Electrical and Electronic Engineering: Theory, Design and Applications

Electrical and Electronic Engineering: Theory, Design and Applications

Edited by
John Fenmore

Willford Press
www.willfordpress.com

Published by Willford Press,
118-35 Queens Blvd., Suite 400,
Forest Hills, NY 11375, USA

ISBN: 978-1-68285-418-1

Cataloging-in-Publication Data

Electrical and electronic engineering : theory, design and applications / edited by John Fenmore.
p. cm.
Includes bibliographical references and index.
ISBN 978-1-68285-418-1
1. Electrical engineering. 2. Electronics. 3. Electronics--Design. I. Fenmore, John.
TK145 .E44 2018
621.3--dc23

For information on all Willford Press publications
visit our website at www.willfordpress.com

Willford Press

Printed in China.

Contents

Preface

This book has been an outcome of determined endeavour from a group of educationists in the field. The primary objective was to involve a broad spectrum of professionals from diverse cultural background involved in the field for developing new researches. The book not only targets students but also scholars pursuing higher research for further enhancement of the theoretical and practical applications of the subject.

Electrical engineering studies electricity and electromagnetism for creating devices to regulate and control electric current and electronic engineering is concerned with the creation of circuits that can contain and transmit electricity. This book on electrical and electronic engineering elucidates new techniques and applications in a multidisciplinary approach. The objective of this book is to give a general view of the different areas of these allied fields, and their applications. It presents the complex subject of electrical and electronic engineering in the most comprehensible and easy to understand language. This book, with its detailed analyses and data, will prove immensely beneficial to professionals and students involved in this area.

It was an honour to edit such a profound book and also a challenging task to compile and examine all the relevant data for accuracy and originality. I wish to acknowledge the efforts of the contributors for submitting such brilliant and diverse chapters in the field and for endlessly working for the completion of the book. Last, but not the least; I thank my family for being a constant source of support in all my research endeavours.

Editor

1

(Lantool) Power Generation Cost Minimization Software Application

Anireh V.I.E.[1,*], John Tarilanyo Afa[2]

[1]Department of Electrical/Computer Engineering, Rivers State University of Science and Technology, Rivers State, Nigeria
[2]Department of Electrical/Electronic Engineering, Niger Delta University, Bayelsa State, Nigeria
*Corresponding author: viaemeka@gmail.com

Abstract Managers of large scale industry like the electricity generation are challenged at many fronts because the tasks involved are complex, influenced by unexpected events and evolve in time. Currently the application of computer methods in this industry has produced tremendous positive results. In this regard, the study presents an effort developed to solve optimal power flow (economic dispatch) problem by minimizing the cost of generation using the lagrangian multiplier method. A decision support system 'Lantool' is presented. The approach is validated by lagrangian method found in technical literatures. The system will assist operators in thermal power plants with the task of planning generation in the most economic way. A result obtained from the application explains the important role decision support systems can play in the management of the electricity generation industry.

Keywords: *optimal power flow, minimizing cost, economic dispatch, lagrangian multiplier method, decision support system*

1. Introduction

The rapid growth in power system industry has made the issue of optimization in power generation and transmission very significant. In the recent time, deregulation in power industry has changed the part played by various units in the electricity industry. Services provided by electric power vendors which originally used to be one single entity are now unbundled into three different entities thus production, transmission and distribution. The service providers' compete in providing good electricity delivery and at the same time strive to make profit. To maintain efficiency and security, reliability of supply are not compromised otherwise customer interest and protection will not be guaranteed. In other to achieve these, best practices in economic dispatch [1] or optimal power flow are employed. The objective is to systematically seek the lowest cost of electricity production that will be consistent with electricity demand. To minimize cost, Optimal power flow (OPF) will increase the use of more efficient generating unit and at the same time addresses two issues of concern - better fuel usage, and reduced green house gas emission [2,3,4] that would result from less efficient generation. OPF therefore seeks to minimize the total cost of generating power (production cost) at various stations while satisfying the loads and the losses in the transmission lines. In the case of developing countries like Nigeria where consumption per capita of electricity is very poor mainly due to high cost of production [5], application of the principles of OPF cannot be over looked. Even when these principles are to be employed the approach to deliver it in a simpler and user friendly manner to enable operators effectively apply it is equally very necessary. It is in view of all these considerations that this paper seeks to present a computer based support system for the electricity generation industry. Computer applications in the form decision support system (DSS) can be a tool used to provide fair and consistent decisions, and at the same time improve the effectiveness of decision making process [6]. A DSS application approach for supporting decision making is typically built for solutions of a certain problem or to evaluate an opportunity. On that note, we present a decision support computer-based tool using Lagrangian multiplier method. The approach was applied to a five-bus network to minimize cost of electricity generation and result is also presented.

Many solution techniques have been proposed and also available to solving optimal power flow problem with varied degree of successes. They can be divided into two main categories, the algorithmic mathematical solution and artificial intelligent solution as reported in several literatures. Among the algorithmic solutions are Interior point (IP) algorithm [7], Simplex algorithm (SA), Quadratic programming (QP), and Dynamic programming (DP) [8]. Langrange relaxation method (LRM) [9,10] Linear programming (LP), Non-linear programming (NLP) and Newton-based methods have also been reported. Currently proposed Artificial Intelligent optimization methods based on heuristics and operational research presented by researchers have emerged with global optimum solution for power system optimization. They include Expert system (ES), Ant Colony search (ACS) [11], Simulated annealing (SA) [12], Artificial Neural networks (ANN) [13,14,15], Fuzzy logic (FL) [16] and Genetic Algorithm (GA) [17]. Others include Meta heuristic methods such as Tabu search (TS) [18], Particle swarm optimization (PSO) [19,20] and Evolutionary programming (EP) [21,22]. Application of these methods depends on the researchers' area of interest as each method has its own advantages and disadvantages. Even as these optimization methods are effective in finding the global optimal solution, however, when adopted in large-scale real-world system, it would take a long computational time. It therefore explains the fact that there exist a gap between theory and practice in the field of optimization. Attempt to bridge this gap is what this paper seek to present by shifting focus to applying analytical method to solving a real-world problem.

1.1. OPF Formulations

Generally, power system can be operated optimally if the principles of equal incremental cost rate are applied to the problem of optimal distribution of loads among the various stations forming a power pool. The most precise way to describe such operation is to use the relation between the input and output parameter as available from the generated data of the individual generators.

The characteristic of this relation is described mathematically as.

$$I = a + bP + cP^2 + dP^3 + \ldots\ldots nP^r \qquad (1)$$

Where I is the input (cost of fuel) and P is the output (power produced by generator) with positive coefficients a, b, c to n.

Alternatively it can be represented graphically as shown in Figure 1:

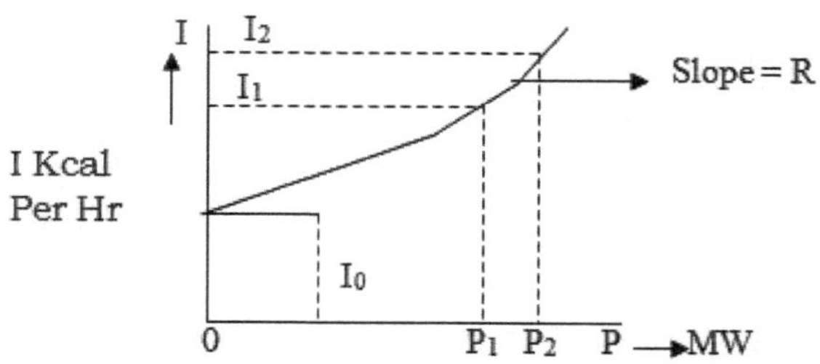

Figure 1. Input – output characteristic of a generator.

I_0 in the above graph represents the amount of input required to keep the generator functioning when there is no load. The slopes of the curve at various load-points give the incremental cost rate. If I_1 and I_2 are the inputs corresponding to the loads P_1 and P_2 respectively, the increase in input required for meeting the increase of load from P_1 to P_2 is given by

$$I_2 - I_1 = \int_{P_1}^{P_2} R_i \, dP \qquad (2)$$

The area under the curve between P_1 and P_2 is the incremental rate. Similarly, from no load to P_1, the increase in input $I_1 - I_0$ is given by the area under the incremental rate curve from $P = 0$ to P_1, and Ri is the incremental rate.

By differentiating the expression in equation 1 to the third power, we get the incremental rate characteristic as:

$$Ri = \frac{dI}{dP} = b + 2c(P) + 3d(P^2) \qquad (3)$$

However, the theory of incremental rate assumes the following conditions:

a. That input – output curves are continuous
b. That first derivatives of the input – output curves are continuous
c. That value of the incremental rate increases with the increase in output

Hence to minimize I (input) the condition will be

$$\frac{dI}{dP} = 0$$

for generators i = 1 --------- n.

Then

$$\frac{dI_1}{dP_1} = 0 \quad \frac{dI_2}{dP_2} = 0 \quad \frac{dIn}{dPn} = 0 \qquad (4)$$

Hence

$$\frac{dI_1}{dP_1} = \frac{dI_2}{dP_2} = \frac{dIn}{dPn} = R \qquad (5)$$

Therefore the input will be a minimum for a specified load when the units have the same incremental rate.

2. Solution to Minimizing Cost of Generation

With the above background information the objective of generation cost minimization is to schedule generation such that input (I) is minimum for the given total power P, subject to restriction that the sum of P_k = P (is the total load received), where P_k is the output of unit k.

Using the lagrangian method, if

$$f\,(P_1, P_2, \text{--------} P_k) = 0 \qquad (6)$$

And

$$\sum_{k=1}^{n} P_k - P = 0 \qquad (7)$$

Then

$$\sum_{k=1}^{n} P_k = P$$

If I represent the cost of input, the minimum input cost is realized when

$$\frac{dI_t}{dP_k} = 0 \text{ where } I_t = \sum_{k=1}^{n} I_k \quad I_t = \text{total input}$$

Applying lagrangian type multiplier where

$I = I_t - \lambda f$ $\qquad \lambda$ = lagrangian type of multiplier

$$\frac{dI}{dP_k} = \frac{dI_t}{dP_k} - \lambda \frac{df}{dP_k} = 0 \qquad (8)$$

Where f is the function of total power received ie

$$f\left(P_1, P_2, ----- P_n\right) = 0 \ or \ \sum_{k=1}^{n} P_k - p = 0$$

$$\text{But RHS } \frac{dI_t}{dP_k} - \frac{\lambda d}{dP_k}\left[\sum_{k=1}^{n} P_k - P\right] = 0$$

$$\frac{dI_t}{dP_k} - \lambda\ [1 - 0] = 0 \qquad (9)$$

$$\therefore \frac{dI_t}{dP_k} = \lambda \quad and \quad \frac{d}{dP_k}\left[\sum_{k=1}^{n} I_k\right] = \lambda$$

Hence

$$\frac{dI_t}{dP_k} = \lambda \qquad (10)$$

The incremental cost of input to k^{th} unit in dollar per hour is equal to the incremental cost of the received power.

The equation 10 may be rewritten as

$$\frac{dI_1}{dP_1} = \frac{dI_2}{dP_2} = \frac{dI_2}{dP_3} ---- = \frac{dI_n}{dP_n} \qquad (11)$$

If the incremental rate of kth unit is written as R_{ik}, then

$$R_{i1} = R_{i2} = R_{i3} ----- R_{ik} = ----- R_{in} = \lambda \qquad (12)$$

Using symbol, C, for incremental production cost in dollar per hour then:

$$Ci_1 = Ci_2 = Ci_3 --- = Ci_k -- Cin = \lambda. \qquad (13)$$

Where λ, the lagrangian multiplier, is the incremental cost of received power in dollar per hour.

In other words, suppose we have a five-bus power system network in a one line diagram as shown in Figure 2, we can develop an computer based algorithm for the optimal distribution cost using the above an analytical solution.

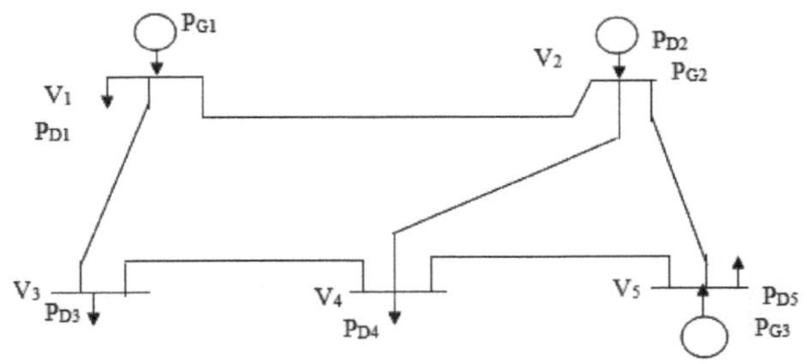

Figure 2. One-Line Diagram of a Five Bus System

2.1. Algorithm for the Optimal Power Flow of the Five-Bus Network

For x MW to be generated from P_{G1}, P_{G2}, and P_{G3} (where the P_{Gi}'s are of equal capacity) that would service the combined external load y in P_{D3}, P_{D4} optimally i.e. x = y, we proceed follows:

1. Obtain the cost functions for each of the generators P_{G1}, P_{G2}, P_{G3} as follows:

$$C_1 = aP_1^2 + bP_1 + c$$
$$C_2 = aP_2^2 + bP_2 + c$$
$$C_3 = aP_3^2 + bP_3 + c$$

2. for optimal economic allocation of load y we apply the lagrangian method, where:

$$C_1 = C_2 = C_3 \quad \text{and} \quad C_1 = dC_1 / dP_1,$$
$$C_2 = dC_2 / dP_2,\ C_3 = dC_3 / dP_3.$$

3. differentiate the C_1, C_2, and C_3 with respect to P_{G1}, P_{G2}, and P_{G3}.
4. solve for P_1, P_2, and P_3 by equating the differential for minimum cost.
5. add P_1, P_2, and P_3 the total will be equal to y, the total load to service i.e $P_1 + P_2 + P_3 = y$.

2.2. Algorithm for the Optimal Cost Savings in the Five-Bus Network

To obtain cost saving effected by the optimal allocation instead of equal division of load we proceed as:

1. let P_{G1}', P_{G2}' and P_{G3}' = x' (x' is one-third of x and (P_{Gi}')'s are of equal capacity).
2. obtain the cost function (C_i')'s of P_1' P_2' P_3' as follows:

$$C_1' = a\left(P_1'\right)^2 + bP_1' + c$$
$$C_2' = a\left(P_2'\right)^2 + bP_2' + c$$
$$C_3' = a\left(P_3'\right)^2 + bP_3' + c$$

3. obtain increased cost of input for P_1' as

$$\Delta C_1 = C_1' - C_1 \text{per unit } (\$/h)$$

4. Saving in cost of input for P_2'

$$\Delta C_2 = C_2 - C_2' \text{ per unit } (\$/h)$$

5. Saving in cost of input for P_3'

$$\Delta C_3 = C_3 - C_3' \text{ per unit } ((\$/h)$$

6. Net increase cost is

$$C_{net} = \Delta C_1 - \Delta C_2 - \Delta C_3 \text{per unit } (\$/h)$$

7. for 24 hours we obtain $C_{day} = C_{net}$ * 24 hours, C_{day} is the daily cost saving for running the three generators.

3. Results

Results obtained from 'Lantool' decision support application is shown in Table 2 for various load demands. As expected the difference between operating the station optimally as against equal distribution is clearly identified in the table as NetSave. The daily cost saving (NetSave) for each of the load demands per day is also shown.

4. Discussion

In the analysis of optimal load division between the various units of a plant, it is expected that the available data on cost is absolutely correct. The implementation of the 'Lantool' program is based on data set for the test cases of Table 1. The table shows generator input cost data as C1, C2, C3, while P1, P2, P3, are power generated data for P_{g1}, P_{g2}, and P_{g3} respectively for the five-bus network in Figure 2. The input cost function is derived from equation 3 with all coefficients positive. (The coefficients for input cost equations are obtained using Matlab Curve Fitting Toolbox). Line plots of cost versus power output is shown in Figure 3. The optimum economy is achieved if every unit (i.e. P_{g1}, P_{g2}, and P_{g3}) operates at the same incremental cost(IC) as seen in Figure 4. At any point on the incremental cost plot, the three generators are operated optimally and fuel utilization is seen to be less. Screen shots of cost with output of simulation, daily cost saving and incremental cost plot/data is shown in Figure 5 and Figure 6. The output of daily net saving per load demand when multiplied by 30 days will give the cost savings per month.

Table 1. Sample of output power and cost of input in $ per hour

P1 (MW)	P2 (MW)	P3 (MW)	C1 ($/hour)	C2 ($/hour)	C3 ($/hour)
24.1	24.1	24.1	598.740 5	581.8886	503.3648
20.8	20.8	20.8	535.232	525.4784	461.8112
21.5	21.5	21.5	548.6125	537.335	470.48
22.3	22.3	22.3	563.9645	550.9574	480.4832
16.6	16.6	16.6	455.978	455.5736	411.4448
22.7	22.7	22.7	571.6645	557.7974	485.5232
22.9	22.9	22.9	575.5205	561.2246	488.0528
22.7	22.7	22.7	571.6645	557.7974	485.5232
25.3	25.3	25.3	622.1045	602.7254	518.9072
23.4	23.4	23.4	585.178	569.8136	494.4048
24.2	24.2	24.2	600.682	583.6184	504.6512
25.1	25.1	25.1	618.2005	599.2406	516.3008
23.6	23.6	23.6	589.048	573.2576	496.9568
22.4	22.4	22.4	565.888	552.6656	481.7408
22	22	22	558.2	545.84	476.72
23.3	23.3	23.3	583.2445	568.0934	493.1312
17.3	17.3	17.3	469.0645	467.0774	419.6432
22	22	22	558.2	545.84	476.72
25.7	25.7	25.7	629.9245	609.7094	524.1392
22.7	22.7	22.7	571.6645	557.7974	485.5232
26.1	26.1	26.1	637.7605	616.7126	529.3968
26.2	26.2	26.2	639.722	618.4664	530.7152
24.3	24.3	24.3	602.6245	585.3494	505.9392
25.9	25.9	25.9	633.8405	613.2086	526.7648

Table 2. Sample result of Optimal vs Equal distribution cost from LanTool per load demand

EQD	L1	L2	L3	L4
Load	55.0001	75.0001	90.0001	100.0001
Gen1cost	488.472	616.25	715	920
Gen2cost	484.167	597.5	686	872
Gen3cost	431.889	515	582	728
OPD				
Load	55.0001	75.0001	90.0001	100.0001
Gen1cost	173.8032	323.5005	439.5947	686.107
Gen2cost	427.0709	539.0336	625.8461	810.2329
Gen3cost	716.8009	791.6816	849.8279	973.0873
	2084.471	1788.825	1625.552	1213.746
NetSave	2084.471	1788.825	1625.552	1213.746

Key: EQD = Equal Distribution of load. OPD = Optimal Distribution of load : L1,L2,L3,L4 = load demand (MW).

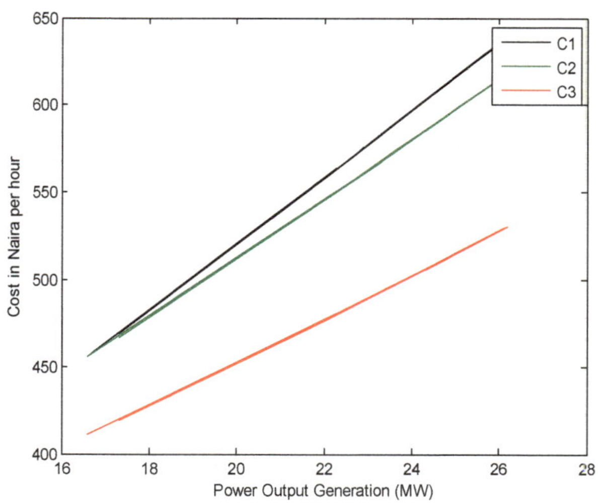

Figure 3. Line Plots of Cost vs Power output Generated for P_{g1}, P_{g2}, and P_{g3}

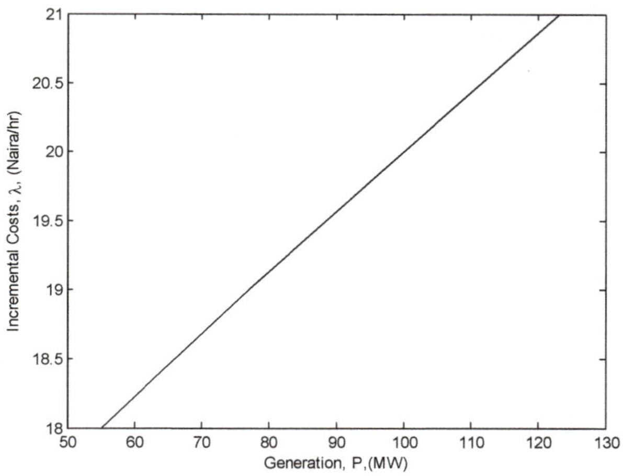

Figure 4. Line Plot of Incremental Cost (Dollar) vs Total Output (MW) P_{g1}, P_{g2}, and P_{g3}

Figure 5. Sample screen for the lagrangian Optimal/Equal distribution cost with output of simulation and daily cost saving data

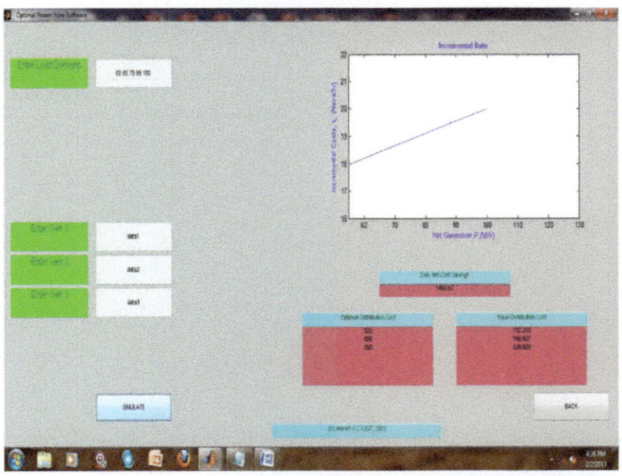

Figure 6. Sample screen for the lagrangian Optimal/Equaldistribution cost with output of simulation, daily cost saving and incremental cost plot/data

5. Conclusion

Minimizing cost of fuel in generating electricity is a real world problem and requires practical solution. The paper presented lantool, a decision support application tool designed to determine the best combination of power generating plant to produce electricity with less fuel cost. It was formulated using the lagrangian multiplier method and developed as an interactive application for operators at gas turbine generating (GTG) plant. The design uses a special object oriented methodology feature of MatLab called handle graphics technology (HGT). This approach provides the application with menu-driven guides that would enable operators with little knowledge of computer navigate through it without difficulty. Lantool can be extended to include an embedded form. When interfaced with a sensor can automatically control the switching of power plant for optimal operation. Simulation data from other stations tested on lantool show no significant difference in result implying that the system can be deployed in a dissimilar geographical location.

References

[1] Dongxiao N, Yanan W. "A novel social-environmental-economic dispatch model for thermal/wind power generation and application" *International Journal of Innovative Computing, Information and Control (ICIC)*. 9(7), 2013.

[2] Alrashidi M.A, Hawary M.E., "Impact of loading conditions on the emission economic dispatch". *Proceedings of World Academy of Science, Engineering and Technology*, 2008.

[3] Wang L, Chanan S., "Reserve constrained multiarea environmental/economic dispatch based on particle swarm optimization with local search". *Engineering Application of Artificial Intelligence Science Direct*, 2008.

[4] Peixoto, H.M., Guerreiro, A.M.G., Dória Neto, A.D., "Image processing for eye detection and classification of the gaze direction", *Proceedings of the 2009 International Joint Conference on Neural Networks*, Atlanta, GA, USA. pp. 2475-2480, 2009.

[5] Sambo A.S., "The Challenges of Sustainable Energy Development in Nigeria". *Presentation to Nigerian Society of Engineers*, Abuja, Nigeria, 2009.

[6] Palma-dos-Reis, A., Zahedi, F. M., "Designing personalized intelligent financial support systems". *Decision Support System*, 26(1): 31-47, 1999.

[7] Martinez L.J., Lora T.A., Santos R.J., "Short term hydrothermal coordination based on interior point nonlinear programing andgenetic Algorithm". *IEEE Porto Power Tech Conference*, 2001.

[8] Shi C.C., Chun H.C., Fomg I-K, Lah P.B., "Hydroelectric generation scheduling with an effective differential dynamic programming algorithm". *IEEE Transactions on Power System*, 1990.

[9] Erion F., Silva E, Laudia S., "Solving the unit commitment problem of hydropower plants via Lagrangian relaxation and sequential quadratic programming". *Journal of Computational & Applied Mathematics*, 2005.

[10] Tkayuki S., Kamu, W., *Lagrangian relaxation method for price based unit commitment Problem*. Engineering Optimization, Taylor Francis, 2004.

[11] Huang J.S., "Enhancement of hydroelectric Generation scheduling using ant colony system based optimization approaches". *IEEE Transactions on Energy Conversion*, 2001.

[12] Mantawy A, Abdel-Magid Y., Selim S., "A simulated annealing algorithm for unit commitment", *IEEE Trans. Power Syst.*, 13(1): 197-204, 1998.

[13] Naresh R, Sharma J., "Short term hydro scheduling using two phase neural network". *International Journal of Electrical Power and Energy Systems*, 2002.

[14] Liu, D., Dong X., Shi, P., "The optimization based on improved membrane algorithm for fuzzy RBF neural network control of ball-plate system", *ICIC Express Letters, Part B: Applications*, 3(5): 1259-1267, 2012.

[15] Lopes, D.C., "Implementation of a Modular Neural Network in a Multiple Processor System on FPGA to Classify Electric Disturbance". *Industrial Electronics. IECON '09. 35th Annual Conference of IEEE*, Porto, Portugal, 2009.

[16] Lee, T.E, Su, J.P., Yu K.W., "Fuzzy gain scheduled alpha-beta-gamma filter design based on particle swarm optimization method", *ICIC Express Letters*, 4(6B): 2305-2310, 2010.

[17] Kazarlis, S., Bakirtzis, A., Petridis, V., "A genetic algorithm solution to the unit commitment problem", *IEEE Trans. Power Syst.*, 11(1): 83-92, 1996.

[18] Lin, W.M, Cheng, F.S, Tsay, M.T., "An improved Tabu search for economic dispatch with multiple minima". *IEEE Transaction on Power System*, 2002.

[19] Jie, X, Deyun, X., "New Metropolis coefficients of particle swarm optimization", *IEEE*, 2008.

[20] Liao, G., "A novel evolutionary algorithm for dynamic economic dispatch with energy saving and emission reduction in power system integrated wind power", *Energy*, 36: 1018-1029, 2011.

[21] Yang, H.T, Yang, P.C, Huang, C. L., "Evolutionary programming based economic dispatch for units with non smooth fuel cost function", *IEEE Transaction on Power System*, 1996.

[22] *Information and Telecommunication Technologies* Peixoto, H.M., Cruz, R. M.S., Guerreiro, A.M.G., Dória Neto, A.D., D'Assunção, A.G., "A comparison of multilayer perceptrons training algorithms for the optimization of frequency selective surfaces", *Proceedings of 8th International Symposium*, Florianopolis, SC, Brazil, 2009.

[23] Gangadhar, K.A. "Electrical Power Systems". Khanna Publishers. Nai,Sarak, Delhi. India, 2006.

2

Quad-Band Dual-Layer Microstrip Antenna Design for Mobile Handset

Gada Mahmood Faisal[1], Kaydar Majeed Quboa[1], Dia Mohamad Ali[2,*]

[1]Elec. & Com. Dept, College of Engineering, Mosul University, Mosul /Iraq
[2]Com. Dept, College of Electronic Engineering, Mosul University, Mosul /Iraq
*Corresponding author: dia_mohamad@yahoo.com

Abstract Due to the rapid acceleration and evolution of the wireless communication and their equipment, integrating more services in the personal devices, Personal Digital Assistance (PDA) devices has to work in multi-band frequencies. In this paper, a quad-band, dual-layer Microstrip Antenna was proposed. The proposed antenna is suggested to be used in a mobile phone handset that covers four frequency bands which include GSM 900 (889-960), GSM 1800 DCS 1800 (1710-1885)W-CDMA (Wideband Code Division Multiple Access); IMT 2000 (International Mobile Telecommunication)1885-2200, Bluetooth and Wi-Fi/WLAN (IEEE 802.11 b/g/n) ISM 2450 (Industrial, Scientific and Medical) (2400-2500). The antenna is fixed on a printed circuit board together with a standard mobile battery and LCD. All other components were packed in a mobile phone handset plastic cover. An extensive analysis of the return loss, radiation pattern, and gain of the designed antenna with and without cover were presented. Packing effects on the antenna parameters were studied, the proposed antenna shows return loss (RL) of -16 dB at 0.92 GHz; -20 dB RL at 1.735 GHz; -37 dB RL at 2.1 GHz; and -16 dB RL at 2.42 GHz which are encouraging results. Designing and simulation of the proposed antenna were based on CST software package.

Keywords: *microstrip antenna, Quad band, mobile, patch*

1. Introduction

Microstrip radiators were first proposed as early as 1953 while the first actual microstrip antenna appeared in 1974 [1]. Microstrip patch antennas (MPAs) consist basically of three layers: metallic layer with antenna radiating patch, dielectric substrate and another metallic layer as the ground plane [2]. The light weight, low cost, low profile, planner configuration and easy in fabrication are the main advantages of MPA which enable it to be used with microwave integrated circuits. In spite of these advantages, MPAs suffer from low gain and relatively large size [3,4].

Due to the development in the field of wireless network and mobile communication, many researches started to develop multi-band MPAs to cover the bands of GSM 900, GSM 1800, GSM 1900, UMTS, Bluetooth, WLAN… etc. Singh et al. [5] designed a conventional rectangular MPA with Ψ-shaped structure suitable to work in two different frequency bands. Chakraborty et al. [6], designed a single feed compact rectangular MPA for quad-band applications using four L-shaped slot structure along the length of the patch. Barun [7], designed a single feed quad-band compact rectangular MPA with L slot on the edge of the patch with size reduction as compared to a conventional MPA. Veereshappa and Mulgi [8], designed a slotted rectangular MPA for quad-band operation and ominidiretional radiation characteristics. Jayasinghe et al. [9], designed a small multi-band MPA for GSM 900, GSM 1800, GSM 1900, UMTS, LTE 2300, and Bluetooth applications using a genetic algorithm. Puri and Tiary [10], designed a simple single layer, single feed triangular slot patch antenna with air dielectric substrate for quad-band frequency operation in different wireless applications. Mendha and Kosta [11], designed and simulated a stacked 3-layer rectangular Ultra Wide Band antenna. AL Noman et al. [12], designed a single layer single feed miniaturized MPA for multi services using multi layers separated by dielectric substrates. Kharade and Patil [13], enhanced the gain of the MPA using multi-layer and multi-dielectric layers separated from patch by air as another dielectric. Sayidmarie et al. [14], designed a compact dual-band; dual ring printed monopole antenna for WLAN applications. Wang et al. [15], designed two broad bands MPA using planer line feed to enhance band width by 25%. AL Mously [16], designed a triple-band; multi-layer patch antenna for 0.9, 1.8 and 2.4 GHz frequency bands.

2. Design Methodolgy

Different methods can be used in the design of multi-band Microstrip antennas. These include single short pin loaded between ground plane and patch, slot-loaded on patch, stacked multi-resonator rectangular patches on thick substrate. Other types such as single feed dual-band using higher order modes of the circular microstrip antenna (CMSA) and triangular microstrip antenna (TMSA) are out of our scope.

In this paper, three of those methods were hybridized to design a quad-band multi-layer Microstrip antenna. The first step is to design a single layer conventional MPA using single feeder. The second step is to stack a parasitic patch on the top of the active patch separated by different dielectric substrate which will increase the required bandwidth. The third step is to add a single short pin at a suitable distance from the center of the patch to add the higher frequency bands with the aid of shifting feeding point. The required quad-band is:

- 1st-band centered at fr1 = 0.92 GHz.
- 2nd-band centered at fr2 = 1.73 GHz.
- 3rd-band centered at fr3 = 2.10 GHz.
- 4th-band centered at f_{r4} = 2.42 GHz.

Step 1: Patch Antenna Design

The antenna design starts with a conventional microstrip antenna MPA1 as shown in Figure 1.

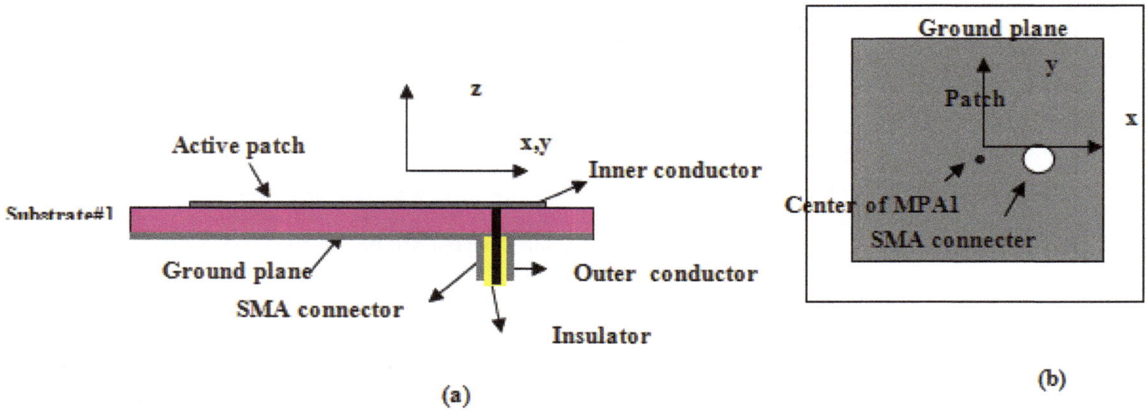

Figure 1. Conventional MPA1 (a)-Side view, (b)-xy cross section

The basic form of a MPA which consists of active radiating patch on one side of a dielectric substrate and a ground plane on the other side the coaxial feed technique is chosen because the feed can be placed at the desired location inside the patch to match its impedance.

The well-known equations used in calculating dimension of this step are listed in Table 1.

Table 1. Used MPA1 design equations

Parameter	Symbol	Equation used
Resonance frequency [17]	f_r	$f_r = \frac{c}{2W\sqrt{(\varepsilon_{r1}+1)/2}}$
Effective dielectric constant [17]	ε_e	$\varepsilon_e = \frac{\varepsilon_{r1}+1}{2} + \frac{\varepsilon_{r1}-1}{2}\left[1+12\frac{h_1}{W}\right]^{-1/2}$
Effective Length [17]	$L_{eff}(mm)$	$L_{eff} = \frac{c}{2f_r\sqrt{\varepsilon_e}}$
Length of patch [17]	$L(mm)$	$L = L_{eff} - 2h$ $*0.412\left[\frac{(\varepsilon_e+0.3)}{(\varepsilon_e-0.258)}\frac{\left(\frac{W}{h_1}+0.26h_1\right)}{\left(\frac{W}{h_1}+0.8\right)}\right]$
Length of ground plane [17]	$L_g(mm)$	$L_g = 6h + L$
Width of ground plane [17]	$W_g(mm)$	$W_g = 6h + W$
Inset feed location [18]	$x_f(mm)$	$x_f = \frac{L}{2\sqrt{\varepsilon_e}}$
	$y_f(mm)$	$y_f = \frac{W}{2}$

where c is the speed of the light, W & L are width and length of patch and ε_r is the relative dielectric constant of the substrate.

The antenna is fed with a coaxial connecter modeled as a standard SMA connector with the following specifications [19]:

- Radius of inner conductor = 0.62 mm.
- Radius of insulator = 2.3 mm.
- Radius of outer conductor = 2.64 mm.
- ε_r of insulator (Teflon-lossy) = 2.1.
- Length underground plane = 2.1 mm.

Step 2: Adding a Parasitic Patch

Adding the parasitic patch will increase the bandwidth of the MPA1 antenna with more than 15%.

Step 3: Adding the Shorting Pin and Shifting the Feeding Point

Adding the short pin to MPA1 and shifting the feeding point to a proper location will lead to the proposed quad-band MPA shown in Figure 2 with specifications and dimensions as given in Table 2.

Table 2. The proposed MPA specifications

Parameter	W_1	L_1	h_1	ε_{r1}	W_2	L_2	h_2	ε_{r2}	W_g	L_g	x_f	y_f	x_s	y_s
Value	34	32	1	4.4	56	47	1.6	1.4	60	50	8	-9	7.6	0

where:

W_1, L_1, and h_1 are the active patch dimension in mm;
W_2, L_2, and h_2 are the parasitic patch dimension in mm;
ε_{r1} is the relative dielectric constant of the substrate #1;
ε_{r2} is the relative dielectric constant of the substrate #2;
(x_s, y_s) is the coordination of the added short pin in mm.

The overall antenna structure is shown in Figure 2. The shorting pin locates at (7.6,0) mm and the feeder at (8,-9) mm.

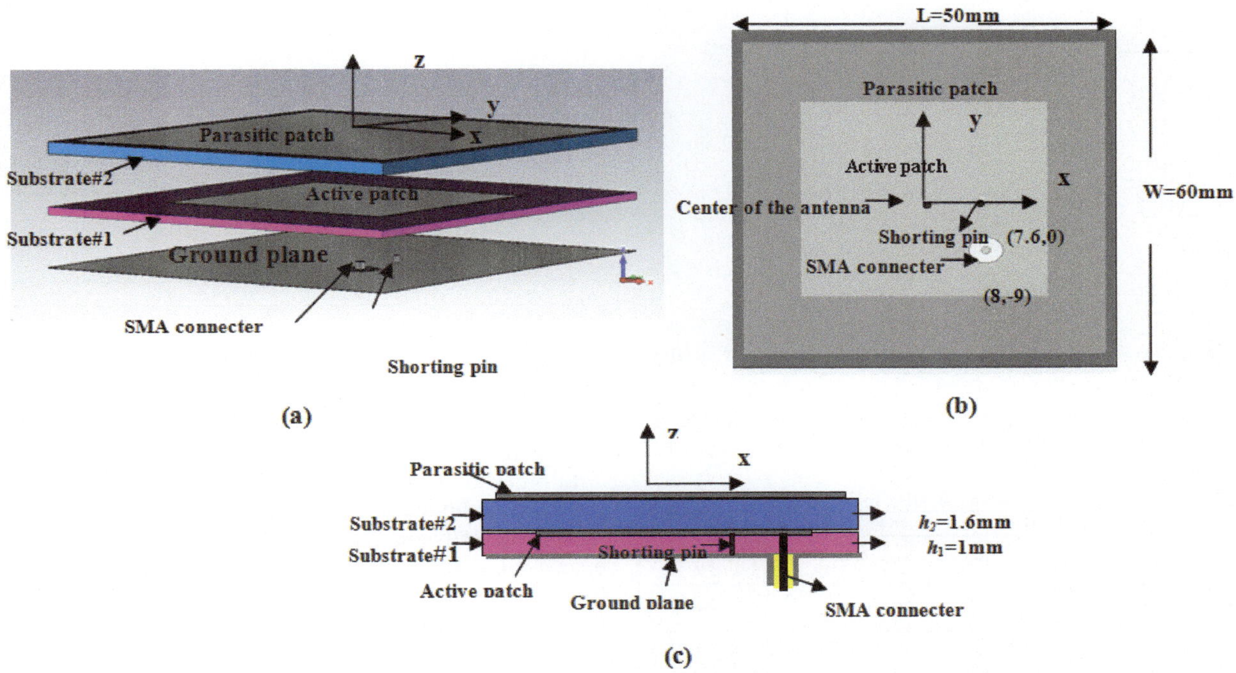

Figure 2. MPA antenna structure: (a) - Layers details, (b) - Top view, and (c) - Connector details

3. Simulation Results

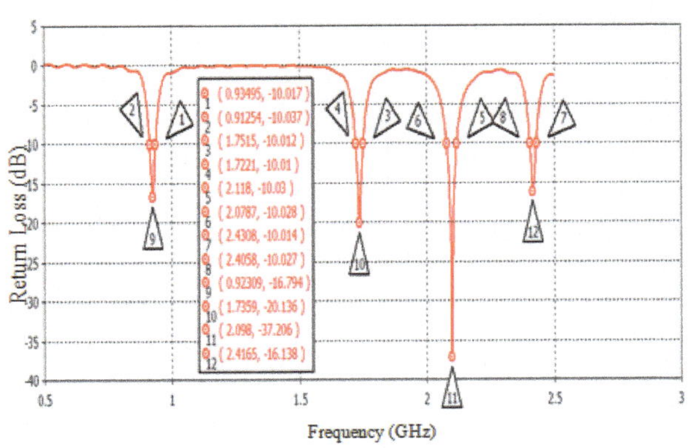

Figure 3. Simulated return loss of quad-band MPA

The proposed quad-band MPA shown in Figure 2 was simulated using CST software package [20]. The detailed return loss (RL) of the final quad-band MPA is shown in Figure 3.

The percent bandwidth *(Bw)* is calculated using the following equation [21]:

$$Bw = \frac{f_h - f_l}{f_c} \times 100\%$$

Table 3. Results of the proposed quad-band MPA

Band	f_c (GHz)	RL (dB)	Bandwidth	VSWR
f_{r1}	0.92	-16.79	2.43	1.4
f_{r2}	1.735	-20	1.7	1.3
f_{r3}	2.098	-37	1.87	1.05
f_{r4}	2.416	-16	1.035	1.4

Where f_h and f_l are the upper and lower cutoff frequencies of the band respectively at a reference return loss of -10 dB and f_c is the center frequency of the assigned band. The RL, bandwidth and voltage standing wave ratio (VSWR) of each band with its center frequency (resonance frequency) are given in Table 3.

The input impedance of the proposed quad-band MPA is investigated as a function of frequency for both real and imaginary parts of antenna's impedance as shown in Figure 4.

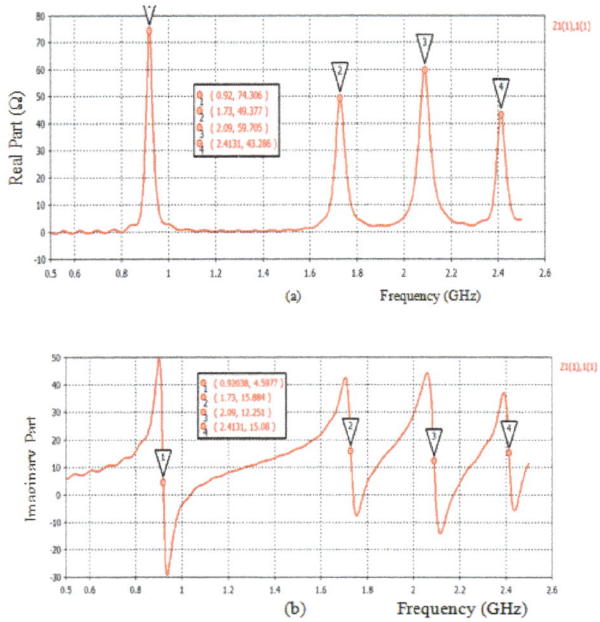

Figure 4. (a) Real part of MPA impedance, (b) Imaginary part of MPA impedance

The input impedance at the resonance frequencies which satisfied the VSWR condition are given in Table 4.

Table 4. Quad-band MPA impedance

Frequency (GHz)	Impedance (Ω)
$f_{r1} = 0.92$	74.3 + j4.6
$f_{r2} = 1.735$	49.4 + j15.9
$f_{r3} = 2.098$	59.7 + j12.3
$f_{r4} = 2.42$	43.3 + j15.1

Table 4 meets the results obtained from drawing the input impedance using Smith chart. As shown in Figure 5, the normalized input impedance approaches the unity (matching point) at four different frequencies. It is clear that these frequencies matched with the VSWR values in Table 3.

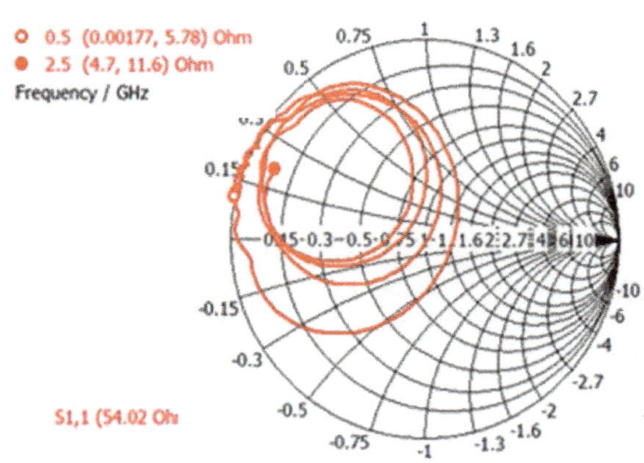

Figure 5. Smith chart for MPA antenna

4. Antenna Packing

The designed quad-band MPA antenna has been assembled on printed circuit board (PCB), with a standard mobile battery, LCD monitor and all are placed in a plastic cover with air gap surrounding all components as shown in Figure 6.

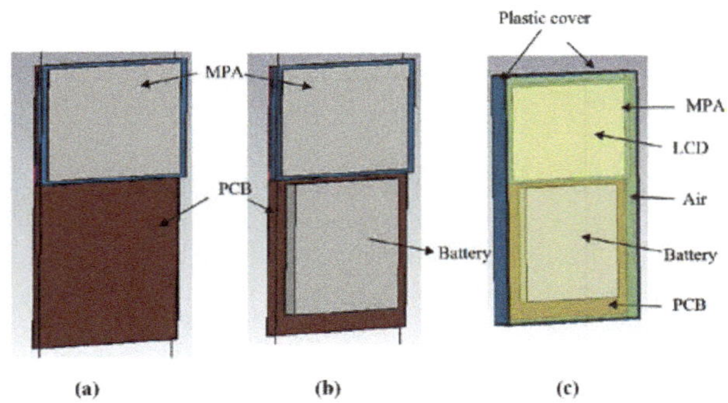

Figure 6. (a) MPA on PCB, (b) MPA on PCB and battery, (c) PCB on PCB battery, LCD monitor and cover

The physical dimensions of PCB, battery, LCD monitor & plastic cover are chosen to be as those of Samsung Galaxy 2 mobile handset as given in Table 5.

Table 5. Handset specifications

Parameter	Material	Dimensions	ε_r
Printed circuit board	PCB	(120 × 60 × 0.2) mm	4.9
Battery	PEC	(58 × 45 × 4) mm	
Cover	plastic	(125 × 65 × 8.1) & 0.8 mm thickness	2.5
Monitor	LCD	(125 × 65 × 0.8) mm	4.78
Surrounding Space	Air		1

In order to evaluate the effect of the packing process, the packed mobile handset has been simulated using CST simulation software and the RL is investigated. The assembled components caused the resonance frequencies to be shifted as given in Table 6. It is clear from Table 6 that small changes in resonance frequencies (1.735 changed to 1.731 and 2.41 GHz changed to 2.42 GHz) occurred.

Table 6. The quad-band MPA with cover

Frequency (GHz)	$f_{r1} = 0.92$	$f_{r2} = 1.731$	$f_{r3} = 2.098$	$f_{r4} = 2.41$
RL	-16	-25	-34	-21

It is worth noting that MPA is covered with a dielectric layer to protect it from the environment, or painted with dielectric paint. This dielectric layer may influence the characteristics of the antenna and cause a shift in resonance frequencies.

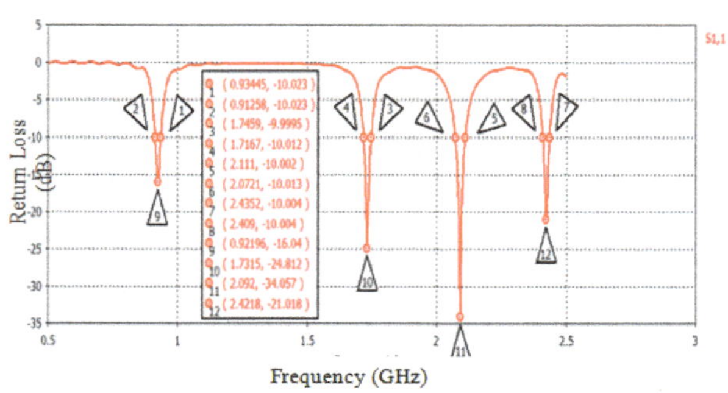

Figure 7. Return loss of quad-band MPA with cover

The return loss of the assembled quad-band MPA antenna with cover shown in Figure 6 is simulated and the result is shown in Figure 7.

The radiation patterns for E and H-plane of the proposed antenna with cover at frequencies 0.92 GHz, 1.731 GHz, 2.09 GHz and 2.41 GHz are shown at Figure 8. In CST software, the mode is chosen to be multi-mode operation which will help in choosing the proper mode for each operating frequency [22]. The radiation pattern of the proposed MPA with cover is shown in Figure 8.a in the mode TM13 at frequency of 0.92 GHz and in the mode TM11 for other frequencies.

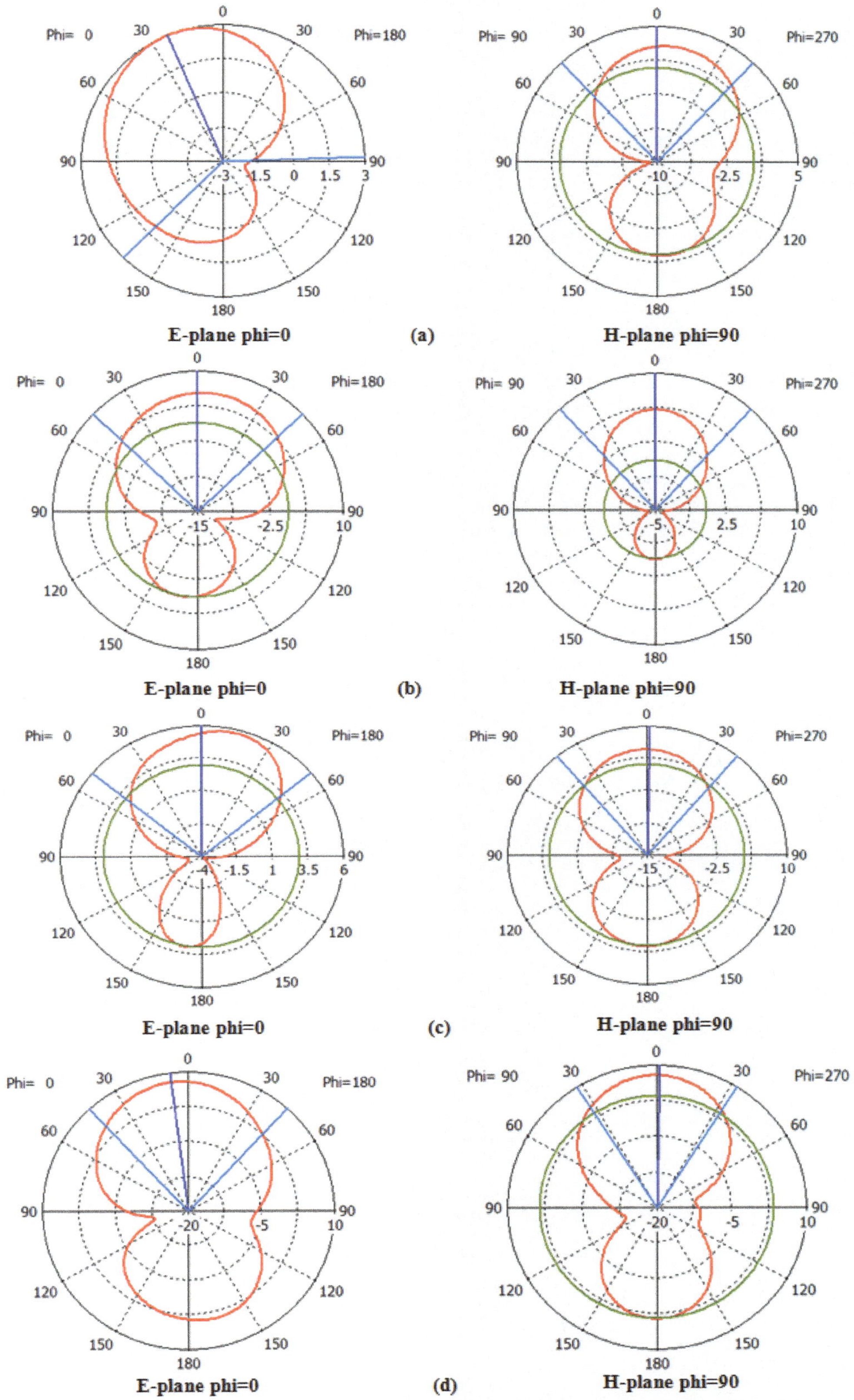

Figure 8. The radiation pattern of the proposed MPA with cover at: (a) 0.92GHz, (b) 1.731GHz, (c) 2.098GHz, and (d) 2.41GHz

From the radiation pattern figure shown in Figure 8 the directivity for the resonance frequencies of the proposed quad-band MPA antenna with cover is given in Table 7.

Table 7. The directivity of quad-band MPA with cover

Frequency (GHz)	$f_{r1} = 0.92$	$f_{r2} = 1.731$	$f_{r3} = 2.098$	$f_{r4} = 2.41$
Directivity (dBi)	3.151	6.096	5.702	7.992

5. Conclusion

A single feed dual layer rectangular microstrip antenna with single short pin has been proposed. It is shown that the proposed quad-band antenna can operate effectively in four frequency bands in the frequency ranges of GSM (900 MHz), GSM (1800 MHz), IMT 2000 (885-2200), Bluetooth and Wi-Fi (2400-2500). Four center frequencies are operating at: 920, 1735, 2098 & 2420 MHz. The bandwidth for each frequency was 22.4 MHz , 29.4 MHz , 40 MHz, and 25 MHz respectively. Maximum gain of the antenna is found to be about 4.39 dBi at 2.42 GHz. The location of the single short pin is optimized in such a way that the antenna can operate in four suitable bands. The antenna with the battery and LCD were fixed together on the printed circuit board and packed in a plastic cover. The dimensions of the designed antenna were 50 × 60 × 2.775 mm.

The proposed antenna characteristics such as return loos, radiation pattern and working bandwidth were acceptable. It is expected that the proposed MPA will be very useful for the future mobile systems and suitable for mobiles having a multi services such as Bluetooth, WIFI WIMAX and of course GSM or 3G services all together.

References

[1] G. Kumar, and K.P. Ray, "Broadband Microstrip Antennas", Artech House, Inc. Boston, London, 2003.

[2] C. A., Balanis, "Modern Antenna Handbook", 1st Edition, John iley & Sons, Inc., 2008.

[3] J. Y. Jan, and L.C. Tseng, "Small Planar Monopole Antenna with a Shoted Parasitic Inverted-L Wire for Wireless Communications in the 2.4, 5.2 and 5.8 GHz Bands", IEEE Trans. Antennas and Propag. Vol. 52, NO. 7, Joly 2004, pp-1903-1905.

[4] U. Chakaraborty, S. Chatterjee, S. K. Chowdhury, and P. P. sarkar, "A Compact Microstrip Patch Antenna for Wireless Communication," Progress In Electromagnetics Research C, Vol. 18, 211-220, 2011.

[5] D. Singh, A. K. Gupta, R. K. Prasad, "Design and Analysis of Dual-Band Ψ-Shape Microstrip Patch Pntenna", International Journal of Advances Engineering & Technology, Mar. 2013. Conference IJAET.

[6] U. Chakaraborty, B. Mazumdar, S.K. Chowdhury, A.K.Bhattacharjee, "A Compact L-slot Microstrip Antenna for Quad band Applications in Wireless Communication", Global Journal of Researches in Engineering Electrical and Electronics Engineering ,vol 12, Issue 2, Version 1.0, Feb. 2012.

[7] B. Mazumdar, "Design Of L-Slit Microstrip Patch Antenna For Wimax & Hiperlan Applications", Journal of computer Networking, Wierless and Mobile Communications (JCNWMC), vol. 2 Issue 1, Sep-2012, 45-50. Conference TJPRC Pvt. Ltd.

[8] M. Veereshappa and S. N. Mulgi, "Design and Development of Quad Band Rectangular Microstrip Antenna with Ominidireciona Radiation Characteristics", / International Journal Of Coputational Engineering Research.

[9] J. W. Jayasinghel, J. Anguera, and D. N. Uduwawala, "A Simple Design of Multiband Microstrip Patch Antenna Robust to Fabrication-Tol-Erancea for GSM ,UMTS, and Bluetooth Applications by Using Genetic Algorithm Optimization", Progress In Electro-magnetics Research M, Vol. 27, 255, 269, 2012.

[10] S. C. Puri, and M.G. Tiary, "A Novel Quad Band Rectangular Microstrip Patch Antenna for Wieless Applications", IJCA Proceedings on International Symposium on Devices MEMS, Intelligent System & Communication, 24-26-2011.

[11] S. E. Mendhe, and Y P. Kosta, "Stacked Three Layer Rectangular Microstrip Patch Antenna for Ultra Wide Band Applications", International journal of Computer Science and Communication, vol. 3, No, 1, January-June 2012, pp. 145-147.

[12] A. AL Nomanovi, M. R. Chowdhury, R. A. Zuboraj and A. Matin, "Design of Miniaturized Dual Band Microstrip Antenna Loaded with Asymmetric J slot", Technol 2012, 2: 1.

[13] A. R. Kharade, V. P. Patil, "Enhancement of Gain of Rectangular Micro strip Antenna Using Multi layer Multi dielectric structure", IOSR Journal of Electronics and Communication Engineering (IOSRJECE) vol. 2, Issue 6 (Sep-Oct 2012), PP 35-40, www.io srjournals.org.

[14] K.H. Sayidmarie and T.A. Nagem, "Compact Dual-Band Dual-Ring Printed Monopole Antennas For WLAN Applications", Progress In Electromagnetic Research B, Vol. 43, 313-331, 2012.

[15] H.H. Wang, K., Yang, Z.Y. Lei and C. I. Li, "Broadband Microstrip Antennas using Coplanar Feed line", Progress In Electromagnetics Research Letters, Vol. 20, 129-136, 2011.

[16] S.I. Al-Mously, "Design and Performance Enhancement of Cellular Handset Antennas and Assessment of Their EM Interaction with a Human", PhD Thesis, Academy of Graduate Studies, School of Applied Science and Engineering Dept. of Electrical and Computer Eng. Libya. Aug., 2009.

[17] C. A. Balanis, "Antenna Theory, Antenna and design", John Wiley & Sons Inc., 2nd Edition, 2005.

[18] G. Frank, "High Frequency Technology: Principles of Mobile Communication Technology", Rectangular Patch Antenna Design Using Coaxial Probe Peed. Htm.

[19] K. Parikh, "Simulation of Rectangular, Single Layer, Coax-Fed Patch Antennas Using Agilent High Frequency Structure Simulator (HFSS)", MSc. Thesis, Faculty of the Virginia Polytechnic. Institute and State University. December 2003.

[20] Computer Simulation Technology: http://www.cst.com/.

[21] L. Warren, and A. Gary, "Antenna Theory and Design" 2nd. Edition, New York, ISBN 0-471-02590-9, 1998.

[22] CST Studio Suite under "Help - > Support Account Settings...". Waveguide Port Overview [CST MWS].

3

DVCC Based K.H.N. Biquadratic Analog Filter with Digitally Controlled Variations

Bilal Arif*, Mohd. Usama Ismail, Ale Imran

Department of Electronics Engineering, Aligarh Muslim University, Aligarh, India
* Corresponding author: arifbilal25@gmail.com

Abstract In this paper, a digitally controlled single input multi output current-mode K.H.N. Biquad Filter is presented. The filter circuit is composed of three DVCCs together with four grounded resistors and two grounded capacitors. The digital control is incorporated using a current-summing network (CSN). Tuning of resonant frequency is carried out by 3–bit digital control word. Block by block replacement has been done to observe the change in the relationship between resonant frequency of the band-pass filter with the control word. The filter circuit showed three different variations when the DVCC blocks were replaced (one by one) with 3-bit DC-DVCC blocks. PSPICE simulations using TSMC 0.25 micron CMOS technology have been performed to validate the theoretical results.

Keywords: *Current-mode, Differential Voltage Current Conveyor (DVCC), multifunctional filter, digitally controlled DVCC (DC-DVCC), Cut off frequency, K.H.N. Biquad filter*

1. Introduction

In the recent times the current mode filters have gained utility in various signal processing applications. These filters have revolutionized the modern day signal processing and have replaced their voltage mode counterparts in several of applications.

Moreover filters which could provide simultaneous realization of the basic filter functions have proved themselves useful in various applications which include touch-tone telephone tone decoder, phase-locked loop FM stereo demodulator and crossover network used in a three-way high-fidelity loudspeaker.

With the inception of Current Conveyors (CC), the filter design has reached a new height and various variants of CC have also gained attraction [1-10]. Various digitally controlled filters have been proposed and realized in the recent times. In [11] H.P. Chen and S.S. Shen presented a DVCC based Universal capacitor grounded voltage mode filter which realized all the five generic filter responses simultaneously. In [12] H. P. Chen presented tunable current mode universal filter, the high output impedance of this filter enable easy cascading in current-mode operation also in different modes of operation it could realize different filter responses simultaneously. Two years back I. A. Khan and A. M. Nahhas presented a CCII based reconfigurable first order multifunction filter whose frequency could be changed with a digital control word [13].

In this paper, the circuit proposed in [14] by Muhammed A. Ibrahim, Shahram Minaei and Hakan Kuntman, has been used to design and implement a digitally controlled current-mode K.H.N. biquad. Using grounded capacitors, the circuit becomes suitable for integration as the grounded capacitor circuit can compensate for the stray capacitances at the respective nodes. PSPICE simulations of the CMOS based programmable filter are performed to demonstrate results.

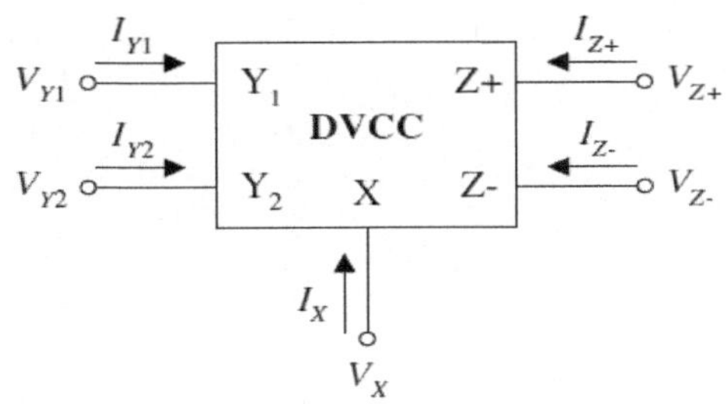

Figure 1. Symbol representing the dual output DVCC [15]

2. DVCC

DVCC is a five-terminal active analog building block illustrated in Figure 1, with terminal characteristics described by the following matrix equation [15].

$$\begin{pmatrix} I_{Y1} \\ I_{Y2} \\ V_X \\ I_{Z+} \\ I_{Z-} \end{pmatrix} = \begin{pmatrix} 0 & 0 & 0 & 0 & 0 \\ 0 & 0 & 0 & 0 & 0 \\ 1 & -1 & 0 & 0 & 0 \\ 0 & 0 & 1 & 0 & 0 \\ 0 & 0 & -1 & 0 & 0 \end{pmatrix} \begin{pmatrix} V_{Y1} \\ V_{Y2} \\ I_X \\ V_{Z+} \\ V_{Z-} \end{pmatrix} \tag{1}$$

DVCC exhibits negligible (ideally zero) input resistance at terminal X, and very high (ideally infinite) resistance at both Y terminals as well as the Z terminal. The output current follows the input current direction with both currents flowing either into or out of the device. The CMOS implementation of DVCC is as shown in Figure 2.

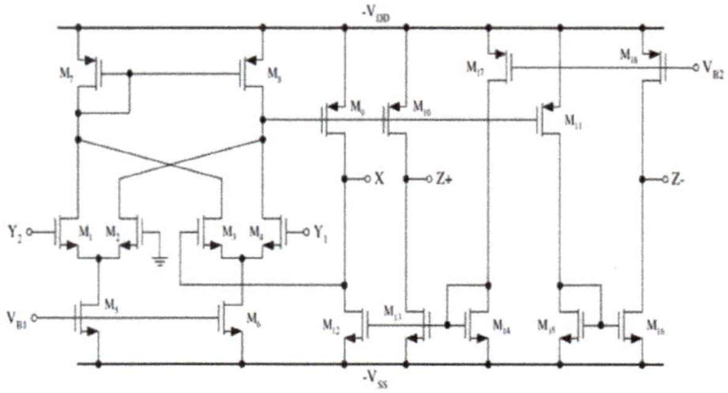

Figure 2. CMOS realization of the dual-output DVCC [15]

IMPLEMENTATION OF KHN BIQUAD

The implemented KHN biquad is illustrated in Figure 3. The analysis of the circuit yields the following equations (2), (3) and (4).

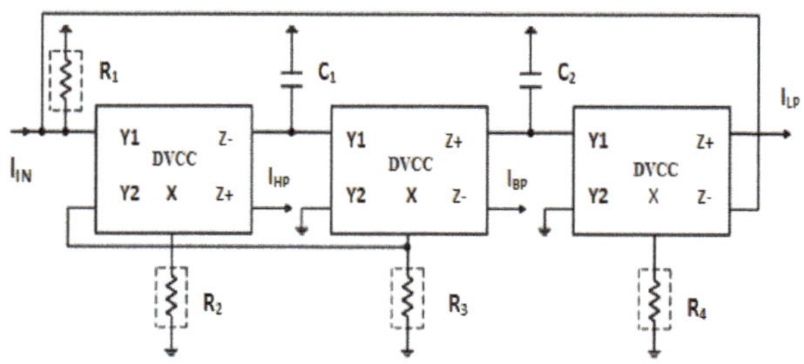

Figure 3. K.H.N. Biquad Filter Realization [14]

$$\frac{I_{BP}}{I_{IN}} = \frac{s\dfrac{R_1}{R_2R_3C_1}}{s^2 + s\left(\dfrac{1}{R_2C_1}\right) + \dfrac{R_1}{R_2R_3R_4C_1C_2}} \quad (2)$$

$$\frac{I_{HP}}{I_{IN}} = \frac{s^2\dfrac{R_1}{R_2}}{s^2 + s\left(\dfrac{1}{R_2C_1}\right) + \dfrac{R_1}{R_2R_3R_4C_1C_2}} \quad (3)$$

$$\frac{I_{LP}}{I_{IN}} = \frac{\dfrac{R_1}{R_2R_3R_4C_1C_2}}{s^2 + s\left(\dfrac{1}{R_2C_1}\right) + \dfrac{R_1}{R_2R_3R_4C_1C_2}} \quad (4)$$

The resonant angular frequency ω_o, and the quality factor, Q, are given by (5) and (6) respectively.

$$= \sqrt{\frac{R_1}{R_2R_3R_4C_1C_2}} \quad (5)$$

$$Q = \sqrt{\frac{R_1C_1R_2}{R_3R_4C_2}} \quad (6)$$

We can see that lowpass, bandpass and highpass functions can be simultaneously realized without changing the topology.

In simulations, using PSPICE the DVCC was realized by the CMOS implementation shown in Figure 2 using TSMC 0.25-μm process parameters. The aspect ratios of the CMOS transistors of the DVCC are presented in Table I. The supply voltages were taken as, $V_{DD} = -V_{SS} = 2$ V and the biasing voltages were assigned values, $V_{B1} = -1.32$ V and $V_{B2} = +0.7$ V. The circuit was designed for $f_o = \omega_o/2\pi = 100$ kHz and Q = 0.707 by choosing $R_1 = R_2 = R_3 = R_4 = 1$ kΩ and $C_2 = 2\ C_1 = 1.125$ nF. The responses of the multifunctional filter are shown in Figure 4 (a) and Figure 4 (b). The results are in full conformity with the theoretical analysis.

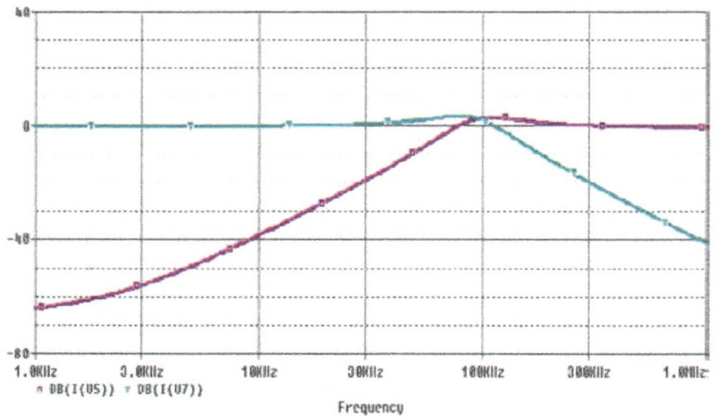

Figure 4 (a). Simulated Lowpass and Highpass responses

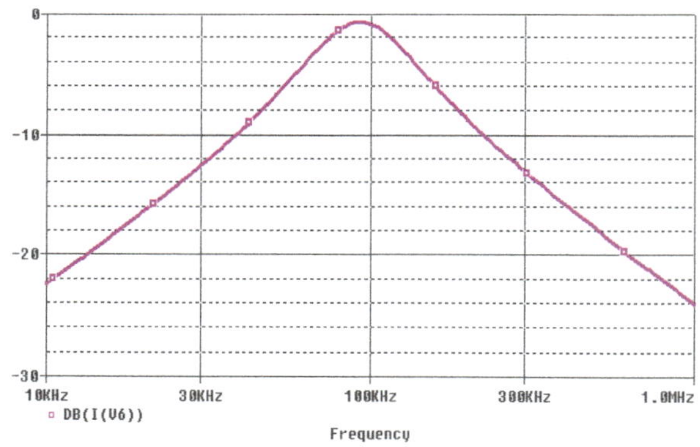

Figure 4 (b). Simulated Bandpass response

Table 1. Aspect ratios of the cmos transistors of the dvcc [15]

Transistors	W (μm)	L (μm)
M_1-M_4	1	0.8
M_5-M_6	24.2	0.8
M_7-M_8	6.8	0.8
M_9-M_{11}, M_{17}	18.6	0.6
M_{12}-M_{14}	25	0.8
M_{15}	19.6	0.8
M_{16}	18	0.8
M_{18}	20	0.6

3. DC-DVCC

To introduce the programmability in the multifunctional filter we have used a digitally controlled DVCC (DC-

DVCC) shown in Figure 5. The modified terminal characteristics for the same are as follows

$$\begin{pmatrix} I_{Y1} \\ I_{Y2} \\ V_X \\ I_{Z+} \\ I_{Z-} \end{pmatrix} = \begin{pmatrix} 0 & 0 & 0 & 0 & 0 \\ 0 & 0 & 0 & 0 & 0 \\ 1 & -1 & 0 & 0 & 0 \\ 0 & 0 & k & 0 & 0 \\ 0 & 0 & -k & 0 & 0 \end{pmatrix} \begin{pmatrix} V_{Y1} \\ V_{Y2} \\ I_X \\ V_{Z+} \\ V_{Z-} \end{pmatrix} \tag{7}$$

Where:

$$k = \frac{I_Z}{I_X} \tag{8}$$

For obtaining the digital control in the DVCC current summing networks (CSNs) are employed at the Z (Z+ and Z-) terminals for controlling the current transfer gain parameter k. The gain parameter k shows a variation from 1 to ($2^n - 1$), where n is the number of transistor arrays. The modified circuit of DVCC with the transistors arrays is as shown in Figure 5. The CSN consists of n transistor pairs, the aspect ratios of whose PMOS and NMOS transistors respectively are given by:

$$\left(\frac{W}{L}\right)_i = 2^i \left(\frac{W}{L}\right)_9 \tag{9}$$

$$\left(\frac{W}{L}\right)_i = 2^i \left(\frac{W}{L}\right)_{12} \tag{10}$$

Furthermore, the current at the Z terminal which is assumed to be flowing out of the DC-DVCC, can be expressed by:

$$I_z = \sum_{i=0}^{n-1} d_i 2^i \left(I_9 - I_{12}\right) \tag{11}$$

Therefore, the proposed DC-DVCC provides a current transfer gain, k equal to:

$$k = \frac{I_z}{I_X} = \frac{\sum_{i=0}^{n-1} d_i 2^i \left(I_9 - I_{12}\right)}{\left(I_9 - I_{12}\right)} = \sum_{i=0}^{n-1} d_i 2^i \tag{12}$$

Where d_i are the bits applied to the i-th branch in the CSN. Now the current flow in a particular branch is enabled or disabled depending upon whether d_i is a logic 1 or logic 0 [16].

4. Comparative Study of Variation in Bandpass Filter Resonant Frequency

In this section the discussion is restricted to the variation in resonant frequency of Band Pass filter only. The circuit shown in Figure 3 was modified by replacing a DVCC block with a DC-DVCC block, one by one so that change in relationship between the resonant frequency and control word can be observed. Each block is assigned an individual gain α, β and γ respectively. The analysis is done and the following expression was obtained for the band pass response.

$$\frac{I_{BP}}{I_{IN}} = -\frac{s\dfrac{\alpha\beta R_1}{R_2R_3C_1}}{s^2 + s\alpha\left(\dfrac{1}{R_2C_1}\right) + \dfrac{\alpha\beta\gamma R_1}{R_2R_3R_4C_1C_2}} \tag{13}$$

The resonant frequency is now defined as

$$= \sqrt{\frac{\alpha\beta\gamma R_1}{R_2R_3R_4C_1C_2}} \tag{14}$$

As could be clearly seen from (14) the resonant frequency of the Bandpass filter can be controlled by changing the values of the gain parameters α, β and γ. This variation will not require any change in the values of the passive components.

In the analysis that follows it is assumed that if the DVCC is replaced by DC-DVCC the gain parameter for respective block takes the value k, however if the DVCC is retained, the gain parameter attains value equal to 1.

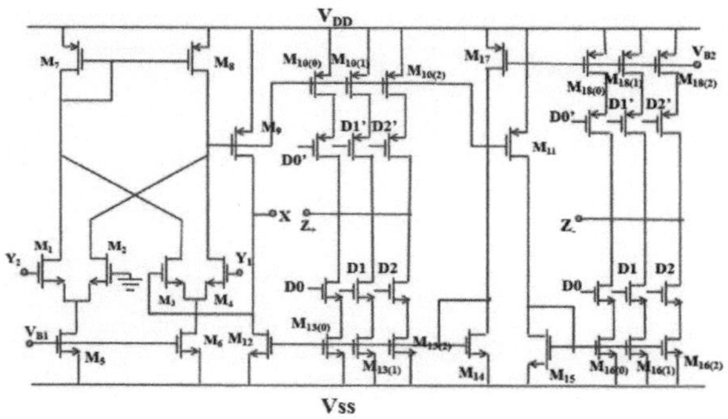

Figure 5. CMOS realization of the DC- DVCC having gain k

Replacing the first block (corresponding to α) by DC-DVCC, therefore for this configuration we have α= k and β=γ= 1, hence the equation for the band-pass response and the expression for the resonant frequency are respectively given by:

$$\frac{I_{BP}}{I_{IN}} = -\frac{s\dfrac{kR_1}{R_2R_3C_1}}{s^2 + sk\left(\dfrac{1}{R_2C_1}\right) + \dfrac{kR_1}{R_2R_3R_4C_1C_2}} \tag{15}$$

$$= \sqrt{\frac{kR_1}{R_2R_3R_4C_1C_2}} \tag{16}$$

The configuration is illustrated in Figure 6.

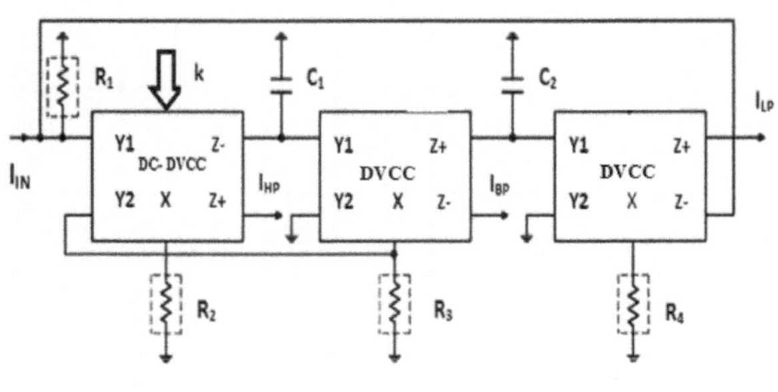

Figure 6. Configuration for α= k and β=γ= 1

From (16) it is evident that the resonant frequency varies with the control word k, in a square root fashion. The configuration shown in Figure 6 is simulated using PSPICE and the Bandpass reses obtained for the control word ([0 0 1] and [1 1 1]) are shown in Figure 7 (a) and Figure 7 (b).

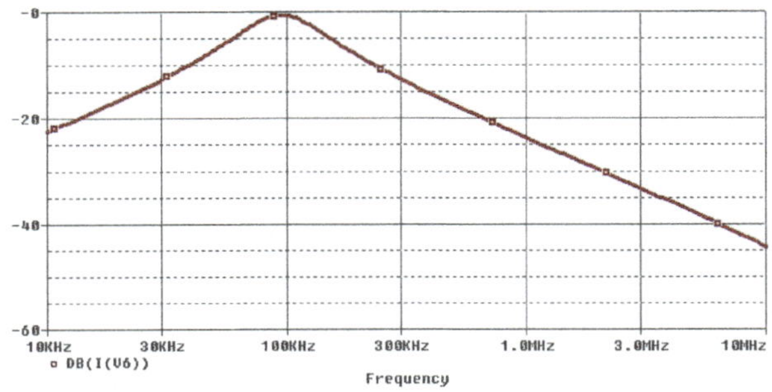

Figure 7(a). Simulated magnitude response (in dB) for band pass filter with control word [$d_2\, d_1\, d_0$ = 0 0 1] selected for circuit of Figure 6

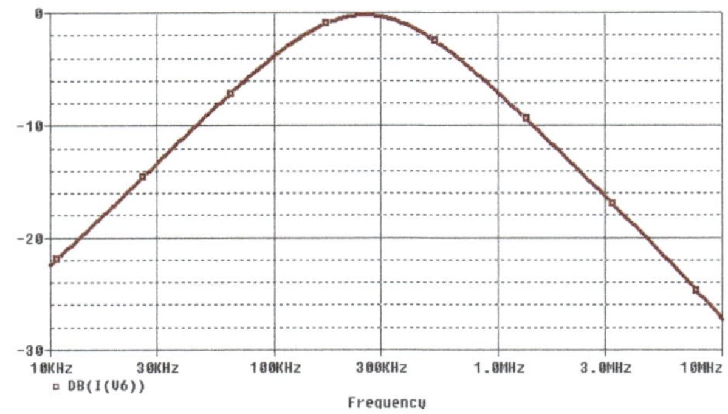

Figure 7(b). Simulated magnitude response (in dB) for band pass filter with control word [$d_2\, d_1\, d_0$ = 1 1 1] selected for circuit of Figure 6.

The simulations show a gradual increase in the resonant frequency when the control word is increased, the observations are recorded in Table II.

Now when the second block is replaced the value of the gain parameters changes to α= β=k and γ= 1 hence the equation for the band-pass response and the expression for the resonant frequency are respectively given by:

$$\frac{I_{BP}}{I_{IN}} = -\frac{s\dfrac{k^2R_1}{R_2R_3C_1}}{s^2 + sk\left(\dfrac{1}{R_2C_1}\right) + \dfrac{k^2R_1}{R_2R_3R_4C_1C_2}} \tag{17}$$

$$= k\sqrt{\frac{R_1}{R_2R_3R_4C_1C_2}} \tag{18}$$

The configuration is illustrated in Figure 8

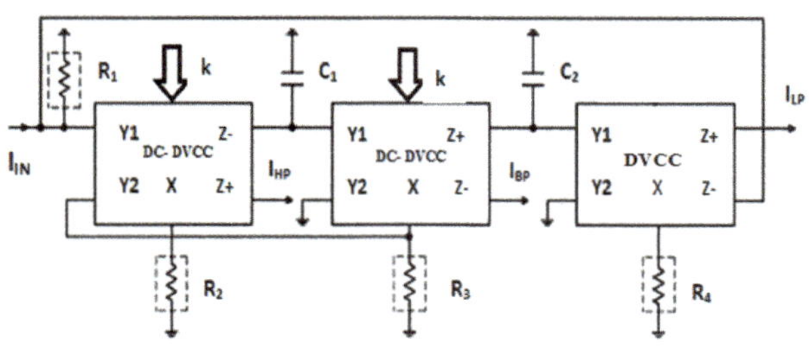

Figure 8. Configuration for α= k and β=γ= k

From (18) it is evident that the resonant frequency varies with the control word k, in a linear fashion. The configuration shown in Figure 8 is simulated using PSPICE and the Bandpass reses obtained for the control word ([0 0 1] and [1 1 1]) are shown in Figure 9 (a) and Figure 9 (b).

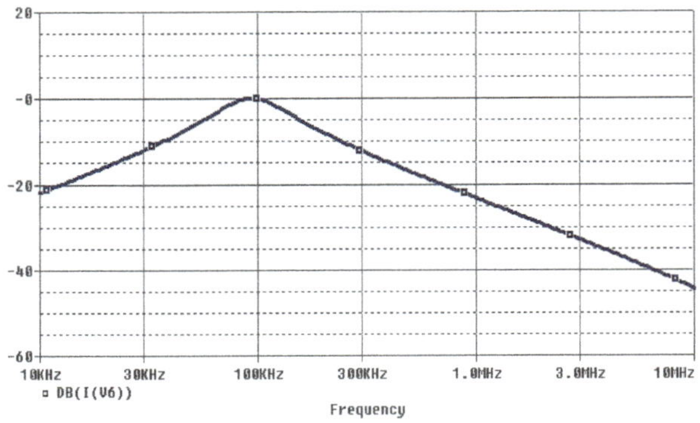

Figure 9(a). Simulated magnitude response (in dB) for band pass filter with control word [$d_2\, d_1\, d_0$ = 0 0 1] selected for circuit of Figure 8

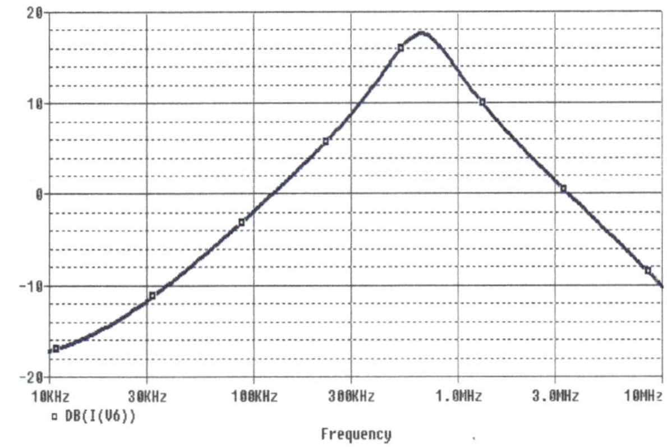

Figure 9(b). Simulated magnitude response (in dB) for band pass filter with control word [$d_2\, d_1\, d_0$ = 11 1] selected for circuit of Figure 6

The simulations show an increase in the resonant frequency when the control word is increased, the observations are recorded in Table 2.

When the third block is replaced the value of the gain parameters changes to α= β=γ=k hence the equation for the band-pass response and the expression for the resonant frequency are respectively given by:

$$\frac{I_{BP}}{I_{IN}} = -\frac{s\dfrac{k^2R_1}{R_2R_3C_1}}{s^2 + sk\left(\dfrac{1}{R_2C_1}\right) + \dfrac{k^3R_1}{R_2R_3R_4C_1C_2}} \tag{19}$$

$$= k^{3/2}\sqrt{\frac{R_1}{R_2R_3R_4C_1C_2}} \tag{20}$$

The configuration is illustrated in Figure 10

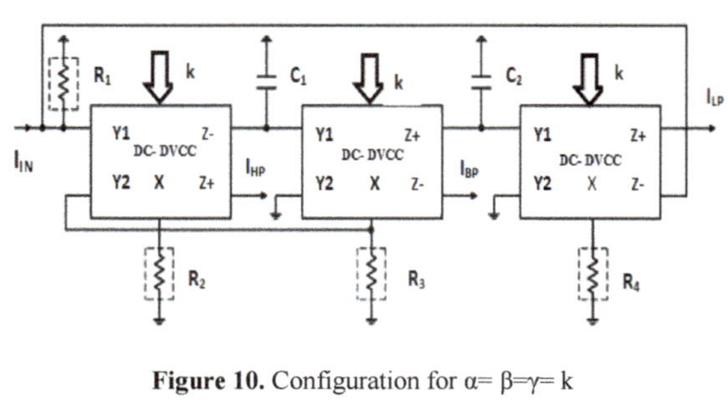

Figure 10. Configuration for α= β=γ= k

From (20) it is evident that the resonant frequency varies with the control word k, in a $k^{3/2}$fashion. The configuration shown in Figure 10 is simulated using PSPICE and the Bandpass reses obtained for the control word ([0 0 1] and [1 1 1]) are shown in Figure 11 (a) and Figure 11 (b).

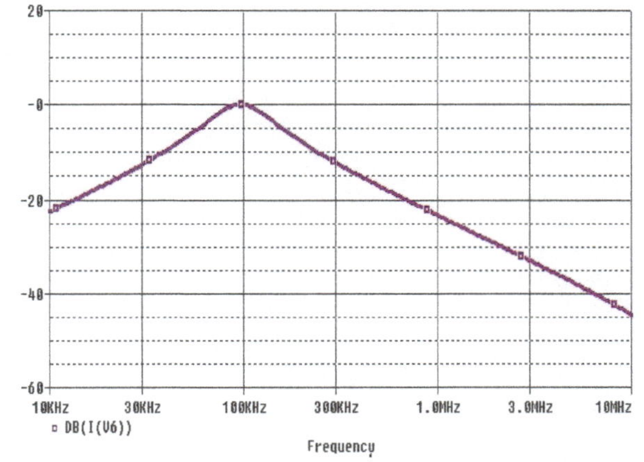

Figure 11(a). Simulated magnitude response (in dB) for band pass filter with control word [d_2 d_1 d_0 = 0 0 1] selected for circuit of Figure 10

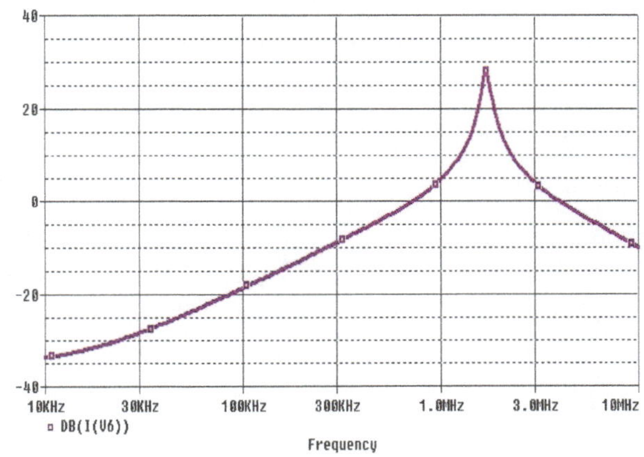

Figure 11(b). Simulated magnitude response (in dB) for band pass filter with control word [d_2 d_1 d_0 = 1 1 0] selected for circuit of Figure 10

The simulations show an increase in the resonant frequency when the control word is increased, the observations are recorded in Table 2.

The circuit presented in [14] for a particular design worked only for a single frequency but using the modification suggested in this paper the utility of the circuit is increased.

Now the circuit when selected for a particular variation and designed for a particular set of values of resistances and capacitances can work for seven different frequencies.

Table 2. Variation in resonant frequency of bandpass responses with the control word

Control word, k	Resonant frequency of BPF of Circuit of Figure 10 (kHz) ($\omega \propto k^{3/2}$)	Resonant frequency of BPF of Circuit of Figure 8 (kHz) ($\omega \propto k$)	Resonant frequency of BPF of Circuit of Figure 6 (kHz) ($\omega \propto \sqrt{k}$)
1	97.79	93.51	95.63
2	270.40	187.00	135.23
3	491.75	282.78	166.29
4	759.70	380.27	191.23
5	1045.13	475.54	213.84
6	1376.03	568.66	235.16
7	1689.10	661.28	257.16

Figure 12, Figure 13 and Figure 14 are the plots showing variation in resonant frequency of the Bandpass filter configurations shown in Figure 6, Figure 8, Figure 10.

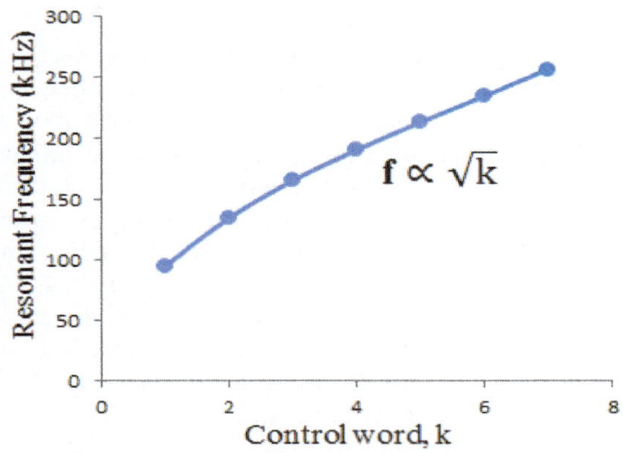

Figure 12. Variation in resonant frequency of BPF with digital control word for circuit of Figure 6

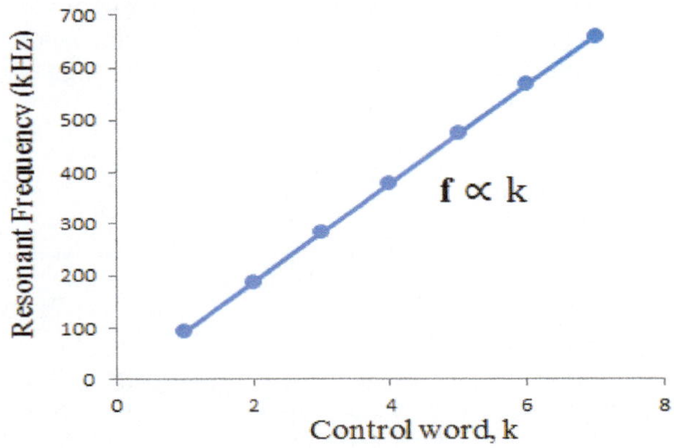

Figure 13. Variation in resonant frequency of BPF with digital control word for circuit of Figure 8

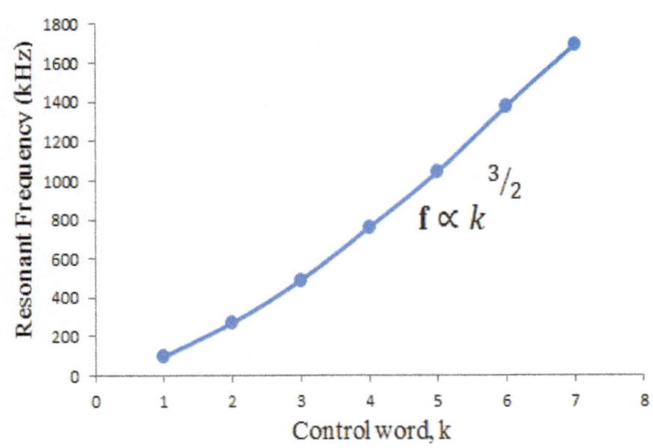

Figure 14. Variation in resonant frequency of BPF with digital control word for circuit of Figure 10

The plots for the variation in resonant frequency obtained by simulation support the theoretical analysis. For the circuit in Figure 6 we have obtained a square root variation then linear variation is observed for the circuit of Figure 8 finally a variation directly proportional to $k^{3/2}$ is observed for circuit of Figure 10.

5. Conclusion

In this paper, a digitally programmable current mode K.H.N. biquad filter based on three DVCCs was presented. Digital control has been achieved with the introduction of CSNs and variation of 3-bit digital control word (k). Now this circuit can be tuned for seven different frequencies for a particular design and variation. This multi frequency tenability within the same design is the contribution for this circuit The circuit was configured in three different ways to provide different relation between resonant frequency of the Bandpass filter and the digital control word (k). The variations obtained were dependent upon the number of DC-DVCCs used. When a single DVCC was replaced with a DC-DVCC square root relation between resonant frequency and k was obtained, this variation became linear on replacement of second DVCC block and when the final block was replaced the resonant frequency became directly proportional to $k^{3/2}$. The observations support the fact that the resonant frequency of Bandpass filter is directly proportional to $k^{a/2}$, where 'a' is the number of DVCC replaced with DC-DVCC. Hence we obtained a circuit whose variation (relationship) of resonant frequency can be controlled. PSPICE simulations were carried out to verify the working of the digitally controlled K.H.N. Biquad Filter.

References

[1] H. O. Elwan, A. M. Soliman. "A novel CMOS current conveyor realization with an electronically tunable current mode filter suitable for VLSI." *Circuits and Systems II: Analog and Digital Signal Processing, IEEE Transactions*, vol. 43, issue. 9, pp. 663-670, Sep. 1996.

[2] C.M. Chang, M.J. Lee. "Voltage-mode multifunction filter with single input and three outputs using two compound current conveyors." *Circuits and Systems I: Fundamental Theory and Applications, IEEE Transactions*, vol. 46, issue. 11, pp. 1364-1365, Nov. 1999.

[3] O. Cicekoglu. "Current-mode biquad with a minimum number of passive elements." *Circuits and Systems II: Analog and Digital Signal Processing, IEEE Transactions*, vol. 48, issue. 2, pp. 221-222, Feb. 2001.

[4] H. Y. Wang, C. T. Lee. "Versatile insensitive current-mode universal biquad implementation using current conveyors." *Circuits and Systems II: Analog and Digital Signal Processing, IEEE Transactions*, vol. 48, issue. 4, pp. 409-413, Apr. 2001.

[5] A.S. Sedra, K.C. Smith. "A second generation current conveyor and its applications". *IEEE Transactions on circuit theory*, vol. 17, pp.132-134, Feb. 1970.

[6] W. Chiu, S. I. Liu, H. W. Tsao, J. J. Chen. "CMOS differential difference current conveyors and their applications." *IEE Proceedings-Circuits, Devices and Systems,* vol. 143, issue. 2, pp. 91-96, Apr. 1996.

[7] H.O. Elwan, A. M. Soliman. "Novel CMOS differential voltage current conveyor and its applications." *IEE Proceedings-Circuits Devices and Systems*, vol. 144, issue. 3, pp. 195-200, Jun. 1997.

[8] T. Dostal, D. Biolek, K. Vrba. "Adjoint voltage-current mode transformation for circuits based on modern current conveyors." *Devices, Circuits and Systems, Proceedings of the Fourth IEEE International Caracas Conference*, 2002, pp. T034-1.

[9] B. Wilson, "Recent developments in current conveyors and current-mode circuits." *Circuits, Devices and Systems, IEE Proceedings G*, vol. 137, issue. 2, pp. 63-77, Apr.1990.

[10] H. Hakan Kuntman. "New Advances and Possibilities in Active Circuit Design." in *Proc. 10th International Conference on Development and Application Systems*, 2010, pp. 9-18.

[11] H. P. Chen and S. S. Shen. "A versatile universal capacitor-grounded voltage-mode filter using DVCCs." *ETRI journal*, vol 29, issue. 4, pp. 470-476, Aug. 2007.

[12] H. P. Chen. "Tunable versatile current-mode universal filter based on plus-type DVCCs." *AEU-International Journal of Electronics and Communications*, vol. 66, issue. 4, pp. 332-339, 2012.

[13] I. A. Khan and A. M. Nahhas. "Reconfigurable Voltage Mode First Order Multifunctional Filter using Single Low Voltage Digitally Controlled CMOS CCII." *International Journal of Computer Applications,* vol. 45, issue. 5, pp. 37-40, May. 2012.

[14] M.A. Ibrahim, S. Minaei, H. Kuntman. "A 22.5 MHz current-mode KHN biquad using differential voltage current conveyor and grounded passive elements." *AEU-International Journal of Electronics and Communications*, vol. 59, issue. 5, pp. 311-318, 2005.

[15] W. Tangsrirat, O. Chaannumsin. "Voltage -mode multifunctional biquadratic filter using single DVCC and minimum number of passive elements". *Indian Journal of Pure and Applied Physics*, vol. 49, pp.703-707, Oct. 2011.

[16] S. A. Mahmoud, M.A. Hashiesh. and A.M. Soliman. "Low-voltage digitally controlled fully differential current conveyor." *Circuits and Systems I: Regular Papers, IEEE Transactions,* vol. 52, issue. 10, pp. 2055-2064, Oct. 2005.

4

Hardware Implementation of the Neural Network Predictive Controller for Coupled Tank System

Ammar A. Aldair*

Electrical Engineering Department, University of Basrah, Basrah, Iraq
*Corresponding author: mmr.ali2@googlemail.com

Abstract In this paper, a neural network based predictive controller is designed for controlling the liquid level of the coupled tank system. The controlled process is a nonlinear system; therefore, a nonlinear prediction method can be a better match in a predictive control strategy. The neural network predictive controller that is discussed in this paper uses a neural network model of a nonlinear plant to predict future plant performance. The simulation results are compared with PID control. The results show that the effectiveness of using the neural predictive controller for the coupled tank system. The Simulink Toolbox in MATLAB has been used to simulate the controlled system with the proposed controller. The VHDL has been used to describe the implementation of neural controller. Xilinx ISE Project Navigator Version 10.1 is used to obtain the compilation and timing test results as well as the synthesized design. The hardware implementation of the neural network predictive controller using FPGA board is proposed. To make sure that the FPGA board works like the simulated neural predictive controller, MATLAB programme is used to compare between the set of the data that are obtained from the ModelSim program and the set of the data that are obtained from the MATLAB Simulink model. Simulation results show that the FPGA board can be used as neural predictive controller for controlling the liquid level of the coupled tank system.

Keywords: *neural network predictive controller, coupled tank system, FPGA*

1. Introduction

The coupled-tank liquid level control system is regarded as the relevant plant to emulate the process control in petrol and chemical industries. The processing plants in these industries largely involve in controlling the liquid level and the flow rate from one tank to another in the presence of nonlinearity and disturbance.

Many authors investigated the problem of controlling liquid flow of single or multiple tanks. The neurofuzzy controller based on the radial basis function neural network is tuned automatically using genetic algorithm to design a control system for the coupled tank [1]. In reference [2], the direct model reference adaptive control is designed for coupled tank system. While, an iterative learning control is used to tune an adaptive neurofuzzy inference system inverse controller for tank system [3]. Khan and Spurgeon presented a second order sliding mode control algorithm for a class of MIMO nonlinear system. This algorithm has been applied for robust control of liquid level in interconnected twin-tanks [4]. Intelligent control including fuzzy control [5,6], neural network control [7], and genetic algorithms [8] have also been applied to the coupled tanks system.

Model predictive control (MPC) is used successfully in some practical application since it was developed in 1980s [9]. The most popular MPC algorithms that found a wide acceptance in industry are: Dynamic Matrix Control (DMC), Model Algorithmic Control (MAC), Predictive Functional Control (PFC), Extended Prediction Self Adaptive Control (EPSAC), Extended Horizon Adaptive Control (EHAC) and Generalized Predictive Control (GPC). GPC has been originally developed with linear plant predictor model, which leads to a formulation that can be solved analytically. But in many industrial processes which usually have nonlinear dynamical behaviours and such complicated systems may be not easily to control by the linear GPC method. In recent years, incorporation of neural networks as intelligent control techniques and adaptive control system has been claimed to be a new method for the control of systems with significant nonlinearities. Those neural network based adaptive control system is called neural network predictive control (NNPC). It is widely used as a controller for many nonlinear systems [10,11,12,13,14]. In this work, a neural network predictive controller is designed for controlling the liquid level of the coupled tank system.

The special-purpose VLSI chips is used to implement the NNPC in real time applications with enhanced processing speed, but due to different neural models and limited use, it has been prohibitively expensive and time consuming to develop such chips. In this paper, the network predictive control is mapped into hardware using field programmable gate array FPGA board to design a control system for coupled tank level. A Very High speed integrated circuit Hardware Description Language (VHDL) codes have been used to describe the operation of the NNPC. The Xilinx Integrated Software Environment (Xilinx ISE 10.1) has been used as environment to type the VHDL codes. The Xilinx Spartan XC3S500E FPGA board has been suggested to use in this work. The Spartan 3E starter Kit board highlights the unique features of the Spartan_3E FPGA family and provides a convenient development board for embedded processing applications. The Hirose 100-pin FX2 Edge connector (J3) can be used as input and output port to receive the digital input error data from the A/D converter (error signal is the difference between the output of the controlled system and desired output) and send the digital control signal to the D/A converter. The analogue control signal is used to force the output of the controlled system to follow the desired output.

For the simulation results, ModelSim XE III 6.4b simulation program will be used with the FPGA-based design. In order to compare the responses of FPGA design with the Simulink design, an M-file will be used to plot the data collected from the ModelSim program and the other data collected from the Simulink design to make sure the FPGA boards work like the neural network predictive controller for coupled tank system. Simulation results show that the FPGA board can be used as neural predictive controller for controlling the liquid level of the coupled tank system.

2. Dynamic Model of the Two Tank System

The coupled tank system is shown in Figure 1. It consists of two identical hold-up tanks coupled by an orifice. The input is supplied by a variable speed pump which supplies liquid to the first tank. The orifice allows the liquid to flow into the second tank and hence out to a reservoir. The control task is to adjust the inlet flow rate *q(t)* by pumping the liquid into the first tank so as to force the level of the liquid in the second tank, h_2k close to a desired set point level, *H*. It is assumed that the back pressure created by the water-head of the second tank does not affect the flow rate of the pump significantly. The controller has been designed based on the neural predictive control.

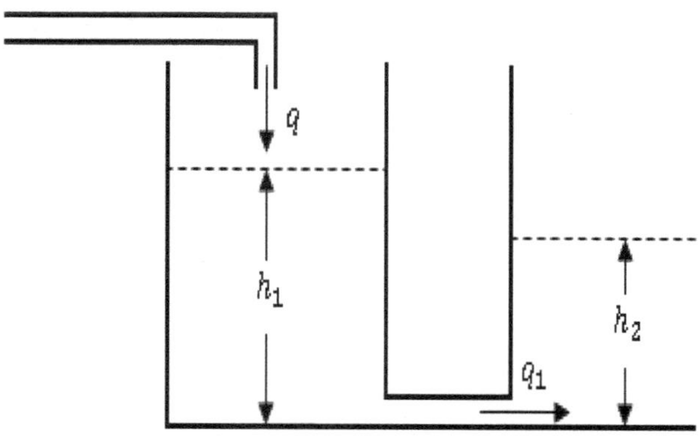

Figure 1. Two Tanks System

2.1. System Modelling

The twin-connected tanks system is a nonlinear dynamical system and the governing dynamical equations can be written as

$$\dot{h}_1 = -\frac{a_{12}}{A}\sqrt{2g\left|(h_1 - h_2)\right|}\,\mathrm{sgn}(h_1 - h_2) + \frac{1}{A}q \quad (1)$$

$$\dot{h}_2 = \frac{a_{12}}{A}\sqrt{2g\left|(h_1 - h_2)\right|}\,\mathrm{sgn}(h_1 - h_2) - \frac{a_2}{A}\sqrt{2g\left|h_2\right|} \quad (2)$$

where h_1 is the level in the first tank;
h_2 is the level in the second tank;
q_1 is the flow rate from the first tank to the second tank;
q_2 is the flow rate from the second tank to a reservoir;
q is the inlet flow rate;
g is the gravitational constant;
A is the cross-section area of first tank and second tank;
a_2 is the cross-section area of the outlet orifice;
a_{12} is the cross-area of the coupled orifice.

2.2. System Constraints

The pump can only pump the water into the first thank, therefore, the inlet flow rate q is always positive. The constraint on the inlet flow rate is given by

$$q \geq 0 \quad (3)$$

In the steady state, for constant water level set points , the respective derivatives must be zero separately i.e.,

$$\dot{h}_1 = \dot{h}_2 = 0 \quad (4)$$

At the steady state, Equation. 1 and Equation 2 cab be rewritten as

$$0 = -\frac{a_{12}}{A}\sqrt{2g\left|(h_1 - h_2)\right|}\,\mathrm{sgn}(h_1 - h_2) + \frac{1}{A}Q \quad (5)$$

$$0 = \frac{a_{12}}{A}\sqrt{2g\left|(h_1 - h_2)\right|}\,\mathrm{sgn}(h_1 - h_2) - \frac{a_1}{A}\sqrt{2g\left|h_2\right|} \quad (6)$$

where Q is the steady state inflow rate. From Equation 5 and Equation 6, to satisfy the constrain in Equation 3, the following constrain should be satisfy as

$$h_1 \geq h_2 \quad (7)$$

3. Neural Network Predictive Control

To design the neural network predictive control, two steps should be carried out:

- Determine the neural network plant model for the given nonlinear system (system identification).
- The neural network plant model is used by the controller to predict future performance.

The following subsection described the steps design of the NNPC.

3.1. Nonlinear System Modeling Using Neural Networks

System identification is the process of developing a mathematical model of a dynamic system based on the input and output data from the actual process. An important stage in control system design is the development of a mathematical model of the system to be controlled. Previous studies in system identification have demonstrated that neural networks are successful in modelling many non-linear systems [13]. Before neural networks are investigated for identification, linear techniques such as auto regressive with exogenous input (ARX) and auto regressive moving average with exogenous input (ARMAX) will be applied to nonlinear system.

The model of any nonlinear system in the form of NARMA is given as

$$y(t+1) = f\begin{pmatrix} y(t-1), y(t-2), \ldots, y(t-n_y), \\ u(t-d-1), \ldots, u(t-d-n_u) \end{pmatrix} \quad (8)$$

where *f(.)* represents an unknown nonlinear system, d represented the time delay of the system, n_y and n_u are the orders of the system. For an unknown model plant, many forms of NN model can be applied. In this work, the multilayer perceptron neural network is used to model the nonlinear system as shown in Figure 2. The neural network design to identify the plant is expressed as

$$\hat{y}(t+1) = \hat{f}\begin{pmatrix} \hat{y}(t-1), \hat{y}(t-2), \ldots, \hat{y}(t-n_y), \\ u(t-d-1), \ldots, u(t-d-n_u) \end{pmatrix} = F\left(\sum_{j=0}^{q} v_j g\left(\sum_{l=1}^{m} w_{lj} u_l + b_{1j}\right) + b_2\right) \quad (9)$$

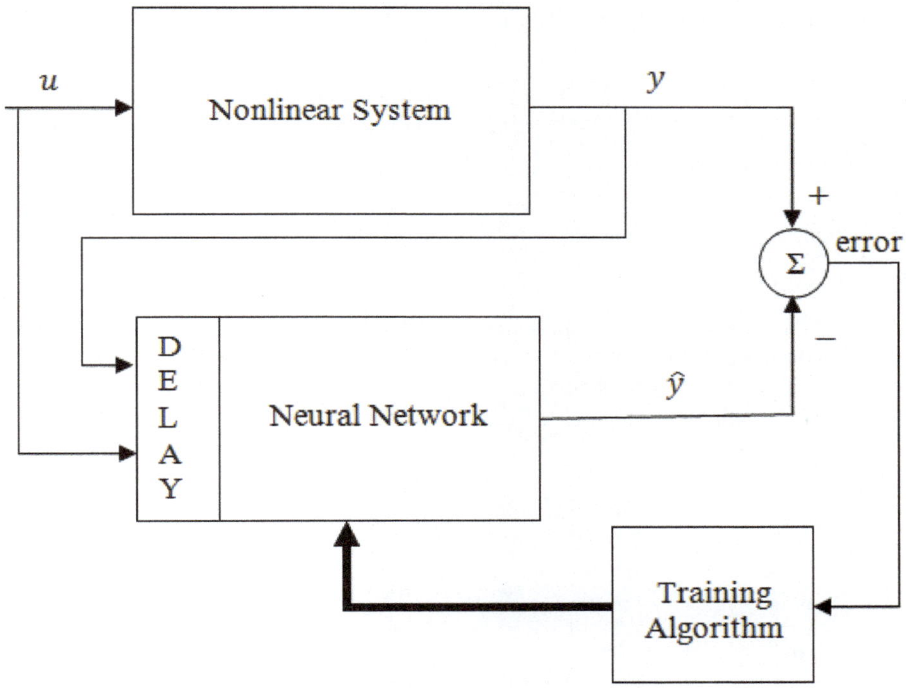

Figure 2. Nonlinear system identification

Where $\hat{y}(t)$ is the output of the neural network system, u_l represents the lth input of the neural network; v_j, w, b_1 and b_2 are the weights and biases of the NN (the adjustable parameters of the NN); *F(.)* and *g(.)* are nonlinear functions ; *q* and *m* are the number of the nodes in the hidden and input layer, respectively.

The adjustable parameters of the NN are determined from a set of examples through the process called training. The examples, or the training data as they are usually called, are a set of inputs, *u(t)*, and corresponding desired outputs, *y(t)*.

Specify the training set by

$$Z^N = \{[u(t), y(t)] | t = 1, \ldots, N\} \quad (10)$$

where *N* is the number of the samples in training set.

The objective of training is then to determine a mapping from the set of training data to the set of possible weights. Therefore, the neural network will produce predictions $\hat{y}(t)$, which in some sense are "close" to the true outputs y(t)

Because the training of neural network is often operated off-line, any training algorithms in batch version can be developed to minimize the following criterion

$$V_N = \frac{1}{2N}\sum_{i=1}^{N}\left[y(t) - \hat{y}(t|\theta)\right]^T \left[y(t) - \hat{y}(t|\theta)\right] \quad (11)$$

where θ is the trainable parameters vector.

The weights are then found as $\theta = \arg\min V_N$. By some kind of iterative minimization scheme:

$$\theta^{(i+1)} = \theta^{(i)} + \mu^{(i)} S^{(i)} \quad (12)$$

Where $\theta^{(i)}$ specified the current iterate; $S^{(i)}$ is the search direction; and $\mu^{(i)}$ the step size.

A large number of training algorithms exist, each of which is characterized by the way in which search direction and step size are selected. The MATLAB Neural toolbox provides a lot of training algorithms such as: Backpropagation algorithm, Variable Learning Rate algorithm, Resilient Backpropagation algorithm, Conjugate Gradient algorithm, Quasi-Newton Algorithm, Levenberg-Marquardt algorithm, and Reduced Memory Levenberg-Marquardt algorithm.

3.2. Predictive Controller

The objective of the predictive control strategy using neural predictors is twofold: (i) to estimate the future output of the plant and (ii) to minimize a cost function based on the error between the predicted output of the processes and the reference trajectory. The cost function, which may be different from case to case, is minimized in order to obtain the optimum control input that is applied to the non-linear plant. In most of the predictive control algorithms a quadratic form is utilized for the cost function:

$$J_P = \sum_{i=N_1}^{N_2} \left[\begin{array}{l} y(k+1) - r(k+i)^2 \\ + p\sum_{i=1}^{Nu} \left(\acute{u}(k+i-1) - \acute{u}(k+i-2) \right)^2 \end{array} \right] \quad (13)$$

with following requirements:

$$\Delta u(k+i-1) = 0, 1 \le N_u < i \le N_2 \quad (14)$$

where N_u is the control horizon; N_I is the minimum prediction horizon; N_2 is the prediction horizon; i is the order of the predictor; r is the reference trajectory; ρ is weight factor; and Δ is the differentiation operator.

The command u may be subject to amplitude constraints

$$u_{\min} \le u(k+i) \le u_{\max} \quad (15)$$

The cost function is often used with the weight factor $\rho = 0$. A very important parameter in the predictive control strategy is the control horizon N_u , which specifies the instant time, since when the output of the controller should be kept at a constant value.

The output sequence of the optimal controller is obtained over the prediction horizon by minimizing the cost function J_P with respect to the vector U. This can be achieved by setting

$$\frac{\partial J_P}{\partial U} = 0 \quad U = \left[u(k-d), \ldots, u\left(k-d+N_u-1\right) \right]^T \quad (16)$$

The analytical approach to the optimization problem needs for the differentiation of the cost function and, finally, leads to a nonlinear algebraic equation; unfortunately this equation cannot be solved by any analytic procedure. This is why a computational method is preferred for the minimization of the cost function, also complying with the typical requirements of the real-time implementations (guaranteed convergence, at least to a sub-optimal solution, within a given time interval). Figure 3 depicts the structure of the NNPC. The controller consists of the neural network plant model and the optimization block. The optimization block determines the values of u' that minimize J_P, and then the optimal u is inputted to the plant (nonlinear system).

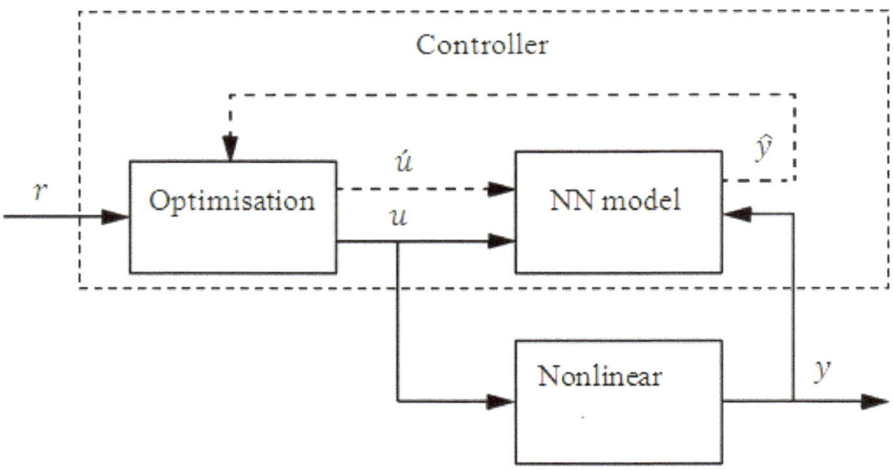

Figure 3. Structure of neural network predictive control

4. Hardware Implementation of the Neural Predictive Control

During the last years, consumer digital devices have been built using either application specific hardware modules (ASICs) or general purpose software programmed microprocessors, or a combination of them. Hardware implementations offer high speed and efficiency but they are tailored for a specific set of computations. If an alternative implementation is needed, a new and expensive design process has to be performed. On the contrary, software implementations can be modified freely during the life-cycle of a device, through patches and updates. However, they are much more inefficient in terms of speed and area. Reconfigurable computing is intended to fill the gap between hardware and software, achieving potentially much higher performance than software, while maintaining a higher level of flexibility than hardware. Reconfigurable devices, including Field-Programmable Gate Arrays (FPGAs), contain an array of computational elements whose functionality is determined through multiple programmable configuration bits. These elements, usually called logic blocks, are connected using a set of routing resources that are also programmable.

Neural networks can be implemented using analogy or digital systems. The digital implementation is more popular as it has the advantage of higher accuracy, better repeatability, lower noise sensitivity, better testability, higher flexibility, and compatibility with other types of pre-processors. Two implementation methods can be used to implement the NNPC: (i) FPGA implementation; and (ii) DSP implementation. FPGA is a suitable hardware for neural network implementation as it preserves the parallel architecture of the neurons in a layer and offers flexibility in reconfiguration.

The VHDL coding developed is tested using Xilinx 10.1 software tool to design the NNPC. This software is synthesised the VHDL codes and generated a configuration file that used to program the FPGA board. The ModelSim 6.4b software tool is used to simulate the output response of the FPGA boards. The same NNPC is designed and simulated using MATLAB software tool. This Software-Based Controller (SBC) will be used to make a comparison with the proposed design. This comparison is important because it tells us to which extent our FPGA-Based Controller (FBC) is close to similar controller designed as a computer program.

5. Simulation and Results

In this work, the Simulink Toolbox is used to simulate the NNPC and coupled tank system. Figure 4 depicts the MATLAB Simulink model of coupled tank system. While, Figure 5 shows the connection between NNPC and coupled tank system.

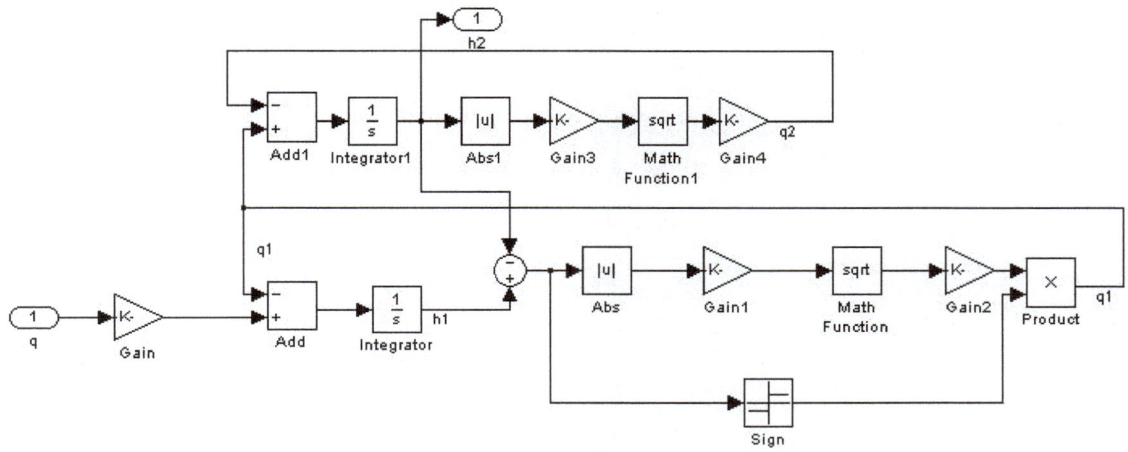

Figure 4. MATLAB Simulink model of coupled tank system

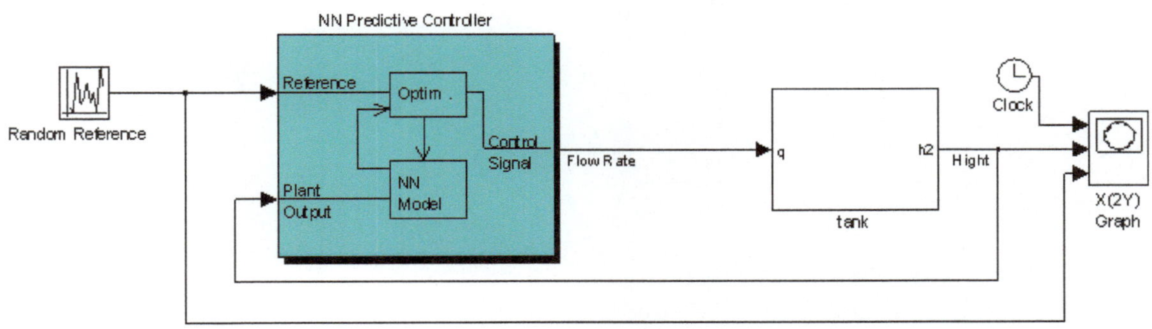

Figure 5. MATLAB Simulink model of NNPC with coupled tank system

As it is mentioned before, to design the neural network predictive control, two steps should be carried out:

- Determine the neural network plant model for the given nonlinear system (system identification).
- The neural network plant model is used by the controller to predict future performance.

To carry out the first step, the neural network is used to design the identification system for the coupled tank system. The training data should be generated by applying a random step inputs to the coupled tank Simulink model. Figure 6 shows the input/output data set that is used to design the neural network identifier.

The Levenberg-Marquardt Algorithm (LMT) is an iterative technique that locates the minimum of a multivariate function expressed as the sum of squares of nonlinear real-valued functions. It has become a standard technique for nonlinear least-squares problems [15], widely adopted in a broad spectrum of disciplines.

To design the NN identifier, we assume that it has four inputs ($y(t-1), y(t-2), u(t-1)$) and $u(t-2)$ and one output $\hat{y}(t)$. The size on hidden layer is assumed seven.

In this work, the Levenberg-Marquardt algorithm is used as training algorithm to adjust the trainable parameters of the NN. After 166 optimization iteration steps, the objective function V_N has been minimized to very small value as shown in Figure 7.

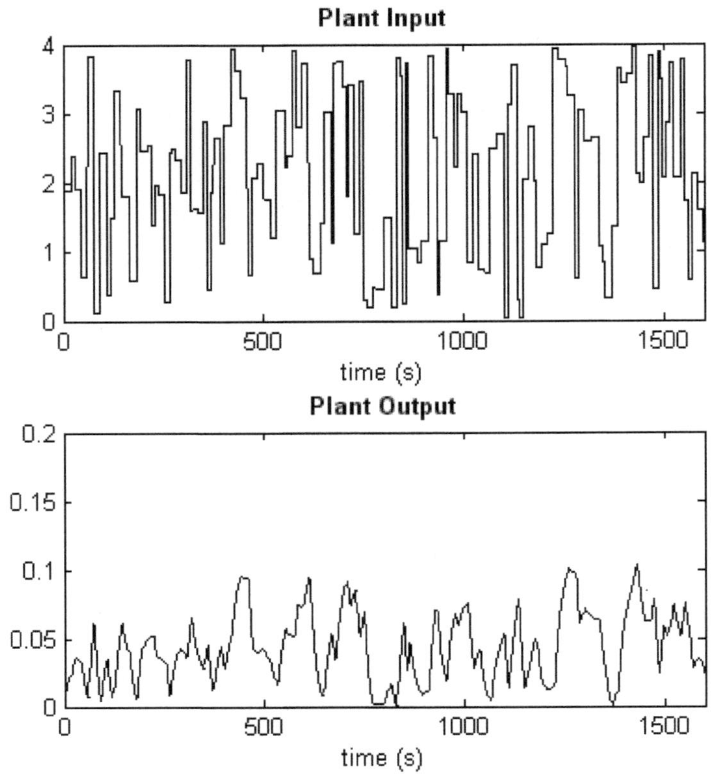

Figure 6. Input/output data training set

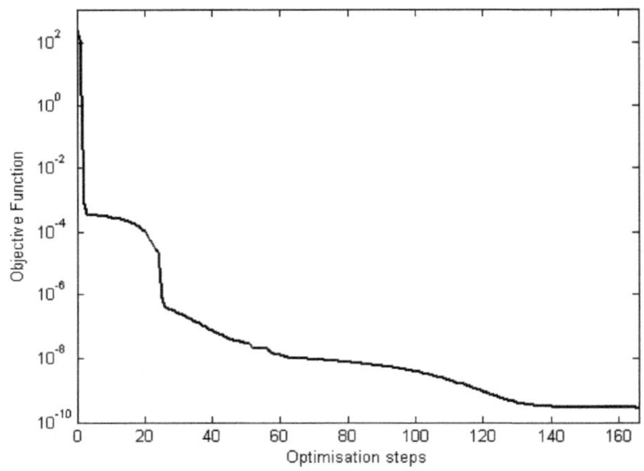

Figure 7. Objective Function during optimization steps

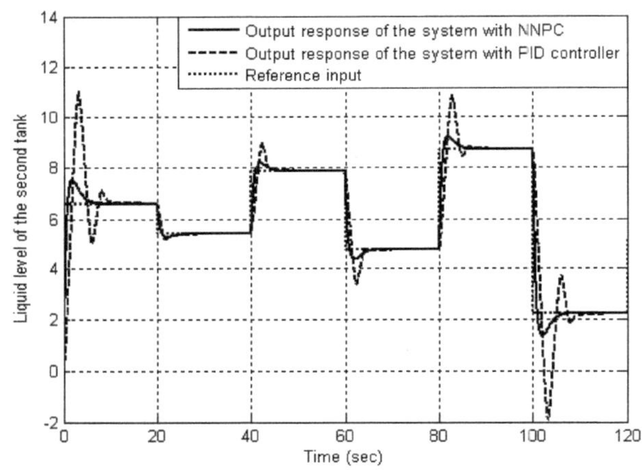

Figure 8. the output response of the coupled tank system with different control system

The optimization problem was addressed in accordance with the computational scenario built in Subsection 3.1. With respect to the notations introduced in Subsection 3.1, the following concrete values were chosen for the tuning parameters of the predictive control algorithm: N1=1, N2=10, Nu=5, and ρ =0.05. The NNPC is generated the suitable control signal to force the liquid level of the second tank to track the reference input signal. In this work, the random step input is used as the reference input signal. In order to investigate the effectiveness of the neural network based predictive controller, the output response of the controlled system with NNPC is compared with output response of the controlled system with PID controller as shown in Figure 8. The Internal model control theory is used to design the PID controller. Figure 8 shows that the overshoot and settle time of the output response are minimized.

After the design of neural network predictive controller has been completed, the FPGA board can be used to implement the proposed control system. First, VHDL codes, which describe the operation of NNP controller, are written (the flowchart of VHDL program is shown in Figure 9). This program is downloaded into the FPGA chip (XILINX Spartan XC3S700AN) by using USB cable. Then, the Hirose 100-pin FX2 Edge connector (one port of the FPGA board) can be used to connect the FPGA board with the tank system. The error between the reference input and the system output can be applied as input to the A/D converter. The digital output of the A/D converter has been applied as input data to the FPGA boards. The FPGA board generates digital inputs to the NNP controller (error and error rate). The NNP controller generates a suitable digital control signal based on the program that is stored in the FPGA chip. The digital control signal is sent to the D/A convertor to generate an analogy control signal which is applied as input to the pump of tank system. Figure 10 shows the control signal obtained from ModelSim program, while, Figure 11 shows the output signal of the NNP controller obtained from MATLAB simulation. By comparing Figure 10 and Figure 11, it can be seen that, the two output signals are identical.

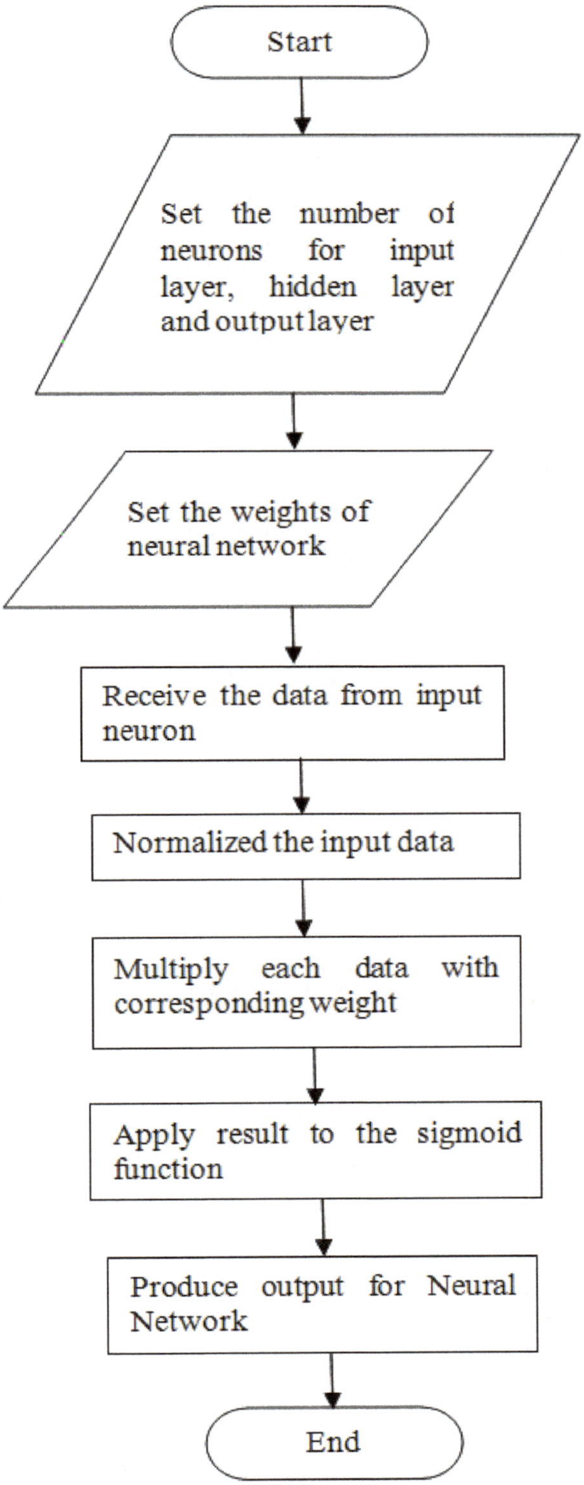

Figure 9. the flowchart of describe the Neural network using VHDL

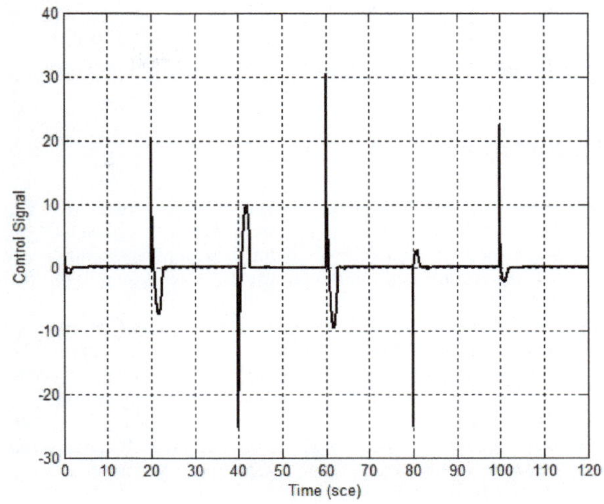

Figure 10. Control signal obtained from FPGA Board

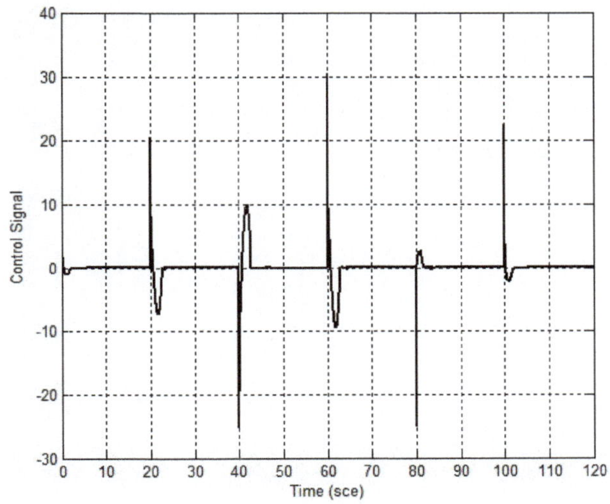

Figure 11. Control signal obtained from MALAB Simulation

6. Conclusion

The neural network based predictive control is proposed in this paper for controlling the liquid level of the nonlinear coupled tank system. Using the neural predictive controller, the liquid level of the second tank is tracked the desired set points by applying the liquid flow rate as a control signal. The Levenberg-Marquardt training algorithm is used to design the neural network identifier which is used to design the predictive control. To investigate the effectiveness of the neural network based predictive controller, the output response of the coupled tank with the NNPC is compared with the response of the controlled system with PID controller. Simulation results show that when the proposed controller is applied, all transient response specification such as: maximum overshoot, settling time delay time, rise time and peak time have been minimized. It means the NNPC can be used effectively to implement the control system for the coupled tank system practically. The obtained adjustable parameters of the NN can be used to implement the controller using an FPGA board. The results show that the control signals obtained from FPGA boards are identical to the control signals which are obtained from design simulation. Therefore, it can be arrived that the FPGA boards can be used to control the coupled tank system.

References

[1] Seng, T.L., M. Khalid, and R. Yusof, *Tuning of a Neurofuzzy Controller by Genetic Algorithms with an Application to A Coupled Tank Liquid Level Control System.* International Journal of Engineering Applications on Artificial Intelligence, 1998. 11: p. 517-529.

[2] Mahyuddin, M.N., M.R. Arshad, and Z. Mohamed. *Simulation of Direct Model Reference Adaptive Control on a Coupled-Tank System Using Nonlinear Plant Model* in *International Conference on Control, Instrumentation and Mechatronics Engineering*. 2007. Johor, Malaysia.

[3] Hassan, K.H., *Self Learning of ANFIS Inverse Control using Iterative Learning Technique.* International Journal of Computer Applications, 2011. 21(8): p. 24-29.

[4] Khan, M.K. and S.K. Spurgeon, *Robust MIMO Water Level Control in Interconnected Twin-tanks Using Second Order Sliding Mode Control.* Control Engineering Practice 2006. 14: p. 375-386.

[5] Rojas, I., et al. *Analysis and Electronic Implementation of a Fuzzy System for the Control of a Liquid Tank.* in *Proceedings of the Sixth IEEE International Conference on Fuzzy Systems* 1997. Barcelona.

[6] Saini, J.S. and Y.P. Singh, *Use of Causal Knowledge in a Real-time Fuzzy Logic Controller.* IEEE Trans. Ind, Application, 1999. 35: p. 554-560.

[7] Evans, J.T., et al., *Implementation and Performance Evaluation of an On-line Neural Network Control Scheme.* International Conference on Control, 1994.

[8] Ng, K.C., et al., *Genetic Algorithms Applied to Fuzzy Sliding Mode Controller Design.* Frst International Conference on Genetic Algorithms in Engineering Systems: Innovations and Applications, 1995.

[9] Clarke, D., C. Mohtadi, and P. Tuffs, *Generalized Predictive Control-Part I: the Basic Algorithm.* Automatica, 1987. 23(2): p. 137-148.

[10] Sedighizadeh, M., M. Rezaei, and V. Najmi, *A Predictive control based on Neural Network for Proton Exchange Membrane Fuel Cell.* World Academy of Science, Engineering and Technology, 2011. 74: p. 461-465.

[11] Lawrynczuk, M., *A Family of Model Predictive Control Algorithms with Artificial Neural Networks.* International Journal of Applied Mathematics and Computer Science, 2007. 17(2): p. 217-232.

[12] Chen, X. and Y. Li, *Neural Network Predictive Control for Mobile Robot Using PSO with Controllable Random Exploration Velocity.* International Journal of Intelligent Control and Systems, 2007. 12(3): p. 217-229.

[13] Li, X., Z. Chen, and Z. Yuan, *Simple Recurrent Neural Network-Based Adaptive Predictive Control for Nonlinear Systems.* Asian Journal of Control, 2002. 4(2): p. 231-239.

[14] Chidrawar, S. and B. Pater, *Generalized Predictive Control and Neural Generalized Predictive Control* Leonardo Journal of Sciences, 2008. 13: p. 133-152.

[15] Wilamowski, B.M., et al., *An Algorithm for Fast Convergence in Training Neural Networks*, in *International Joint Conference on Neural Networks.* 2001: Washington, DC p. 1778-1782.

Design of Robust PID Power System Stabilizer for Multimachine Power System Using HS Algorithm

D. K. Sambariya[1,*], R. Prasad[2]

[1]Department of Electrical Engineering, Rajasthan Technical University, Kota, India
[2]Department of Electrical Engineering, Indian Institute of Technology Roorkee, Roorkee, India
*Corresponding author: dksambariya_2003@yahoo.com

Abstract In this paper, the design of a proportional, derivative and integral (PID) based power system stabilizer (PSS) is carried out using a new Meta heuristic harmony search algorithm (HSA) to optimize the parameters. The design of proposed PID controller is considered with an objective function based on eigenvalue shifting to guarantee the stability of nonlinear plant for a wide range of conditions using HSA. The HSA optimized PIDPSS (HSPIDPSS) controller is applied to the standard IEEE ten-machine thirty nine-bus test power system model in decentralized manner and the performance is compared with a robust fuzzy controller. The robustness is tested by considering four plant conditions with change of active power, active load and faults at different buses of the power system to establish, the superior performance with HSPIDPSS over the FPSS.

Keywords: *Harmony Search Algorithm optimized PID Power System Stabilizer (HSPIDPSS), Fuzzy Power System Stabilizer, IEEE ten-machine thirty nine-bus test power system, proportional-derivative and integral controller*

1. Introduction

Power systems are complex multi-component dynamic systems in which the system characteristics fluctuate with varying loads and varying generation schedules. These power systems suffer by low frequency oscillations on sudden changes in load or occurrence of fault. The transfer of bulk power across weak transmission lines is hindered due to continuous persistence of such a low frequency oscillation (0.2–3.0 Hz) [1].

In early 1960s, the fast acting, high-gain automatic voltage regulators (AVR) were applied to the generator excitation system which in-turn invites the problem of low frequency electromechanical oscillations in the power system. To reduce the low-frequency oscillations, the PSS adds a stabilizing signal to AVR that modulates the generator excitation to damping electrical torque component in phase with rotor speed deviation, which increases the generator damping. The uniformly adopted type of PSS is known as conventional PSS (CPSS), which consists with the lead-lag type components. The recent development in CPSS design is carried out in [2], using Support vector machine and Harmony search algorithm for IEEJ Western Japan ten-machine power system. Similar to CPSS a Proportional, Integral Derivative (PID) controller many be connected to modulate the signal of the AVR to damp-out the small signal oscillations. The tuning of the PID gains is based on the conventional methods such as Zeigler/Nichols method, gain and phase margin method, Cohen/Coon pole placement, gain scheduling and minimum variance methods. However, these methods suffer with some limitations as (a) extensive methods to set gains, (b) difficulty to deal with gains for a large, complex and nonlinear power system, and (c) poor performance in a closed loop because of changing conditions [3].

To mitigate the shortcomings of these conventional methods; many optimization based algorithms have been proposed. The methods available in literature are as Tabu search [4], Evolutionary algorithm [5], Differential Evolution (DE) algorithm [6], Simulated Annealing, Genetic Algorithm [7], Fuzzy logic [8-11], Interval type-2 fuzzy logic [12,13], Artificial Bee Colony (ABC) [14], Particle swarm optimization [15], iterative linear matrix inequalities algorithm [16]. The above optimization methods works well but fails with the objective function as highly epistetic with large number of parameters. To such objective function, these methods may give degraded results with large computational burden.

The Harmony Search (HS) Algorithm is proposed by Geem in [17], is inspired from the process of the improvisation used by musicians to achieve the harmony. The HS algorithm [18,19] is a meta-heuristic optimization algorithm which is similar to the PSO [20] and GA [21]. It has been implemented extensively in the fields of engineering optimization [18], in recent years. It became an alternative to other heuristic algorithms like PSO [20]. It is a derivative free, meta-heuristic optimization (which doesn't use trial-and-error), inspired by the way musicians improvise new harmonies, and it uses higher-level techniques to solve problems efficiently [22]. HS can be classified as a population-based evolutionary algorithm such as GA and the PSO algorithm [23].

The article in [2]; presents the application of support vector machines (SVM) to adaptive PSS design in a multi-machine power system based on the HSA. In SVM, data from power system model as input and the SVM parameters along with model features optimized by harmony search based on the k-fold cross-validation technique. It presents a hybrid technique to use SVM and HSA in the design of PSS for IEEJ Western Japan ten machine power systems. However, the paper incorporate some limitations as (i) The parameters of CPSS are optimized to satisfy the D-stability region with damping ratio as 0.02 which is not an effective way of damping because the preferred damping ratio is always more than 0.1 as per dynamics of generators in small signal stability analysis as in [20,24] for New England ten multi-machine power system, for single-machine infinite-bus power system as in [25] and considered as even 0.2 in [26]. However, the paper has not demonstrated the plot of eigenvalue for all electromechanical modes of oscillations on s-plane concerning to all generators. Furthermore, the proposed EMOs are ($-0.2989 \pm j4.1377$, $\zeta = 0.0720$; $-0.2713 \pm j3.4281$, $\zeta = 0.0788$; $-0.3719 \pm j3.4837$, $\zeta = 0.1062$; $-0.1331 \pm j2.3868$, $\zeta = 0.0557$; $-0.1129 \pm j3.8358$, $\zeta = 0.0294$; $-0.0753 \pm j3.1278$, $\zeta = 0.0241$) very weak in nature. As per these EMOs, it is very clear that the location is very near to imaginary axis of the s-plane, which cannot be suggestive to guarantee the stability of the power system (ii) It is observed that some of the results for speed variation are settling at about 50 seconds, which is due to lesser damping factor selection otherwise the settling time should be around 5-10 seconds as in [20,25] and clearly mentioned in [26] (iii) No quantitative analysis of the response such as the performance indices (ITAE, IAE and ISE) are calculated for the proposed and the comparing responses. (iv) The power system considered in this article is IEEJ Western Japan ten-machine power system which is lesser a complex system as compared to the IEEE New England ten-machine thirty nine-bus power system model in terms of interconnection and design as well. As per above limitations, it seems that a very basic model of harmony search algorithm may be used to tune the CPSS parameters of the power system model.

The limitations observed as above are mitigated in this article by plot of dominant electromechanical mode of oscillation on the s-plane and verifying in D-shape sector drawn with maximum value of real part of EMO as -1 ($\sigma < -1$) and minimum value of damping factor as 0.1 ($\xi > 0.1$). The robustness is tested by considering four plant conditions with change of active power, active load and faults at different buses of the power system to establish, the superior performance with HSPIDPSS over the FPSS. The superiority of HSPIDPSS is validated in terms of settling time and the performance indices (ITAE, IAE and ISE) as lesser in magnitude as compared to FPSS.

In the organization of paper, the problem is formulated in section 2. A review on test system, PID based PSS design fundamental and the objective function are introduced in this section. An overview on Harmony Search Algorithm which is used to optimize the PID controller parameters is introduced in section 3. The performance analysis is carried out in section 4 for IEEE ten-machine thirty nine-bus test power system model. Lastly the analysis is concluded in section 5, followed by references.

2. Problem Formulations

2.1. Test System Representation

The general representation of a power system using nonlinear differential equations can be given by

$$\dot{X} = f(X, U) \tag{1}$$

Where, X and U represents the vector of state variables and the vector of input variables. As in [21], the power system stabilizers can be designed by use of the linearized incremental models of power system around an operating point. The system representation based on differential equations and used data is given in [13,27]. The state equations of a power system can be written as

$$\Delta\dot{X} = A\Delta X + BU \tag{2}$$

Analysis of ten-machine thirty nine-bus power system as in Figure 1 can be carried out by simultaneous solution of equations consisting of prime movers, synchronous machines with excitation systems, transmission line network, dynamic and static loads, and other devices like static VAR and HVDC converters based compensators. The dynamics of generator rotors, excitation, prime movers, and other related devices are being represented by differential equations. Thus, the complete multi-machine model consists of large numbers of ordinary differential equations (ODE) and algebraic equations. These are linearized about an operating point (nominal) to derive a linear model for the small signal oscillatory behaviour of power systems [28,29]. The range of variation in operating point can generate a set of linear models corresponding to each operating point/condition.

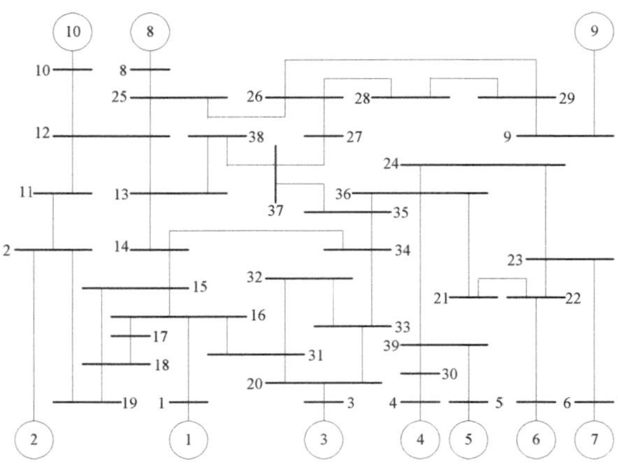

Figure 1. Line diagram of New England 39 Bus test Power System

The state equations of a power system, consisting 'N' number of generators and N_{pss} number of power system stabilizers can be written as in Eqn. (2). Where, A is the system matrix with order as $4N\times 4N$ (40×40) and is given by $\partial f / \partial X$, while B is the input matrix with order $4N\times N_{pss}$ (40×10) and is given by $\partial f / \partial U$. The order of state vector ΔX is $4N\times 1$ (40×1), the order of ΔU is $N_{pss}\times 1$

(10×1). Here, the well known Heffron-Phillip linearized model is used to represent the large multimachine power system as in Figure 2.

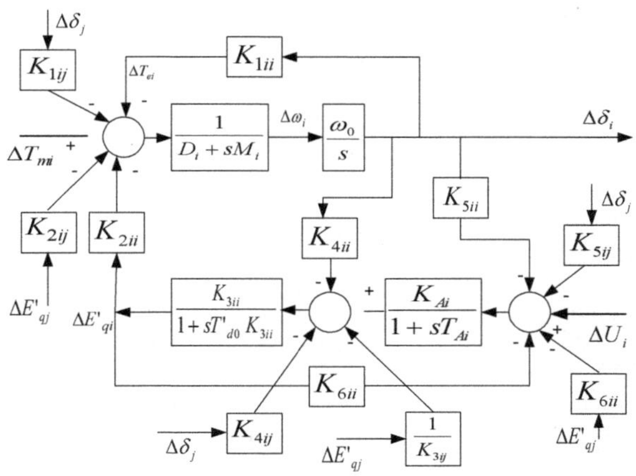

Figure 2. Heffron-Philips Model for Multimachine Power System

2.2. PID tuning Scheme

Since last 60-70 years, the Proportional, Integral and derivate (PID) controller is most popularly known and used feedback controller in the field of complex process industries. It can provide excellent robust control performance over a wide range of operating conditions of a power system because of its three different modes of operation [1,16]. The proportional controller mode can reduce the rise time but unable to reduce the steady-state error of the response. The higher value of proportional gain may cause the system to become an unstable but lower value make in-sensitive or lesser sensitive to even large value of error [1]. The derivative control mode improves the system stability by reducing overshoot and improving transient response. The integral control mode of operation may eliminate the steady-state error but may worsen the transient response of the system. The lower value of integral gain value makes a system sluggish while the higher value causes the random increase in the overshoot [4,16]. Therefore, to design the PID controller, all three gains require special attention to get the control signal by the trial-and-error method based on the experience and plant behaviour. The block diagram representation of PID controller for a closed-loop system is shown in Fig. 3 and mathematically represented by Eqn. (3).

$$Upid = K_p \Delta\omega(t) + K_i \int_0^{T_{SIM}} \Delta\omega(t)\, dt + K_d \frac{\partial \Delta\omega(t)}{\partial t} \tag{3}$$

2.3. Objective Function

To increase the system damping over a wide range of operating conditions and configuration of a power system, a robust tuning must be incorporated. Therefore, PSS design is formulated as an eigenvalue-based objective function. The two sub-objective functions based on minimization of real part of eigenvalue and maximization of damping factor are taken into consideration for optimization of PID parameters [24,26].

The optimization constraints are the limits/bounds on the optimized parameters such as gain and the time constants. Thus, the optimization problem is subjected to.

$$\begin{cases} K_p^{\min} \le K_p \le K_p^{\max} \\ K_d^{\min} \le K_d \le K_d^{\max} \\ K_i^{\min} \le K_i \le K_i^{\max} \end{cases} \tag{4}$$

Typical ranges of the optimized parameters are selected as -5.00 to 50.00 for $K_{p,}$ K_d and Ki. Considering one of the above objectives, the proposed approach employs HS algorithm to solve this optimization problem for an optimal set of PSS parameters. Considering i[th] eigenvalue representation as $\lambda_i = \sigma_i \pm j\omega_i$ and the associated damping ratio expression as $\xi_i = -\sigma_i / \sqrt{\sigma_i^2 + \omega_i^2}$ [26].

$$J = \sum_{i=1}^{n} (\sigma_0 - \sigma_i)^2 + \alpha \sum_{i=1}^{n} (\xi_0 - \xi_i)^2 \tag{5}$$

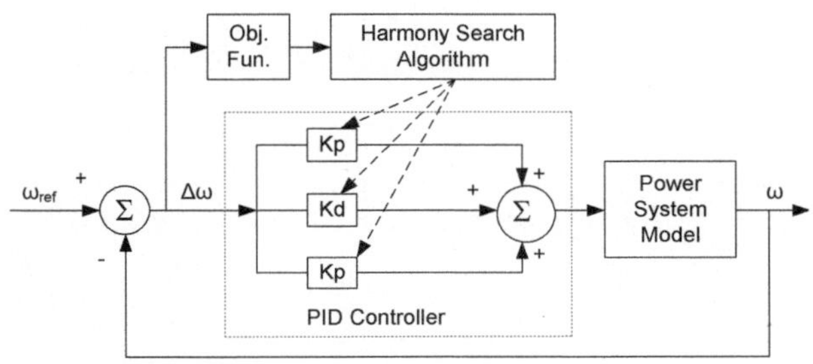

Figure 3. PID controller tuned by Harmony Search Algorithm

3. Overview of Tuning Methodology: HSA

This algorithm is an evolutionary meta-heuristic which is inspired by a natural phenomenon. It is inspired by the method used by musicians to improvise new optimal harmonies. In this algorithm, an analogy in between optimization and improvisation is designed such that to mimic the process used by musicians to get the note that results to a finest pleasing harmony, whenever, musicians are synchronized with other musicians [30,31].

3.1. Procedural Steps of HS Algorithm

Each row of harmony memory (HM), consists of *N* decision variables and the fitness score *w* ([x^1, x^2... *w*]). The HM is initialized with HMS randomly generated solution vectors [32,33].

Step 1: Initialization Process: Let in an optimization problem, the objective function is represented by Minimization of *F(x)* which is subjected to $x_i \in X_i$, *i=1, 2, 3 ... N.* Where, x is the set of design variables (x_i), ($x_i^L \le x_i \le x_i^U$), and N is the count of design variables.

1. Define the variable limits as lower (x_i^L) and upper (x_i^U) or $x_i^L \le x_i \le x_i^U$.

2. Deciding the value of harmony memory size (HMS), from the range $10 \le \text{HMS} \le 100$.

3. Decide value of HMCR (harmony memory consideration rate) within the range $0.0 \le \text{HMCR} \le 1.0$

$$x'_1 \leftarrow \begin{cases} x'_i \in \{x_i^1, x_i^2, \ldots, x_i^{HMS}\} & : prob.\ HMCR \\ x'_i \in X_i & : prob.(1-HMCR) \end{cases} \quad (6)$$

4. Decide the value of PAR (pitch adjustment rate) from the range $0.0 \le \text{PAR} \le 1.0$

$$x_j^i = x_j^L + rand(0,1) \times (x_j^U - x_j^L) \quad (7)$$

$$x'_i \leftarrow \begin{cases} yes & with\ probability\ PAR \\ no & with\ probability\ (1-PAR) \end{cases} \quad (8)$$

5. Compute the step size (b_i) as $b(i) = (x_i^U - x_i^L)/N$

6. Specify the maximum limit of iteration number.

Step 2: HM Initiation: The HM matrix as in Eqn. (9) is filled with randomly generated possible solution vectors for HMS and is sorted by the values of the objective function f(x).

$$HM = \begin{bmatrix} x_1^1 & x_2^1 & \cdots & x_{N-1}^1 & x_N^1 \\ x_1^2 & x_2^2 & \cdots & x_{N-1}^2 & x_N^2 \\ \vdots & \vdots & \vdots & \vdots & \vdots \\ x_1^{HMS-1} & x_2^{HMS-1} & \cdots & x_{N-1}^{HMS-1} & x_N^{HMS-1} \\ x_1^{HMS} & x_2^{HMS} & \cdots & x_{N-1}^{HMS} & x_N^{HMS} \end{bmatrix} \Rightarrow \begin{bmatrix} f(x^1) \\ f(x^2) \\ \vdots \\ f(x^{HMS-1}) \\ f(x^{HMS}) \end{bmatrix} \quad (9)$$

Step 3: Improvisation Process: A New Harmony vector $x' = (x'_1, x'_2, \ldots, x'_n)$ is generated based on three criteria: random selection, memory consideration, and pitch adjustment.

Random Selection: To decide the value of x'_1 for the New Harmony $x' = (x'_1, x'_2, \ldots, x'_n)$, the HS algorithm randomly selects a value from range with a probability of "*1 - HMCR*".

Memory Consideration: To decide the value of x'_1, the HS algorithm randomly selects a value x_1^j from the HM with a probability of "*HMCR*", where *j = 1, 2,…, HMS*. It can be represented as in Eqn. (9).

Pitch Adjustment: Each element of the New HM vector $x' = (x'_1, x'_2, \ldots, x'_n)$ is subjected to determine whether it should be pitch-adjusted or not. The selected x'_i is further adjusted by adding an amount to the value with a probability of PAR. The PAR parameter is called the probability of pitch adjustment and is represented by Eqn. (8).

The *'no'* with probability *'1 - PAR'* represents that the probability of not adding any amount. On the other hand, if the pitch adjustment decision for x'_i is yes, then x'_i is replaced by $x'_i \leftarrow x'_i \pm bw$; where, bw is distance bandwidth in case a continuous variable. The pitch adjustment is performed on every variable of the New HM vector.

Step 4: Updating the HM: Let the HM vector is $x' = (x'_1, x'_2, \ldots, x'_n)$, which is resolved by minimization of objective function is better than the worst harmony present in the HM. Therefore, the New Harmony is inserted into the HM, while, the worst harmony is removed from the HM.

Step 5: Checking the stopping criterion: If the maximum count of improvisations is reached and the stopping criterion as maximum number of iterations is satisfied, then the process of computation is terminated. Otherwise, go to steps 3 and 4 to repeat the process. The HS algorithm is shown in the Figure 4.

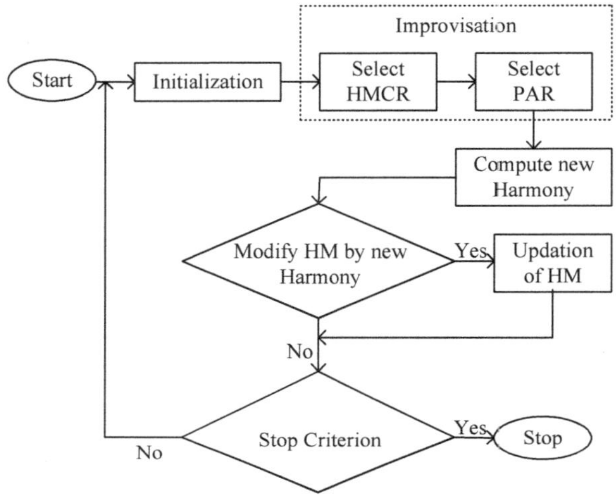

Figure 4. Harmony Search Algorithm

4. Results and Discussions

4.1. Experimental Plant Creation

The line diagram of ten-machine thirty nine-bus power system (New England thirty nine bus test system) as in Figure 1, while the system data are as in [21]. A linear representation of the system without PSSs is formed around a nominal point and the power system models (plants) are generated by applying change in active power, active load and fault at different bus numbers as in Table 1. The plants generated are subjected to operate without PSS (i.e. under open loop) and the speed responses of generators (ten in number) in each plant are recorded. It is clear that none of generator of any plant is showing stable operation. The proposed HSPIDPSS and FPSS would be implemented on these plants and the performance analysis is carried out.

4.2. HSA Optimized PID Parameters

The above generated plants as in Table 1 are equipped with the PID controller as in Figure 3 and optimized by Harmony Search algorithm subjected to the objective function (as in section 2.3) with the parametric bounds such as $-5 \le K_p \le 50$, $-5 \le K_d \le 50$ and $-5 \le K_i \le 50$. The optimized parameters of controllers connected to generators for plant-1 are enlisted in the Table 2. The performance of the HSA during optimization for 200 iterations is shown in Figure 5 under function minimization and in terms of fitness function value. This figure incorporates the maximum function value at starting as 0.0018 and minimum at the end of 200 iterations as 9.4803×10^{-4}.

Table 1. Ten-machine thirty nine-bus power system plants

Plant	Active Power	Fault Location
1	[5.519816;10.00; 6.5000; 5.080; 6.3200; 6.500; 5.60000; 5.400; 8.3000; 2.500]	Bus No. – 16, Base case for active power and active load
2	[5.519816; 10.0; 6.6000; 5.08000; 6.1200; 6.5000; 5.60000; 5.4000; 8.30000; 2.5000]	Bus No. – 13, Active power of Gen -3 and 5 are changed., Active load of Gen -1 and 2 are changed.
3	[5.419816; 10.10; 6.4000; 5.18000; 6.32000; 6.5000; 5.60000; 5.4000; 8.3000; 2.50000]	Bus No. – 11, Active power of Gen – 1, 2, 3 and 4 are changed, Active load of Gen -2 and 4 are changed.
4	[5.519816; 10.0; 6.5000; 5.08000; 6.32000; 6.5000; 5.50000; 5.5000; 8.3000; 2.50000]	Bus No. – 9, Active power of Gen – 7 and 8 are changed, Active load of Gen -8 and 9 are changed.

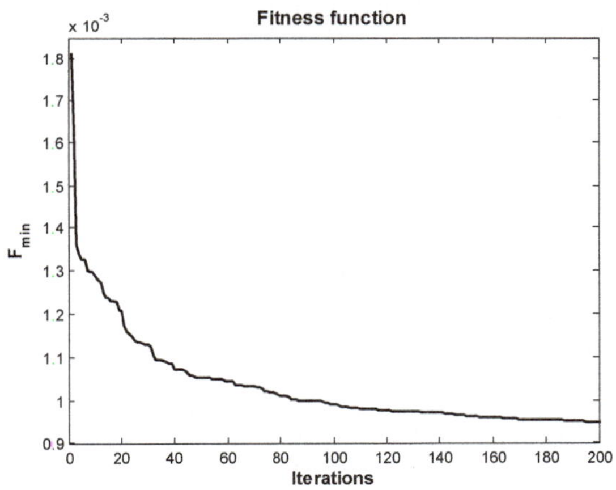

Figure 5. Plot of fitness function with iterations

Table 2. The harmony search tuned PID parameters for ten generators of plant 1-4

Generators	Kp	Kd	Ki
Gen-1	49.9948	15.5289	38.1243
Gen-2	48.9951	49.9127	49.2617
Gen-3	46.9101	23.3802	48.6788
Gen-4	36.2787	48.5096	46.2030
Gen-5	49.2328	35.6628	46.8553
Gen-6	48.6677	39.2584	49.7246
Gen-7	45.7080	11.1073	49.6559
Gen-8	47.4491	33.3796	44.4684
Gen-9	46.3637	13.7397	43.7974
Gen-10	46.9797	19.5897	45.1682

The initializing parameters of the harmony search algorithm to optimize the controller parameters are as the number of variables (NVAR) is 20, maximum number of iterations as 200, memory size (HMS) as 40 (equal to the order of the plant matrix), harmony consideration rate (HMCR) as 0.9, minimum pitch adjusting rate (PAR_{min}) as 0.2, maximum pitch adjusting rate (PAR_{max}) as 0.5, minimum (bw_{min}) and maximum (bw_{max}) value of bandwidth as 0.0001 and 1, respectively.

4.3. Eigenvalue Analysis

The test system without PSS and equipped with HSPIDPSS is subjected to linear mode for determination of connected eigenvalue for different plant conditions as mentioned in Table 1. The nonlinear simulation of the test system without PSS reveals that none of the generators are giving a stable response; therefore, the related damping factor is negative and the plot of the eigenvalue is incorporated in Figure 6 with circular red colour for all four plant conditions. The eigenvalue of the test system with the HSPIDPSS associated to different plant conditions are represented in Figure 6 and tabulated in Table 3. The eigenvalue plot of the system with HSPIDPSS satisfies the d-shape sector condition ($\sigma < -1$ and $\xi > 0.1$) as mentioned in section 2.3. No eigenvalue having real part (σ) as > -1 and associating damping factor (ξ) as < 0.1 can prove system stability within the D-shape sector. The test system without PSS has 20 electromechanical modes of oscillations (EMOs) out of total 40 eigenvalue. On application of HSPIDPSS, one eigenvalue to the real axis of s-plane is added for a generator, and the EMO is being adjusted to giving stable operation of the generators. Therefore, the total eigenvalue of the test system becomes as 50 with HSPIDPSS. On the other hand, application of conventional PSS as in [2] adds one extra EMO and one eigenvalue at real axis, making total eigenvalue as 60.

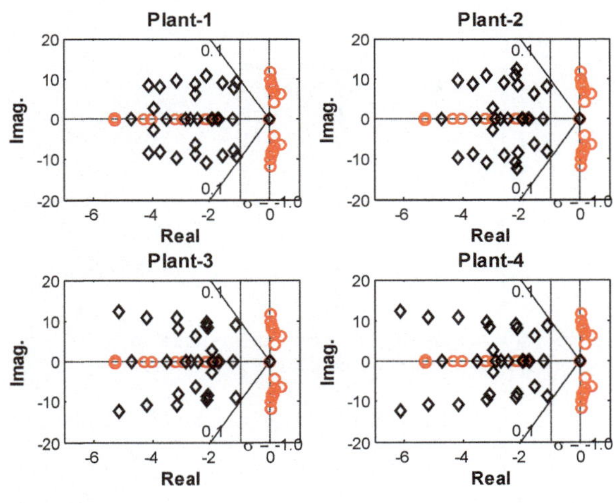

Figure 6. Plot of dominant EMOs for plant 1 to 4 of test system without PSS and with HSPIDPSS

4.4. Speed Response Analysis

Since the objective of the paper is to optimize the PID controller parameters using Harmony Search Algorithm, therefore, the system responses without PSS, with FPSS as in [19] and with HSPIDPSS are compared. Since four plants consists forty generator responses, therefore, it is not possible to show generator wise response analysis in graphical form. It is clear that the speed response with HS optimized PIDPSS is quickly settling as compared to the response with fuzzy PSS for all generators. To be considerate of space constraint, the speed response with FPSS and with HSPIDPSS for plant 1, 2, 3 and 4 are shown in Fig. 7(a), 7(b), 8(a), 8(b), 9(a), 9(b), 10(a), and 10(b), respectively. However, to four plants in terms of settling time with FPSS and with HSPIDPSS is provided in Table 4. The oscillations are damped out much faster of the power system with HSPIDPSS as compared to system equipped with FPSSs. This illustrates the potential and superiority of the proposed design approach to obtain an optimal set of PID parameters using HSA (HSPIDPSS).

Table 3. Representation of least damped EMOs with damping factor and frequency in Hz of all ten generators for plant 1 to 4

Gen	Plant-1	Plant-2	Plant-3	Plant-4
1	-1.1300+9.3983i [0.1194, 1.4958]	-2.1302 +12.3967i [0.1694, 1.9730]	-5.1302 +12.3956i [0.3824, 1.9728]	-6.1316 +12.3960i [0.4434, 1.9729]
2	-1.1869 + 7.9126i [0.1483, 1.2593]	-2.1843 +10.9059i [0.1964, 1.7357]	-4.1876 +10.9123i [0.3583, 1.7367]	-5.1861 +10.9142i [0.4292, 1.7370]
3	-2.1620 +10.7767i [0.1967, 1.7152]	-3.1619 +10.7789i [0.2815, 1.7155]	-3.1626 +10.7747i [0.2816, 1.7149]	-4.1610 -10.7757i [0.3602, 1.7150]
4	-3.1536 + 9.7621i [0.3074, 1.5537]	-4.1542 + 9.7615i [0.3916, 1.5536]	-2.1535 + 9.7610i [0.2154, 1.5535]	-3.1542 - 9.7666i [0.3073, 1.5544]
5	-1.5712 - 9.1341i [0.1695, 1.4537]	-2.5717 + 9.1386i [0.2709, 1.4545]	-1.1706 + 9.1343i [0.1271, 1.4538]	-2.1716 - 9.1336i [0.2313, 1.4537]
6	-2.5303 - 8.8194i [0.2758, 1.4037]	-3.6301 + 8.8220i [0.3805, 1.4041]	-2.1320 + 8.8187i [0.2350, 1.4035]	-1.1298 - 8.8204i [0.1270, 1.4038]
7	-3.7228 - 8.1134i [0.4170, 1.2913]	-1.1231 + 8.1171i [0.1371, 1.2919]	-3.1223 + 8.1169i [0.3590, 1.2919]	-2.1230 - 8.1138i [0.2531, 1.2914]
8	-4.1008 - 8.3493i [0.4409, 1.3288]	-2.1014 + 8.3562i [0.2439, 1.3299]	-2.1010 + 8.3492i [0.2440, 1.3288]	-3.1015 - 8.3544i [0.3480, 1.3296]
9	-2.5359 - 6.3300i [0.3719, 1.0075]	-1.5355 + 6.3284i [0.2358, 1.0072]	-2.5367 - 6.3356i [0.3717, 1.0083]	-1.5357 - 6.3290i [0.2358, 1.0073]
10	-3.9570 + 2.7781i [0.8184, 0.4422]	-2.9627 + 2.7840i [0.7287, 0.4431]	-1.9562 - 2.7756i [0.5761, 0.4417]	-2.9573 - 2.7791i [0.7287, 0.4423]

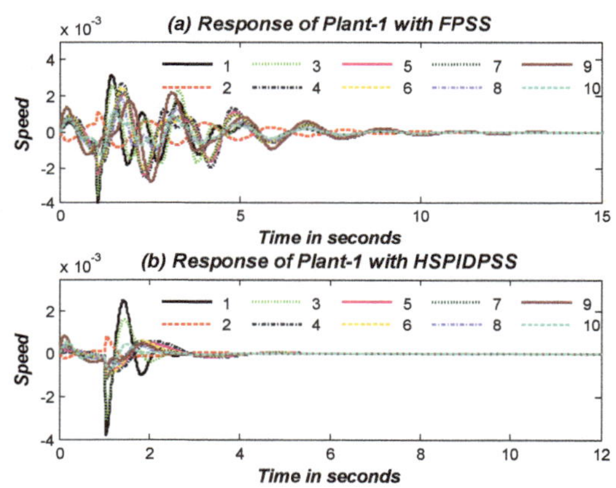

Figure 7. *Speed response of* ten generators for (a) Plant-1 with Fuzzy PSS, and (b) Plant-1 with HSA tuned PID based PSS

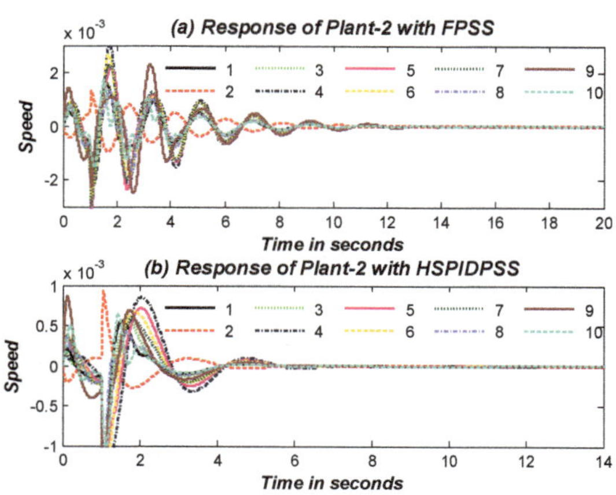

Figure 8. Speed response of ten generators for (a) Plant-2 with Fuzzy PSS, and (b) Plant-2 with HSA tuned PID based PSS

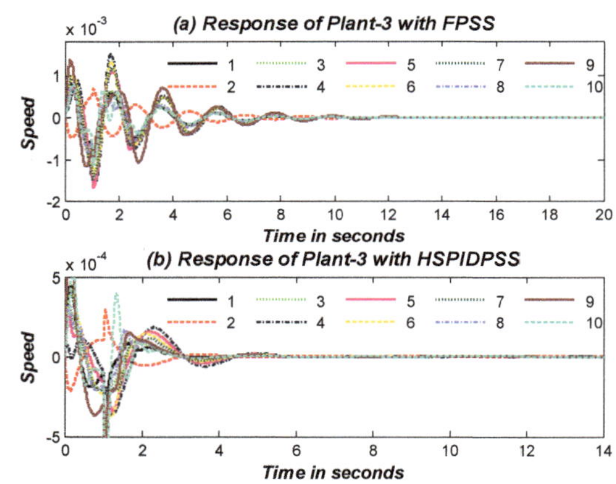

Figure 9. Speed response of ten generators for (a) Plant-3 with Fuzzy PSS, and (b) Plant-3 with HSA tuned PID based PSS

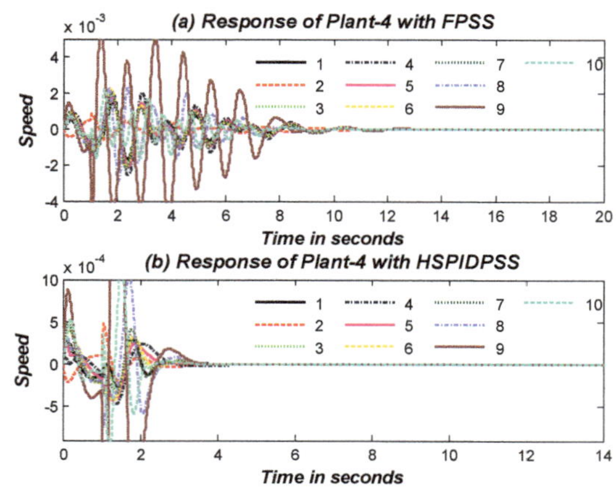

Figure 10. Speed response of ten generators for (a) Plant-4 with Fuzzy PSS, and (b) Plant-4 with HSA tuned PID based PSS

4.5. Performance Index Based Analysis

For completeness and clear perceptiveness about the system response for all the system conditions, three performance indices (PIs) that reflect the settling time and overshoot are introduced and evaluated. These indices are defined as.

ITAE: Integral of the Time-Weighted Absolute Error

$$ITAE = \int_{t=0}^{T=t_{sim}} t\left|\Delta\omega(t)\right|dt \tag{10}$$

ISE: Integral Square Error

$$ISE = \int_{t=0}^{T=t_{sim}} \left|\Delta\omega(t)\right|^2 dt \tag{11}$$

IAE: Integral of the Absolute Error

$$IAE = \int_{t=0}^{T=t_{sim}} \left|\Delta\omega(t)\right|dt \tag{12}$$

where, t_{sim} is the simulation time of the system. The performance indices as ITAE calculated for speed response of plants 1 to 4, without PSS (open), with FPSS and with HSPIDPSS are mentioned in Table 5. Here, it should be cleared that the least value of performance index represents the best performance of the system. As in plant 1 to 4, the PIs with HSPIDPSS as compared to without PSS (open) and with PIDPSS are least. The response with least value of performance index as compared to others represents best speed response performance.

Table 4. Settling time of Speed response of different generators of different plants 1-4

Generators of a plant	Plant-1		Plant-2		Plant-3		Plant-4	
	With FPSS	With HSPIDPSS	With FPSS	With HSPIDPSS	With FPSS	With HSPIDPSS	With FPSS	With HSPIDPSS
Gen-1	9.80	2.73	9.05	4.18	8.27	1.62	10.8	2.50
Gen-2	11.60	4.30	12.60	4.54	9.22	3.02	11.82	2.94
Gen-3	9.71	3.64	11.46	3.84	8.32	4.14	10.97	2.80
Gen-4	12.52	4.47	12.74	5.54	10.37	5.60	13.04	2.83
Gen-5	11.98	4.32	13.10	5.17	10.61	4.24	13.28	2.45
Gen-6	10.85	4.38	11.52	4.53	11.67	3.13	14.15	2.64
Gen-7	11.78	4.26	13.89	4.40	12.66	5.06	13.24	2.58
Gen-8	11.85	4.43	11.45	4.30	10.71	1.86	11.09	3.26
Gen-9	11.75	4.11	13.83	4.28	12.62	2.14	13.36	2.96
Gen-10	9.54	3.54	10.34	4.11	9.31	2.09	10.14	2.42

Table 5. Performance indices (ITAE, IAE & ISE) based analysis of test system with controllers (FPSS, HSPIDPSS) & with No-PSS

Power System Model	Performance Index	No-PSS	FPSS	HSPIDPSS
Plant-1	ITAE	10.1103	0.1835	0.0162
	ISE	33.7468	7.0380E-06	6.0821E-06
	IAE	14.2518	1.0153E-02	9.7441E-03
Plant-2	ITAE	9.1624	0.1763	0.0205
	ISE	28.0987	5.4935E-06	4.1689E-06
	IAE	12.8673	1.0022E-02	1.0918E-03
Plant-3	ITAE	8.7297	0.0656	0.0063
	ISE	26.3265	1.0135E-06	1.0039E-06
	IAE	12.0369	4.3459E-03	3.3941E-04
Plant-4	ITAE	9.9514	0.2134	0.0114
	ISE	33.0254	6.8345E-06	95.2511E-06
	IAE	14.1362	8.5675E-03	7.7710E-04

5. Conclusions

In this paper, the proportional integral derivative based power system stabilizer is designed for IEEE ten-machine thirty nine–bus test power system model. The constants of PID controller are optimized by using Harmony Search Algorithm with an objective function based on eigenvalue shifting. The speed response of the generator for the multimachine power system without PSS, with fuzzy PSS and with the harmony search algorithms based optimized PIDPSS are compared. It is found that the response with FPSS can stabilize the system under adverse operating conditions (under several faults) with prolonged settling time as compared against a few of seconds with HSPIDPSS. The superior performance of HSPIDPSS is further illustrated by using performance indices associated to speed response as the value is least as compared to the system without PSS and with FPSS. The linear mode of the test system with HSPIDPSS is analyzed to guarantee system stability in D-shape sector on s-plane.

References

[1] M. Soliman, A. L. Elshafei, F. Bendary, and W. Mansour, "Robust decentralized PID-based power system stabilizer design using an ILMI approach," *Electric Power Systems Research,* vol. 80, pp. 1488-1497, 2010.

[2] J. Pahasa and I. Ngamroo, "Adaptive Power System Stabilizer Design Using Optimal Support Vector Machines Based on Harmony Search Algorithm," *Electric Power Components and Systems,* vol. 42, pp. 439-452, 2014/04/04 2014.

[3] P. Kundur, M. Klein, G. J. Rogers, and M. S. Zywno, "Application of power system stabilizers for enhancement of overall system stability," *IEEE Transactions on Power Systems,* vol. 4, pp. 614-626, 1989.

[4] A. Bagis, "Tabu search algorithm based PID controller tuning for desired system specifications," *Journal of the Franklin Institute,* vol. 348, pp. 2795-2812, 2011.

[5] S. Panda, "Multi-objective evolutionary algorithm for SSSC-based controller design," *Electric Power Systems Research,* vol. 79, pp. 937-944, 2009.

[6] S. Panda, "Differential evolution algorithm for SSSC-based damping controller design considering time delay," *Journal of the Franklin Institute,* vol. 348, pp. 1903-1926, 2011.

[7] J. Zhang, J. Zhuang, H. Du, and S. Wang, "Self-organizing genetic algorithm based tuning of PID controllers," *Information Sciences,* vol. 179, pp. 1007-1018, 2009.

[8] D. K. Sambariya and R. Prasad, "Optimal Tuning of Fuzzy Logic Power System Stabilizer Using Harmony Search Algorithm," *International Journal of Fuzzy Systems,* pp. 1-14, 2015/06/17 2015.

[9] D. K. Sambariya, R. Gupta, and A. K. Sharma, "Fuzzy Applications to Single Machine Power System Stabilizers," *Journal of Theoretical and Applied Information Technology,* vol. 5, pp. 317-324, 2009.

[10] R. Gupta, D. K. Sambariya, and R. Gunjan, "Fuzzy Logic based Robust Power System Stabilizer for Multi-Machine Power System," in *IEEE International Conference on Industrial Technology, ICIT 2006.*, 2006, pp. 1037-1042.

[11] D. K. Sambariya, "Power system stabilizer design using compressed rule base of fuzzy logic controller," *Journal of Electrical and Electronic Engineering,* vol. 3, pp. 52-64, July 1, 2015 2015.

[12] D. K. Sambariya and R. Prasad, "Evaluation of interval type-2 fuzzy membership function & robust design of power system stabilizer for SMIB power system," *Sylwan Journal,* vol. 158, pp. 289-307, April 2014 2014.

[13] D. K. Sambariya and R. Prasad, "Power System Stabilizer design for Multimachine Power System using Interval Type-2 Fuzzy Logic Controller," *International Review of Electrical Engineering (IREE),* vol. 8, pp. 1556-1565, October 2013 2013.

[14] H. Gozde and M. C. Taplamacioglu, "Comparative performance analysis of artificial bee colony algorithm for automatic voltage regulator (AVR) system," *Journal of the Franklin Institute,* vol. 348, pp. 1927-1946, 2011.

[15] W.-D. Chang and S.-P. Shih, "PID controller design of nonlinear systems using an improved particle swarm optimization approach," *Communications in Nonlinear Science and Numerical Simulation,* vol. 15, pp. 3632-3639, 2010.

[16] H. Bevrani, T. Hiyama, and H. Bevrani, "Robust PID based power system stabiliser: Design and real-time implementation," *International Journal of Electrical Power & Energy Systems,* vol. 33, pp. 179-188, 2011.

[17] Zong Woo Geem, Joong Hoon Kim, and G. V. Loganathan, "A New Heuristic Optimization Algorithm: Harmony Search," *Simulation,* vol. 76, pp. 60-68, February 1, 2001 2001.

[18] Z. W. Geem, "Harmony Search Applications in Industry," in *Studies in Fuzziness and Soft Computing, Volume 226, Soft Computing Applications in Industry.* vol. 226, B. Prasad, Ed., ed: Springer-Verlag Berlin Heidelberg, 2008, pp. 117-134.

[19] D. K. Sambariya and R. Prasad, "Design of Harmony Search Algorithm based tuned Fuzzy logic Power System Stabilizer," *International Review of Electrical Engineering (IREE),* vol. 8, pp. 1594-1607, October 2013 2013.

[20] S. M. Abd-Elazim and E. S. Ali, "A hybrid Particle Swarm Optimization and Bacterial Foraging for optimal Power System Stabilizers design," *International Journal of Electrical Power & Energy Systems,* vol. 46, pp. 334-341, 2013.

[21] H. Alkhatib and J. Duveau, "Dynamic genetic algorithms for robust design of multimachine power system stabilizers," *International Journal of Electrical Power & Energy Systems,* vol. 45, pp. 242-251, 2013.

[22] L. Wang, R. Yang, Y. Xu, Q. Niu, P. M. Pardalos, and M. Fei, "An improved adaptive binary Harmony Search algorithm," *Information Sciences,* vol. 232, pp. 58-87, 2013.

[23] J. Yu and P. Guo, "Improved PSO Algorithm with Harmony Search for Complicated Function Optimization Problems," in *Advances in Neural Networks – ISNN 2012.* vol. 7367, J. Wang, G. Yen, and M. Polycarpou, Eds., ed Shenyang, China: Springer-Verlag Berlin Heidelberg, 2012, pp. 624-632.

[24] M. A. Abido, "Robust Design of Power System Stabilizers for Multimachine Power Systems Using Differential Evolution," in *Computational Intelligence in Power Engineering.* vol. 302, B. Panigrahi, A. Abraham, and S. Das, Eds., ed: Springer Berlin Heidelberg, 2010, pp. 1-18.

[25] D. K. Sambariya and R. Prasad, "Robust tuning of power system stabilizer for small signal stability enhancement using metaheuristic bat algorithm," *International Journal of Electrical Power & Energy Systems,* vol. 61, pp. 229-238, 2014.

[26] A. Khodabakhshian and R. Hemmati, "Multi-machine power system stabilizer design by using cultural algorithms," *International Journal of Electrical Power & Energy Systems,* vol. 44, pp. 571-580, 2013.

[27] K. R. Padiyar, *Power System Dynamics Stability and Control.* Hyderabad, India: BS Publications, 2008.

[28] D. K. Sambariya and R. Prasad, "Robust Power System Stabilizer Design for Single Machine Infinite Bus System with Different Membership Functions for Fuzzy Logic Controller," *Proceedings of 7th International Conference on Intelligent Systems and Control (ISCO 2013) , at Korpagam College of Engineering, Coimbatore, Tamilnadu, India,* vol. IEEE Catalog number (Print): CFP1387T-PRT, ISBN (Print): 978-1-4673-4601-6, pp. 13-19, January 4-5, 2013 2013.

[29] D. K. Sambariya and R. Prasad, "Design of PSS for SMIB System Using Robust Fast Output Sampling Feedback Technique," *Proceedings of 7th International Conference on Intelligent Systems and Control (ISCO 2013) at Karpagam College of Engineering, Coimbatore, Tamilnadu, India,* vol. IEEE Catalog number (Print): CFP1387T-PRT, ISBN (Print): 978-1-4673-4601-6, pp. 166-171, January 4-5, 2013 2013.

[30] P. Yadav, R. Kumar, S. K. Panda, and C. S. Chang, "An Intelligent Tuned Harmony Search algorithm for optimisation," *Information Sciences,* vol. 196, pp. 47-72, 2012.

[31] D. Manjarres, I. Landa-Torres, S. Gil-Lopez, J. Del Ser, M. N. Bilbao, S. Salcedo-Sanz, *et al.*, "A survey on applications of the harmony search algorithm," *Engineering Applications of Artificial Intelligence,* vol. 26, pp. 1818-1831, 2013.

[32] J. Fourie, S. Mills, and R. Green, "Harmony filter: A robust visual tracking system using the improved harmony search algorithm," *Image and Vision Computing,* vol. 28, pp. 1702-1716, 2010.

[33] J. Fourie, R. Green, and Z. W. Geem, "Generalised Adaptive Harmony Search: A Comparative Analysis of Modern Harmony Search," *Journal of Applied Mathematics,* vol. 2013, pp. 1-13, 2013.

6

Design Proposed Features Extraction Recognition System of Latin Handwritten Text Based on 3D-Discrete Multiwavelet Transform

Laith Ali Abdul-Rahaim*

Electrical Engineering Department, Babylon University, Babil, Iraq
*Corresponding author: drlaithanzy@yahoo.com

Abstract On-line handwriting recognition is the task of determining what letters or words are present in handwritten text. It is of significant benefit to man-machine communication and can assist in the automatic processing of handwritten documents. It is a subtask of the Optical Character Recognition (OCR), whose domain can be machine-print only.The introduced system is a character-based recognition and it is a writer independent system. The recognition responsibility of the proposed system is for 52 character classes [uppercases (A-Z) and the lowercases (a-z)]. The suggested system includes the essential stages needed for most of the pattern recognition systems. These stages are the preprocessing stage, the features extraction stage, the pattern matching and classification stage and the postprocessing stage. The proposed method employs the 3 Dimensional Multiwavelet transform 3D-DMWTCS using multiresolution image decomposition techniques working together with multiple classification methods as a powerful classifier. The classification stage is designed by using a minimum distance classifier depending on Euclidean Distance which has a high speed performance. The system design also includes a modest postprocessing stage that makes a consistency between the recognized characters within the same word in relation to their upper and lower cases.The overall classification accuracy of proposed systems can be obtained are 95.305 percent with 3D-DMWTCS based on the Rimes database.

Keywords: *3D-DMWTCS,2D-DMWTCS, dpi, HWR, RR, OCR, ED, MDC*

1. Basic Concepts of the Handwriting Recognition

Handwriting (HW) is one of the most important ways in which civilized people communicate. It is used for both personal (e.g. letters, notes, addresses on envelopes, etc.) and business communications (e.g. bank cheques, business forms, etc.). The writing is a physical process where the brain sends an order through the nervous system to the arm, hand and fingers, where together they manipulate the writing tool. Therefore, a person's handwriting is as unique as human fingerprints and facial features. However, it varies depending upon many factors (age, education, temper, left or right handed writer, etc.).With the advent of the computer, it became possible that machines could also reduce the amount of mental labor needed for many tasks. One of these tasks is recognizing a human's handwriting. Of course, much progress has been made in the way of computer handwriting recognition, but a computer will never be able to read a human's handwriting as good as a human. Even so, it doesn't hurt to try to develop technology which can approach the recognition ability of humans. Since the handwriting is one of the most important ways in which people communicate, it would provide an easy way of interacting with a computer [1,2]. A recognition system can be either "on-line" or "off-line." It is "online" if the temporal sequence of points traced out by the pen is available, such as with electronic personal data assistants that require the user to "write" on a screen where the path of the pen is measured by a device such as a digitizing tablet. It is "off-line" if it is applied to previously written text, such as any image scanned by a scanner. The on-line problem is usually easier than the off-line problem since more information is available. So far, most of the off-line handwriting recognition systems are applied to reading letters, postal addresses and then automatic sorting of postal mail, processing forms like bank cheques or discrimination of the different scripts for individual writers (Handwriting identification) [3]. The real progress in character recognition was achieved in the advancement age (after 1990). In early 1990s, image processing and pattern recognition were efficiently combined with artificial intelligence techniques. Efficient tools such as Neural Networks (NN), Support Vector Machines (SVM), Hidden Markov Model (HMM), fuzzy set reasoning, and natural language combined with more powerful computers and more accurate electronic equipment's have provided quite satisfactory results for restricted applications [4]. The challenge is to recognize the HW texts that are written by people in real life situations and makes an automatic transcription by computer, where only the image of the handwriting is available. It becomes more important to make the transfer of information between people and machines simple, fast and reliable. Chevalier et al [5] presented a two-dimensional approach of the processing of handwriting. It combines a Markovian model, an efficient decoding algorithm, a windowed spectral features extraction scheme and a rigorous evaluation methodology. They applied this principle to a digit recognition task and to a word recognition task [6].

2. Model for on-line Handwriting Recognition

Handwriting Recognition is interpretation of data which describes handwritten objects to generate a description of that interpretation in a desired format. Or in other words it is a determination what letters or words are present in a digital image of handwritten text. HWR is of significant benefit to man-machine communication and can assist in the automatic processing of the handwritten documents. A wide variety of techniques are used to perform off-line handwriting recognition. To convert this image into information understandable by computers requires the solution to a number of challenging problems. Firstly, pre-processing steps are achieved on the image to reduce some undesirable variability that only contributes to complicate the recognition process. Operations like binarization, noise removal, skew, slant and slope corrections, thinning, smoothing, normalization, etc. are carried out at this stage. The second step is the segmentation of the word into a sequence of basic recognition units such as characters or pseudo-characters. However, segmentation may not be present in all systems. Recognition approaches can be either "Holistic" or segmentation-based. "Holistic" means that words or sentence are processed as a whole without segmentation into characters or strokes [6,7]. In segmentation-based approaches, whole or partial characters are recognized individually after they have been extracted from the text image. The final step is to extract discriminated features from the input pattern to either build up a feature vector or to generate graphs, string of codes or sequence of symbols. However, the characteristics of the features depend on the preceding steps; say whether segmentation of words into characters was carried out or not. The pattern recognition model to handwriting recognition consists of pattern training, that is, one or more patterns corresponding to handwritten words or characters of the same known class are used to create a pattern representative of the features of that class. The recognition includes a comparison of the test pattern with each class reference pattern and measuring a similarity score (e.g. distance, probability) between the test pattern and each reference pattern [8,9]. The pattern similarity scores are used to decide which reference pattern best matches the unknown pattern. The post-processing or verification may also be included in some systems. However, for meaningful improvements in recognition, it is necessary to incorporate the recognition process other sources of knowledge such as language models [10]. A limited vocabulary is one of the most important aspects of systems that rely on large vocabularies because it contributes to improve the accuracy as well as to reduce computation. In the case of systems that deal with large vocabularies, other additional modules may be included such as pruning or lexicon reduction mechanisms [11,12]. If we desire a system to distinguish objects of different types, it must be first decided which characteristics of the objects should be measured to produce descriptive parameters called (features) of the object, and the resulting parameters values comprise the feature vector for each object. Proper selection of the features is important, since only these will be used to identify the objects.Good features have four characteristics [13,14]; Discrimination, Reliability, Independence and Small Numbers. The complexity of a pattern recognition system increases rapidly with dimensionality of the system. More importantly, the number of objects required to train the classifier and to measure its performance increases exponentially with the number of features. And it is necessary to avoid the redundancy.The features extraction is an important step in achieving good performance for the recognition. For off-line HWR systems the feature extraction methodology is based on one or more of Global features and Geometrical and topological features. The geometrical and topological (structural) features describe the geometry and topology characteristic of a character. Some examples of extracted features are strokes and bays (kerning) in various directions, dots, end points, intersections of line segments (junctions), loops, curves (turnings), etc as shown in Figure 1. Each of these features can be encoded by a single number. Geometrical and topological features have a high tolerance to distortion and style variations. Due to complexity of extracting the geometrical and topological features and the great variations in local properties of HW characters, it is rather difficult to generate feature masks. But once they are implemented, they can process characters at high speed independently in [15] are examples of using the geometrical features.

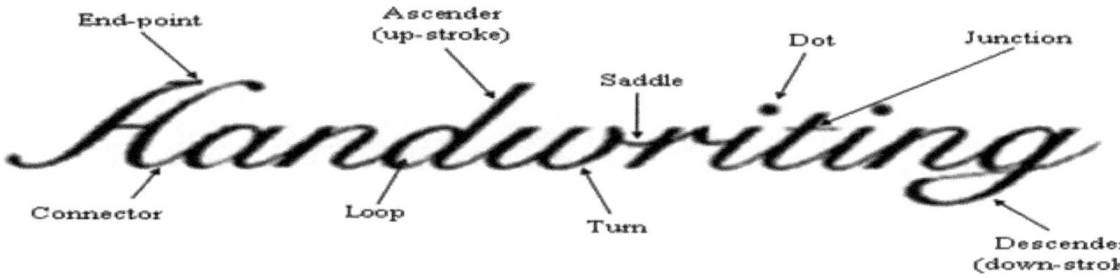

Figure 1. Geometrical and topological features

Statistical features: The statistical features are derived from the statistical distribution of pixels of a character. They are numerical measures computed over images or regions of images [16,17]. They include, but are not limited to, pixel densities, histograms of chain code directions, moments, Fourier descriptors, the aspect ratio of the character, characteristic loci, crossing and distances. The statistical features take some topological and dynamic information into account and consequently can tolerate minor distortions and style variations [18,19]. A key question in handwriting recognition is how test and reference patterns are compared to determine their similarity. Pattern comparison can be done in a wide variety of ways. The goal of a classifier is to match a sequence of observations derived from an unknown handwritten word against the reference patterns that were previously trained, and obtain confidence scores (distance, cost, or probabilities) to further decide which model best represents the test pattern [20]. Contextual approaches lead to excessive growth in the number of models since one model is needed for each pair of adjacent characters. Parameter estimation may be unreliable since, for practical applications, a restricted set of training data is generally available. It is thus desirable to reduce the number of models and model parameters while preserving model refinement. Hence, model sharing and parameter tying are necessary to reduce the number of parameters. Schussler and Niemann [21] describe a context-dependent system using HMMs, where all sub word units (from monographs to the whole word) are modeled within a word hierarchy. Models with not enough training samples are eliminated. The system is tested on a small and dynamical lexicon. The state-based tying proposed by Natarajan et al.[22] uses a mixture of 128 Gaussians associated to each state position of contextual models (trigraphs) corresponding to the same base character. The total number of models can also be reduced as in [23,24] by clustering all trigraphs according to contexts described not as characters but as ascending or descending strokes. Fink and Plotz [25] also proposed a system based on contexts, these contexts being defined as broad categories. A data-driven clustering of the 1,500 Gaussian densities of the mixture is performed at each state position and for each category. It is worth noting that clustering and tying the contextual models may offer, in addition to reducing the number of parameters, the possibility to automatically capture common contextual effects on handwriting a given letter [26].

3. Feature Extraction Using Multiwavelet Transform and Classification

Multiwavelet transform is a new concept in the framework of wavelet transform but has some important differences. In particular, whereas wavelet has an associated one scaling function and wavelet function, Multiwavelet has two or more scaling functions and wavelet functions. One of the well-known Multiwavelet was constructed by Donovan, Geronimo, Hardin, and Massopust (DGHM). DGHM Multiwavelet simultaneously possesses orthogonality, compact support, approximation order 2 and symmetry [13]. Next, we give a brief overview of the Multiwavelet transform [14]. Unlike scalar wavelet, even though the Multiwavelet is designed to have approximation order *p*, the filter bank associated with the Multiwavelet basis does not inherit this property. Furthermore, since the Multiwavelet have more than one scaling function; the dilation equation becomes dilation with matrix coefficients. Thus, in applications, one must associate a given discrete signal into a sequence of length r vectors (where r is the number of scaling functions) without losing certain properties of the underlying Multiwavelet. Similar to the traditional scalar wavelet transform, the two-dimensional Multiwavelet transform can be achieved by applying the one-dimensional transform on the rows by treating each row as a one-dimensional signal and afterward on columns. However, for the applications using Multiwavelet, profiteering process must be applied to each row and each column to initiate the vector sequence $c0$ to the filter bank.. There is growing interest in using wavelet features for images in many applications, including object identification, medical-images retrieval, and texture analysis. These features are generally extracted from the two-dimensional discrete Multiwavelet (2D-DMWTCS) coefficients of the image under processing. Wavelets are localized basis functions which are translated and dilated versions of some fixed mother wavelet. The main feature of wavelets is that they are able to provide localized frequency information about a function or signal. Such information is particularly beneficial for classification. There exists an abundant variety of wavelets, and the fundamental problem to overcome is deciding which wavelet will produce the best results for a particular application. The following procedure must be doing to calculate a single level 2-D Discrete Multiwavelet Transform using GHM four multifilter and a critical sampled scheme of preprocessing (approximation row preprocessing):

1. *Checking phoneme Signal Dimensions:* Phoneme matrix should be a square matrix, N×N matrix, where N must be power of 2. So that the first step of the transform procedure is checking input phoneme dimensions. If the phoneme matrix is not a square matrix, some operation must be done to the adding rows or column of zeros to get a square matrix.
2. *Constructing A Transformation Matrix:* Using the transformation matrix, such as given in the following matrix format [10]:

$$\begin{aligned}&\text{new odd-row}\\&=(0.373615)[\text{same odd-row}]\\&+(0.11086198)[\text{next even-row}]\\&+(0.11086198)[\text{previous even-row}]\end{aligned} \tag{1}$$

$$\text{new even-row} = (\sqrt{2}-1)[\text{same even-row}] \tag{2}$$

An N/2×N/2 transformation matrix should be constructed using GHM low-and high-pass filtersmatrices given below [9]:

$$\begin{bmatrix} H_0 & H_1 & H_2 & H_3 & 0 & 0 & \dots \\ G_0 & G_1 & G_2 & G_3 & 0 & 0 & \dots \\ 0 & 0 & H_0 & H_1 & H_2 & H_3 & \dots \\ 0 & 0 & G_0 & G_1 & G_2 & G_3 & \dots \\ \dots & \dots & \dots & \dots & \dots & \dots & \dots \end{bmatrix} \tag{3}$$

$$\left.\begin{aligned}
&H_0 = \begin{bmatrix} \frac{3}{5\sqrt{2}} & \frac{2}{5} \\ -\frac{1}{20} & -\frac{3}{10\sqrt{2}} \end{bmatrix}, H_1 = \begin{bmatrix} \frac{3}{2\sqrt{2}} & 0 \\ \frac{9}{20} & \frac{1}{\sqrt{2}} \end{bmatrix} \\
&H_2 = \begin{bmatrix} 0 & 0 \\ \frac{9}{20} & -\frac{3}{10\sqrt{2}} \end{bmatrix}, H_3 = \begin{bmatrix} 0 & 0 \\ -\frac{1}{20} & 0 \end{bmatrix} \\
&G_0 = \begin{bmatrix} -\frac{1}{20} & -\frac{3}{10\sqrt{2}} \\ \frac{1}{10\sqrt{2}} & \frac{3}{10} \end{bmatrix}, G_1 = \begin{bmatrix} \frac{9}{20} & -\frac{1}{\sqrt{2}} \\ -\frac{9}{10\sqrt{2}} & 0 \end{bmatrix} \\
&G_2 = \begin{bmatrix} \frac{9}{20} & -\frac{3}{10\sqrt{2}} \\ \frac{9}{10\sqrt{2}} & -\frac{3}{10} \end{bmatrix}, G_3 = \begin{bmatrix} -\frac{1}{20} & 0 \\ -\frac{1}{10\sqrt{2}} & 0 \end{bmatrix}
\end{aligned}\right\} \quad (4)$$

3. After substituting GHM matrix filter coefficients values as given in eq. (3), an N×N transformation matrix results with the same dimensions of input image dimensions after preprocessing. Preprocessing Rows: Approximation-based row preprocessing can be compute by applying Eqs. (1), and (2) to the odd- and even-rows of the input N′N matrix respectively. Input matrix dimensions after row preprocessing are the same N′N.
4. *Transformation of image Rows :* The procedure can be done as follows:

a. Apply matrix multiplication between the N×N constructed transformation matrix by the N×N row preprocessed input phoneme matrix.

b. Permute the resulting N×N matrix rows by arranging the row pairs 1, 2 and 5, 6 …, N–3, N–2 after each other at the upper half of the resulting matrix rows, then the row pairs 3, 4 and 7, 8… N–1, N below them at the next lower half.

5. *Preprocess Columns:* To repeat the same procedure used in preprocessing rows:

a. Transpose the row transformed N×N matrix resulting from step 4.

b. Repeat step 3 to the N×N matrix (transpose of the row transformed N×N matrix), which results in N×N column preprocessed matrix.

6. *Transformation of Columns :* Transformation of phoneme columns is applied next to the N×N column preprocessed matrix as follows:

a. Apply matrix multiplication between the N×N constructed transformation matrixes by the N×N column preprocessed matrix.

b. Permute the resulting N×N matrix rows by arranging the row pairs 1, 2 and 5, 6 …, N–3, N–2 after each other at the upper half of the resulting matrix rows, then the row pairs 3, 4 and 7, 8… N–1, N below them at the next lower half.

7. *The Final Transformed Matrix:* The following procedure must be doing to get final transformation matrix:

a. Transpose the resulting matrix that get from column transformation step.

b. Apply coefficients permutation to the resulting transpose matrix. Coefficients permutation is apply to each of the basic four subbands of the resulting transpose matrix so that each subband permutes rows then permutes columns. Finally, a N×N DMWT matrix results from the N×N original matrix using approximation-based preprocessing.

The main features of this Multiwavelet type are the ability to provide localized frequency information about a character image. Such information is particularly beneficial for classification. For a given binary image containing a single character, there are many pre-processing procedures performed prior to feature extraction. The most important thing is to make our system independent of each character concerning its *shape*, *position (location in the word)* and *size*. In relation to its shape normalization, this can be achieved by slant and skew corrections steps. Concerning stroke width normalization, this is done by skeletonization (thinning) approach and successive steps of stroke thickening. These steps leave each stroke with approximately the same width. Related to character position normalization, it is achieved by first character segmentation and then a bounding rectangle of each character is found. This removes any differences due to location of character within each image. Next this bounding rectangle is scaled to a (32 × 32) pixel image (A^{j+1}), in order to scale (size) normalization. The wavelet decomposition is applied at one level of resolution, yielding four subband images {Approximations (A^j), Horizontal details (D^j_H), Vertical details (D^j_V) and Diagonal details (D^j_D)} each containing 16 X 16 pixels. Therefore, the feature vector is formed by these subband images with (*1xd*) dimensions, where *d*= 4 X 16 X 16. Figure 2 illustrates the 1-level of 3D-(DMWTCS or DWT) step. For each subband image, the values of the (wavelet or Multiwavelet) coefficients are normalized to the range [0, 1]. Figure 3 shows the 3D-(DMWTCS or DWT) coefficients for all subbands [19]. The main information is concentrated at the approximation subband and the other are distributed in the other subband images. Other experiments were done to find out which subband contains more important characteristics (important features for recognition).These experiments used the same test data set and the same data base. The results show the features relevance to the recognition process. The features that are extracted from the approximation subband contribute by about 53% of the importance in the recognition task, while the features extracted from the all other subbands contribute with about 47% of the significance in the recognition process. But this conclusion does not mean that the correct recognition is about 53 % by using the approximation only.

2.1. A Proposed Computation Method of 3D-DMWTCS

Discrete Multiwavelet transform given a good indication in applications of signal processing. Recent work on Multiwavelet have been studies the basic theory, methods of constructing new multifilters and the denoising and compression applications in of video and image [12,17,21,22]. The algorithms for computing three dimensional discrete Multiwavelet transform Critically

Sampled (3D-DMWTCS) have been described in this section in a simple and easy way to verify procedure using matrix multiplication and addition. To compute 3D-DMWTCS one must know how compute 1, 2D- discrete Multiwavelet transforms Critically Sampled and its inverse 1,2D-IDMWTCS. In 3D- discrete Multiwavelet transformation Critically Sampled algorithm is defined in 3D, so the transformation procedure will done successively in x-, y- and z-directions.

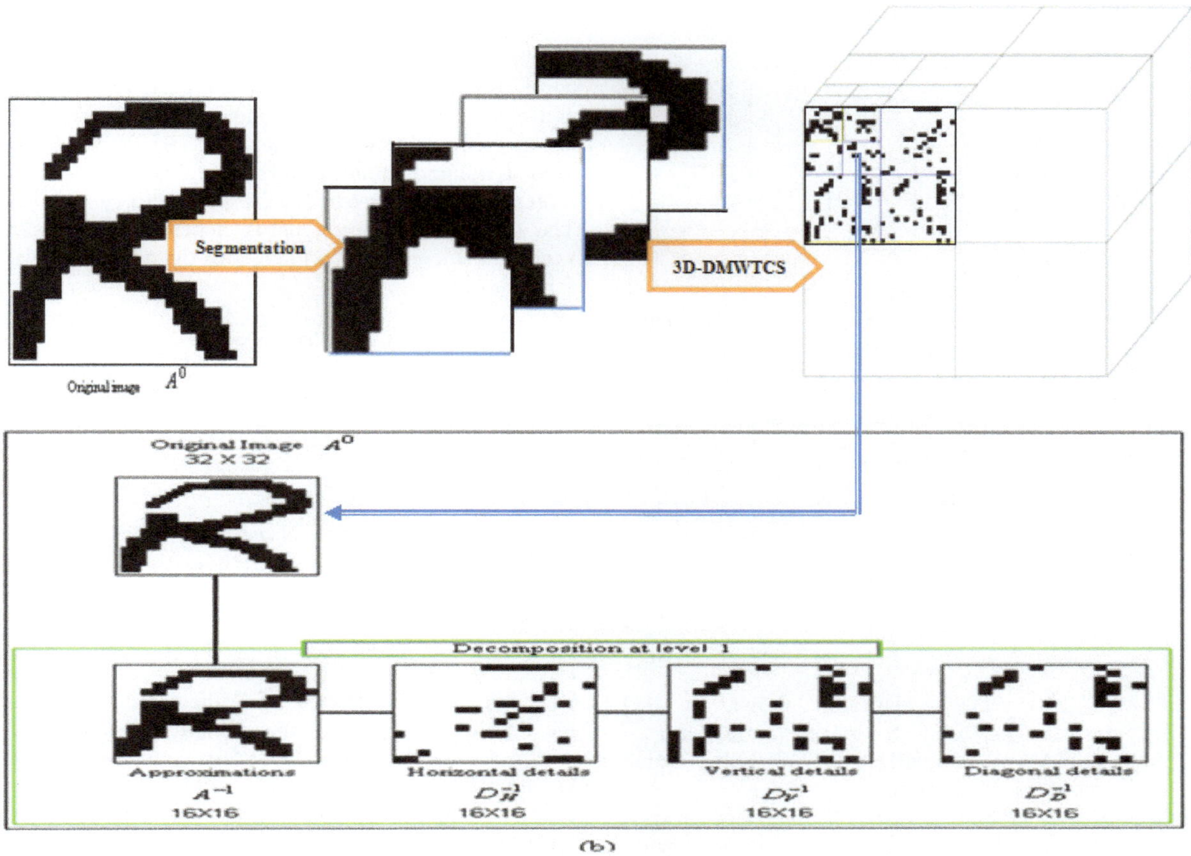

Figure 2. decomposition of 3D-DMWTCS on character imagea-square view mode b- Tree view mode

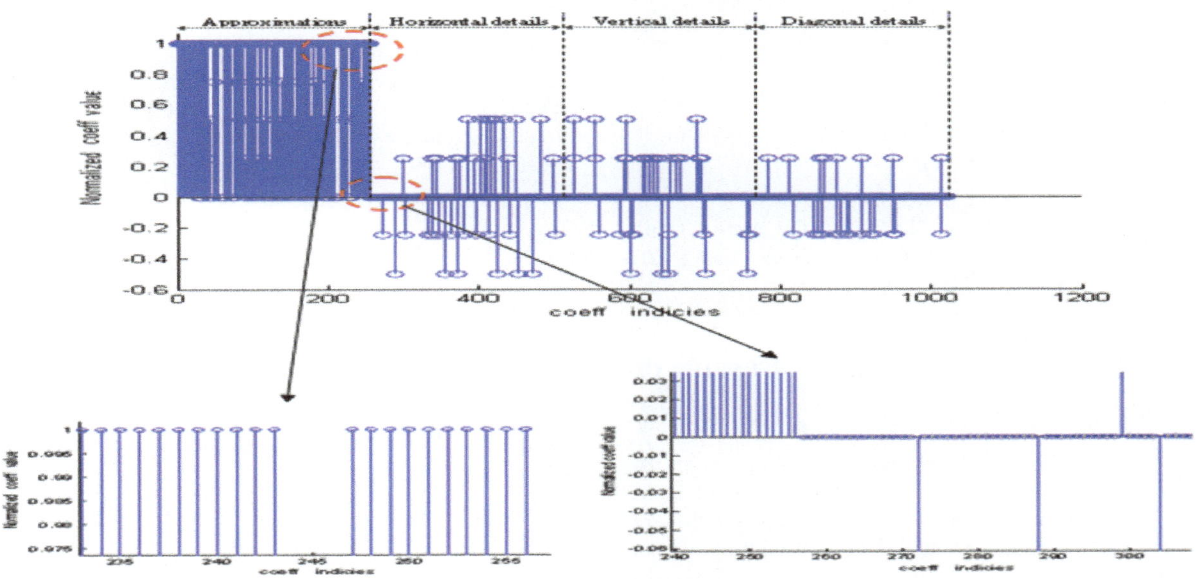

Figure 3. 3D-DMWTCS normalized coefficients

For a 2D-DMWTCS, the procedure was applied to each vector in x-direction first, and then to each vector in y-direction. Similarly, in 3D- discrete Multiwavelet transformation Critically Sampled the procedure is defined in 3D and the transformation algorithm is applied successively in x-, y- and z-direction.

Let's take a general 3D signal, for example any NxNxM matrix. The computation 3D-DMWTCS need the following procedure:

1. construct 3D-matrix A to represent the 3D input signal,

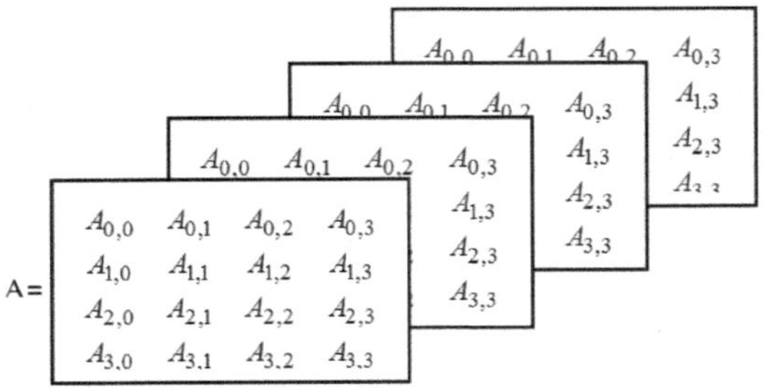

(5)

2. Using 2D DMWTCS algorithm to each NxN input matrix, which result in a Y matrix ($N\times N\times M$).

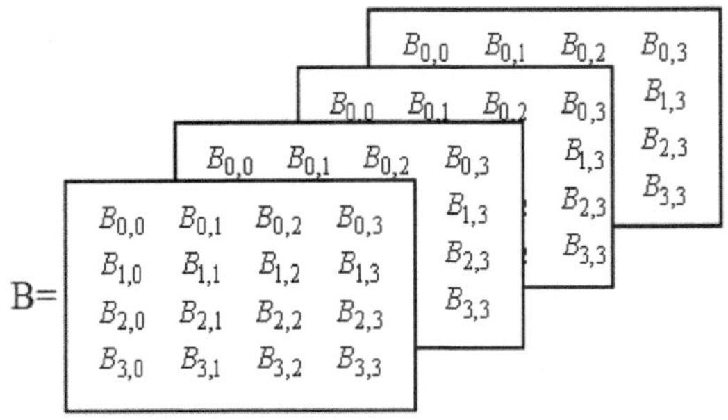

(6)

3. Using 1D-DMWTCS algorithm shown in [11] tocompute 1D-DMWTCS to each element of NxNmatrix in all M matrices in z-direction, which can summarized as follows:

a. For each i,j element in the 1st matrix construct a vector of $Mx1$ for each element in z-direction output matrices from 2D-DMWTCS in step 2, this operation is done as below:

$$V(i,j)=\begin{bmatrix} B_{i,j,0} & B_{i,j,1} & B_{i,j,2} & B_{i,j,3} \end{bmatrix}^T_{1\times M} \quad (7)$$

Where $i,j=0,1,2,\cdots,N$.

b. Applying 1D-DMWTCS algorithm to each the construct vector $V(i,j)$.

4. Repeat step 3 for all construct vector i, j.
5. Finally, a$N\times N\times M$ of 3D-DMWTCS matrix results from the N×N×M original matrix using 3D-DMWTCS.

4. Training and Testing Phases of the Recognition Task

The goal of the training phase is to extract and prepare the best parameter values (features) of the character models. The training phase deals with the handwritten characters (some times are called letters) with known and defined letters classes. Each input letter image is adequately pre-processed and its relevant features are then extracted from the preprocessed image forming a feature vector (f_{dx1}), where d is the features number. For each character class, the feature vectors is generated which are also known as the class reference feature vectors. These vectors have the goal of representing its corresponding character classes. This class reference feature vectors are generated by the application of 2D-DMWTCS. As a result, the system has reference feature vectors which are forming the feature matrix $\boldsymbol{F}$ as stated in equation (8). $\boldsymbol{F}$contains$Mx52$ feature vectors f where$M$ is the number of training samples and (52) is the number of letter classes (upper lower cases of Latin alphabet). For example: $f_{1,1}$ and $f_{2,1}$are the feature vectors of the letter (A) of the first and second training samples and so on while $f_{1,2}$ and $f_{2,2}$ are the feature vectors of the letter (B) of the first and second training samples and so on. Each column in $\boldsymbol{F}$ represents the features of one class for M training samples. The size of the feature matrix depends on the feature vector dimension and the number of the available training samples M.

$$F=\begin{bmatrix} f_{1,1} & f_{1,2} & \cdots & \cdots & f_{1,52} \\ f_{2,1} & f_{2,2} & \cdots & \cdots & f_{2,52} \\ . & . & . & . & . \\ . & . & . & . & . \\ f_{M,1} & f_{M,2} & \cdots & \cdots & f_{M,52} \end{bmatrix} \quad (8)$$

An extended training phase, i.e., more samples of handwritten letters with various styles would improve the system performance. The training data set have to be with various styles rather than to be of large quantity. The testing phase of the classification, an image from an unknown class is initially pre-processed and then having its feature vector extracted using the same techniques in the training phase. From this initially extracted feature vector, for each distinct letter class, a new and different feature vector is then generated. Next, the classifier is used to assign the input handwritten character to the class that best accommodates the input image.

4.1. The Minimum Distance Classifier (MDC)

The implemented minimum distance classification based on calculating and comparing the Euclidean Distances (ED). The Euclidean Distances are between the feature vector of the unknown input character (to be classified) and the reference feature vectors as shown in equation (9) [19].

$$ED_n=\left\|f_i-f_{m,n}\right\| = \sqrt{\sum_{m=1}^{M}\left|f_i-f_{m,n}\right|^2} \quad n=1,\ 2\ \ldots\ 52 \quad (9)$$

Where $\left\|f_i-f_{m,n}\right\|$ denotes the Euclidean Distance between the vectors f_i and $f_{m,n}$. f_i is the feature vector of the unknown input character pattern, the subscript i is denotes to the word "input". $f_{m,n}$is the mth feature vector of the nth character class that belongs to the feature matrix $\boldsymbol{F}$.

$$CED_n=\sum_{m=1}^{M} ED_{m,n} \qquad n=1,\ 2\ldots\ 52 \quad (10)$$

$$\text{CED}=[CED_1\ CED_2\ \ldots\ldots CED_{52}] \quad (11)$$

where CED_n is the Class Euclidean Distance of the nth character class for M training samples and **CED** is the Class Euclidean Distance vector. The chosen class will be the one that achieves the smallest *CED* in **CED** In other words; it is the smallest distance between the input feature vector and the most representative (nearest) vector of the reference feature vectors. Figure 4 and Figure 5 show some examples of how minimum distance classifier work to classify the HW letters T&Z which appear at the left bottom corner of each plot. The x-axes of the plots represent the 52 characters classes of upper and lower cases. The y-axes are the *EDs* between the input features vector and the reference features vectors as shown in equation (11).

Concerning Figure 4, the CED_{20} is the smallest among all *CEDs*. The CED_{20} is the sum of *M* Euclidean Distances of the 20th class that belongs to the letter " T ", therefore, the input letter will be classified as T and so on for the other inputs. Figure 5 shows the letter "s-lowercase" classification. It is clear there are two smallest *CEDs*(CED_{19} and CED_{45}). The letter "s-lowercase" will be classified as "S-uppercase" since $CED_{19}<CED_{45}$. The recognition between letters with approximately having similar patterns of their upper and lower cases is still an open problem till this point of the proposed recognition system. Such letters are "s" and "S", "w" and "W", "z" and "Z" "c" and "C" etc. The unique difference between their upper and lower patterns is the size. The size difference is lost by the size normalization step at the preprocessing stage which may waste this feature between upper and lower cases of the letters above. This problem will be discussed and partially treated at the postprocessing stage.The false classification may come from calculating the smallest *ED* which may give an assurance to a wrong class as shown in Figure 6. Figure 7 shows the avoiding of the false classification for the HW letter "Y" as letter V. It is clear that there is some similarity between the HW "Y" with the letter V which is the reason for this very small *ED*.

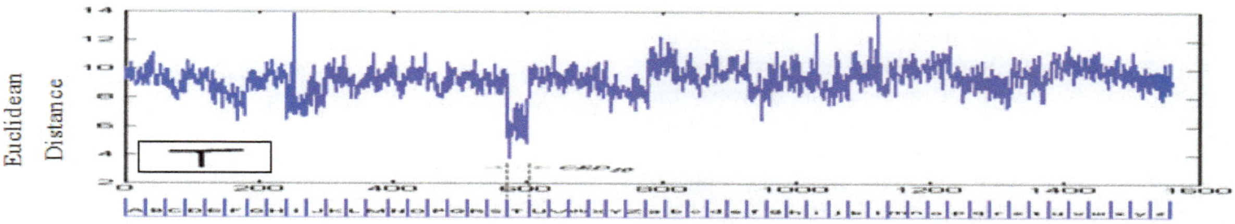

Figure 4. Letter "T" classification

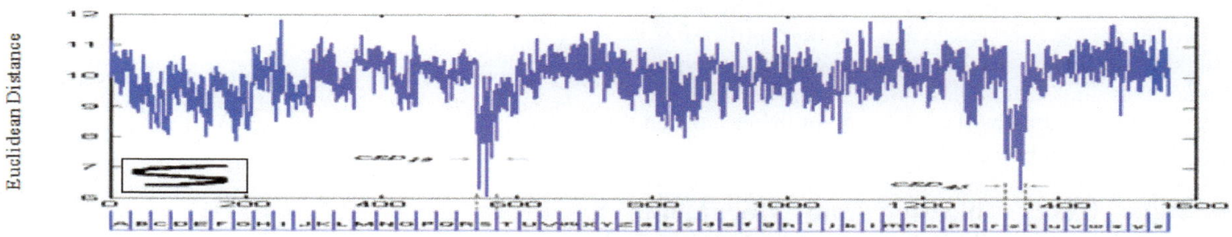

Figure 5. Letter "S" classification

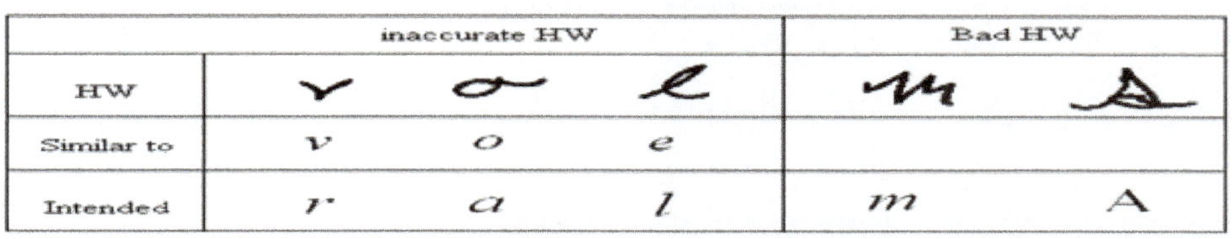

Figure 6. Inaccurate and bad HW

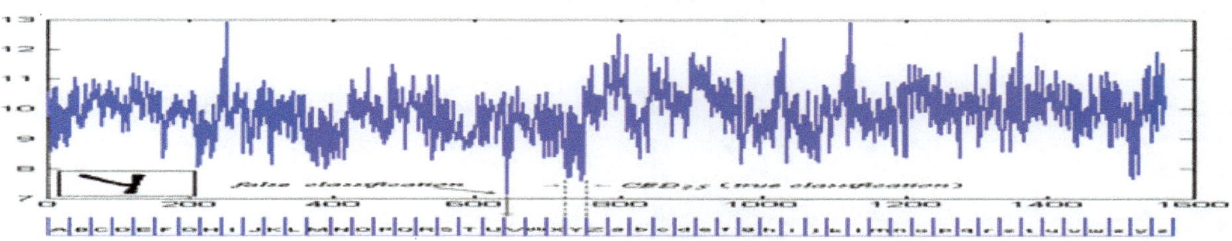

Figure 7. True and false classification

Sometimes the false classification could not be avoided due to the great similarity between the input HW letter and the unintended character class as shown in Figure 8. The input HW letter is " *r* ", but it is classified as " v " due to the great similarity between them. The solution of this problem is beyond the scope of this work (out of the proposed system ability). The Recognition process will be character by character and the designed program preserves the recognized characters to their words.

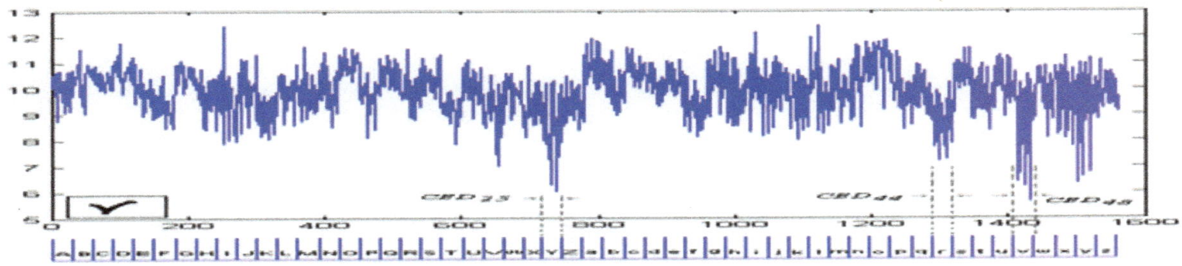

Figure 8. false classification due to not good HW

4.2. Post processing

The postprocessing step includes all processes that may be made to enhance the recognition and make the decision by different ways such as the prior contextual knowledge, integration of grammatical and syntactic knowledge, spelling, punctuation mark … etc. It focuses on solving the problem of the bad recognition between the characters between their upper and lower cases. The principle of its processing depends upon the comparison between the recognized characters within the same word. The normal word may contain all letters with their lowercases as "*university*" or may have only the first one with its uppercase especially if this word at the beginning of the sentence as "*University of Babylon*" or it represents a name as or "*Babylon*". It is not proper that the word is written in lowercase letters except one or more letters at its middle or last in uppercase as in "*uniVersitY*". Sometimes, the terms may be written with all letters in uppercases as in "*Handwriting Recognition*". This case can be easily discovered by the inspection of the recognized letters with the same word and the adjacent ones.

4.3. Data Base Collection and Document Image Acquisition

The proposed handwriting recognition system is with two phases; the first is the training phase that uses training handwritten samples while the second is the testing and recognition phase that needs test samples. The collected database for training must include different handwritten styles related to the scope of the proposed system. The data base was collected locally from various right hand and left hand writers with different ages, educations, temper etc. Characters were written by writers using specific forms on plain, white paper sheets with black ink pen to give clear strokes with sharp edges. Each filled form contains 52 Latin characters including the upper (A-Z), lower (a-z) cases and (0-9) numerical digits as shown in Figure 9. The collected data base depends upon the variety not on the quantity. The selection of the training samples must avoid as possible the redundant HW styles. The proposed handwriting recognition system processes data that were captured from a flatbed scanner. They were scanned at *150-dots-per-inch* resolution, in 256 levels of gray to produce one file per writer. The next task is to segment each form into its component characters. The pixel histogram calculations based segmentation algorithm takes a simple approach, looking for the gaps between lines and characters. Figure 10 shows the segmentation results.

Figure 9. HW data collection form

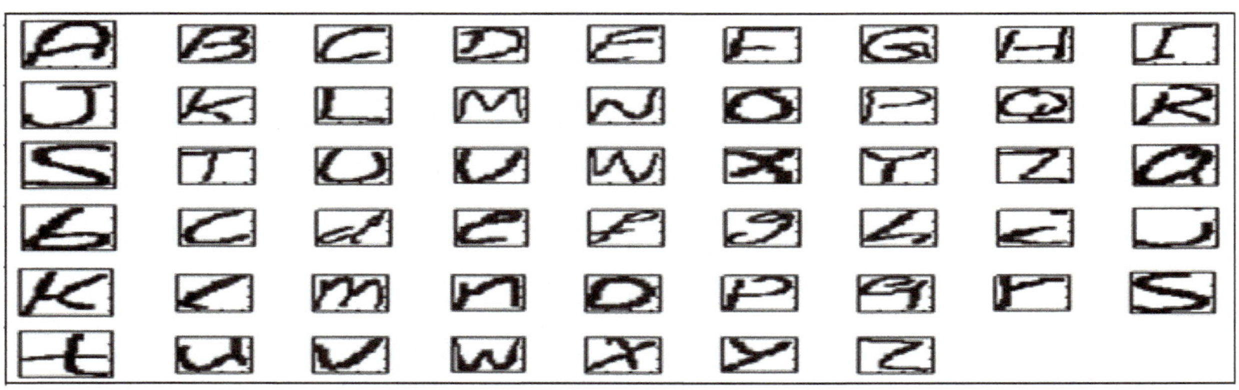

Figure 10. HW characters after the segmentation process from the form

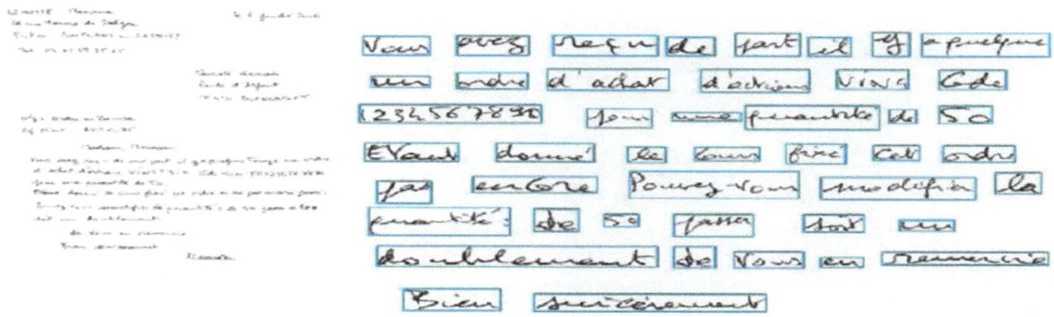

Figure 11. some mail letters handwritten of the Rimes of the training database and extractedwords data set classified as their classes and processed by the normalization steps

Each segmented character will pass through some steps (preprocessing steps) to be under the effects of same steps that the test samples will pass. These steps are binarization, thinning, thickening the strokes in order to smooth them and make all strokes approximately with the same thickness and characters resizing (size normalization) to be (32x32) pixels. These steps will make processing independent of strokes thickness and characters sizes. Figure 11 shows some of the characters of the training data set classified as their classes processed by the above steps. The attention during the data base collection was upon the quality not on quantity of the collected HW samples. The selection of the training samples process was avoiding the redundancy of characters samples. The redundancy is useless or unavailing. A large number of training HW samples may include bad HW writing styles and may make a negativity recognition process. Figure 12 shows the steps of the proposed HWRS. Figure 13 shows the block diagram of a complete proposed Handwriting Recognition System.

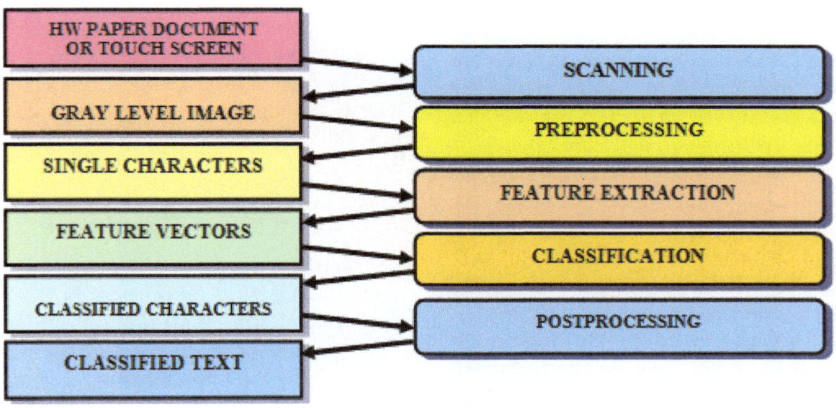

Figure 12. Steps of the proposed HWRS

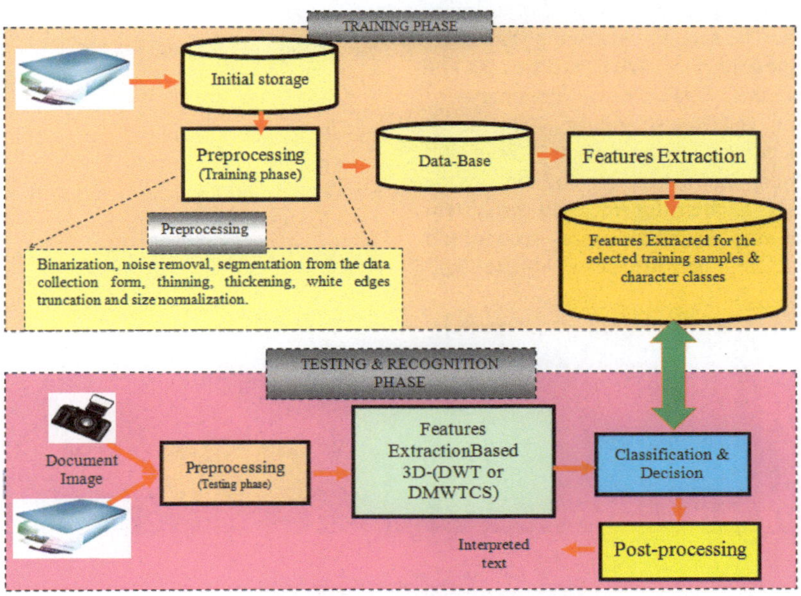

Figure 13. Block diagrams of the proposed HWRS

Figure 14 shows the raw handwritten texts written by different writers that under test in column (a) and the same texts after the recognition in column (b) by the proposed Handwriting Recognition System before the postprocessing step.

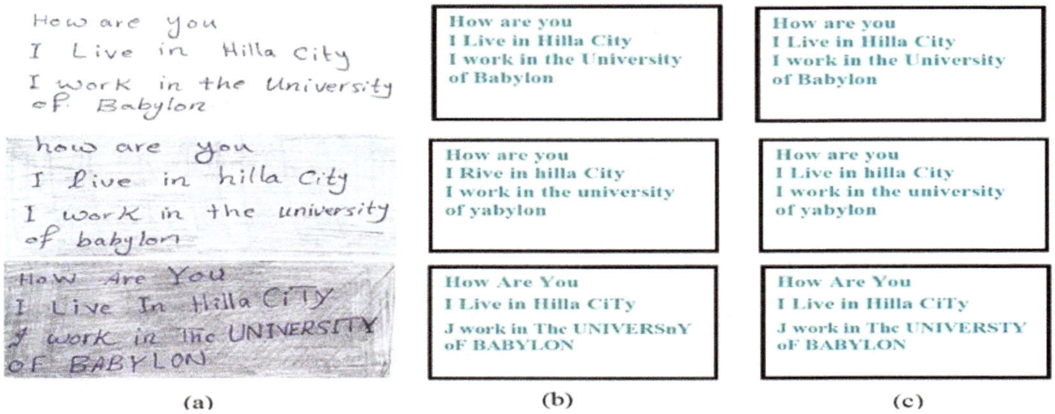

Figure 14. HW Recognition examples (a) Raw Handwritten text. (b) The recognized text Using 2D-DMWTCS (c) The recognized text Using 3D-DMWTCS

5. Performance Evaluation of the Proposed HWRS

The Rimes database [26,27] was presented in 2006 and gathers more than 12,500 handwritten documents written by about 1,300 volunteers. It was created to provide data relative to advanced mailrooms, and the panel of documents offers large variability and makes the database a challenging one. Based on the Rimes database, it is possible to proceed logo recognition, document structure retrieval, and word and character recognition. Since 2007, evaluation campaigns have occurred [19] which enable participants to compare their results. In our work, we use the presegmentedword images given by the database to perform isolatedword recognition. The database consists of 59,203 wordimages divided into three subsets: 44,197 images for training, 7,542 for validation, and 7,464 for testing. The first experiment considers the full dictionary of size 5,334, including all words from the training, validation, and test sets. The second experiment considers only the reduced test dictionary of size 1,612 words. The last experiment uses the 4,943-words dictionary of training and validation databases, giving 392 out-of-vocabulary words (5.5 percent of the testset). About 300 new trigraphs are present in the test set and not in the other sets. The proposed handwriting recognition system is designed to be on-line character-based recognition system. Therefore; the system performance evaluation will be concerned with character recognition. This section focuses on the evaluation of the recognition task, accuracy rate, and the recognition time as well as the proposed integrated recognition system. The following evaluation experiments were made using a Laptop Toshiba of 2.2 GHz,4 dual cores, 8Gbyte internal memory RAM and the system has 2Mbyte cash memory. The proposed system was built and tested using Matlab2014a. The histograms and tables of this evaluation were achieved by using the locally collected HW samples.

5.1. Recognition Rate Evaluation

In order to evaluate the proposed system, two types of experiments were performed. In the first one the system was trained with 1560 training characters of 30 different HW styles. The Recognition Rate (RR %) was inspected by letters of the Latin alphabetic characters written by 100 writers. The inspection was for the uppercase (A-Z) and the lowercase (a-z) separately as shown in Figure 15 and Figure 16. It is clear that the characters having simple patterns like C, O can be recognized more accurately (high RR) than characters with more complexity (having multiple strokes and junctions) like R, K, q etc. By comparing the results shown in Figure 15 and Figure 16, the RR of the uppercase characters is higher than of their lowercases. The reason behind this result is that the uppercase letters are always more obvious than the lowercases since the first have more right-angled straight strokes than the curved ones.

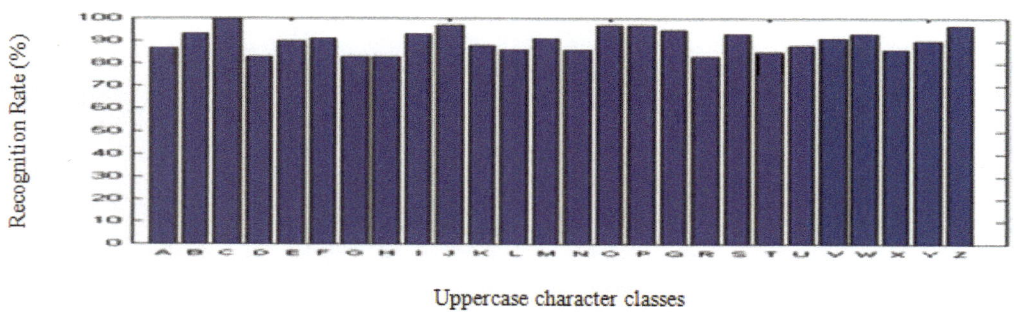

Figure 15. The Recognition Rate (%) inspected by letters of 100 writers for uppercase letters

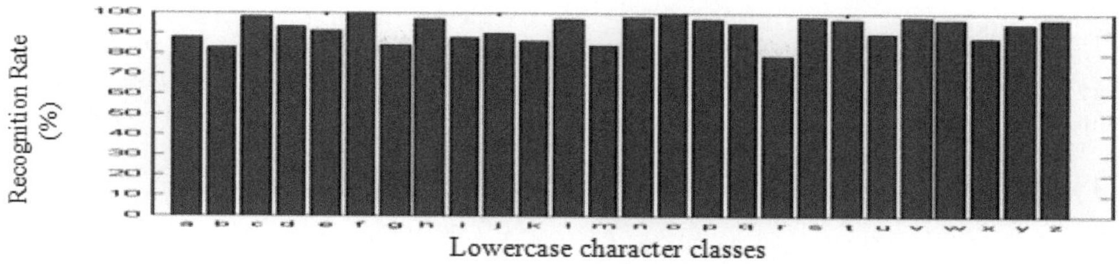

Figure 16. The Recognition Rate (%) inspected by letters of 100 writers for lowercase letters

The second experiment was the Recognition Rate (%) inspection for 100 testing samples for both upper and lower characters cases. Two of the test handwritten samples were the same as used for the training and they showed a RR of 100% (circled ones). Other test samples were written by the same persons who contributed in the training samples and these samples showed very high RR. The other samples with neat, very clear and fine handwriting. The other test samples were new HW samples. The weak or bad HW test samples gave the worst results as shown in Figure 17. Table 1 summarizes the results of the two inspection experiments for 3D-Multiwavelet [17].

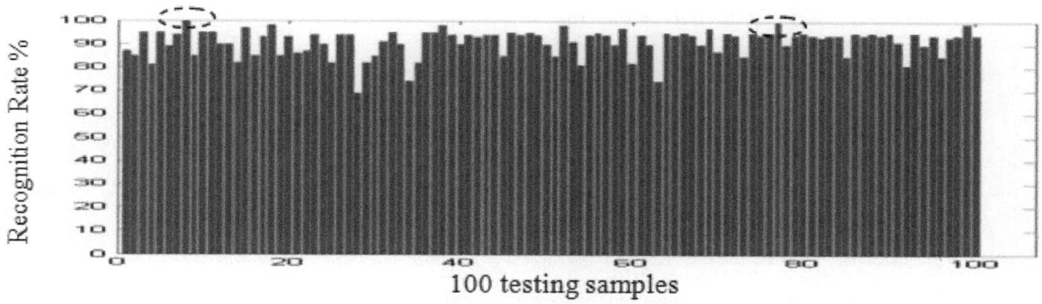

Figure 17. the Recognition Rate (%) inspected for 100 testing samples for both upper and lower characters cases

Table 1. Recognition Rate (%) of the proposed system for experiments 1 & 2

Characters Classes	2D-DMWTCS RR%	3D-DMWTCS RR%
Uppercase (26 classes) inspected by letters	94.28	96.94
Lowercase (26 classes) inspected by letters	91.63	93.67
Upper/lower (52 classes) inspected by writers	92.995	95.305

5.2. Recognition Time Computations

The main challenge is to speedup the recognition process and to improve the recognition accuracy. However, these two aspects are in mutual conflict. It is relatively easy to improve recognition speed while trading away some accuracy. But it is much harder to improve the recognition speed while preserving the accuracy. The recognition time of the proposed handwriting recognition system depends on many factors, The features and the specifications of the processing system like microprocessor speed and the available processing memory; size of the HW document to be recognized (number of sentences, words and characters); The scanning resolution (dpi); The degree of the HW document noise; The degree of the lines skew; degree of the characters slant; HW style (discrete or mixed styles), which affect on characters segmentation step; type and the decomposition level of the 3D-DMWT; number of the training data samples (size of the data-base); It is worth mentioning that the designed system uses some Matlab functions and subroutines that may slow down the processing speed. The goal is the minimization of the recognition time as possible. The recognition time that will be computed is per one character since the proposed HWRS is a character-based recognition. The HW document that is under test was a moderate noisy document and consists of six sentences (lines) with different skews namely L1, L2 … L6. These lines contain about 28 individual words (wo) with about 131 characters with different slants. The HW doc was scanned by flat scanner at 256 gray scales with 150-dpi resolution. The 3D-DMWTCS was by using Daubechies family at level 1 of decomposition that extracts 1024 features per character. Concerning the skew estimation and correction step, it is clear that the time required for long sentences is more than that for shorter ones. For the word segmentation, the time required to segment a line into individual words depends upon the number of the words and the characters which form it (line size). The time needed for the slant estimation and correction also depends upon the word size and its slant degree. As regard the thinning and thickening step, many experiments were made to compute the time required for this process and find the average time per character. There is interference between the segmentation and the recognition steps of the proposed system. Once the character is segmented, it will be delivered to the recognition step that includes the features extraction and the classification steps. Therefore; the needed time will be counted for these steps together.

6. General Discussion

The primary goal of the work is to design a complete modular on-line recognition system (writer-independent) for the handwritten text. This system deals with discrete and mixed HW styles in the upper and lower cases of Latin script letters and it is based on character recognition. The proposed system is a 3D-Discrete Multiwavelet Transform-based as a features extractor. The classifier was trained by a locally collected data base during the training phase. Patterns matching and classification is during the recognition phase. However, in every case the presented algorithm significantly improves the word and character segmentation and then recognition enhancement. With the application of the proposed approach one avoids any effect on the characters connectivity in the word and makes a negligible change in their aspect ratios and on shape nature. Some problems may arise when a word itself includes characters with different slants but rarely the writer himself writes a word with different characters orientations. However, in every case the presented algorithm significantly improves the character segmentation and then recognition enhancement. The reported results can, however, be considerably improved by training the system for a specific writer. Since the extracted rules are based on vectors consisting of human comprehensible information rather than opaque numerical data, a further improvement on the reported accuracy by refining the automatically extracted rules manually is possible.

7. Conclusions

The work developed in this paper aims to design an online system of high recognition accuracy of Latin handwritten types using (3D-DMWTCS). Multiwavelet representation has the advantage that the variations in the character shapes caused by the writing styles of different persons will cause only minor changes in the wavelet representation. The main information and features are concentrated at the approximation subband and the others are distributed in the other subband images. The recognition process depends on the features included in the approximation subband by about 53%. Using (3D-DMWTCS) for features extraction makes the recognition system need only a small training set to achieve high recognition accuracy regarding to the training samples (Data-Base), the collected data base depends upon the variety not on the quantity for the proposed HWRS based on (3D-DMWTCS). The selection of the training samples must avoid the redundant HW styles as possible. The relation between the number of the training samples and the Recognition Rate (RR) is not in direct proportional. The random increase in the training samples will not of necessity increase the RR; it may confuse the recognition process. Low resolution scanning (less than 100 dpi) will give erosion characters patterns. The high resolution scanning (more than 200 dpi) will increase the pixel size and improve the recognition but increase the time of the processing steps. The suitable resolution was 150 dpi. The capturing color mode has an effect on the recognition. The experiments show that the scanning with 256 gray scales is better than black/white or color mode. The proposed HWRS based on (3D-DMWTCS) system achieve an overall classification accuracy of 95.305 percent with 3D-DMWTCS using the Rimes database. These results have a gain of 2.31dB for 3D-DMWTCS compare with (2D-DMWTCS).

References

[1] Cun-Zhao Shi; Chun-Heng Wang; Bai-Hua Xiao; Song Gao; Jin-Long Hu, "Scene Text Recognition Using Structure-Guided Character Detection and Linguistic Knowledge," Circuits and Systems for Video Technology, IEEE Transactions on , vol.24, no.7, pp.1235,1250, July 2014

[2] A.-L. Bianne, C. Kermorvant, and L. Likforman-Sulem, "Context-Dependent HMM Modeling Using Tree-Based Clustering for the Recognition of Handwritten Words," Proc. SPIE Document Recognition and Retrieval, 2010..

[3] Assabie Y. and Bigun J" Writer-independent offline recognition of handwritten Ethiopic characters", In: Proc. 11th ICFHR, August 19-21, Montreal, 2008.

[4] Ballesteros J., Travieso C.M., Alonso J.B. and Ferrer M.A, " Slant estimation of handwritten characters by means of Zernike moments", IEE electronics letters vol.41 no.20, 29th September,2005

[5] Bruce K., "Multiwavelets for Quantitative Pattern Matching" Proceedings of the 42nd Hawaii International Conference on System Sciences, pp 1- 10,2009.

[6] Cheng-Lin L., Jaeger S. and Nakagawa M. "Online recognition of Chinese characters: the state-of-the-art", IEEE transactions on pattern analysis and machine intelligence, vol. 26, no. 2, February 2004.

[7] Gao X., Fen X., and Bodong L., "Construction of Arbitrary Dimensional Biorthogonal Multiwavelet Using Lifting Scheme" IEEE Transactions On Image Processing, VOL. 18, NO. 5, MAY, pp 942-955,2009.

[8] Deselaers, T.; Keysers, D.; Hosang, J.; Rowley, H.A., "GyroPen Gyroscopes for Pen-Input With Mobile Phones," Human-Machine Systems, IEEE Transactions on , vol.45, no.2, pp.263,271, April 2015.

[9] Shahriarpour, E.; Sadri, J., "Recognition of legal amount words on Persian bank checks using Hidden Markov Model," Intelligent Systems (ICIS), 2014 Iranian Conference on , vol., no., pp.1,5, 4-6 Feb. 2014

[10] Breuel, T.M.; Ul-Hasan, A.; Al-Azawi, M.A.; Shafait, F., "High-Performance OCR for Printed English and Fraktur Using LSTM Networks," Document Analysis and Recognition (ICDAR), 2013 12th International Conference on , vol., no., pp.683,687, 25-28 Aug. 2013

[11] El-Anzy L., " Design and Simulation of STBC-(OFDM and CDMA) Transceivers based on Hybrid Transforms ", Ph.D. Thesis, Univ. of Technology, Electrical and electronic engineering, Dep., Oct.2006.

[12] Rodolfo P., Gabriela S., Tsang I. and George D "Text Line Segmentation Based on Morphology and Histogram Projection" 10th International Conference on Document Analysis and Recognition, pp 651-655., 2009.

[13] Suriya K. , Kitti A., Thanatchai K. "Recognition of power quality events by using Multiwavelet-based neural networks" Elsevier Ltd., Electrical Power and Energy Systems Vol 30 ,pp 254–260, 2008.

[14] K. Ubul, A. Adler and M. Yasin, "Multi-Stage Based Feature Extraction Methods forUyghur Handwriting Based Writer Identification," In Genetic Algorithms in Applications, InTech , 2012.

[15] Shanker A., Tajagopalan A. "Off-line signature verification using DTW", Pattern Recognition Lett. 28, pp1407–1414, 2007.

[16] Xiangqian Wu; Youbao Tang; Wei Bu, "Offline Text-Independent Writer Identification Based on Scale Invariant Feature Transform," Information Forensics and Security, IEEE Transactions on , vol.9, no.3, pp.526,536, March 2014

[17] Yousri K., Thierry P., AbdelMajid B., "A Multi-Lingual Recognition System for Arabic and Latin Handwriting" 10th International Conference on Document Analysis and Recognition, pp 1196-1200, 2009.

[18] Matheel E., Rabab F., "Novel Video Denoising Using 3-D Transformation Techniques," International Journal of Engineering and Advanced Technology (IJEAT), Volume-2, Issue-5, June 2013.

[19] R. Cruz, G. Cavalcanti, and T. Ren, "An ensemble classifier for offline cursive character recognition using multiple feature extraction techniques," in The 2010 International Joint Conference on Neural Networks (IJCNN),pp. 1 –8, July 2010.

[20] AbuzaraidaM. A., A. M. Zeki and A. M. Zeki, "Segmentation Techniques for Online Arabic Handwriting Recognition: Asurvey," in 3rd International Conference on Information and Communication Technology for the Moslem World, Jakarta,Indonesia, , pp. D37-D40,2010.

[21] R. Smith, "History of the Tesseract OCR engine: what worked and what didn't ," in DRR XX, San Francisco, USA, Feb. 2013.

[22] KN. Sriraam and R. Shyamsunder. "3-D Medical Image Compression Using 3-DWavelet Coders", Digital Signal Processing, 21(1):100–109, January 2011.

[23] Ibrayim, M., and Askar H., "Design and implementation of prototype system for online handwritten Uyghur character recognition," Wuhan University Journal of Natural Sciences,vol. 17 no. 2, pp. 131-136, April 2012.

[24] Dreuw, P., Rybach, D., Heigold, G., and Ney, H., Guide to OCR for Arabic Scripts Chp. Part II: RWTH OCR: A Large Vocabulary Optical Character Recognition System for Arabic Scripts, Springer, London, UK, pp. 215-254,July, 2012.

[25] Abuzaraida, M.A.; Zeki, A.M.; Zeki, A.M., "Feature extraction techniques of online handwriting arabic text recognition," Information and Communication Technology for the Muslim World (ICT4M), 2013 5th International Conference on , vol., no., pp.1,7, 26-27 March 2013

[26] http://rimes.it sudparis.eu/, 2010.

[27] E. Augustin, M. Carre, E. Grosicki, J.-M. Brodin, E. Geoffrois, and F. Preteux, "Rimes Evaluation Campaign for Handwritten Mail Processing," Proc. Int'l Workshop Frontiers in Handwriting Recognition, pp. 231-235, 2006.

New Very Low Frequency Oscillator Using only a Single CFOA

D. K. Srivastava[1], V. K. Singh[1,*], R. Senani[2]

[1]Electronics Engineering Department, Institute of Engineering and Technology, Lucknow, 226021 India
[2]Division of Electronics and Communication Engineering, NetajiSubhas Institute of Technology, Sector 3, Dwarka, New Delhi, 110078, India
*Corresponding author: vksingh@ietlucknow.edu

Abstract Some time back, Elwakil presented a systematic method of realizing Very Low frequency (VLF) oscillators using current feedback operational amplifiers (CFOA) and demonstrated that the classical Wienbridge oscillator, employing the concept of composite resistor (containing two positive and one negative resistance) can be tailored to generate VLF oscillations. The circuit proposed by Elwakil, however, required two CFOAs along with six resistors and two capacitors. The object of this paper is to report a new VLF oscillator circuit which, in contrast to Elwakil's circuit, requires onlya single CFOA. The workability of the new circuit has been demonstrated by experimental results using commercially available AD844type CFOAs.

Keywords: *Oscillators, Current-feedback-operational amplifiers, current-mode circuits, single element controlled oscillators, very low-frequency oscillators*

1. Introduction

Very low frequency (VLF) oscillators (capable of generating frequencies < 20Hz) are required in many applications such as bio-medical instrumentation, geo-physical instrumentation and in some measurements in control system applications.

Owing to the well-known advantages of current feedback op-amps (CFOA) (for instance, see [1,2,3,4]) and their wide spread acceptability as alternative building blocks, there have been numerous studies in recent literature on realizing sinusoidal oscillators using CFOAs, for instance, see [5,6,7,8] and references cited therein. In [8], Elwakil had reported a systematic method of designing VLF oscillators using CFOAs. The circuit of [8], however, requires two CFOAs along with six resistors and two capacitors (one of which is floating).

The main object of this paper is to introduce a new CFOA-based VLF oscillator, which, by contrast to the circuit of [8], requires no more than a single CFOA and employs both grounded capacitors, as preferred for integrated circuit implementation.

2. Proposed VLF oscillator using a single CFOA

CFOA is a four terminal building block which internally contains a plus type second generation current conveyor [9] and a voltage follower and is characterized by the following terminal equations

$$i_y = 0,\ v_x = v_y,\ i_z = i_x,\ v_w = v_z \tag{1}$$

The symbolic notation of the CFOA and its equivalent circuit are shown in Figure 1.

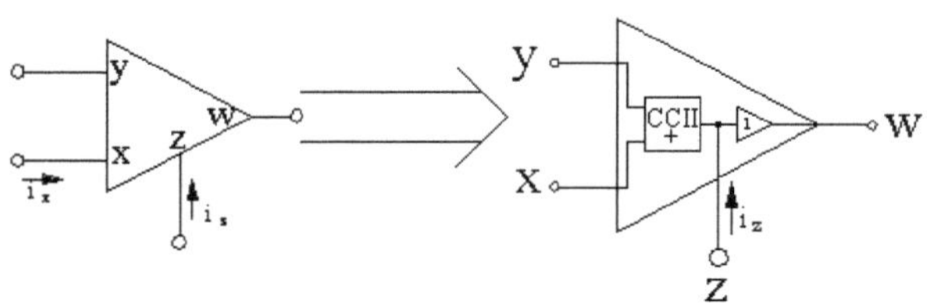

Figure 1. Symbolic notation and the equivalent circuit of the CFOA

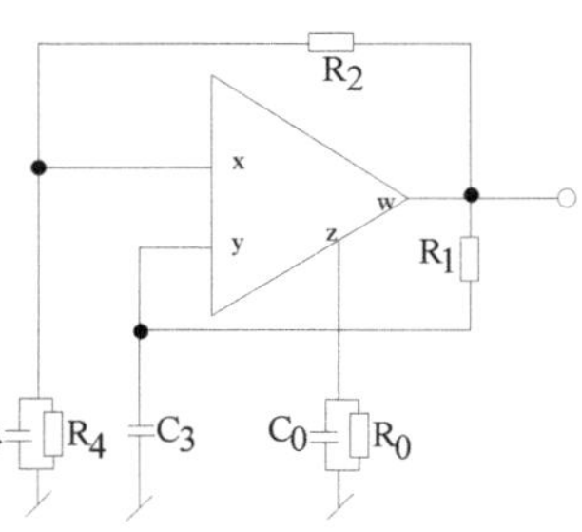

Figure 2. The proposed VLF oscillator

The proposed circuit configuration is shown in Figure 2. Using equation (1) for the CFOA, a straight forward analysis of the circuit gives the condition of oscillation as

$$\frac{C_4 - C_0}{C_3} = \frac{R_1}{R_0} + \frac{R_1}{R_2} \quad (2)$$

whereas the frequency of oscillation is given by

$$f_0 = \frac{1}{2\pi}\sqrt{\frac{\left(\frac{R_4}{R_0} - 1\right)}{\left(R_1 R_4 C_0 C_3\right)}} \quad (3)$$

From (2) and (3) it is seen that frequency of oscillation can be adjusted by varying R_4 (while, ensuring $\frac{R_4}{R_0} > 1$) whereas the condition of oscillation can also be independently adjusted by C_4 and/or R_2, out of which using a variable resistance R_2 is obviously more preferable. From equation (3), one can see that keeping the difference term in the numerator of FO as small as feasible, generation of VLF oscillations should be possible. It may be mentioned that while designing the circuit, the resistor values should be chosen such that they are much larger than the parasitic x port input impedance R_x (typically around 50Ω) but smaller than the parasitic z port resistance R_p (typically around 3MΩ) of the CFOA. Similarly, the external capacitors have to be chosen such that their values are larger than the parasitic output capacitance C_p (looking into the z terminal) of the CFOA (typically of the order of 4-5pF). This will ensure that the parasitic impedances of the CFOA will not have much effect on the realized frequency of oscillation.

3. Experimental Results

The workability of the proposed circuit has been checked by bread-boarded versions of the circuit realized by using AD844 type of CFOAs biased with ±12V DC power supplies, taking R_0=R_1=1MΩ, C_0=C_3=100 nF, C_4=200 nF with the CO adjusted by variable resistance R_2 (1MΩ fixed + 0.0.5M Ω variable). Figure 3 (a) shows the variation of FO with R_4, while typical waveform generated from this circuit is shown in Figure 3 (b). The experimental results, thus, confirm the practical workability of the new oscillator circuit.

4. Comparison with Previously Known Oscillators

A comparison of the proposed new oscillator with those previously published in literature [5,6,7,8,10] is now in order. This is shown in Table 1.

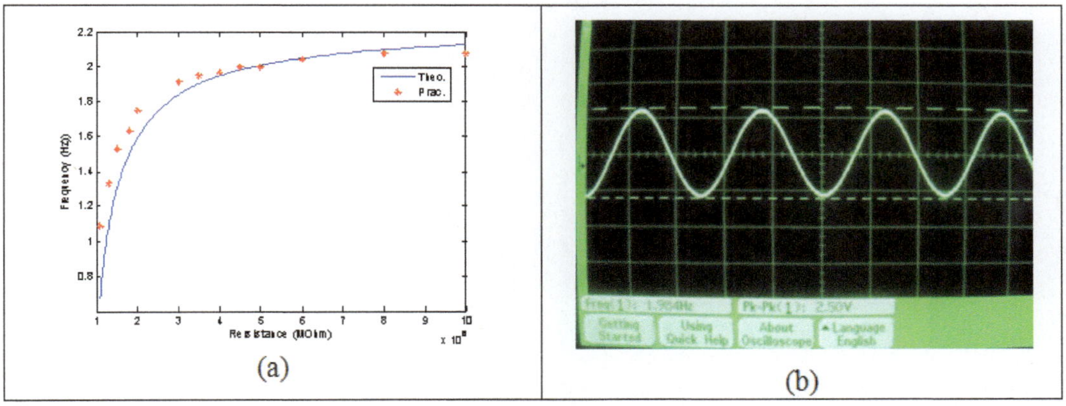

Figure 3. Experimental results of the proposed oscillator (a) variation of oscillation frequency with variable resistance R_4 (b) a typical waveform generated by the circuit (1.984Hz, 2.5V p-p)

Table 1. Comparison with relevant previously published oscillator circuits

Ref. No.	Active components	Whether circuit uses all grounded capacitors	No. of resistors	Whether independent control of CO is available	Whether independent control of FO is available
[5]	1 OP-AMP	NO	06	YES	YES
[6]	2 CFOA	NO	03	YES	YES
[7]	2 CFOA	NO	04	YES	YES
[8]	2 CFOA	NO	06	YES	YES
[10]	1 CFOA	YES	04	NO	YES
Proposed Circuit	**1 CFOA**	**YES**	**04**	**YES**	**YES**

From Table 1, it is seen that the proposed circuit is the only single-CFOA oscillator which possesses all the desirable features while using a reasonable number (only four) of the resistors.

5. Concluding Remarks

Anew CFOA-based VLF oscillator circuit configuration has been presented which employs only a single CFOA in contrast to an earlier CFOA-based VLF oscillator proposed by Elwakil [8] which requires two CFOAs. The workability of the new circuit has been confirmed by experimental results using AD844 type CFOAs and it has been found that the circuit can generate a sinusoidal waveforms of frequency as low as 2 Hz. Lastly, it must be mentioned that although a somewhat similar single CFOA oscillator has been reported earlier in [10], however, the circuit described here has the advantage of providing independent control of CO also, which is not available in the quoted circuit of [10].

Acknowledgements

This work was conducted partly at Analog Signal Processing Research Lab, NSIT, New Delhi and partly at Analog Signal Processing Research Lab, IET, Lucknow. Thanks are due to Prof. D R Bhaskar and Prof. A. K. Singh for their support.

References

[1] C.Toumazou and F.J.Lidgey, "Current feedback op-amps: a blessing indisguise?," *IEEECircuits Devices Mag.,* vol. 10, no.1, pp. 34-37, 1994.

[2] F.J.Lidgeyand K.Hayatleh, "Current feedback operational amplifiers and applications, *Electron. Commun. Eng,* vol. 9, no. 4, pp. 176-182, 1997.

[3] A.M. Soliman, "Applications of current feedback amplifier," *Anal. Intregr. Circuits Signal Process.* vol. 11, pp. 265-302, 1996.

[4] R. Senani, "Realization of a class of analog signal processing/signal generation circuits: novel configurations using current feedback op-amps", Frequenz, vol. 52, no. 9-10, pp. 196-206, 1998.

[5] R. Senani and D. R. Bhaskar, "Single op-amp sinusoidal oscillators suitable for generation of very low frequencies," IEEE Trans. Instrum. Meas., vol. 40, no. 4, pp. 777-779, 1991.

[6] D.R. Bhaskar and R. Senani, "New CFOA-based Single-Element-Controlled Sinusoidal Oscillators," IEEE Trans. Instrum. Meas. vol. 55, no. 6, pp. 2014-2021, 2006.

[7] M. T.Abuelma'atti, "Identification of a class of two CFOA-based sinusoidal RC oscillators," Analog Integrated Circuits Signal Process., vol. 65, pp. 419-428, 2010.

[8] A. S. Elwakil, "Systematic realization of low-frequency oscillators using composite passive-Active resistors", IEEE Trans. Instrum. Meas., vol. 47, no. 2, pp. 584-586, 1998.

[9] A. S. Sedra and K.C. Smith, "A Second generation Current Conveyor and its applications", IEEE Transaction on Circuit Theory, vol. CT-17, pp. 132-134, February 1970.

[10] A. Toker, O. Cicekoglu and H.Kuntman, "On the oscillator implementation using a single current feedback op-amp", Computers and Electrical Engg., vol. 22, pp. 375-389, 2002.

8

Physical Substantiation of Huygens Principle and the Reciprocity Theorem

F. F. Mende*

B.I. Verkin Institute for Low Temperature Physics and Engineering, NAS Ukraine, 47 Lenin Ave., Kharkov, Ukraine
*Corresponding author: mende_fedor@mail.ru

Abstract The Huygens principle says, that each element of wave front can be examined as the center of the second disturbance, which generates second spherical waves, and the resulting light field at each point of space will be determined by the interference of these waves. This principle is the basic postulate of geometric optics; however, it does not reveal physical nature of this phenomenon. Are examined the laws of the self-induction of such reactive elements as capacity and inductance, which made possible to base the physical bases of Huygens's principle. It is shown that with the connection to the sources of direct current and voltage they have the effective resistance, which depends on the time. Is examined the parametric self-induction of the elements indicated and it is shown that in the regime of parametric self-induction the capacity and inductance can not only derive energy from the power sources, but also return it into the external circuits. The new method of determining the velocity of propagation of constant stresses and currents in the long lines is developed. This task cannot be solved, using standard wave equations for the long lines. The physical special features of the work of the receiving directional Yagi antennas are examined and it is shown that the reciprocity theorem for such antennas is not carried out.

Keywords: *Huygens principle, antennas, reciprocity theorem, self-induction*

1. Introduction

Many specialists consider that the capacity and inductance are reactive elements and cannot take away energy in the alternating current circuits or voltage [1]. In order to be convinced of the fact that this is not always carried out, it suffices to conduct simple experiment. If we connect capacitor to the electric brush of alternating current, to and then open it, then it is possible to reveal that the capacitor is charged. If we repeat this experiment repeatedly, then it is possible to establish that a voltage drop across the terminals of capacitor varies in the limits from the zero values, to the amplitude value of a potential difference in the network. This fact attests to the fact that the capacitor can take away energy from the electric brush, since the energy, accumulated in the capacitor, is determined by the equation

$$W_C = \frac{1}{2}CU^2$$

where U is voltage drop across the terminals of capacitor.

To load capacitor is possible and from the source of direct current, and in this case the energy, accumulated in the capacitor, will be obtained from this source. Idea about the fact that the reactive elements cannot take away energy in the alternating current circuits, was formed for that reason, that with the examination of the properties of reactive elements in such chains are examined the periodic processes of infinite duration and the moment of turning off of reactive element from the chain is not considered.

If we charge capacitor from the source of direct current, then a voltage drop across it will be grow, and it will derive energy from the source, accumulating it similar to storage battery. In this case the capacity can be represented as the effective resistance, which depends on the time. The same relates also to the inductance, connected to the dc power supply. But capacity and inductance can return energy into the external circuit, if their value changes.

The special features of capacity and inductance indicated make it possible to solve a question about the velocity of propagation of constant voltage and currents in the long lines, what cannot be made, using telegraphic of equations.

Theory and different constructions of antennas are in detail examined in monographs [2,3]. Primary attention in these works is given to the calculation of the transmitting antennas, at basis of which lies the principle of interference. However, up to now are not clear the physical the bases of the work of some types of the receiving directional antennas. With incomprehensibly what means the receiving Yagi antennas, which received very wide acceptance, ensure this high mu-factor. For guaranteeing the high mu-factor this antenna must gather energy from the lateral space, which surrounds antenna, but, as this it makes, the pill box of their times remains not clear.

2. Laws of the Self-induction

To the laws of the self-induction of reactive elements should be carried those laws, which describe the reaction of these elements with the galvanic connection to them of the sources of current or voltage. These laws are the basis of the theory of electrical chains. The results of this theory can be postponed also by the electrodynamics of material media, since such media can be represented in the form equivalent diagrams with the use of such elements [2].

By self-induction we will understand the reaction of material structures with the constant parameters to the connection to them of the sources of voltage or current. to the self-induction let us carry also that case, when its parameters can change with the presence of the connected power source or the energy accumulated in the system. This self-induction we will call parametric[4,5,6,7].

Let us refine some terms, utilized in the article. We will call such generator, which ensures lumped voltage on any external circuit constant-potential generator. The internal resistance of the ideal voltage generator is equal to zero. We will call such generator, which provides the assigned current in any external circuit direct-current generator. The internal resistance of ideal current generator approaches infinity.

Ideal capacity cannot be connected to the ideal voltage source, since. at the moment of connection the current of the charge of capacity will approach infinity. Capacity can be connected only to the current source. Ideal inductance cannot be connected to the ideal current source, since. at the moment of connection a voltage drop across inductance will approach infinity. Inductance can be connected only to the dc power supply.

3. Capacitive Self-induction

If the capacity C is charged to a potential difference U, then the charge Q, accumulated in it, is determined by the equation

$$Q_{C,U} = CU$$

Magnitude of the charge can change with the method of changing the potential difference with a constant capacity, or with a change in the capacity with a constant potential difference.

If capacitance value or voltage drop across it depend on time, then the current strength is determined by the equation:

$$I = \frac{dQ_{C,U}}{dt} = C\frac{\partial U}{\partial t} + U\frac{\partial C}{\partial t}$$

Consequently, current in circuit can be obtained by two methods: changing voltage on the capacity with a constant capacity or changing capacity with constant voltage on it.

For the case, when the capacity C_0 is constant, we obtain

$$I = C_0\frac{\partial U}{\partial t} \tag{3.1}$$

Let us connect to the capacity direct-current generator and we will support in it the direct current I_0. Then, after integrating Eq. (3.1) with respect to the time, we obtain

$$U = \frac{I_0 t}{C_0} \tag{3.2}$$

This equation, which connects the direct current, which flows through the capacity, and voltage on it is the analog of Ohm law. For this reason the value

$$R = \frac{t}{C_0}$$

plays the role of effective resistance, which depends on time. The it should be noted that obtained result is completely obvious; however, such properties of capacity, which customary to assume by reactive element they were for the first time noted in the work [3].

The power, output by current source, is determined in this case by the equation:

$$P(t) = \frac{I_0^2 t}{C_0} \tag{3.3}$$

The energy, accumulated by capacity in the time t, we will obtain, after integrating Eq. (3.3) with respect to the time:

$$W_C = \frac{I_0^2 t^2}{2C_0}$$

Substituting here the value of current from Eq. (3.2), we obtain the dependence of the value of the accumulated in the capacity energy from the instantaneous value of voltage on it:

$$W_C = \frac{1}{2}C_0 U^2$$

If we support at the capacity constant voltage U_0, and to change capacity, then the current, which flows through it, is written down

$$I = U_0\frac{\partial C}{\partial t} \tag{3.4}$$

This case relates to the parametric self-induction. This equation, which connects the direct current, which flows through the capacity, and voltage on it is the analog of Ohm law. For this reason the value

$$R_C = \left(\frac{\partial C}{\partial t}\right)^{-1}$$

plays the role of effective resistance, which depends on time.

The power, expended in this case by source, is determined by the equation:

$$P = U_0^2\frac{\partial C}{\partial t}$$

From this equation is evident that depending on the sign of derivative the expendable power can have different signs. When the derived positive, expended by source power proceeds with an increase in the energy, stored up in the capacity. If derived negative, then work carries out the external source, which ensures the decrease of capacity. The energy, spent by this source, is separated in the external circuit.

4. Inductive Self-induction

Let us examine the processes, proceeding in the inductance. Let us introduce the concept of the flow

$$\Phi_{L,I} = LI$$

Voltage on the inductance is equal to the derivative of the flow of current induction on the time:

$$U = \frac{d\Phi_{L,I}}{dt} = L\frac{\partial I}{\partial t} + I\frac{\partial L}{\partial t}$$

Let us examine the case, when the inductance L_0 is constant:

$$U = L_0\frac{\partial I}{\partial t} \quad (4.1)$$

Will maintaina constant voltage U_0 . Then, after integrating Eq. (4.1) with respect to the time, we obtain

$$I = \frac{U_0 t}{L_0} \quad (4.2)$$

This equation, which connects the direct current, which flows through the capacity, and voltage on it is the analog of Ohm law. Therefore the value of

$$R = \frac{L_0}{t}$$

plays the role of effective resistance, which depends on time.

The power, expended in this case by source, is determined by the equation:

$$P(t) = \frac{U_0^2 t}{L_0} \quad (4.3)$$

After integrating Eq. (4.3) on the time, we will obtain the energy, accumulated in the inductance

$$W_L = \frac{1}{2}\frac{U^2 t^2}{L}$$

After substituting in this equation the value of voltage from Eq. (4.2), we obtain the energy, stored up in the inductance,

$$W_L = \frac{1}{2}L_0 I^2$$

If we support in the inductance the direct current I_0, and to change inductance, then the current, which flows through it, will be written down

$$U = I_0\frac{\partial L}{\partial t} \quad (4.4)$$

This case relates to the parametric self-induction. This equation, which connects the direct current, which flows through the capacity, and voltage on it is the analog of Ohm law. Therefore the value

$$R = \frac{dL}{dt}$$

plays the role of the effective resistance.

The power, expended in this case by source, is determined by the equation:

$$P = I_0^2\frac{\partial L}{\partial t}$$

As in the case with the capacity, effective resistance can be both the positive and negative. This means that the inductance can how derive energy from without, so also return it into the external circuits.

5. Propagations of Constant Voltage and Current in the Long Lines

The processes of the propagation of voltages and currents in the long lines determine the wave equations:

$$\frac{\partial^2 U}{\partial z^2} = \frac{1}{v^2}\frac{\partial^2 U}{\partial t^2}, \quad \frac{\partial^2 I}{\partial z^2} = \frac{1}{v^2}\frac{\partial^2 I}{\partial t^2}$$

which are obtained from the telegraphic equations:

$$\frac{\partial U}{\partial z} = -L_0\frac{\partial I}{\partial t}, \quad \frac{\partial I}{\partial z} = -C_0\frac{\partial U}{\partial t}$$

where L_0 and C_0 are linear inductance and the capacity of line.

The velocity of propagation in this line is determined by the equation

$$v = \frac{1}{\sqrt{L_0 C_0}}$$

The knowledge second derivative voltages and currents is required during the solution of wave equations.

But wave equations do not give the answer to the question, as one should enter when to incoming line it is connected dc power supply. The results, obtained in two previous divisions, give answer to these questions. The processes, examined in two previous paragraphs, concern chains with the lumped parameters, when the distribution of potential differences and currents in the elements examined can be considered uniform. However, there are chains, for example the long lines, in which the spatial distribution of voltages and currents they are not uniform.

Let us examine processes in the long line, whose capacity and inductance are the distributed parameters. We will consider that the linear capacity and the inductance of line is equal C_0 and L_0 respectively. If we to this line connect the dc power supply o U_0, then its front will be extended in the line some by the speed v, and the moving coordinate of this front will be determined by the equation $z = vt$. In this case the total quantity of the charged capacity and the value of the summary inductance, along which it flows current, calculated from the beginning lines to the location of the front of voltage, will change according to the law:

$$C(t) = zC_0 = vtC_0, L(t) = zL_0 = vtL_0$$

The source of voltage U_0 will in this case charge the being increased capacity of line, for which from the source to the charged line in accordance with Eq. (3.4) must leak the current:

$$I = U_0 \frac{\partial C(t)}{\partial t} = vU_0C_0$$

This current there will be the leak through the conductors of line, that possess inductance. But, since the inductance of line in connection with the motion of the front of voltage, also increases, in accordance with Eq. (4.4), on it will be observed a voltage drop:

$$U = I \frac{\partial L(t)}{\partial t} = IvL_0 = v^2U_0C_0L_0 \qquad (5.1)$$

But a voltage drop across the conductors of line in the absolute value is equal to the voltage, applied to its entrance; therefore in the last expression should be placed $U = U_0$. Then from Eq. (5.1) we obtain

$$v = \frac{1}{\sqrt{L_0C_0}}$$

This equation coincides with the velocity of propagation, obtained from the wave equations.

However, in the electrodynamics, based on Maxwell equations, there are no such concepts as capacity and inductance, and there are concepts of the electrical and magnetic permeability of medium. In the carried out examination such concepts as electrical and magnetic fields also was absent. Let us show how to pass from such categories as linear inductance and capacity, current and voltage in the line to such concepts as dielectric and magnetic constant, and also electrical and magnetic field. For this let us examine the simplest construction of line, located in the vacuum (Figure 1)

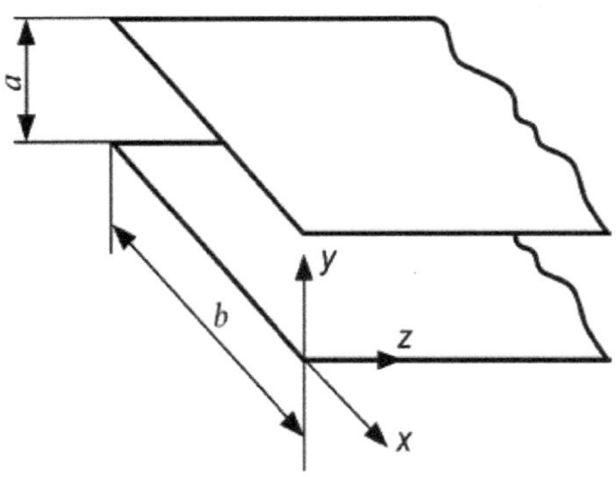

Figure 1. The two-wire circuit, which consists of two ideally conducting planes

We will consider that $b >> a$ and edge effects it is possible not to consider. Then the following connection will exist between the linear parameters of line and the magnetic and dielectric constants:

$$L_0 = \mu_0 \frac{a}{b}, C_0 = \varepsilon_0 \frac{b}{a}$$

where μ_0 and ε_0 is dielectric and magnetic constant of vacuum.

The velocity of propagation in this line is determined by the equation

$$v = \frac{1}{\sqrt{L_0C_0}} = \frac{1}{\sqrt{\mu_0\varepsilon_0}} = c$$

where c is velocity of propagation of light in the vacuum.

The wave drag of the line examined will be equal

$$Z = \frac{a}{b}\sqrt{\frac{\mu_0}{\varepsilon_0}} = \frac{a}{b}Z_0$$

where $Z_0 = \sqrt{\frac{\mu_0}{\varepsilon_0}}$ is wave drag of free space.

This with the observance of the condition $a = b$ we obtain the equality $L_0 = \mu_0$. This means that magnetic permeability μ_0 plays the role of the longitudinal specific inductance of vacuum. In this case is observed also the equality $C_0 = \varepsilon_0$. This means that the dielectric constant ε_0 plays the role of the transverse specific capacity of vacuum. In this interpretation both μ_0 and ε_0 acquire clear physical sense.

If we to the line of infinite length, connect the dc power supply U_0, then the field strength in the line will be equal

$$E_y = \frac{U_0}{a}$$

and the current, which flows into the line from the power source, will be determined by the equation:

$$I = \frac{U_0}{Z} = \frac{aE_y}{Z}$$

Magnetic field in the line will be equal to the specific current, flowing in the line

$$H_x = \frac{I}{b} = \frac{aE_y}{bZ}$$

Substituting here the value of Z, we obtain

$$H_x = \frac{E_y}{Z_0}$$

The same connection between the electrical and magnetic field exists also for the case of the transverse electromagnetic waves, which are extended in the free space.

Comparing expressions for the energies, it is easy to see that the specific energy can be expressed through the electrical and magnetic fields

$$\frac{1}{2}\mu_0 H_x^{\ 2} = \frac{1}{2}\varepsilon_0 E_y^{\ 2} \qquad (5.2)$$

This means that the specific energy, accumulated in the magnetic and electric field in this line is identical.

The special feature of this line will be the fact that in it, in contrast to the free space, the stationary magnetic and electric fields can be extended, but this case cannot be examined by the method of solution of Maxwell equations.

Consequently, conditionally it is possible to consider that the long line is the device, which with the connection to it of dc power supply is filled up with two forms of the energy: electrical and magnetic. The specific densities of these energies are equal, and since and electrical and magnetic energy fill identical volumes, the general energy, accumulated in these fields is identical. The distribution of electrical and magnetic pour on in the section from the beginning of line and to the front of voltage in it is uniform.

Is obvious $I(t)U(t)$ represents the power P, transferred through the cross section of line in the direction of z. If in this equation current and voltage was replaced through the tensions of magnetic and electrical pour on, then we will obtain $P = abE_yH_x$. The work of E_yH_x represents the absolute value of Poynting vector.

6. Physical substantiation of Huygens principle and the reciprocity theorem

Huygens principle says, that each element of wave front can be examined as the center of the second disturbance, which generates second spherical waves, and the resulting light field at each point of space will be determined by the interference of these waves. This principle is the basic postulate of geometric optics; however, it does not reveal physical nature of this phenomenon. From geometric optics it is known that any beam of light is dispersing and that the area of its section in the process of propagation always increases. This phenomenon is subordinated to Huygens principle. But is there any physical explanation of this principle? It occurs, there exists.

Let us examine flat monochromatic TEM the wave, passing through the slot, whose width is considerably longer more than wavelength (Figure 2).

Figure 2. Passage of the plane wave through the slot

After passage through the slot wave begins to be enlarged in the transverse direction, and this expansion is subordinated to Huygens principle. In this case in the expanding wave the ends of the paths of constant phase in the process of their motion move with the speed of light still, also, in the transverse direction. Difference is only the fact that in the line the wave of transverse electric field is propagated, and self-expansion occurs in the direction of propagation. In this case occurs the self-expansion of the vector of electric field still, also, in the transverse direction [8]. In the long line there is no such expansion, since wave in the transverse direction I limit the conductors of line. The transverse transformation of wave is accompanied by the fact that, beginning from the center of path of constant phase along it begins leak bias current. This process is very similar to the expansion of the compressed elastic, when all its sections begin evenly to be enlarged. In this case the energy density of electromagnetic wave begins to decrease, being evenly distributed in the increasing volume, occupied by the expanding wave.

This simple examination, he indicates the physical causes for Huygens postulate and is in fact new physical law.

With this phenomenon is connected the so-called reciprocity theorem for the antenna systems, which does not up to now have its physical substantiation. This theorem says, that the mu-factor of the directional antennas is identical both with the emission and with the reception of signal. The strangeness of this theorem consists in the fact that the directional antenna can form the narrowly directed beam, when the radiated energy is concentrated in some direction. This means that the energy density is concentrated in the space limited from the lateral sides. This one can see well based on example of laser beam. However, receiving directional antenna is located in the fields of transmitter, which are evenly distributed in the space, and in order to increase its mu-factor to the directional receiving antenna it is necessary to know how to gather energy from the lateral space. This actually so, but as this it makes, until now, it remains riddle. A question does consist in that, is it possible to find some physical causes for this strange behavior of the directional receiving antennas.

Is known that the laws of geometric optics, when ray can be considered practically rectilinear, work when beam width considerably more than wavelength. In this case Huygens principle works. When the width of the slit becomes comparable with the wavelength, the slit is strongly divergent beam, and when the size of the gap becomes smaller the wavelength, the slit is radially divergent waves.

Let us examine two cases. In Figure 3 is depicted the case, when the width of obstacle is considerably lower than the wavelength.

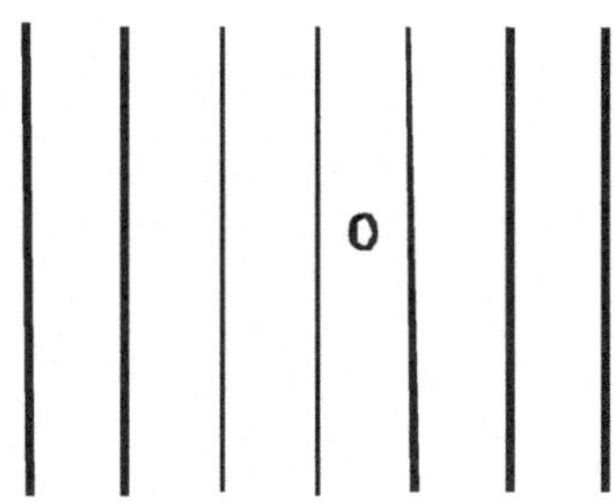

Figure 3. Diffraction by the waves of the obstacle, whose dimensions are considerably lower than the wavelength

Bend around obstacles such waves practically do not change their shape.

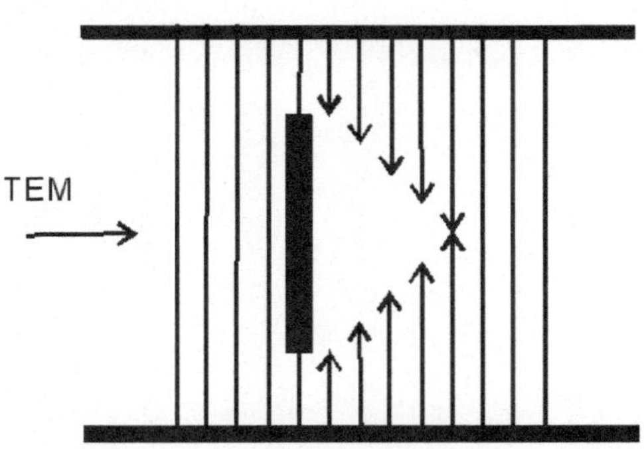

Figure 4. The case, when on the oscillation loop is located the obstacle, whose dimensions are greater wavelength

Quite different is the process of diffraction when the dimensions of the barrier longer wavelength (Figure 4).

Let us examine wave process in the long line, when on the way of propagation TEM wave is located the absorbing partition, whose dimensions are greater wavelength.

It is evident that from the right side from the absorbing obstacle there is a shadow zone, where the waves absent. But passing obstacle, the ends of the paths of constant phase begin to converge, bringing additional energy into the shadow zone from the lateral space. This process obeys the law of diffraction and has obvious special feature. Since the absorbing partition absorbed the part of the energy of the wave incident to it, the energy density of waves to the right beyond the limits of shadow, will be less than the energy density of waves to the partition. Direction the motions of energy in the expanding section of wave are shown by pointers. This means that in order to liquidate shadow and to restore normal wave process, wave energy begins to be pumped over from the lateral sections of those removed from the shadow zone, enlarging the sections of waves torn by obstacle. In this case Poynting vector to the right of shadow will be less than to the left of the partition. This means that as a result this diffraction process the specific energy again evenly was distributed on the perimeter of line, but its density will become less. If we on the route waves place consecutively several absorbing partitions of TEM wave, then after passage by the wave of sequential partition the Poyting vector will always decrease. For the realization of this process the wave energy always will be redistributed from the lateral regions line to the central region. This process is analogous to the process of the propagation of the wave front in the transmission line, only expansion of the lateral front of paths of constant phase occurs in the transverse direction and in the same direction is transferred the energy.

If the cross section of line is considerably greater than the section of the absorbing partitions, then a change in the vector of the Poyting vector will be insignificant, and each subsequent partition will absorb almost this energy as the previous partition. This connected with the fact that cross section fields on in the line considerably more than the cross section of partitions. But this process is connected with the fact that wave energy from the lateral sections, situated beyond the limits of the absorbing partitions, it is pumped over into the region of the arrangement of partitions.

If we instead of the absorbing partitions place the director of a Yagi antenna, then process will be repeated. But this will mean that this antenna uses not only wave energy, which directly falls to the directors, but also assembles energy from the surrounding space.

Let us make one refinement. We examined the case, when the directors of antenna were located at a distance, considerably larger than wavelength. But in a Yagi antenna the distance between the directors composes 0.25λ. In this case such shadow, which is depicted on Figure 4., it cannot be formed after the directors. But, since each vibrator absorbs the specific share of energy of the wave incident to it, wave energy after the passage of this vibrator must decrease. But this it does not occur for that reason, that into the region of the arrangement of the following vibrator energy to be redistributed from the lateral regions. With this is connected the high mu-factor of the antenna examined.

However, to say that for the type of Yagiantennas performed reciprocity theorem, it is impossible, since the formation of directional transmitting antennabeam obeys the laws of interference, and the gain of the receiving antennais happening according to the laws of diffraction.

Reference

[1] I. S. Gonorovsky. *Radio Circuits and Signals*. M.: Soviet Radio, 1977.

[2] S. Ramo, John. Winner. *Fields and Waves inmodernelectronics*. OGIZ, 1948.

[3] I. V. Goncharenko, *Antennas HF and VHF. Part m5. Directional HF antenna. Shortened, phased, multi-band.* Radiosoft, 2010.

[4] F. F. Mende. *Great errors and mistakes of physicists XIX-XX centuries. The revolution in modern physics.* Kharkiv, NTMT 2010.

[5] F. F. Mende New electrodynamics. Revolution in the modern physics. Kharkov, NTMT, 2012.

[6] F. F. Mende, New approaches in contemporary classical electrodynamics. Part I, Engineering Physics, № 1, 2013.

[7] F. F. Mende, New Properties of Reactive Elements and the Problem of Propagation of Electrical Signals in Long Lines, *American Journal of Electrical and Electronic Engineering*, vol. 2, no. 5 (2014): 141-145.

[8] F. F. Mende. *Problems of modern physics and their solutions* Palmarium Academic Publishing, 2010.

9

Harmonic Aggregation Techniques: Methods to Compensate for Interaction Effects

Mohammad Mahdi Share Pasand*

Department of Electrical and Electronics Engineering, Standard Research Institute- SRI, Alborz, Iran
*Corresponding author: sharepasand@standard.ac.ir

Abstract The interaction effect in aggregation of harmonic currents is investigated. Several approaches including IEC 61000-3-6 recommended method are utilized and compared. A simple practical method is proposed to account for/compensate interaction effect. The proposed technique is computationally simple and does not require the knowledge of current harmonics in each branch. A rectifier is considered and simulated as an example to show the interaction effect and performance of the proposed method.

Keywords: *electromagnetic compatibility, harmonic distortion, harmonic aggregation, power quality, current harmonics, interaction effect*

1. Introduction

Power quality has been attracting increasing attention during recent years. This is mostly due to increasing use of nonlinear power electronic/ switching devices. On the other hand, much more number of sensitive electronic components such as computers is being used each day. This results in a growing importance of power quality issues. Harmonic pollution (i.e. levels of high order harmonics in the current/voltage waveform with a magnitude which causes unacceptable performance degradation) may cause compatibility problems between different components connected to a shared bus, distribution losses, damage to several components such as transformers, power switches and electric motors, accidental operation of remotely controlled switches and breakers(false tripping), equipment malfunction due to excess voltage, metering errors in power distribution and distributed measurement and control systems, fires in wiring, penalties on monthly bill units, generator failures etc. Permissible levels of harmonic pollution in IEC and IEEE international standards are provided as static indices measured at PCCs [1][3,4]. However, since it is not always practical to measure power quality indices in PCCs, especially when the network behavior is time varying (e.g. inserting new loads, time varying components, etc.), some approximate methods are recommended in IEC 61000 series for evaluating the total harmonic current based on single component behavior [3].

To have a proper estimation of harmonic current based on individual harmonic currents is that one can determine whether it is possible to add new components without violating the THD requirements or not. If the new component violates the THD level requirements, it has to be integrated with a harmonic filter. Also in the situation when one or a few loads are to be provided with a harmonic filtering means to reduce THD, it is desired to be able to determine the harmonic level after adding the filter at the design stage. Also appropriate aggregation formulae will be useful for optimal placement of harmonic filters in a network. (i.e. whether to provide each load with its harmonic filter or to design a single filter for a group of loads consisting n loads, and to determine n to yield best results).

Two major phenomena may impact aggregation of harmonic currents. First is the magnitude aggregation error which is caused by the phase differences between several harmonic components. This issue has been addressed in [1,2,3,4] and references therein. Another complication which is to be addressed in this paper is the interaction effect. Interaction is the effect of adding new loads on the harmonic components of the previously installed loads. This phenomenon is caused because of line impedance which results in harmonic voltages due to harmonic currents. As a result, a nonlinear load could pollute the distribution network via line impedances carrying harmonic currents. Interaction effect has not been issued in previous works, since the common de-facto method for harmonic aggregation studies is to model harmonic loads as harmonic current sources. [8,9,10,13,14] However in real world, harmonic loads may exhibit different harmonic currents depending on applied voltages. As a result, a load drawing harmonic current may cause another load to draw more harmonic current through harmonic voltage drop in the line.

[1] Point of Common Coupling

2. Approximation of Harmonic Current

The most conservative method to sum up individual harmonics is to add the harmonic currents magnitudes (peak or RMS) to estimate the aggregated harmonic current. This method, namely linear summation, presumes that all I_{hn} s are in phase as stated in equation (1). If the components (usually load currents) are in phase, a scalar summation would be accurate; however this will lead to a conservative yet unrealistic approximation of THD [3]. It has been represented that in many cases, arithmetic summation is too pessimistic and actual harmonics cancels each other due to phase differences, resulting in a very smaller aggregated harmonic level. (See [6] and references therein.)

Another method is to vector-sum the harmonic currents which lead to an exact value for the aggregated harmonic current. This method is accurate, however the actual phases are not constant [6] and the so called vector sum may result inaccurate estimation if the phase angles are not measured for a long enough period. An approximate method is to use statistical methods as [6] which require a probability density function for harmonic phases or sums. The probability that a phase equals a specific value depends on switching time which is human dependant as well as internal circuitry of the device.

The third method is to establish formulae based on empirical data. IEC 61000-3-6 recommends to add up harmonics based on Table 2 and equation 2, which is arithmetic summation for harmonic orders below 5 and a root of sums for higher orders (2).

$$I_{hTOT} = \sum_{j=1}^{n} I_{hj} \tag{1}$$

In which the following notation is used:

I_{hTOT}: Aggregated harmonic current (order h) for all components

I_{hj}: Harmonic current (order h) for j^{th} component

IEC 61000 recommended formulation is:

$$I_{hTOT} = \sqrt[\alpha]{\sum_{j=1}^{n} I_{hj}^{\alpha}} \tag{2}$$

In which:

Table 1. IEC 61000-3-6 method for harmonic aggregation [5]

α	Harmonic Order
1	h<5
1.4	5<h<10
2	h>10

Using root sum of squares –RSS– is also possible, assuming 90 degrees phase differences. Other aggregation techniques include summation of vector components presuming random phase differences driven by probability densities of different types. These methods require experimental data to estimate phase angle difference PDF, from the experimental histograms. Statistical approaches are more logical to be applied especially when an appropriate density function is available. Actual PDFs can be modeled only through complicated analytical functions. [6] In addition, because of fixed limits recommended for THD levels [4,9], it is more practical to derive conservative deterministic models based on statistical analysis. One approach is to define a percentile in the statistical model, not to be violated by the system. For instance [10] sets 5% limit for the THD (i.e. The THD level may not exceed 95% of the recommended limit) which is recommended by IEC as well [3].

The recommended methods –namely harmonic aggregation techniques or methods of harmonic summation are not accurate due to the fact that harmonic components have different- usually time variant - phases. A statistical analysis of harmonic phases is performed in [7], in which probability density functions are estimated on the basis of simplifying assumptions. The PDF of harmonic phases is considered to be uniform in [6,7], which are too optimistic. In fact, the statistical analysis of harmonics is valid only if experimental data for individual appliance harmonics is used to model phase behaviors. In this study, experimental data are drawn to model the phase behavior, examine the current aggregation method, evaluate the statistical assumptions of [7] and to derive statistical indices for THD levels which is optimistically considered to be in steady state in standards IEEE 519, IEC 61000 (CIGRE is one exception [6] which admits that permissible harmonics could be time varying).

3. Interaction Effects on Aggregated and Individual Harmonic Currents

Definition: Interaction drift is defined as follows:

$$\delta \triangleq \frac{THD_{i-single} - THD_{i-multiple}}{THD_{i-single}}$$

In which $THD_{i-multiple}$ represents the total harmonic distortion of current drawn by a load paralleled with a number of identical loads and $THD_{i-single}$ represents the total harmonic distortion of current drawn by a single load connected directly to the voltage source.

Interaction drift is caused because of the voltage drop between parallel loads due to resistance / impedance existing in wirings. It may also model the change in harmonic distortion due to current reduction in each load being paralleled with other loads.

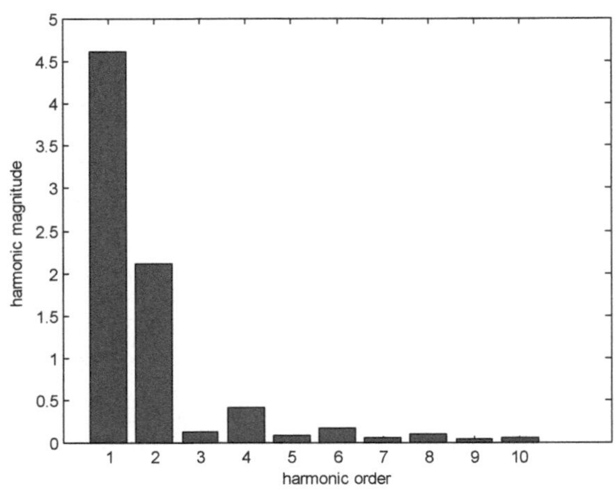

Figure 1. Harmonic spectrum of a rectifier

Consider N identical loads with similar phase behaviors. A proper example of this situation is a bank of paralleled voltage controlled switching devices like a bank of rectifiers. In this situation, phase of harmonic current does not affect the aggregated current and therefore the interaction effect could be analyzed independently.

Figure 1 depicts harmonic components of a rectifier.

A line impedance of 0.1Ω is considered to model line impedance and its effect on individual harmonic levels.

Table 2. THD levels for different number of loads

Number of loads	THD for each load (%)	Aggregated Current THD (%)
1	46.93	46.93
2	46.93,48.1	47.47
3	46.93,48.15,47.45	47.47
4	46.93,48.11,47.24,49.01	47.69
5	46.93, 48.13,47.22,49.28,47.55	47.68

As shown by Table 2, increasing the number of identical loads, does not severely affect the THD level. This is due to the fact that harmonic currents of this type of load have almost equal phases in all times. The following figure depicts phase behavior of half wave rectifiers.

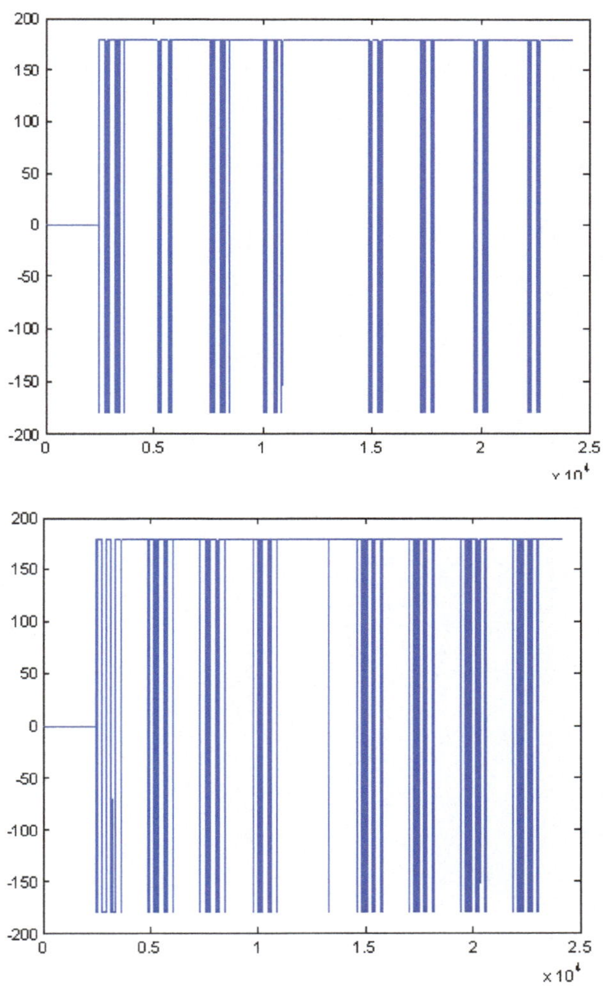

Figure 2. Phase behavior of 5th harmonic current in a rectifier

Unlike THD level, current harmonics are affected when additional loads are inserted into the bus. The flow of harmonic currents through line impedance induces harmonic voltage drops and causes harmonic pollution to distribute through the network. This along with the first harmonic voltage drop causes a load to draw different current compared to the situation in which a single load is directly connected to the bus. The following table depicts the difference cause by interaction effect. Interaction effect is of great importance especially when power electrical loads are located far from each other and line impedances are significant.

Table 3. Current Harmonics of order 3, for different number of loads

Num. of loads	3rd CH[2] for each load	Aggregated 3rd CH (A)	Approximation of aggregated 3rd CH (RSS) (A)	Approximation of aggregated 3rd CH[3] (IEC) (A)	Practical Approximation of aggregated 3rd CH (A)
1	0.1264	0.1264	0.1264	0.1264	0.1264
2	0.1264, 0.1527	0.2791	0.1982	0.2791	0.2528
3	0.1264, 0.1427, 0.1107	0.3798	0.2204	0.3798	0.3792
4	0.1264, 0.1335, 0.09147, 0.1204	0.4717	0.2380	0.4717	0.5056
5	0.1264, 0.1293, 0.082, 0.1061, 0.0689	0.5134	0.2357	0.5134	0.6320

It is not practical to measure harmonics of all branches, therefore it is desired to estimate total harmonic current based on harmonic current of a single load and phase behavior of loads. Practical approximation (left column of Table 3 and Table 4) is based on the measured current harmonic for a single load which could be approximated from THD level or as a worst case scenario, by the fact that a single device complies with an emission EMC standard as in [7].

In Table 4, two approximate values are included due to the difference between approximations given in Table 1 for harmonic orders below and above 5.

The following observations are made:

- Harmonic currents of diode rectifiers are mostly in phase and summation method is the most appropriate estimation. This is due to the fact that change in forward voltages of diodes does not impact the phase of harmonic currents. (Phase difference is about 0.001rad per 1volt forward voltage difference).

- In presence of line resistance, harmonic currents of loads are generally decreased. Therefore, estimation of total current harmonic based on harmonic current of a

[2] Current Harmonic

[3] Based on IEC 61000-3-6 given in Table 1

single load is not accurate. (Compare 3rd and 7th columns of Table 3 and Table 4)

Assuming identical loads with slight differences in phase and magnitude of harmonics, one may obtain the following approximate relations:

- As line impedance affects harmonic currents, j^{th} harmonic current of the kth branch could be approximated as:

$$I_{hj}^{k} \approx I_{hj}^{1} - \frac{I_{h1}^{1}}{v_1}\left(r_l . I_{hj}^{1} . \frac{k(k+1)}{2} \right)$$

In which I_{hj}^{1} and I_{h1}^{1} represent the jth and 1st harmonic currents of a single load.

Table 4. Current Harmonics of order 5, for different number of loads

Num. of loads	5th CH for each load	Aggregated 5th CH (A)	Approximation of aggregated 5th CH (RSS) (A)	Approximation of aggregated 5th CH[4] (IEC) (A)	Approximation of aggregated 5th CH[5] (IEC) (A)	Practical Approximation of aggregated 5th CH (A)
1	0.0751	0.0751	0.0751	0.0751	0.0751	0.0751
2	0.0751, 0.09006	0.1652	0.1173	0.1652	0.1357	0.1502
3	0.0751, 0.0841, 0.06559	0.2248	0.1499	0.2579	0.1889	0.2253
4	0.0751, 0.07872, 0.05432, 0.07047	0.2786	0.1406	0.2786	0.1882	0.3004
5	0.0751, 0.07625, 0.04915, 0.06198, 0.04085	0.3031	0.1392	0.3033	0.1936	0.3755

- The aggregated harmonic current for N branches, could be therefore computed as:

$$I_{hj}^{agg} \approx NI_{hj}^{1} - \frac{\frac{I_{h1}^{1}}{v_1} r_l . I_{hj}^{1} \sum k(k+1)}{2}$$

$$= NI_{hj}^{1} - \frac{I_{h1}^{1} r_l . I_{hj}^{1}}{2v_1}\left(\underbrace{\frac{N(N+1)}{2}}_{\sum k} + \underbrace{\frac{N^3}{3} + \frac{N^2}{2} + \frac{N}{6}}_{\sum k^2} \right)$$

$$= NI_{hj}^{1} - \frac{I_{h1}^{1} r_l . I_{hj}^{1}}{2v_1}\left(\frac{N^3}{3} + N^2 + \frac{2N}{3}\right)$$

Where loads do not exhibit similar phase behavior, the aforementioned approximation will be too conservative and the first component of approximation could be replaced considering appropriate approximation methods like the one given in [4].

[4] Based on IEC 61000-3-6 for h<5

[5] Based on IEC 61000-3-6 for h>5

The advantage of this proposed method is that it does not require any measurement in individual branches. The only measurements needed are the first and j^{th} harmonic current components, number of similar branches (loads or PCCs) and bus voltage.

The following tables compare practical approximations for aggregated current harmonic in 3rd and 5th harmonic orders.

Table 5. Comparison of practical approximation methods

Num. of loads	Actual aggregated value for 3rd CH	Practical approximation of aggregated value for 3rd CH	Proposed approximation for 3rd CH
1	0.1264	0.1264	0.1264
2	0.2791	0.2528	0.2470
3	0.3798	0.3792	0.3559
4	0.4717	0.5056	0.4474
5	0.5134	0.6320	0.5157

Table 6. Comparison of practical approximation methods

Num. of loads	Actual aggregated value for 5th CH	Practical approximation of aggregated value for 5th CH	Proposed approximation for 5th CH
1	0.0751	0.0751	0.0751
2	0.1652	0.1502	0.1467
3	0.2248	0.2253	0.2115
4	0.2786	0.3004	0.2658
5	0.3031	0.3755	0.3064

The following figure shows the approximations and actual values of aggregated harmonic current. The proposed method is situated above the actual value when the number of loads increases. Note that using approximate values is inevitable in most applications because of practical limitations in measuring all harmonic values.

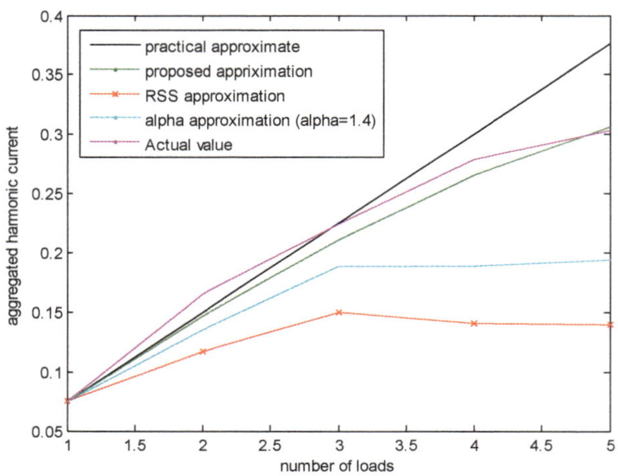

Figure 3. Practical approximations for aggregated 5th harmonic current

Figure 3 shows that the proposed method results in acceptable aggregated value which is below the actual value for small number of loads. Because loads are similar, the best approximation for the aggregated harmonic is linear summation; however, linear summation requires the value of harmonic currents in all branches which is not practical. Other methods including RSS and alpha approximation are less accurate as there is no cancellation due to phase difference in harmonic currents of this example.

4. Conclusion

The interaction effect which may adversely impact the accuracy of harmonic aggregation computations is studied. It is shown that existing approximations are not accurate or require extensive measurement data to calculate the aggregated harmonic. A formulation is proposed for a set of similar nonlinear loads to calculate the aggregated harmonic accurately and without need to measure harmonic currents at each branch. Examples are investigated with in-phase harmonic loads using different methods including IEC 61000-3-6 aggregation formulae.

References

[1] Allen, George W, "Design of Power-Line Monitoring Equipment", IEEE Transactions on Power Apparatus and Systems, Vol.PAS-90, Issue: 6, Nov. 1971.

[2] S Chattopadhyay, M Mitra, S Sengupta ,"Electric power quality", Springer, 2011.

[3] G.D.Castro, A., M. Bollen and A. Moreno-Monuz, "Street Lamps Aggregation Analysis Through IEC 61000-3-6 Approach", International Conference on Electricity Distribution, Stockholm, June 2013.

[4] Pasand, Mohammad Mahdy Share, and Zahra Rahmatian. "Harmonic aggregation techniques for power quality assessment a standard framework." *International Journal of Engineering & Technology* 3.3 (2014): 365-371.

[5] Pasand, MM Share. "Harmonic Aggregation Techniques for Power Quality Assesment A review of different methods." *International Journal of Engineering Science and Technology* 6.7 (2014): 423.

[6] IEEE std. 519, "Recommended Practices and Requirements for Harmonic Control in Electrical Power Systems", 1992.

[7] IEC 61000, "Electromagnetic compatibility (EMC) – Part 3-6: Limits – Assessment of emission limits for the connection of distorting installations to MV, HV and EHV power systems", Basic EMC publication, ed.2, 2008. (IEC/TR 61000-3-6).

[8] Y. Baghzouz, et.al., "Time-Varying Harmonics: Part II—Harmonic Summation and Propagation" , IEEE Transactions On Power Systems, Vol. 17, No. 1, 2002.

[9] Y. Baghzouz, Task Force on Probabilistic Aspects of Harmonics: "Time-varying harmonics: Part I—Characterizing measured data," IEEE Trans. Power Delivery, vol. 13, pp. 938-944, 1998.

[10] A. Cavallini, R. Langella, et.al., "Gaussian modeling of harmonic vectors in power systems," in IEEE 8th Int. Conf. Harmonics and Quality of Power, Athens, Greece, Oct. 14–16, 1998, pp. 1010-1017.

[11] IEC 61000-3-2: 2009, Electromagnetic Compatibility (EMC) - Part 3-2: Limits-Limits for harmonic current emissions (Equipment input current ≤ 16A per phase) (IEC 61000-3-2).

[12] CIGRE WG 36-05, "Assessing voltage quality with relation to harmonics, flicker and unbalance,", France, CIGRE Rep. 36-203, 34th Session, 1992.

[13] Medeiros, F. ; Brasil, D.C. ; Ribeiro, P.F. ; Marques, C.A.G. ; Duque, C.A. "A new approach for harmonic summation using the methodology of IEC 61400-21", 14th International Conference on Harmonics and Quality of Power (ICHQP), 2010.

[14] Hsiao, Y. "Design of Filters for Reducing Harmonic Distortion and Correcting Power Factor in Industrial Distribution Systems." Tamkang Journal of Science and Engineering, 4(3), 193-199, 2001.

[15] Wakileh, G. J. "Power systems harmonics: fundamentals, analysis, and filter design", Berlin: Springer, 2001.

[16] Persson, Jonas,"Comparing Harmonics Mitigation Techniques." (2014).

[17] Buccella, Concettina, et al. "Harmonic mitigation technique for multilevel inverters in power systems." Power Electronics, Electrical Drives, Automation and Motion (SPEEDAM), 2014 International Symposium on. IEEE, 2014.

[18] Aye, Thet Mon, and Soewin Naing. "Analysis of Harmonic Reduction by Using Passive Harmonic Filters." (2014).

[19] Sharma, Vandana, and Anurag Singh Tomer. "Comparative Analysis on Control Methods of Shunt Active Power Filter for Harmonics Mitigation." International Journal of Science and Research 3 (2014).

[20] Yang, Kai, Math HJ Bollen, and E. O. Anders Larsson. "Aggregation and amplification of wind-turbine harmonic emission in a wind park." Power Delivery, IEEE Transactions on 30.2 (2015): 791-799.

[21] Yang, Kai, Math HJ Bollen, and E. O. Larsson. "Wind power harmonic aggregation of multiple turbines in power bins." Harmonics and Quality of Power (ICHQP), 2014 IEEE 16th International Conference on. IEEE, 2014.

10

A Bibliography of the Work Done on Externally-Linear-Internally-Nonlinear Circuits during 1979-2014

Manoj Kumar Jain[1], V. K. Singh[1], Raj Senani[2,*]

[1]Department of Electronics and Communication Engineering, Institute of Engineering and Technology, Lucknow, India
[2]Division of Electronics and Communication Engineering, Netaji Subhas Institute of Technology, Sector 4, Dwarka, New Delhi, India
*Corresponding author: senani@ieee.org

Abstract There is a continued demand of analog circuits which are capable of operating at low supply voltages and have low-power consumption and large dynamic range. There are several techniques of designing circuits satisfying such requirements a majority of which fall into the general category of *externally-linear-internally-nonlinear* (ELIN) circuits. In this paper, we present a bibliography of ELIN circuits which covers log-domain, exponential state-space domain, square root domain and other related circuits, based upon the work done during 1979 to 2014. It is hoped that this compilation (with a brief overview) should be useful to research scholars, educators, students, practicing engineers and anybody who is interested in knowing about the current state-of-the- art of log domain, translinear and square root domain circuits.

Keywords: *log domain circuits, dynamic translinear circuits, square root domain circuits, companding, filters, oscillators*

1. A Brief Overview of ELIN Circuits

The log-domain circuits are based on natural logarithmic relation between the voltage and current of diodes and between base-emitter voltage and collector current of bipolar junction transistors. The first log-domain filter circuit was proposed by Adams in 1979 [1]. Adams conceived the first ever log-domain filter as *a circuit, composed of both linear and non-linear elements, which, when placed between a log converter and an anti-log converter (in the 'log-domain'), caused the system to act as a linear filter*. Thus, Adams circuit was also the first externally linear and internally nonlinear (ELIN) circuit, demonstrating the compressing and expanding (companding) technique.

The log-domain filters are also recognized as translinear (TL) filters. The TL filters are essentially dynamic translinear circuits, which are based upon the *dynamic translinear principle* which is inherited from the conventional *static translinear principle*, formulated by Gilbert in 1977 [2].

The last few decades have witnessed considerable interest of researchers in realising linear analog signal processing employing ELIN circuits. A number of ELIN circuits are realized by using companding techniques. Tsividis [3] showed the advantages of syllabic companding in signal processing. The log-domain circuits, dynamic translinear circuits, square-root-domain circuits and syllabically companding filters all are the subsets of ELIN circuits.

The TL filter concept was reinvented in 1990 by Seevinck [4] who coined the term *Current-mode companding* and also introduced the first class-AB integrator. Both Adams [1] and Seevinck [4] however, proposed first-order filters and hence, these ideas unfortunately, did not attract the required attention of the researchers for many years. The idea of log-domain filtering actually took off in 1993, when Frey [5] demonstrated that higher-order log-domain filters can be systematically synthesized using state-space techniques, in which BJTs were directly used to realize the log-domain filters by mapping from state-space linear differential equations. Furthermore, Frey also proposed 'Exponential State-space Filters' in [6], and showed that this technique is also suitable for high frequency applications [9]. After that many researchers started taking interest in log-domain/translinear circuits. In 1994, Toumazou, Ngarmnil and Lande [7] proposed the first log-domain filters for implementation in MOS technology. However, the first experimental results of log-domain filters were published by Perry and Roberts in [12] in 1995 who also proposed an alternative synthesis method based on the simulation of LC ladder filters. The first experimental results of MOSFET-based circuit were presented by Ngarmnil, Toumazou and Lande [11] in 1995. On the other hand, the first application of the underlying design principle for the realization of an oscillator was proposed in 1995 by Pookaiyaudom and Mahattanakul [10].

In syllabic companding, the gains of the compressor and expander are adjusted according to slowly varying characteristic of the signal envelope or power [3,14,87].

Log-domain or dynamic translinear circuits are the examples of instantaneous companding circuits but the true instantaneous companding in log domain circuits is required to operate in class B or class AB [4,8,23,31,49,67, 69,72,74,77,81,84,91,92,94,125,126,40,153,171,181,187, 191]. Both the type of circuits are suitable to implement linear and nonlinear differential equations and several synthesis approaches have been advanced by the various researchers on their own preferences such as, the state-space methods as in [5,8,13,19,50,51,64,66,77,156,160, 171,186,181,188], wave active filter concepts as in [131, 164, 173, 203], signal flow graph based approaches as in [12, 22, 52, 53, 110], leap frog designs as in [38,158,173], synthesis based on Bernoulli operator as in [33,64,68,79], modular design approaches as in [35,70], linear transformation based methods as in [21,152,175], approaches based upon simulation of LC ladders as in [22,52,97,142,154,164,166,195], unified matrix method as in [96,133], Taylor series based design as in [73], image parameter methods as in [44] and Impedance scaling based method as in [195].

Initially, log-domain circuits employed only BJTs and capacitors but subsequently it was recognized that such circuits may also be realized by using MOS transistors as well when they are operated in subthreshold region or weak inversion, as in [7,47,65,69,71,74,80,86,89,94,97, 118,136,137,139], BiCMOS technology as in [11,16,31, 32,44,46,69,138,177,178], using lateral bipolar transistors as in [95], using floating gate MOS (FGMOS) transistors as in [125,126,129,165,191] and using multiple input translinear elements as in [99,135]. Researchers have also found scope of using Darlington topology in log-domain circuits to improve gain [151].

Many different approximation techniques have so far been used to realize log domain/translinear filters such as Chebyshev [12,16,22,31,46,53,84,86,97,154,157], Elliptic [21,22,110,181], Bessel [162,179] and Butterworth [38,58,132]. These filters have been realized by using the cascaded-biquads approach, methods based upon simulation of LC ladders or using multiple-feedback topologies. Besides filters and oscillators, work has also been done on realizing integrators [15,17,26,38,43,72,194] and differentiators [15,108] which are the basic building blocks of analog circuit synthesis.

Apart from normal filter realization techniques, researchers have also given attention on filters using common mode feedback topology [130], multiple feedback *follow the leader feedback* (FLF) topology [49, 168], inverse-follow-the leader-feedback approach [53] and wavelet based filter designs [113,143,170].

Other than conventional filters, such as those presented in [146,187], the log-domain/dynamic translinear techniques have also been used in a number of other applications, such as realization of oscillators [10,13,27,32,56,67,92,101,119,169,189,197,205,211], PLL [32,60], RF applications [19], syllabic companding filters [36,42,58,102], RMS detectors [18,147], RMS-DC converters [24,30,41,62,177,204], ΣΔ A/D convertors [122], pulse duration modulation (PDM) circuits [114] and adaptive filters [28,111,129,140,165,201].

A number of researchers have also carried out noise analysis and the analysis of other second order effects/transistor non-idealities; for instance, see [34,37,45, 48,50,55,59,61,75,78,81,83,106,109,116,182,199,202].

The circuits containing MOS transistors operating in the quasi-quadratic law (or strong inversion) are called square-root domain circuits [25,57,124]. Exponential State-Space circuits [6,9,23], Sinh domain circuits [6,23,167], Tanh domain circuits [6,23,167], circuits with Piecewise-Linearization [63] and the circuits having voltage dependent capacitors [63] are few more examples of ELIN circuits.

In conclusion, the log-domain/translinear circuits have attracted a great deal of technological interest. Such circuits are useful for realizing high performance current mode linear and nonlinear signal processing and are highly suitable for VLSI implementation because of employing only transistors and capacitors and offering several of the following advantages such as electronic tunability of the parameters of the realized functions, use of low-voltage supply voltages, low-power consumption, extended dynamic range, low noise, low third order intermodulation distortion and low total harmonic distortion, etc.

It is hoped that the bibliography presented in this communication should be useful to research scholars, educators, students, practicing engineers and anybody who is interested in knowing about the current state of the art of log domain, translinear and square root domain circuits and their applications.

References

[1] Adams, R.W., "Filtering in the Log-domain", *Preprint #1470 presented at the 63rd AES conference*, New York, NY, 1979.

[2] Gilbert, B., "Translinear circuits: A proposed classification", *Electronics Letters*, 11 (1), 14-16, 1975.

[3] Tsividis, Y.P., Gopinathan, V. and Tóth, L., "Companding in signal processing", *Electronics Letters*, 26 (17), 1331-1332, 1990.

[4] Seevinck, E., "Companding current-mode integrator: A new circuit principle for continuous-time monolithic filters", *Electronics Letters*, 26 (24), 2046-2047, 1990.

[5] Frey, D.R., "Log-domain filtering: An approach to current-mode filtering", *IEE Proceedings G Circuits, Devices and Systems*, 140 (6), 406-416, 1993.

[6] Frey, D.R., "A general class of current mode filters", IEEE *International Symposium on Circuits and Systems (ISCAS'93)*, 2, 1435-1438, 1993.

[7] Toumazou, C., Ngarmnil, J. and Lande, T.S., "Micropower log-domain filter for electronic cochlea", *Electronics Letters*, 30 (22), 1839-1841, Oct. 1994.

[8] Frey, D., "Current-mode class AB second order filter", *Electronics Letters*, 30 (3), 205-206, 1994.

[9] Frey, D., "A 3.3 volt electronically tunable active filter usable to beyond 1 GHz", *International Symposium on Circuits and Systems (ISCAS'94)*, 5, 493-496, 1994.

[10] Pookaiyaudom, S. and Mahattanakul, J., "A 3.3 volt high-frequency capacitor-less electronically-tunable log-domain oscillator", *International Symposium on Circuits and Systems (ISCAS'95)*, 2, 829-832, 1995.

[11] Ngarmnil, J., Toumazou, C. and Lande, T.S., "A fully tuneable micropower log-domain filter", *Twenty-first European Solid-State Circuit Conference (ESSCIRC'95), Lille, France*, 86-89, 19-21 Sep., 1995.

[12] Perry, D. and Roberts, G.W., "Log-domain filters based on LC ladder synthesis", *Proceeding of the IEEE International Symposium on Circuits and Systems (ISCAS '95), Seattle, WA*, 1, 311-314, 30 Apr.-3May, 1995.

[13] Thanachayanont, A., Pookaiyaudom, S. and Toumazou C., "State-space synthesis of log-domain oscillators", *IET Electronics Letters*, 31 (21), 1797-1799, 1995.

[14] Tsividis, Y., "General approach to signal processors employing companding", *Electronics Letters*, 31 (18), 1549-1550, 1995.

[15] Tsividis, Y., "On linear integrators and differentiators using instantaneous companding", *IEEE Transactions on Circuits and Systems II: Analog and Digital Signal* Processing, 42 (8), 561-564, 1995.

[16] Punzenberger, M. and Enz, C., "A new 1.2 V BiCMOS log-domain integrator for companding current-mode filters", *IEEE International Symposium on Circuits and Systems (ISCAS '96) Connecting the World, Atlanta, GA*, 1, 125-128, 12-15 May, 1996.

[17] Fried, R., Python, D. and Enz, C.C., "Compact log-domain current mode integrator with high transconductance-to-bias current ratio", *Electronics Letters*, 32 (11), 952-953, 1996.

[18] Frey, D.R., "Explicit log domain root-mean-square detector", *United States Patent*, 1996.

[19] Frey, D.R., "Log domain filtering for RF applications", *IEEE Journal Solid-State Circuits*, 31 (10), 1468-1475, 1996.

[20] Ngarmnil, J. and Toumazou, C., "Micro power log-domain active inductor", *Electronics Letters*, 32 (11), 953-955, 1996.

[21] Liu, S.I. and Liao, Y.H., "Table-based log-domain linear transformation filter", *Electronics Letters*, 32 (19), 1771-1772, 1996.

[22] Perry, D. and Roberts, G.W., "The design of log-domain filters based on the operational simulation of LC ladders", *IEEE Transactions on Circuits and Systems II: Analog and Digital Signal Processing*, 43 (11), 763-774, 1996.

[23] Frey, D., "Exponential state space filters: A generic current mode design strategy", *IEEE Transaction on Circuits and Systems I: Fundamental Theory and Applications*, 43 (1), 34-42, 1996.

[24] Mulder, J., van der Woerd, A.C., Serdijn, W.A. and van Roermund, A.H.M., "An RMS-DC converter based on the dynamic translinear principle", *Proceeding of the 22nd European Solid-State Circuits Conference (ESSCIRC'96), Neuchatel, Switzerland*, 17-19 Sept., 1996.

[25] Mulder, J., van der Woerd, A.C., Serdijn, W.A. and van Roermund, A.H.M., "Current-mode companding √x-domain integrator", *Electronics Letters*, 32 (3), 198-199, 1996.

[26] Mahattanakul, J., Toumazou, C. and Pookaiyaudom, S., "Low-distortion current-mode companding integrator operating at f_T of BJT", *Electronics Letters*, 32 (21), 2019-2021, 1996.

[27] van der Woerd, A.C., Mulder, J., Serdijn, W.A. and van Roermund, A.H.M., "Recent trends in translinear circuits", *in Proc. Electronics-ET'96, Sozopol, Bulgaria*, 1, 14-21, 1996.

[28] Frey, D.R. and Steigerwald, L., "An adaptive analog notch filter using log filtering", *IEEE International Symposium on Circuits and Systems (ISCAS'96) Connecting the World, Atlanta, GA*, 1, 297-300, 12-15 May, 1996.

[29] Frey, D.R., "Log-filtering using gyrators", *Electronics Letters*, 30 (1), 26-28, 1996.

[30] Mulder, J., Serdijn, W.A., van der Woerd, A.C., and van Roermund, A.H.M., "Dynamic translinear RMS-DC converter", *Electronics Letters*, 32 (22), 2067-2068, 1996.

[31] Punzenberger, M. and Enz, C.C., "A 1.2-V low-power BiCMOS class AB log-domain filter", *IEEE Journal Solid-State Circuits*, 32 (12), 1968-1978, 1997.

[32] Thanachayanont, A., Payne, A. and Pookaiyaudom, S., "A current-mode phase-locked loop using a log-domain oscillator", *Proceedings of the IEEE International Symposium on Circuits and Systems (ISCAS '97)*, 1, 277-280, 9-12 Jun., 1997.

[33] Drakakis, E.M., Payne, A.J. and Toumazou, C., "Bernoulli operator: a low-level approach to log-domain processing", *Electronics Letters*, 33 (12), 1008-1009, 1997.

[34] Leung, V.W., El-Gamal M. and Roberts, G.W., "Effects of transistor non-idealities on log-domain filters", *Proceedings of IEEE International Symposium on Circuits and Systems (ISCAS '97)*, 1, 109-112, 9-12 Jun., 1997.

[35] Mahattanakul, J. and Toumazou, C., "Modular log-domain filters", *Electronics Letters*, 33 (13), 1130-1131, 1997.

[36] Frey, D.R. and Tsividis, Y.P., "Syllabically companding log domain filter using dynamic biasing", *Electronics Letters*, 33 (18), 1506-1507, 1997.

[37] Mulder, J., van der Woerd, A.C., Serdijn, W.A. and van Roermund, A. H. M., "General current-mode analysis method for translinear filters", *IEEE Transaction on Circuits and Systems I: Fundamental Theory and Applications*, 44 (3), 193-197, 1997.

[38] Serdijn, W.A., Broest, M., Mulder, J., van der Woerd, A.C. and van Roermund, A.H.M., "A low-voltage ultra-low-power translinear integrator for audio filter applications", *IEEE Journal of Solid-State Circuits*, 32 (4), 577-581, 1997.

[39] Payne, A. and Thanachayanont, A., "Translinear circuit for phase detection", *Electronics Letters*, 33 (18), 1507-1509, 1997.

[40] Tsividis, Y., "Externally linear, time-invariant systems and their application to companding signal processors", *IEEE Transactions on Circuits and Systems II: Analog and Digital Signal Processing*, 44 (2), 65-85, 1997.

[41] Mulder, J., van der Woerd, A.C., Serdijn, W.A. and van Roermund, A.H.M., "An RMS-DC converter based on the dynamic translinear principle", *IEEE Journal of Solid-State Circuits*, 32 (7), 1146-1150, 1997.

[42] Mulder, J., Serdijn, W.A., van der Woerd, A.C. and van Roermund, A. H. M., "A syllabic companding translinear filter", *Proceeding of IEEE International Symposium on Circuits and Systems (ISCAS'97)*, 1, 101-104, 9-12 Jun., 1997.

[43] Mahattanakul, J. and Toumazou, C., "Instantaneous companding and expressing: A dual approach to linear integrator synthesis", *Electronics Letters*, 33 (1), 4-5, 1997.

[44] Ruymbeke, G. van, Enz, C., Krummenacher, F. and Declerq, M., "A BiCMOS programmable continuous-time filter using image-parameter method synthesis and voltage-companding technique", *IEEE Journal of Solid-State Circuits*, 32 (3), 377-387, 1997.

[45] Mulder, J., Kouwenhoven, M.H.L. and van Roermund, A.H.M., "Signal x noise intermodulation in translinear filters", *Electronics Letters*, 33 (14), 1205-1207, 1997.

[46] Punzenberger, M. and Enz, C.C., "A compact low-power BiCMOS log-domain filter", *IEEE Journal Solid-State Circuits*, 33 (7), 1123-1129, 1998.

[47] Germanovix, W., O'Neill, G., Toumazou, C., Drakakis, E.M., Kitney, R.I. and Lande, T.S., "Analogue micro powered log-domain tone controller for auditory prostheses", *Electronics Letters*, 34 (11), 1051 - 1052, 1998.

[48] Fox, R.M., "Design-oriented analysis of log-domain circuits", *IEEE Transactions Circuits and Systems II: Analog and Digital Signal Processing*, 45 (7), 918-921, 1998.

[49] Punzenberger, M. and Enz, C., "Log-domain filters for low-voltage low-power applications", *IEEE International Conference on Electronics, Circuits and Systems, Lisboa*, 1, 41-44, 1998.

[50] Frey, D.R., "State-space synthesis and analysis of log-domain filters", *IEEE Transactions on Circuits and Systems II: Analog and Digital Signal Processing*, 45 (9), 1205-1211, 1998.

[51] Frey, D.R., "Synthesis of distortion compensated log-domain filters using state space techniques", *Proceedings of the IEEE International Symposium on Circuits and Systems (ISCAS '98), Monterey, CA*, 1, 321-324, 31 May-3 Jun.,1998.

[52] El-Gamal, M. and Roberts, G.W., "Very high-frequency log-domain band pass filters", *IEEE Transactions on Circuits and Systems II: Analog and Digital Signal Processing*, 45 (9), 1188-1198, 1998.

[53] Wu, J. and El-Masry, E.I., "Log-domain synthesis of an nth-order filter", *International Journal of Electronics*, 84 (4), 359-369, 1998.

[54] Brandtner, T., Papathanasiou, K. and Hamilton, A., "Palmo cell using sampled data log-domain integrators", *Electronics Letters*, 34 (8), 733-735, 1998.

[55] Mulder, J., Kouwenhoven, M.H.L., Serdijn, W.A., van der Woerd, A.C. and van Roermund, A.H.M., "Noise considerations for translinear filters", *IEEE Transaction on Circuits and Systems II: Analog and Digital Signal Processing*, 45 (9), 1199-1204, 1998.

[56] Serdijn, W.A., Mulder, J., van der Woerd, A.C. and van Roermund, A.H.M., "A wide-tunable translinear second-order oscillator", *IEEE Journal of Solid-State Circuits*, 33 (2), 195-201, 1998.

[57] Mulder, J., Serdijn, W.A., van der Woerd, A.C. and van Roermund, A.H.M., "A 3.3 V current-controlled √-domain oscillator", *Analog Integrated Circuits and Signal Processing*, 16 (1), 17-28, 1998.

[58] Mulder, J., Serdijn, W.A., van der Woerd, A.C. and van Roermund, A.H.M., "An instantaneous and syllabic companding translinear filter", *IEEE Transactions on Circuits and Systems I: Fundamental Theory and Applications*, 45 (2), 150-154, 1998.

[59] Tóth, L., Tsividis, Y.P. and Krishnapura, N., "On the analysis of noise and interference in instantaneously companding signal processors", *IEEE Transactions on Circuits and Systems II: Analog and Digital Signal Processing*, 45 (9), 1242-1249, 1998.

[60] Payne, A., Thanachayanont, A. and Papavassilliou, C., "A 150-MHz translinear phase-locked loop", *IEEE Transactions on*

Circuits and Systems II: Analog and Digital Signal Processing, 45 (9), 1220-1231, 1998.

[61] Kouwenhoven, M.H.L., Mulder, J. and van Roermund A.H.M., "Noise analysis of dynamically nonlinear translinear circuits", *Electronics Letters*, 34 (8), 705-706, 1998.

[62] Surakampontorn, W. and Kumwachara, K., "A dual translinear based true RMS to DC convertor", *IEEE Transactions on Instrumentation and* Measurement, 47 (2), 459-464, 1998.

[63] Tsividis, Y.P., "Externally linear integrators," *IEEE Transactions on Circuits and Systems II: Analog and Digital Signal Processing*, 45 (9), 1181-1187, 1998.

[64] Drakakis, E.M., Payne, A.J. and Toumazou, C., ""Log-domain state-space": a systematic transistor-level approach for log-domain filtering", *IEEE Transactions on Circuits and Systems II: Analog and Digital Signal Processing*, 46 (3), 290-305, 1999.

[65] El-Masry, E.I. and Wu, J., "CMOS micropower universal log-domain biquad", *IEEE Transactions Circuits and Systems I: Fundamental Theory and Applications*, 46 (3), 389-392, 1999.

[66] Frey, D., "Distortion compensation in log-domain filters using state-space techniques", *IEEE Transactions Circuits and Systems II: Analog and Digital Signal Processing*, 46 (7), 860-869, 1999.

[67] Ferrer, E. and Fox, R.M., "Limit-cycle oscillations in a log-domain-based filter", *IEEE Transactions on Circuits and Systems II: Analog and Digital Signal Processing*, 46 (6), 832-836, 1999.

[68] Drakakis, E.M., Payne, A.J. and Toumazou, C., "Log-domain filtering and the Bernoulli cell", *IEEE Transactions on Circuits and Systems I: Fundamental Theory and Applications*, 46 (5), 559-571, 1999.

[69] Enz, C., Punzenberger, M. and Python, D., "Low-voltage log-domain signal processing in CMOS and BiCMOS", *IEEE Transactions on Circuits and Systems II: Analog and Digital Signal Processing*, 46 (3), 279-289, 1999.

[70] Mahattanakul, J. and Toumazou, C., "Modular log-domain filters based upon linear Gm-C filter synthesis", *IEEE Transactions on Circuits and Systems I: Fundamental Theory and Applications*, 46 (12), 1421-1430, 1999.

[71] Fox, R.M. and Nagarajan, M., "Multiple operating points in a CMOS log-domain filter", *IEEE Transactions on Circuits and Systems II: Analog and Digital Signal Processing*, 46 (6), 705-710, 1999.

[72] Poort, P.J., Serdijn, W.A., Mulder, J., van der Woerd, A.C. and van Roermund, A.H.M., "A 1-V Class-AB Translinear Integrator for Filter Applications", *Analog Integrated Circuits and Signal Processing*, 21 (1), 79-90, 1999.

[73] Abuelmáatti, M.T. and Abed, S.M., "A translinear circuit for analogue function synthesis based on a Taylor series", *International Journal Electronics*, 86 (11), 1341-1348, 1999.

[74] Serdijn, W.A., Kouwenhoven, M.H.L., Mulder, J. and van Roermund, A.H.M., "Design of high dynamic range fully integratable translinear filters", *Analog Integrated Circuits and Signal Processing*, 19 (3), 223-239, 1999.

[75] Mulder, J., Kouwenhoven, M.H.L., Serdijn, W.A. and van der Woerd, A.C., "Nonlinear analysis of noise in static and dynamic translinear circuits", *IEEE Transactions on Circuits and Systems II: Analog and Digital Signal Processing*, 46 (3), 266-278, 1999.

[76] Seevinck, E., Vittoz, E.A., du Plessi, M. and Joubert, T.H., "CMOS translinear circuits for minimum supply voltage", *IEEE Transactions on Circuits and systems II: Analog and Digital Signal Processing*, 47 (12), 1560-1564, 1999.

[77] Frey, D.R. and Tola, A.T., "A state-space formulation for externally linear class AB dynamical circuits," *IEEE Transactions on Circuits and Systems II: Analog and Digital Signal Processing*, 46 (3), 306-314, 1999.

[78] Leung, V.W. and Roberts, G.W., "Analysis and Compensation of Log-Domain Biquadratic Filter Response Deviations due to Transistor Non-idealities", *Analog Integrated Circuits and Signal Processing*, 22 (2-3), 147-162, 2000.

[79] Drakakis, E.M. and Payne, A.J., "A Bernoulli Cell-Based Investigation of the Non-Linear Dynamics in Log-Domain Structures", *Analog Integrated Circuits and Signal Processing*, 22 (2-3), 127-146, 2000.

[80] Fragnie`re, E., Vittoz, E. and van Schaik, A., "A Log-Domain CMOS Transcapacitor: Design, Analysis and Applications", *Analog Integrated Circuits and Signal Processing*, 22 (2-3), 195-208, 2000.

[81] Tola, A.T. and Frey, D.R., "A Study of Different Class AB Log Domain First Order Filters", *Analog Integrated Circuits and Signal Processing*, 22 (2-3), 163-176, 2000.

[82] Germanovix, W. and Toumazou, C., "Design of a micropower current-mode log-domain analog cochlear implant", *IEEE Transactions on Circuits and Systems II: Analog and Digital Signal Processing*, 47 (10), 1023-1046, 2000.

[83] Leung, V.W. and Roberts, G.W., "Effects of transistor nonidealities on high-order log-domain ladder filter frequency responses", *IEEE Transactions on Circuits and Systems II: Analog and Digital Signal Processing*, 47 (5), 373-387, 2000.

[84] El-Masry, E.I. and Wu, J., "Fully Differential Class-AB Log-Domain Integrator", *Analog Integrated Circuits and Signal Processing*, 25 (1), 35-46, 2000.

[85] Frey, D., "Future implications of the log domain paradigm", *IEE Proceedings - Circuits, Devices and Systems*, 147 (1), 65-72, 2000.

[86] El-Masry, E.I. and Wu, J., "Low Voltage Micropower Log-Domain Filters", *Analog Integrated Circuits and Signal Processing*, 22 (2-3), 209-220, 2000.

[87] Krishnapura, N., Tsividis, Y. and Frey, D.R., "Simplified technique for syllabic companding in log-domain filters", *Electronics Letters*, 36 (15), 1257-1259, 2000.

[88] Edwards R.T. and Cauwenberghs, G., "Synthesis of Log-Domain Filters from First-Order Building Blocks", *Analog Integrated Circuits and Signal Processing*, 22 (2-3), 177-186, 2000.

[89] Masmoudi, D., Serdijn, W.A., Mulder, J., van der Woerd, A.C., Tomas, J. and Dom, J.P., "A new current-mode synthesis method for dynamic translinear filters and its applications in hearing instruments", *Analog Integrated Circuits and Signal Processing*, 22, 221-229, 2000.

[90] Mulder, J., Serdijn, W.A., van der Woerd, A.C. and van Roermund, A.H.M., "Dynamic Translinear Circuits-An Overview", *Research Perspectives on Dynamic Translinear and Log-Domain Circuits, Analog Integrated Circuits and Signal Processing*, 22 (2-3), 5-20, 2000.

[91] El-Gamal, N.M., Baki, R.A. and Bar-Dor, A., "30-100-MHz NPN-only variable-gain class-AB instantaneous comapanding filters for 1.2-V applications", *IEEE Journal of Solid-State Circuits*, 35 (12), 1853-1864, 2000.

[92] Mahattanakul, J. and Toumazou, C., "Instantaneous companding current-mode oscillator based on class-AB transconductor", *Analog Integrated Circuits and Signal Processing*, 23 (1), 57-64, 2000.

[93] Tóth, L., Efthivoulidis, G. and Tsividis, Y.P., "Noise analysis of externally linear systems", *IEEE Transactions on Circuits and Systems II: Analog and Digital Signal Processing*, 47 (12), 1365-1377, 2000.

[94] Python, D. and Enz, C.C., "A micropower class-AB CMOS log-domain filter for DECT applications", *IEEE Journal of Solid-State Circuits*, 36 (7), 1067-1075, 2001.

[95] Krishnapura, N. and Tsividis, Y., "A micropower log-domain filter using enhanced lateral PNPs in a 0.25 μm CMOS process", *IEEE Symposium on VLSI Circuits, Digest of Technical Papers, Kyoto, Japan*, 179-182, 14-15 Jun., 2001.

[96] Ng, A.E.J., Sewell, J.I., Drakakis, E.M., Payne and A.J., Toumazou, C., "A unified matrix method for systematic synthesis of log-domain ladder filters", *Proceedings of the IEEE International Symposium on Circuits and Systems (ISCAS 2000)Sydney, NSW*, 1, 149-152, 6-9 May, 2001.

[97] Li S., Wang Y. and Wu, J., "Design of low-voltage and low-power fully integrated filter based on log-domain current-mode integrator", *Journal of Electronics (China)*, 18 (4), 346-350, 2001.

[98] Fox, R.M., Ko, H.J. and Eisenstadt, W.R., "Dynamic Current Requirements in Single-Ended Log-Domain Filters", *Analog Integrated Circuits and Signal Processing*, 28 (1), 73-81, 2001.

[99] Minch, B.A., "Multiple-input translinear element log-domain filters", *IEEE Transactions on Circuits and Systems II: Analog and Digital Signal Processing*, 48 (1), 29-36, 2001.

[100] Mahon, A. and Feely, O., "Nonlinear dynamics of log-domain circuit", *Electronics Letters*, 37 (15), 929-930, 2001.

[101] Özoğuz, S., "Simple log-domain chaotic oscillator", *Electronics Letters*, 37 (23), 1378-1379, 2001.

[102] Frey, D., Tsividis, Y.P., Efthivoulidis, G. and Krishnapura, N., "Syllabic-companding log domain filters", *IEEE Transactions on Circuits and Systems II: Analog and Digital Signal Processing*, 48 (4), 329-339, 2001.

[103] Mulder, J., Serdijn, W.A., van der Woerd, A.C. and van Roermund, A.H.M., "A generalized class of dynamic translinear circuits", *IEEE Transactions on Circuits and Systems II: Analog and Digital Signal Processing*, 48 (5), 501-504, 2001.

[104] Diepstraten, F.M., Kuijstermans, F.C.M., Serdijn, W.A., van der Kloet P., van Staveren, A., Neerhoff, F.L., Verhoeven, C.J.M. and van Roermund, A.H.M., "Dynamic Behavior of Dynamic Translinear Circuits: the Linear Time-Varying Approximation", *IEEE Transaction on Circuits and Systems I: Fundamental Theory and Applications*, 48 (11), 1333-1337, 2001.

[105] López-Martin, A.J. and Carlosena, A., "Systematic Design of Companding Systems by Component Substitution", *Analog Integrated Circuits and Signal Processing*, 28 (1), 91-106, 2001.

[106] Efthivoulidis, G., Tóth, L. and Tsividis, Y., "Noise in externally linear filters", *Analog Integrated Circuits and Signal Processing*, 28 (1), 63-72, 2001.

[107] El-Gamal, M.N. and Roberts, G.W., "A 1.2-V n-p-n-only integrator for log-domain filtering", *IEEE Transactions Circuits and Systems II: Analog and Digital Signal Processing*, 49 (4), 257-265, 2002.

[108] Vlassis S. and Psychalinos. C., "A Novel Log-Domain Differentiator", *Analog Integrated Circuits and Signal Processing*, 32 (3), 285-287, 2002.

[109] Ng, A.E.J. and Sewell, J.I., "Direct noise analysis of log-domain filters", *IEEE Transactions on Circuits and Systems II: Analog and Digital Signal Processing*, 49 (2), 101-109, 2002.

[110] Psychalinos, C. and Vlassis, S., "On the exact realization of LOG-domain elliptic filters using the signal flow graph approach", *IEEE Transactions on Circuits and Systems II: Analog and Digital Signal Processing*, 49 (12), 770-774, 2002.

[111] McDonald, E.J. and Minch, B.A., "Synthesis of a translinear analog adaptive filter", *Proceeding of the IEEE International Symposium on Circuits and Systems (ISCAS 2002), Phoenix-Scottsdale, AZ*, 3, III-321-III-324, 26-29 May, 2002.

[112] López-Martin, A.J. and Carlosena, A., "1.5 V CMOS companding filter", *Electronics Letters*, 38 (22), 1346-1348, 2002.

[113] Haddad, S.A.P., Houben, R. and Serdijn, W.A., "First derivative Gaussian wavelet function employing dynamic translinear circuits for cardiac signal characterization", *Proceedings of the ProRISC Workshop on Circuits, Systems and Signal Processing, Veldhoven, the Netherlands*, 288-291, 28-29 Nov., 2002.

[114] Serra-Graells, F. and Huertas, J.L., "1 V CMOS Subthreshold Log Domain PDM", *Analog Integrated Circuits and Signal Processing*, 34 (3), 183-187, 2003.

[115] Baki, R.A. and El-Gamal, M.N., "A low-power 5-70-MHz seventh-order log-domain filter with programmable boost, group delay, and gain for hard disk drive applications", *IEEE Journal of Solid-State Circuits*, 38 (2), 205-215, 2003.

[116] Baki, R.A., Beainy, C. and El-Gamal, M.N., "Distortion analysis of high-frequency log-domain filters using Volterra series", *IEEE Transactions on Circuits and Systems II: Analog and Digital Signal Processing*, 50 (1), 1-11, 2003.

[117] Serrano-Gotarredona, T. and Linares-Barranco, B., "Log-domain implementation of complex dynamics reaction-diffusion neural networks", *IEEE Transactions on Neural Networks*, 14 (5), 1337-1355, 2003.

[118] Grech, I., Micallef, J. and Vladimirova, T., "Low-Power Log-Domain CMOS Filter Bank for 2-D Sound Source Localization", *Analog Integrated Circuits and Signal Processing*, 36 (1-2), 99-117, 2003.

[119] Özoğuz, S. and Sengör, N.S., "On the realization of NPN-only log-domain chaotic oscillators", *IEEE Transactions on Circuits and Systems I: Fundamental Theory and Applications*, 50 (2), 291-294, 2003.

[120] Drakakis, E.M. and Burdett, A.J., "Operational DC constraints for a class-A, third-order, observer canonical-form log-domain filter", *IEEE Transactions on Circuits and Systems I: Fundamental Theory and Applications*, 50 (10), 1337-1342, 2003.

[121] Drakakis, E.M. and Burdett, A.J., "Spectral impact of the modulation index upon internal log-domain currents", *IEEE Transactions on Circuits and Systems I: Fundamental Theory and Applications*, 50 (10), 1343-1347, 2003.

[122] Serra-Graells, F., "1 V all-MOS ΣΔ A/D converters in the log-domain", *Analog Integrated Circuits and Signal Processing*, 35, 47-57, 2003.

[123] Minch, B.A., "Construction and transformation of multiple-input translinear element networks", *IEEE Transactions on Circuits and Systems I: Fundamental Theory and Applications*, 50 (12), 1530-1537, 2003.

[124] De La Cruz-Blas, C.A., Lopez-Martin, A. and Carlosena, A., "1.5-V MOS translinear loops with improved dynamic range and their applications to current-mode signal processing", *IEEE Transactions on Circuits and Systems II: Analog and Digital Signal Processing*, 50 (12), 918-927, 2003.

[125] López-Martín, A.J., Fernández, R., de la Cruz Blas, C.A. and Carlosena, A. "A 1 V Micropower FGMOS Class AB Log-Domain Filter", *Analog Integrated Circuits and Signal Processing*, 41 (2-3), 137-145, 2004.

[126] Rodriguez-Villegas, E., Yufera, A. and Rueda, A., "A 1-V micropower log-domain integrator based on FGMOS transistors operating in weak inversion", *IEEE Journal of Solid-State Circuits*, 39 (1), 256-259, 2004.

[127] Gerosa, A., Maniero, A. and Neviani, A., "A fully integrated dual-channel log-domain programmable preamplifier and filter for an implantable cardiac pacemaker", *IEEE Transactions on Circuits and Systems I: Regular Papers*, 51 (10), 1916-1925, 2004.

[128] Serra-Graells, F., Gómez, L. and Huertas, J.L., "A true-1-V 300-μW CMOS-subthreshold log-domain hearing-aid-on-chip", *IEEE Journal of Solid-State Circuits*, 39 (8), 1271-1281, 2004.

[129] Abshire, P.A., Wong, E.L., Yiming, Z. and Cohen, M.H., "Adaptive log domain filters using floating gate transistors", *Proceedings of the IEEE International Symposium on Circuits and Systems (ISCAS '04)*, 1, I-29 - I-32, 23-26 May, 2004.

[130] Fox, R.M., Ko, H.J. and Eisenstadt, W.R., "Differential log-domain filters with high-gain common-mode feedback", *IEEE Transactions on Circuits and Systems I: Regular Papers*, 51 (2), 254-263, 2004.

[131] Psychalinos, C., Fragoulis, N. and Haritantis, I., "Log-domain wave filters", *IEEE Transactions on Circuits and Systems II: Express Briefs*, 51 (6), 299-306, 2004.

[132] Fragoulis, N., Psychalinos, C. and Haritantis, I., "Modular log-domain filters realized using wave port terminators", *IEEE Transactions Circuits and Systems I: Regular Papers*, 55 (11), 2235-2244, 2004.

[133] Ng, A.E.J., Teplechuk, M.A. and Sewell, J.I., "Unified matrix method for systematic synthesis of log-domain ladder filters", *IEE Proceedings - Circuits, Devices and Systems*, 151 (4), 285-293, 2004.

[134] De La Cruz-Blas, C.A., Lopez-Martin, A.J. and Carlosena, A., "Low-voltage CMOS nonlinear transconductors and their application to companding current-mode filters", *Analog Integrated Circuits and Signal Processing*, 38 (2-3), 137-147, 2004.

[135] Minch, B.A., "Synthesis of static and dynamic multiple-input translinear element networks", *IEEE Transactions on Circuits and Systems I: Regular Paper*, 51 (2), 409-421, 2004.

[136] Serra-Graells, F., Rueda, A. and Huertas, J.L., "Low-voltage CMOS log-companding techniques for audio applications", *Analog Integrated Circuits and Signal Processing*, 38 (2-3), 121-135, 2004.

[137] López-Martín, A.J., De La Cruz-Blas, C.A. and Carlosena, A., "1.2-V 5-μW class-AB CMOS log-domain integrator with multidecade tuning", *IEEE Transactions on Circuits and Systems II: Express Briefs*, 52 (10), 665-668, 2005.

[138] Halvorsrød, T., Luzi, W. and Lande, T.S., "A log-domain μbeamformer for medical ultrasound imaging systems", *IEEE Transactions on Circuits and Systems I: Regular Papers*, 52 (12), 2563-2575, 2005.

[139] Shoemaker, P., "A Methodology for Long Time Constant Log-Domain Filters in CMOS", *Analog Integrated Circuits and Signal Processing*, 42 (2), 161-178, 2005.

[140] Zaghloul, K.A. and Boahen, K.A., "An ON-OFF log domain circuit that recreates adaptive filtering in the retina", *IEEE Transactions on Circuits and Systems I: Regular Papers*, 52 (1), 99-107, 2005.

[141] Tola, A.T., Arslanalp, R. and Yilmaz, S.S., "Design of current mode, low voltage and electronically tunable class AB differential type KHN filter in the log domain", *Proceedings of the IEEE 13th Signal Processing and Communications Applications Conference*, 131-134, 16-18 May, 2005.

[142] Kontogiannopoulos, N. and Psychalinos, C., "Log-domain filtering by simulating the topology of passive prototypes", *IEEE Transactions on Circuits and Systems I: Regular Papers*, 52 (10), 2043-2054, 2005.

[143] Haddad, S.A.P., Bagga, S. and Serdijn, W.A., "Log-domain wavelet bases", *IEEE Transactions on Circuits and Systems I: Regular Papers*, 52 (10), 2023-2032, 2005.

[144] Serra-Graells, F. and Huertas, J.L., "Low-Voltage CMOS subthreshold log-domain filtering", *IEEE Transactions on Circuits and Systems I: Regular Papers*, 52 (10), 2090-2100, 2005.

[145] Ascoli, A., Mahon, A. and Feely, O., "Nonlinear dynamics of first- and second-order log-domain circuits", *IEEE Transactions on Circuits and Systems I: Regular Papers*, 52 (7), 1372-1381, 2005.

[146] Tola, A.T., Arslanalp, R. and Yilmaz, S.S., "A design of differantial type class AB Tow-Thomas filter in the log domain", *International conference on electrical and electronics engineering, Bursa, Turkey*, 105-109, 7-11 Dec., 2005.

[147] Qizhang, Y., Eisenstadt, W.R., Fox, R.M. and Tao, Z., "A translinear RMS detector for embedded test of RF ICs", *IEEE Transaction on Instrument and Measurement*, 54 (5), 1708-1714, 2005.

[148] Kircay, A. and Cam, U., "A novel first-order log-domain allpass filter", *AEU - International Journal of Electronics and Communications*, 60 (6), 471-474, 2006.

[149] Hamilton, T.J., Jin, C. and van Schaik, A., "An analysis of matching in the Tau cell log-domain filter", *Proceedings of the IEEE International Symposium on Circuits and Systems (ISCAS 2006), Island of Kos*, 421-424, 21-24 May, 2006.

[150] Teplechuk, M.A. and Sewell, J.I., "Approximation of arbitrary complex filter responses and their realisation in log domain", *IEE Proceedings Circuits, Devices and Systems*, 153 (6), 583-590, 2006.

[151] Arslanalp, R. and Tola, A.T., "By Using Darlington Topology Improvement of In-Band Gain for the Log Domain Filters," *IEEE 14th Signal Processing and Communications Applications, Antalya*, 1-4, 17-19 Apr., 2006.

[152] Psychalinos, C., "Improved building blocks for log-domain linear transformation filters", *Proceedings of the IEEE International Symposium on Circuits and Systems (ISCAS 2006), Island of Kos*, 1535-1538, 21-24 May, 2006.

[153] Kircay, A., Cam U. and Tola, A.T., "Novel first-order differential class-AB log-domain allpass filters", *AEU - International Journal of Electronics and Communications*, 60 (10), 705-712, 2006.

[154] Ascoli, A., Feely O. and Curran, P., "On the Chaotic Behavior of a Third-Order Log-Domain Filter", *Nonlinear Dynamics*, 44 (1-4), 45-54, 2006.

[155] Teplechuk, M.A. and Sewell, J.I., "Realisation of asymmetrical complex filters in log-domain", *Proceedings of the IEEE International Symposium on Circuits and Systems (ISCAS 2006), Island of Kos*, 4, 21-24 May, 2006.

[156] Kircay, A. and Cam, U., "State-Space Synthesis of Current-Mode First-Order Log-Domain Filters", *Turk J Elec Engin*, 14 (3), 399-415, 2006.

[157] Drakakis, E.M., "Systematic derivation of explicit design formulae for log-domain: A 3rd-order lowpass example", *Microelectronics Journal*, 37 (7), 646-656, 2006.

[158] Psychalinos, C., "On the transposition of G_m–C filters to DC stabilized log-domain filters", *International Journal of Circuits Theory and Applications*, 34, 217-236, 2006.

[159] Psychalinos, C., "Realization of log-domain high-order transfer functions using first-order building blocks and complementary operators", *International Journal of Circuits Theory and Applications*, 35, 17-32, 2006.

[160] Kircay, A. and Cam, U., "A Novel Log-Domain First-Order Multifunction Filter", *Electronics and Telecommunications Research Institute (ETRI) Journal*, 28 (3), 401-404, 2006.

[161] Groza, R., Festila, L. and Fazakas, A., "A Log-Domain summing-amplifier for serial signal flows", *Inter-Ing 2007*, IV-4-2-4, 2007.

[162] Groza, R., Festila, L. and Szopos, E., "Bessel Filter implementation in Log-Domain", *Inter-Ing 2007*, IV-3-1-6, 2007.

[163] Feştilă, L., Groza, R., Cirlugea, M. and Fazakas, A., "Modular analysis and design of log-domain circuits based on LIN↔ELIN transformations", *Analog Integrated Circuits and Signal Processing*, 50 (3), 231-249, 2007.

[164] Teplechuk, M.A. and Sewell, J.I., "Wave realisations of filters and equalisers in log-domain", *IET Circuits, Devices & Systems*, 1 (6), 433-443, 2007.

[165] Zhai, Y. and Abshire, P.A., "Adaptive log domain filters for system identification using floating gate transistors", *Analog Integrated Circuits and Signal Processing*, 56 (1-2), 23-36, 2007.

[166] Ascoli, A., Curran, P. and Feely, O., "Modelling the dynamics of log-domain circuits", *International Journal Of Circuits Theory and Applications*, 35 (1), 33-70, 2007.

[167] Tongkulboriboon, S., Pawarangkoon, P. and Kiranon, W., "Externally linear current amplifiers", *International Journal of Electronics*, 94 (6), 587-596, 2007.

[168] Shah N.A. and Khanday, F.A., "A DC stabilized log-domain nth-order multifunction filter based on the decomposition of nth-order HP filter function to FLF topology", *International Journal of Circuit Theory and Applications*, 37 (10), 1075-1091. 2008.

[169] Psychalinos, C. and Souliotis, G., "A log-domain multiphase sinusoidal oscillator", *AEU - International Journal of Electronics and Communications*, 62 (8), 622-626, 2008.

[170] Hongmin, L., Yigang, H. and Sun, Y., "Detection of Cardiac Signal Characteristic Point Using Log-Domain Wavelet Transform Circuits", *Circuits, Systems, and Signal Processing*, 27 (5), 683-698, 2008.

[171] Kircay, A. and Cam, U., "Differential Type Class-AB Second-Order Log-Domain Notch Filter", *IEEE Transactions on Circuits and Systems I: Regular Papers*, 55 (5), 1203-1212, 2008.

[172] Groza, R., Festila, L., Hintea, S. and Cirlugea, M., "Log-Domain Binary SVM Image Classifier", *Knowledge-Based Intelligent Information and Engineering Systems (Lecture Notes in Computer Science)*, 5179, 368-375, 2008.

[173] Psychalinos, C., "Low-Voltage Log-Domain Complex Filters", *IEEE Transactions Circuits and Systems I: Regular Papers*, 55 (11), 3404-3412, 2008.

[174] Shah, N.A., Iqbal, S.Z. and Parveen, N., "Log-Domain low pass high pass first-order filter", *Indian Journal Pure & Applied Physics*, 46 (9), 667 -670, 2008.

[175] Psychalinos, C., "Log-domain linear transformation filters revised: Improved building blocks and comparison results", *International Journal of Circuit Theory and Applications*, 36 (2), 119-133, 2008.

[176] Vercauteren, T., Pennec, X., Perchant, A. and Ayache, N., "Symmetric Log-Domain Diffeomorphic Registration: A Demons-based Approach", *Medical Image Computing and Computer-Assisted Intervention-MICCAI-2008, Lecture Notes in Computer Science*, 5241, 754-761, 2008.

[177] Farshidi, E. and Sayedi, S.M., "A 1.2 V current-mode true RMS-DC converter based on the floating gate MOS translinear principle", *Microelectronics Journal*, 39 (2), 293-298, 2008.

[178] Loizos, D.N., Sotiriadis, P.P. and Cauwenberghs, G., "A translinear SiGe BiCMOS current-controlled oscillator with 80Hz-800MHz tuning range", *Analog Integrated Circuits and Signal Processing*, 57 (1-2), 107-115, 2008.

[179] Ip, H.M.D., Drakakis, E.M. and Bharath, A.A., "A 19 nW analogue CMOS log-domain 6th-order Bessel filter without E-minus cells", *Microelectronics Journal*, 40 (8), 1170-1174, 2009.

[180] Katsiamis, A.G., Drakakis, E.M. and Lyon, R.F., "A Biomimetic, 4.5 μW, 120+ dB, Log-Domain Cochlea Channel with AGC", *IEEE Journal of Solid-State Circuits*, 44 (3), 1006-1022, 2009.

[181] Kircay, A. and Cam, U., "A new 5th-order differential type class-AB log-domain elliptic lowpass filter for video frequency applications", *Analog Integrated Circuits and Signal Processing*, 60 (3), 221-229, 2009.

[182] Fragoulis, N., Psychalinos, C. and Haritantis, I., "Analysis and compensation of transistor non-idealities in log-domain wave active filters", *AEU - International Journal of Electronics and Communications*, 63 (2), 132-138, 2009.

[183] Georgiou, P. and Toumazou, C., "Chemical log-domain filter", *Electronics Letters*, 45 (8), 391-392, 2009.

[184] Tola, A.T., Arslanalp, R. and Yilmaz, S.S., "Current mode high-frequency KHN filter employing differential class AB log domain integrator", *International Journal of Electronics and Communications (AEÜ)*, 63 (7), 600-608, 2009.

[185] Shah, N.A. and Khanday, F.A., "Log-domain synthesis of nth order universal filter", *Analog Integrated Circuits and Signal Processing*, 59 (3), 309-315, 2009.

[186] Frey, D.R. and Drakakis, E.M., "Unifying perspective on log-domain filter synthesis", *Electronics Letters*, 45 (17), 861-863, 2009.

[187] Tola, A.T., Arslanalp, R. and Yilmaz, S.S., "Current mode Tow-Thomas biquadratic differential class AB log domain filter", *International Review of Electrical Engineering*, 4 (6), 1426-1432, 2009.

[188] Arslanalp, R. and Tola, A.T., "State space representation for log domain filtering synthesis", *Indian Journal of Pure and Applied Physics*, 47, 745-752, 2009.

[189] Prommee, P., Sra-ium, N. and Dejhan, K., "High-frequency log-domain current-mode multiphase sinusoidal oscillator", IET Circuits, Devices & Systems, 4 (5), pp. 440-448, 2010.

[190] Psychalinos, C. and Vlassis, S., "Low-voltage log-domain differentiators", *Analog Integrated Circuits and Signal Processing*, 63 (2), 299-306, 2010.

[191] Farshidi, E., "Synthesis of class-AB log-domain filters based on nonlinear transconductance", *Microelectronics Journal*, 41 (5), 266-276, 2010.

[192] Fragoulis, N., "Universal log-domain biquad", *Electronics Letters*, 46 (7), 487-488, 2010.

[193] Tola, A.T., Yilmaz, S.S. and Arslanalp, R., "Current Mode Log Domain Notch Filter Design Based on Adding Filtering Blocks", *Electronics World*, 116 (1896), 42, 2010.

[194] Shah, N.A., Iqbal, S.Z. and Parveen, N., "Log-domain all pass filter based on integrators", *Analog Integrated Circuits and Signal Processing*, 67 (1), 85-88, 2011.

[195] Manetti S. and Piccirilli, M.C., "Log-Domain Filter Design Based on Impedance Scaling of LC Ladder Structures", *Circuits, Systems, and Signal Processing*, 30 (2), 371-389, 2011.

[196] Psychalinos, C., "Log-domain SIMO and MISO low-voltage universal biquads", *Analog Integrated Circuits and Signal Processing*, 67 (2), 201-211, 2011.

[197] Prommee, P., Prapakorn, N. and Swamy, M.N.S., "Log-Domain Current-mode Quadrature Sinusoidal Oscillator", *Radioengineering*, 20 (3), 600-607, 2011.

[198] Shah, N.A. and Khanday, F.A., "A Generic Current Mode Design for Multifunction Grounded Capacitor Filters Employing Log-Domain Technique", *Active and Passive Electronic Components*, 1-10, 2011.

[199] Farshidi, E., "Generalized Noise Analysis of Log Domain Static Translinear Circuits", *International Scholarly and Scientific & Innovation, World Academy of Science, Engineering and Technology*, 5 (4), 177-180, 2011.

[200] Fernández, D., Martínez-Alvarado, L. and Madrenas, J., "A Translinear, Log-Domain FPAA on Standard CMOS Technology", *IEEE Journal of Solid-State Circuits*, 47 (2), 490-503, 2012.

[201] Kasimis, C., Souliotis, G. and Psychalinos, C., "Novel log-domain frequency-adaptive filter", *International Journal of Electronics*, 99 (2), 197-209, 2012.

[202] Duduk, N. and Tola, A.T., "A Study about Effects of Transistors' Nonideal Characteristics on Log Domain Filters," *International Conference on Applied Electronics (AE), Pilsen*, 69-74, 5-7 Sept., 2012.

[203] Fragoulis, N., "Wave log-domain filters using lattice sections", *International Journal of Circuit Theory and Applications*, 40 (4), 385-394, 2012.

[204] Farshidi, E. and Asiaban, H., "A new true RMS-to-DC converter using up-down translinear loop in CMOS technology", *Analog Integrated Circuits and Signal Processing*, 70 (3), 385-390, 2012.

[205] Prommee, P. and Wongprommoon, N., "Log-domain All-pass Filter-based Multiphase Sinusoidal Oscillators", *Radioengineering*, 22 (1), 14-23, 2013.

[206] Prommee, P. and Dejhan, K., "Single-input Multiple-output Tunable Log-domain Current-mode Universal Filter", *Radioengineering*, 22 (2), 474 - 484, 2013.

[207] Giagkoulovits, C. and Psychalinos, C., "0.5V Cardiac Sense Amplifier Realization Using Log-Domain Filtering", *ISRN Biomedical Engineering*, 2013, 11 pages, 2013.

[208] Vargas-Bernal, R. and De La Cruz-Blas, C.A., "A survey on the static and dynamic translinear paradigm", *Integrated Circuits for Analog Signal Processing*, 289-306, 2013.

[209] Tsirimokou, G., Laoudias, C. and Psychalinos, C., "0.5-V fractional-order companding filters", *International Journal of Circuit Theory and Applications*, 2014.

[210] Shaheem, M.A., Savaria, Y. and Hamoui, A.A., "Design and modeling of high-resolution multibit log-domain ΔΣ modulators", *Analog Integrated Circuits and Signal Processing"*, 79 (3), 569-582, 2014.

[211] Papadimitriou, K.I. and Drakakis, E.M., "CMOS weak-inversion log-domain glycolytic oscillators: A cytomimetic circuit example", *International Journal of Circuit Theory and Applications*, 42 (2), 173-194, 2014.

[212] Eskiyerli, M.H., Payne, A.J. and Toumazou, C., "State space synthesis of integrators based on the MOSFET square law," *Electronics Letters*, 32 (6), 505-506, 1996.

[213] Eskiyerli, M. and Payne, A., "Square Root Domain" Filter Design and Performance," *Analog Integrated Circuits and Signal Processing*, 22(2-3), 231-243, 2000.

[214] Psychalinos, C. and Vlassis, S., "A Systematic Design Procedure for Square-Root-Domain Circuits Based on the Signal Flow Graph Approach," *IEEE Transactions on Circuits and Systems-I: Fundamental Theory and Applications*, 49(12), 1702-1712, 2002.

[215] Psychalinos, C. and Vlassis, S., "A High Performance Square-Root Domain Integrator," *Analog Integrated Circuits and Signal Processing*, 32(1), 97-101, 2002.

[216] De La Cruz-Blas, C.A., López-Martín, A.J. and Carlosena, A., "1.5 V tunable Square-Root Domain filter," *Electronics Letters*, 40(4), 213-214, 2004.

[217] Mohammed, K.O. and Soliman, A.M., "Realization of Square Root Domain Filters from Passive Filters," *Frequenz*, 58(11-12), 272-276, 2004.

[218] Vlassis, S. and Psychalinos, C., "A Square-Root Domain Differentiator Circuit," *Analog Integrated Circuits and Signal Processing*, 40(1), 53–59, 2004.

[219] De La Cruz-Blas, C.A., López-Martín, A.J. and Carlosena, A., "1.5-V Square-Root Domain Second-Order Filter With On-Chip Tuning," *IEEE Transactions on Circuits and Systems-I: Regular Papers*, 52(10), 1996-2006, 2005.

[220] Yu, G.J., Huang, C.Y., Liu, B.D. and Chen, J.J., "Design of Square-Root Domain Filters," *Analog Integrated Circuits and Signal Processing*, 43(1), 49-59, 2005.

[221] Mohammed, K.O. and Soliman, A.M., "A Tunable Square Root Domain Oscillator," *Analog Integrated Circuits and Signal Processing*, 43(1), 91-95, 2005.

[222] Yu, G.J., Huang, C.Y., Chen, J.J. and Liu, B.D., "Design of Current-Mode Square-Root Domain Band-Pass Filter with Reduced Voltage," *Analog Integrated Circuits and Signal Processing*, 44(3), 239–250, 2005.

[223] Al-Gahtani, M.A., "Class-AB Square Root Domain Filter," *WSEAS Transaction on Circuits and Systems*, 5(8), 1381-1387, 2006.

[224] Ozoguz, S., Abdelrahman, T.M. and Elwakil, A.S., "Novel Approximate Square-Root Domain All-Pass Filter with Application to Multiphase Oscillators," *Analog Integrated Circuits and Signal Processing*, 46(3), 297-301, 2006.

[225] Farshidi, E. and Sayedi, S.M., "Class-AB Square root Domain Filters Based on the Floating Gate MOS Translinear Principle," *Iranian Journal of Electrical & Electronic Engineering*, 2(2), 62-70, 2006.

[226] Ragheb, T.S.A. and Soliman, A.M., "New Square-Root Domain Oscillators," *Analog Integrated Circuits and Signal Processing*, 47(2), 165-168, 2006.

[227] Kontogiannopoulos, N. and Psychalinos, C., "Switched-current filters revisited: Square-root domain sampled-data filters," *IEEE Transactions on Circuits and Systems Ii-Express Briefs*, 53(12), 1373-1377, 2006.

[228] Psychalinos, C., "Square-root domain wave filters," *Int. J. Circuit Theory and Applications*, 35(2), 131-148, 2007.

[229] Psychalinos, C., "Square-Root Domain Operational Simulation of LC Ladder Elliptic Filters," *Circuits, Systems & Signal Processing*, 26(2), 263-280, 2007.

[230] Menekay, S., Tarcan, R.C. and Kuntman, H., "The Second Order Low Pass Filter Design with A Novel Higher Precision Square-Root Circuit," *IU-Journal of Electrical & Electronics Engineering*, 7(1), 323-329, 2007.

[231] De La Cruz-Blas, C.A., López-Martín, A., Carlosena, A. and Sarmiento, A., "Multiple operating points in a square-root domain first-order filter," *Int. J. Circ. Theor. Appl.*, 35(1), 71-91, 2007.

[232] Psychalinos, C., "Design of square-root domain filters by substituting the passive elements of the prototype filter by their equivalents," *Int. J. Circuit Theory and Applications*, 36(2), 185-204, 2008.

[233] Stoumpou, E., Khanday, F.A., Psychalinos, C. and Shah, N.A., "A low-voltage square-root domain n-th order multifunction FLF filter topology," *Analog Integrated Circuits and Signal Processing*, 61(3), 315-322, 2009.

[234] Yu, G.J., Tsao, Y.J. and Lin, Y.S., "1V Square-Root Domain Low-Pass Filter using Translinear Loop Technology in Biomedical Engineering," *Life Science Journal*, 6(4), 92-96, 2009

[235] Keserlioglu, M.S. and Kircay, A., "The Design of Current-Mode Electronically Tunable First-Order Square-Root-Domain Filters Using State-Space Synthesis Method," *International Review on Modelling and Simulations*, 2(2), 124-128, 2009.

[236] Ölmez, S. and Çam, U., "A novel square-root domain realization of first order all-pass filter," *Turk J Elec Eng & Comp Sci*, 18(1), 141-146, 2010.

[237] Yu, G.J., "A Systematic Synthesis for High-Order Square-Root Domain Filters with Reduced Voltage," *Life Science Journal*, 7(1), 15-29, 2010.

[238] Yu, G.J. and Lin, Y.S. "Low Voltage Tunable Square-Root Domain Band-Pass Filter with Translinear Loop Technique in Biomedical Engineering," *Life Science Journal*, 7(1), 30-33, 2010.

[239] Asiaban, H. and Farshidi E., "A New True RMS-to-DC Converter in CMOS Technology," *World Academy of Science, Engineering and Technology*, 4(11), 234-237, 2010.

[240] Jelodarian, P. and Farshidi, E., "A Current-mode Continuous-time Sigma-delta Modulator based on Translinear Loop Principle," *World Academy of Science, Engineering and Technology*, 4(6), 422-425, 2010.

[241] De La Cruz-Blas, C.A. and Feely, O., "Limit cycle behavior in a class-AB second-order square root domain filter," *Analog Integrated Circuits and Signal Processing*, 68(2), 175-181, 2011.

[242] Yilmaz, S.S. and Tola, A.T., "Fifth Order Butterworth Low Pass Square-Root Domain Filter Design," *Elektronika Ir Elektrotechnika*, 115(9), 55-58, 2011.

[243] Stoumpou, E. and Psychalinos, C., "Square-root domain linear transformation filters," *International Journal of Circuit Theory and Applications*, 39(7), 719-731, 2011.

[244] Yilmaz, S.S., Tola, A.T. and Arslanalp, R., "A Novel Second-Order All-Pass Filter Using Square-Root Domain Blocks," *Radioengineering*, 22(1), 179-185, 2013.

[245] Laoudias, C., Psychalinos, C. and Stoumpou, E., "1.5V square-root domain universal biquad filters," *Int. J. Circ. Theor. Appl.*, 41(3), 307-318, 2013.

[246] Khanday, F.A., Psychalinos, C. and Shah, N.A., "Square-Root-Domain Realization of Single-Cell Architecture of Complex TDCNN," *Circuits Systems and Signal Processing*, 32(3), 959-978, 2013.

[247] Shah, N.A. and Khanday, F.A., "Synthesis of High-order all pass filters using Novel Low Voltage Current-Mode Square-Root-Domain low-order all pass Filters", *Maejo International Journal of Science and Technology*, 7(03), 385-395, 2013.

[248] Farshidi, E. and Manoochehri, S., "A New Approach for Current-Mode SRD Filters," *International Journal of Electronics Communications and Electrical Engineering*, 3(12), 1-10, 2013.

[249] Kircay, A., Keserlioglu, M.S. and Çam, U., "A new current-mode square-root-domain general notch filter," *Journal of Circuits, Systems, and Computers*, 22(1), 1250072-1-1250072-10, 2013.

[250] Khanday, F.A. and Shah, N. A., "Realization of low-voltage square-root-domain all-pass filters," *Maejo Int. J. Sci. Technol.*, 7(3), 422-432, 2013.

[251] Kircay, A., "Electronically tunable current-mode square-root-domain first-order multifunction filter," *Int. J. Electronics*, 101(2), 212-219, 2014.

[252] Khanday, F.A., Psychalinos, C. and Shah, N.A., "Universal filters of arbitrary order and type employing square-root-domain technique". *Int. J. Electronics*, 101(7), 894-918, 2014.

[253] Shaterian, M., Twigg, C.M. and Azhari, J., "MTL-based implementation of current-mode CMOS RMS-to-DC converters," *Int. J. Circ. Theor. Appl.*, 2014.

11

Design and Implementation of Controlled Zeta Converter Power Supply

Ali H. Ahmad[1], Nashwan Saleh Sultan[2,*]

[1]Department of Electrical, Eng. College, University of mosul, Mosul, Iraq
[2]Technical Eng.College of Mosul, Mosul, Iraq
*Corresponding author: nashwansaleh86@yahoo.com

Abstract This work includes Zeta voltage converter controller design and implementation. The mathematical model of the zeta converter circuit operating in the continuous conduction mode in state-space form is presented. Fuzzy Logic controller and hybrid Fuzzy Logic controller/ particle swarm optimization techniques are used to designed controller. Analysis and comparison between simulation and practical responses of open loop, close loop fuzzy logic controller and hybrid fuzzy logic/particle swarm optimization controller results are performed for different, working conditions such as sudden changes in the load resistance and reference voltages. The results show that there are significant improvement in the results for the proposed hybrid FLC/PSO control technique.

Keywords: *Fuzzy Logic Controller (FLC), Particle Swarm Optimization (PSO), Zeta Converter*

1. Introduction

A DC-DC converter is widespread in modern portable electronic equipments and systems. The batteries are providing constant input voltage to the converter, then the converter converts it into wide range of values depending on the charge level. At low charge level, the voltage may drop below the battery voltage for continuously supplying the load with constant voltage [1,2]. There for it needs to be regulated. There are many research works dealing with the direct voltage converters performances and their control. One of those works was that of (A. H. Ahmed ..et, al) in 2006 A.C [3]. They designed and analyzed a regulated controller of Cuk converter using H_∞ / μ technique. O. A. Taha in 2007A.C[4] had studied CUK converter circuit performances, its designed and implement a robust controller for the cuk converter using H_∞ synthesis technique. He studied the effects of CUK converter parameters changes on the circuit stability. S. S.Sabri in 2008 A.C, he studied the CUK converter circuit performance and designed controller for it using Fuzzy Logic Controller based genetic algorithm to improve its performance [5]. E. Vuthchhay and C. Bunlaksanusorn in 2010A.C [6]. They studied zeta converter circuit performance, Modeling and Control of the converter performance. R. Suresh Kumar, had studied BOOST converter circuit performance and its control using PID controller. He used a particle swarm optimization technique for the design and improvement of its performance [7]. Because of the lack of the studies about the ZETA converter circuit, the present work is dealing with the design of the Fuzzy Logic Controller and hybrid Fuzzy Logic/ Particle Swarm Optimization Controller to control converter circuit output voltage and improve its output performance.

2. Zeta Converter Mathematic Model

The dc-dc voltage Zeta converter is assumed to operate in the continuous conduction mode(CCM).There exist two circuit states within one switch period T. First state is when switch is turned on (DT),and another when it is turned off [(1-D)T].The general state space mathematical model of the zeta converter is given by:

$$x' = Ax + Bu \tag{1}$$

$$y = Cx + Eu \tag{2}$$

Where

x: is $n \times 1$ state vector; A: is $n \times n$ system matrix; u:is $m \times 1$ input vector;
B: is $n \times m$ input matrix; E: is $L \times m$ matrix; y : is $L \times 1$ output vector; C: is $L \times n$ output matrix;

The zeta converter circuit is shown in Figure 1. It consists of IGBT transistor as a switch, Diode, two capacitors C_1 and C_2, two inductors L_1 and L_2 with internal resistances r_1 and r_2 respectively and load resistor R_L. In the first mode of operation, the converter circuit(switch is ON) shown in Figure 2. During the interval (DT), the inductors L_1 and L_2 are in charging state[6].

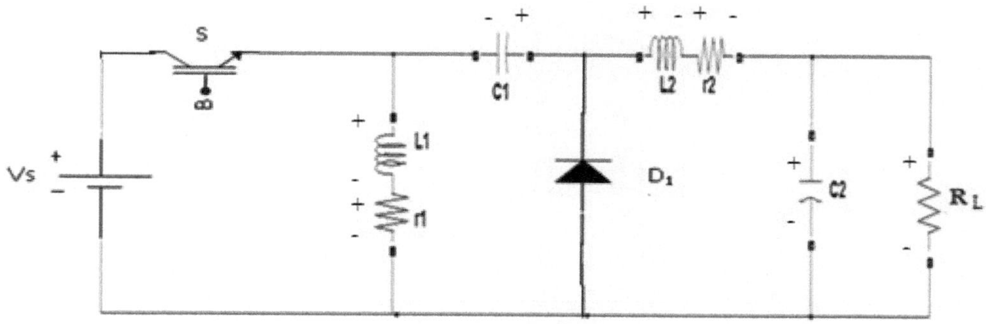

Figure 1. Zeta converter circuit

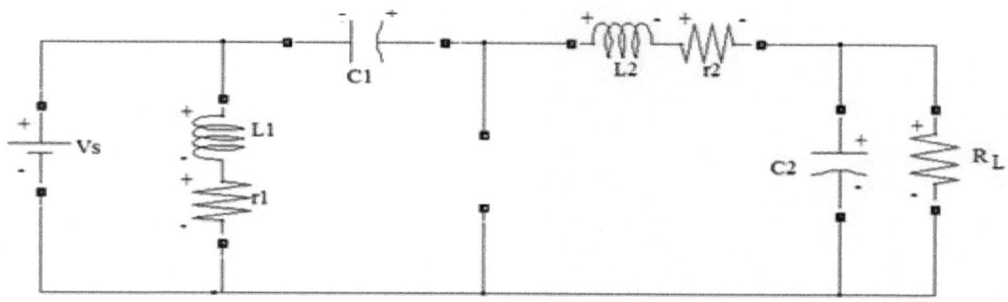

Figure 2. Mode one of Zeta converter circuit

Considering :

$$i_{L_1} = x_1, i_{L_2} = x_2$$
$$V_{C_1} = x_3, V_{C_2} = x_4$$

Applying the Kirchhoff's voltage law to the circuit in Figure 2 and writing the voltage equations for the open loop circuit and rearranging the equations to be in the state space form. The system equations became to be:

$$\frac{di_{L_1}}{dt} = \frac{1}{L_1}.V_s - \frac{r_1}{L_1}.i_{L_1} \quad (3)$$

$$\frac{dv_{C_1}}{dt} = -\frac{1}{C_1}.i_{L_2} \quad (4)$$

$$\frac{di_{L_2}}{dt} = \frac{1}{L_2}.V_{C_1} + \frac{1}{L_2}V_s - \frac{r_2}{L_2}.i_{L_2} - \frac{1}{L_2}.vc_2 \quad (5)$$

$$\frac{dv_{C_2}}{dt} = \frac{1}{C_2}.i_{L_2} - \frac{1}{RC_2}.vc_2 \quad (6)$$

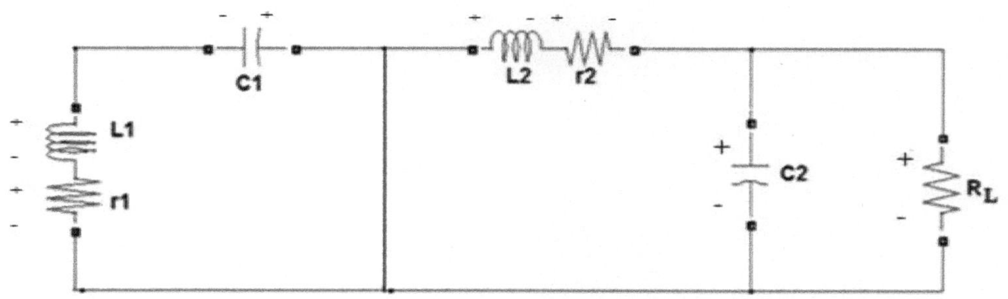

Figure 3. Mode two of Zeta converter circuit

In the second mode of operation (1-D)T the IGBT transistor is OFF. The converter equivalent circuit is shown as in Figure 3. In this mode the inductors (L_1, L_2) are in the discharging state. L_1 is discharging its stored energy into the capacitor C_1 , and the inductor L_2 transform energy to output section [6,8]. Then by the same way the system equations became to be:

$$\frac{di_{L_1}}{dt} = \frac{r_1}{L_1}.i_{L_1} - \frac{1}{L_1}.vc_1 \quad (7)$$

$$\frac{dvc_1}{dt} = \frac{1}{C_1}i_{L_1} \quad (8)$$

$$\frac{di_{L_2}}{dt} = -\frac{1}{L_2}.vc_2 - \frac{r_2}{L_2}.i_{L_2} \quad (9)$$

$$\frac{dv_{C_2}}{dt} = \frac{1}{C_2}.i_{L_2} - \frac{1}{RC_2}.vc_2 \quad (10)$$

Using on state and off state equation, the system state space equivalent equation became to be:

$$\begin{bmatrix} \dot{x}_1 \\ \dot{x}_2 \\ \dot{x}_3 \\ \dot{x}_4 \end{bmatrix} = \begin{bmatrix} -\frac{r_1}{L_1} & 0 & \frac{D-1}{L_1} & 0 \\ 0 & -\frac{r_2}{L_2} & \frac{D}{L_2} & -\frac{I}{L_2} \\ \frac{1-D}{C_1} & -\frac{D}{C_1} & 0 & 0 \\ 0 & \frac{1}{C_3} & 0 & -\frac{1}{RC_3} \end{bmatrix} \begin{bmatrix} x_1 \\ x_2 \\ x_3 \\ x_4 \end{bmatrix} + \begin{bmatrix} \frac{D}{L_1} \\ \frac{D}{L_2} \\ 0 \\ 0 \end{bmatrix} u \quad (11)$$

$$y = \begin{bmatrix} 0 & 0 & 0 & 1 \end{bmatrix} \begin{bmatrix} x_1 \\ x_2 \\ x_3 \\ x_4 \end{bmatrix} \quad (12)$$

The relation between input and output voltages of the zeta converter is given by:

$$Vo = \frac{D.Vs}{(1-D)} \quad (13)$$

3. Zeta Converter Circuit Design

Applying Kirchhoff's voltage law on Zeta converter circuits: first and second modes. The produced ripple inductors currents and ripple capacitors voltages are given by the equations below:

$$\Delta I_1 = \frac{DV_S}{FL_1} \quad (14)$$

$$\Delta I_2 = \frac{DV_S}{FL_2} \quad (15)$$

$$\Delta V_{C1} = \frac{DV_S}{8F^2C_1L_1} \quad (16)$$

$$\Delta V_{C2} = \frac{DV_S}{8F^2C_2L_2} \quad (17)$$

Where: F is the switching frequency (F=5Khz), $\Delta I_1 = 2.5A$, $\Delta I_2 = 2.5A$, $\Delta V_{C1} = 0.06V$, $\Delta V_{C2} = 0.06V$, D=0.5.

From the equations above the critical values of the inductors and capacitors of the Zeta converter circuit may be introduced as the follows: [6]

$$L_1 \geq \frac{(1-D)^2 R_L}{2DF} \quad (18)$$

$$L_2 \geq \frac{(1-D)R_L}{2F} \quad (19)$$

$$C_1 \geq \frac{D}{8F(1-D)R_L} \quad (20)$$

$$C_2 \geq \frac{1}{8FR_L} \quad (21)$$

Whereas the above components became to be: L_1 = 0.5mH, L_2 = 0.5mH, C_1 =900µF, C_2 = 1000µF.

4. Fuzzy Logic Controller Design

Fuzzy logic controller has been designed for writing its inputs as : The error e(t)and the error change Δe(t) of the output voltage. The linguistic variables are defined as (N, NS, Z, PS, P)where N means negative, NS negative small, Z zero, PS positive small, P positive. Triangular membership functions of the fuzzy logic controller are considered. The fuzzy rules are summarized in Table 1. The surge no type of fuzzy inference engine is considered [5]. The error range are taken between (-30 and 30) as shown in Figure 4a, and range of error change are taken between(-12 and 12) as shown in Figure 4b.

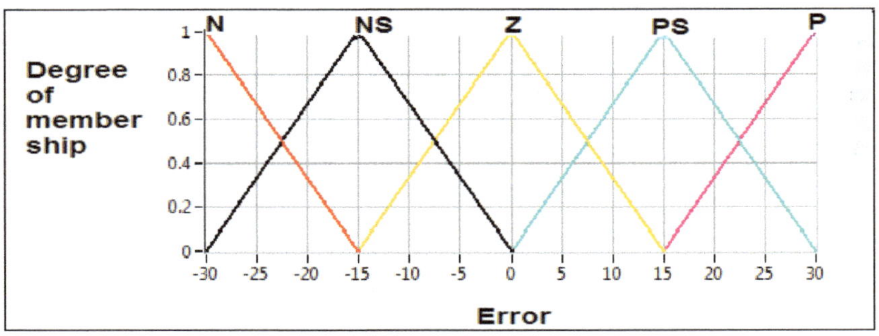

Figure 4a. The error membership function

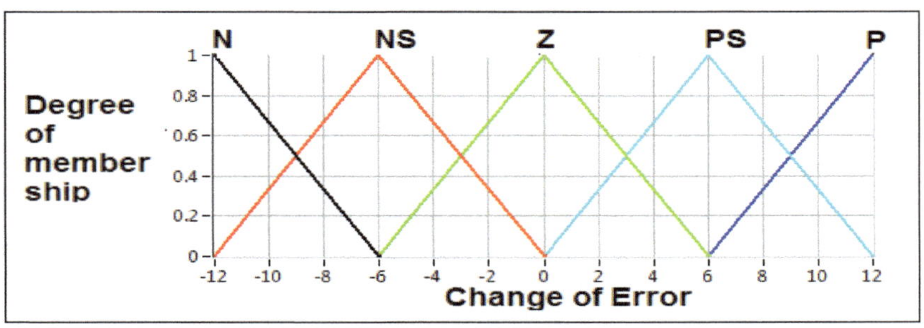

Figure 4b. The change of error membership functio

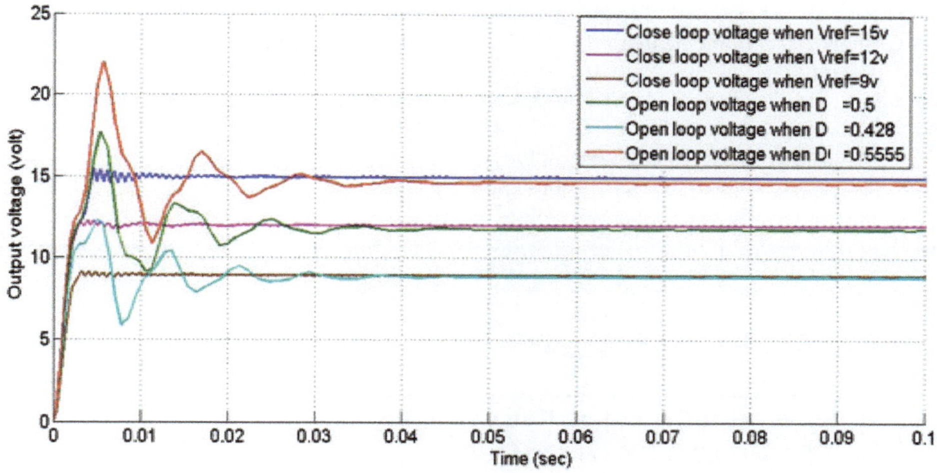

Figure 5. output voltage responses of the open loop and closed loop systems with FLC when Vref=9,12 and15 volts and R_L (10) Ω

Table 1. Fuzzy Logic Controller rule base

e/Δe	N	NS	Z	PS	P
NB	N	N	N	NS	Z
NS	N	N	NS	Z	PS
ZE	N	NS	Z	PS	P
PS	NS	Z	PS	P	P
PB	Z	PS	P	P	P

The output voltage responces for the open loopand closed loop systems with Fuzzy Logic Controller are shown on the same graph in Figure 5 for referance voltages (9,12 and15) volts and load resistance (10) Ω.

5. Particle Swarm Optimization Algorithm

Particle swarm optimization technique, first developed by (J. Kennedy and R. Eberhart, 1995) as one of the modern heuristic algorithms. It was inspired by the social behavior of the bird and fish schooling and has been found to be robust in solving continuous nonlinear optimization problems [7,9]. This algorithm is based on the following scenario: a group of birds are randomly searching for food in an area and there is only one piece of food. All birds are unaware where the food is, but they do know how far the food is at each time instant. The best and most effective strategy to find the food would be to follow the bird which is nearest to it. Based on such scenario, the PSO algorithm is used to solve the optimization problem.

In PSO, each single solution is a "bird" in the search space; this is referred to as a "particle". The swarm ismodeled as particles inamulti-dimensional space, which have positions and velocities. These particles have two essential capabilities: their memory of their own best position and knowledge of the global best. Members of the swarm communicate good positions to each other and adjust their own position and velocity based on good positions according to following equations (22,23): [7,10].
Let:
i to be the number of particles
j to be the number of iterations

$$v(k+1)_{i.j} = w.v(k)_{i.j} + C_1 r_1 \left(\text{gbest} - x(k)_{i.j}\right) + C_2 r_2 \left(\text{pbest}_i - x(k)_{i.j}\right) \tag{22}$$

$$x(k+1)_{i.j} = x(k)_{i.j} + v(k+1)_{i.j} \tag{23}$$

Where as:

$v(k)_{i.j}$ is velocity of the particle i

$x(k)_{i.j}$ is position of the particle i

C_1, C_2 are the acceleration constants

W is the ineratia weight facter

r_1, r_2 are random numbers between 0 and 1

Pbest is the best position of the specific particle
Gbest is the best particle of the group.

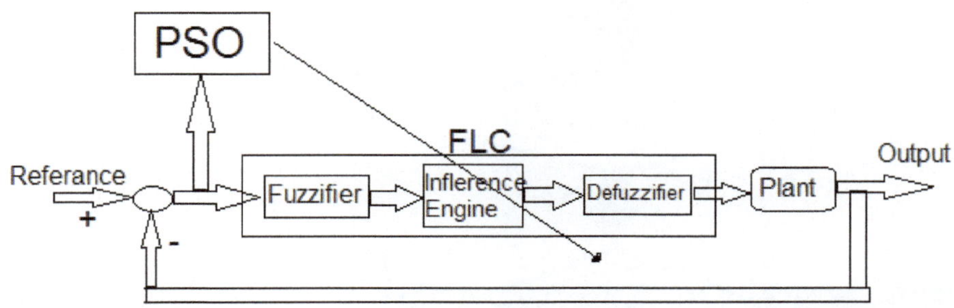

Figure 6. system block daigram

6. Hybrid Fuzzy Logic /Particle Swarm Optimization Controller Design

The design of an optimal fuzzy logic controller is performed using the PSO algorithms to search globally its optimal gains. The structure of the fuzzy logic controller with PSO algorithms is shown in Figure 6. The optimization algorithm is implemented by using MATLAB m-file program and linked with the system simulation program in MATLAB/SIMULINK, to check the system performance in each particle. The PSO produces the FLC controller gains which give optimal performance for the Zeta converter. The minimum absolute integral error is adopted in this work, which is given by the following equation (20).

$$ISE_{min} = \frac{1}{N}\sum_{i=1}^{N}\left|Vr - Vact(i)\right| \qquad (24)$$

When n: is number of semples considered.

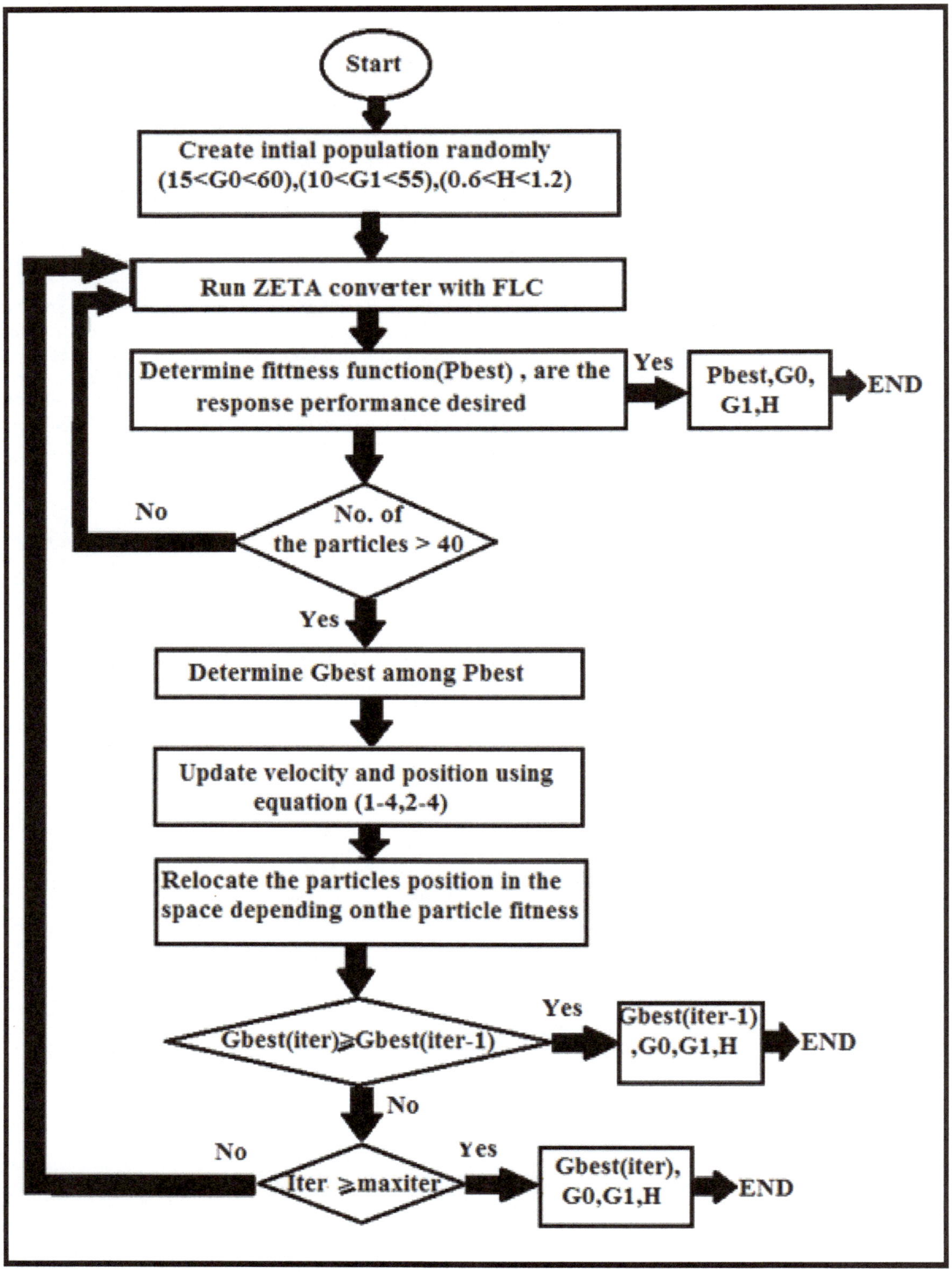

Figure 7. FLC / PSO flowchart

The PSO technique has been programed using MATLAB M-file. PSO algorithm is used in this work to compute the fuzzy controller gains G,G1,H during the system disturbance, the error information reaches the PSO algorithm. Then, it startes to update the fuzzy gains values for mantianing the desired system performance..The hybrid FLC/ PSO flowchart is shown in Figure 7.

7. System Realizationz

The system is implemented practically as open loop, closed-loop with FLC and hybrid FLC/ PSO as shown in Figure 8. The open-loop voltage response when duty ratio (D=0.5555) and load resistance (R_L = 10-40-10Ω) is shwon in Figure 9. The close loop voltage response with FLC when referance voltage (Vref=15 volt) and load resistance (R_L = 10Ω) is shwon in Figure 10. The close loop voltage response with hybrid FLC/ PSO when referance voltage (Vref=15 volt) and load resistance (R_L = 10Ω) is shwon in Figure 11.

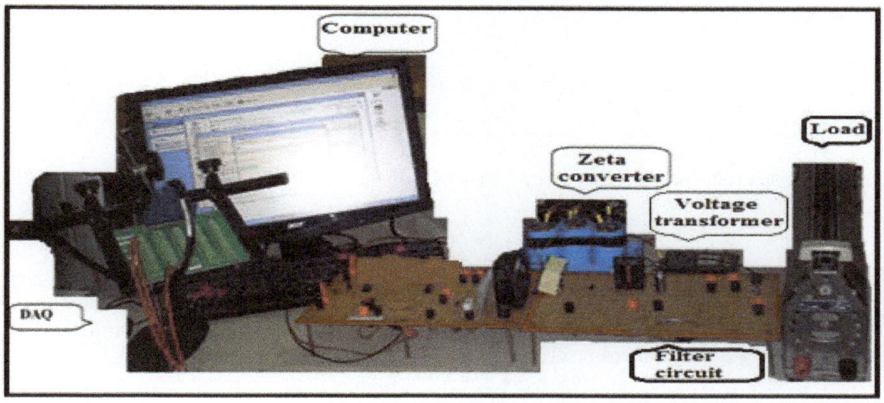

Figure 8. Practical system

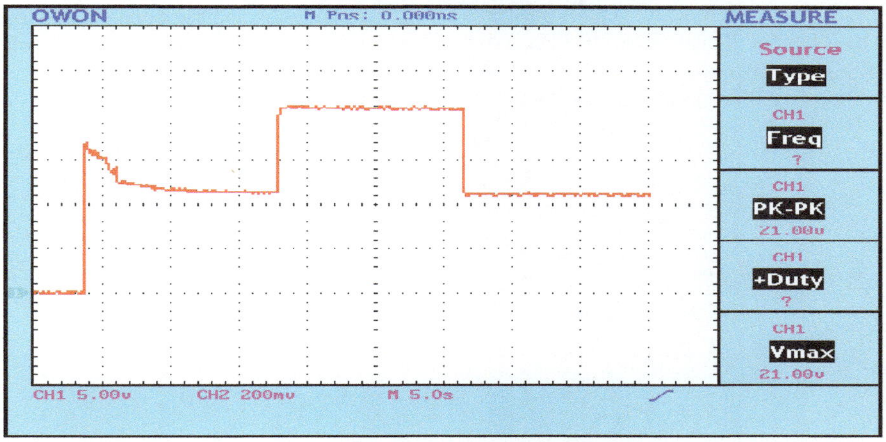

Figure 9. output voltage response of the open loop when (D=0.5555) and (R_L =10-40-10 Ω)

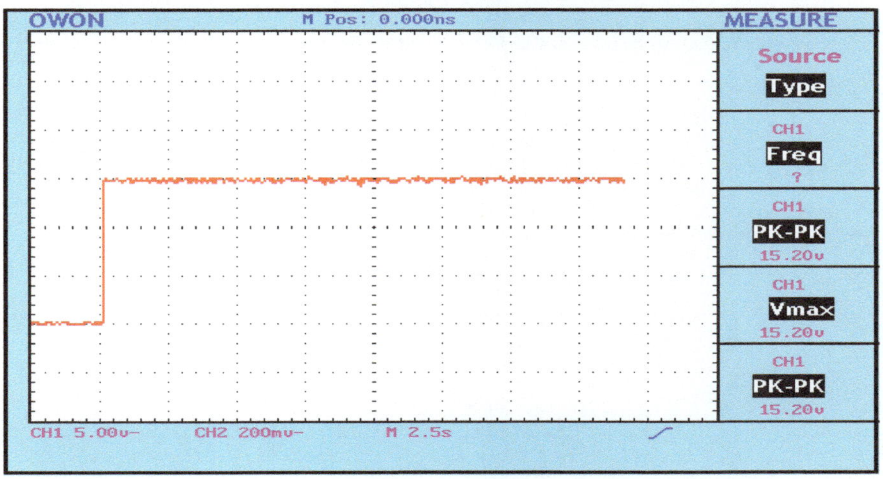

Figure 10. output voltage response of the closed loop system with FLC for V_{ref} 15 volt and (R_L = 10-40-10) Ω

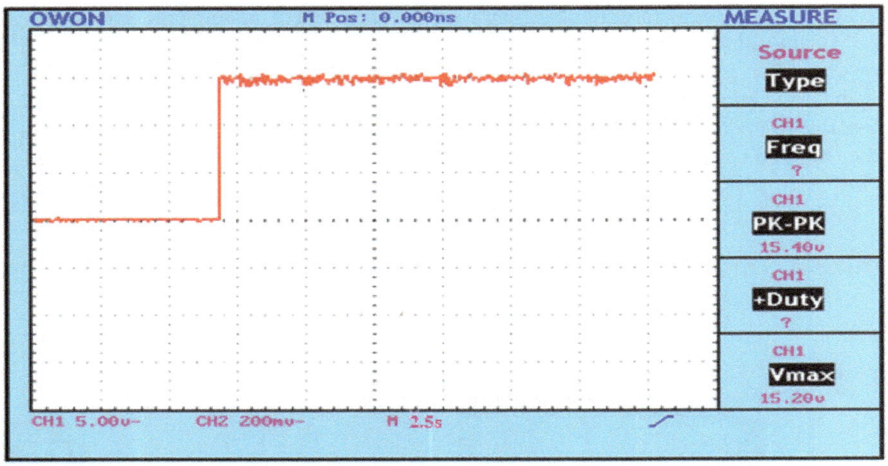

Figure 11. output voltage response of the closed loop system withhybrid FLC/PSOfor V_{ref} = 15 volt and (R_L = 10-40-10) Ω

8. Results and Discussion

The output voltages for the simulated openloop and closed loop systems with fuzzy logic controller are shwon in Figure 12, for referance voltages (9, 12 and 15) volt, for varying the load resistance (10-40-10)Ω at time (t=0.05-0.1-0.15 sec) respectively. The fuzzy logic controller gains are chosen by trail and error method to get the bestzeta converteroutput performance. Thisprocess needed long time. The closed loop system with the FLC/PSO controller response shown in Figure 13 has agreat improvement in it performance of the absolute itegral error was minimum. It oovercome the probem of long time mannal tunning. The output voltage of theclosed loop system with FLC had beenadaptedusing PSOtechniquewhen referance voltages (9,12 and 15) voltsand varying load resistances equal to (10-40-10Ω) as shwon in Figure 13. The closed looppractical system response with FLC/ PSO when referance voltages (9-12-15volt) and load resistance (10Ω) is shwon in Figure 14. Comparision between theoritical and practicalresults for open loop, close loop with FLC and hybrid FLC/PSO controllerfor voltage (15 volt) are shwon in the Table 2.

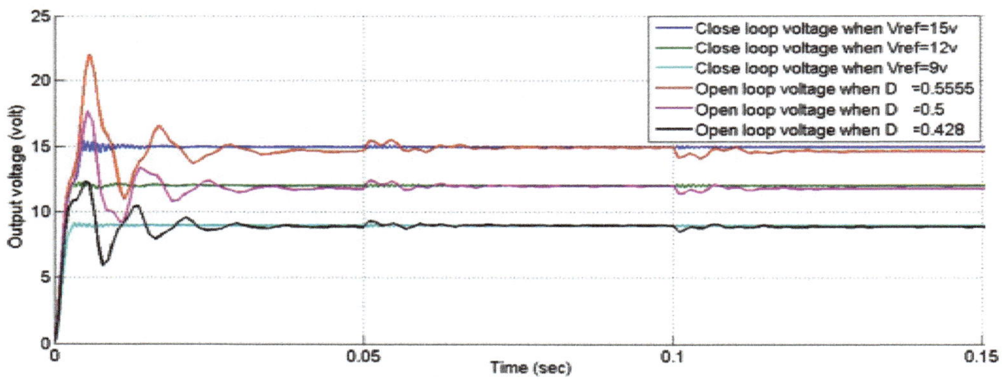

Figure 12. open loop system and close loop system with FLC controller responsesfor V_{ref} = 9,12,15 volts and R_L = 10-40-10 Ω

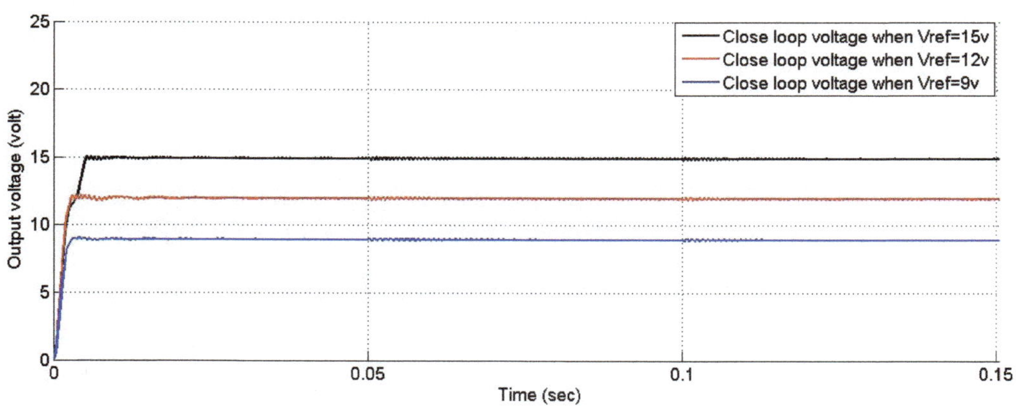

Figure 13. close loop system with FLC/ PSO controller responsesfor V_{ref} = 9,12,15 volts and R_L =10-40-10 Ω

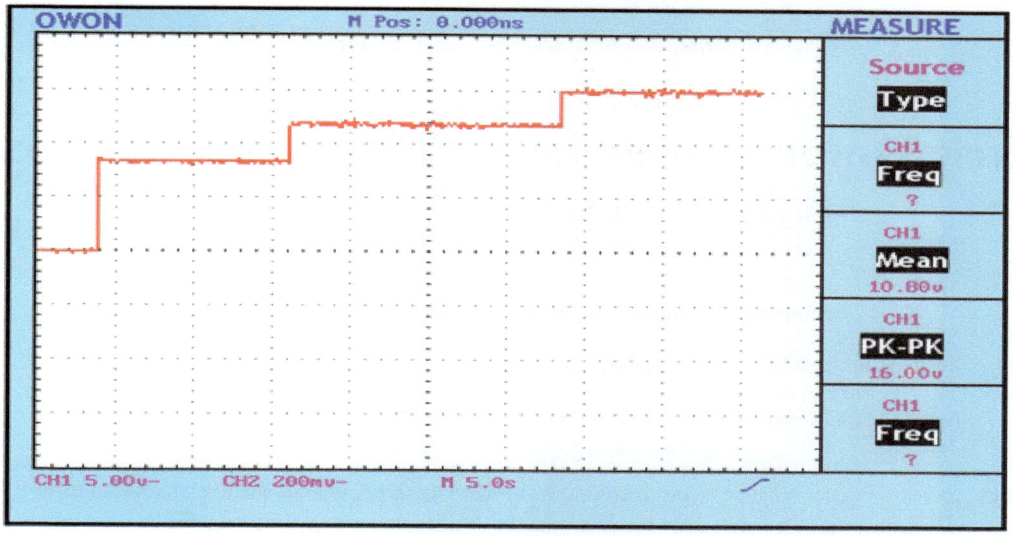

Figure 14. output voltage response of the closed loop system for V_{ref} = 9,12,15 volts and R_L (10) Ω.withFLC /PSO controller

Table 2. Comparision of the system performance for the different techniques when referance voltage (Vref=15 volts)

Design system	Steady state error[%]	Peak Overshoot [%]	Setlling time[ms]
Zeta converter circuit theoretically	6.7	51.7	38
Close loop system with FLC	0.47	2.7	7
Close loop system with FLC /PSO	0.4	0.91	5
Zeta converter circuit practically	16.7	49	312
Close loop system with FLC practically	0.49	2.9	22
Close loop system with FLC / PSO practically	0.43	0.95	13

From Table 2 above, it is shown that the performance of the converter for open loop system has high peak overshoot, the steady state time and state steady error. The system manually optimized fuzzy controller gains response performance (P.O.S, t_{ss} and e_{ss}) have been improved. Hybrid FLC/PSO (P.O.S, t_{ss} and e_{ss}) has improvements. The realized system response performance appears verified the theoretical ones except the steady state time witch was slightly higher. Comparison of the results show that the proposed(FLC/PSO)control is the best.

9. Conclusion

Zeta converter circuit has been designed. Its mathematical model in state space form has been developed. The circuit output voltage response as an open-loop system has been analyzed. Two techniques has been use to regulate the output voltage of the zeta converter circuit. Fuzzy logic controller technique has been used. Hybrid fuzzy logic /particle swarm optimization technique has been proposed. The output voltage for the Zeta converter with the feedback controller and two technique has been studied. It has been seen that the response of the Zeta converter circuit with the proposed technique was better than that of FLC technique. The computation time for the FLC/PSO gains were very short. The designed circuit with the controllers was implemented. The practical results was verified the simulated results. Therefore one can conclude that the proposed method is the best one.

References

[1] J. Falin," Designing DC/DC converters based on ZETA topology", Analog Applications Journal, Texas Instruments Incorporated, pp, 16-20, 2010.

[2] D. C. Martins, "Zeta Converter with High Power Factor Operating in Continuous Conduction Mode", Federal University of Catarina, Industrial Electronics, Control, and Instrumentation, IEEE pp. 1802-1807, 1996.

[3] A.H. Ahmed, M.K. Alkhatat, A.A. Abdullah, "Hinf/Muo - SYNTHESIS for CUK CONVERTOR CIRCUIT CONTROLLER", AL RAFIDAIN Engineering Joyrnal, vol. 14, no. 4, 2006.

[4] O.A.Taha "Cuk Converter Circuit Controller Design and Implemementation "M.Sc Thesis Mosul University, Mosul Iraq, 2007.

[5] S.S.Sabri "Optimal Fuzzy Controller Design For Cuk Converter Circuit Using Genetic Algorithm", M.Sc Thesis Mosul University, Mosul Iraq, 2008.

[6] E. Vuthchhay and C. Bunlaksanusorn," Modeling and Control of a Zeta Converter", International Power Elecronics Conference, AUNSEED-NET project, JICA.IEEE, pp. 612-619. 2010.

[7] K.E. Khoshmardan, M.R. Dastranj, M.O. Taleghani, A. Hajipoor "Design a Fuzzy Logic Based Speed Controller for DC Motor with Particle Swarm Optimization "PSO" Algorithm", Australian Journal of Basic and Applied Sciences. Azad University, Sabzevar, Iran, pp. 1283-1290, 2011.

[8] E. Vuthchhay, C. Bunlaksananusorn, and H. Hirata "Dynamic Modeling and Control of a Zeta Converter", International Symposium on Communications and Information Technologies, Oct. 2008.

[9] H. T. Hassan, R. Zafar, S. A. Mohsin, "Fully Informed Particle Swarm Optimization Technique for Electrical Power System", Canadian Journal on Electrical and Electronics Engineering Vol. 3, No. 3, pp. 121-127., March 2012.

[10] K.y.Lee and. M. A.El-Sharkawi, "MODERN HEURISTIC OPTIMIZATION TECHNIQUES", A JOHN AND SONS.INC. 2008.

12

Electric Power Transmission Enhancement: A Case of Nigerian Electric Power Grid

Emmanuel U. Oleka[1,*], **Samuel N. Ndubisi**[2], **Gerald K. Ijemaru**[3]

[1]Electrical and Computer Engineering Department, North Carolina A&T State University, Greensboro NC, USA
[2]Electrical Engineering Department, University of Port Harcourt, PortHarcourt, Nigeria
[3]Electrical and Electronic Engineering Department, Federal University of Oye-Ekiti, Ekiti State Nigeria
*Corresponding author: eoleka@aggies.ncat.edu

Abstract Increase in economic activities resulting from increase in population and social advancement has led to increase in electrical energy demands. This has increased the burdens on the existing transmission assets and in some cases, has caused the loading of the transmission assets beyond their design limits with consequent reduction in power quality and power outages in extreme cases. Many techniques have been developed to enhance the capabilities of the power grids. This paper looks into these techniques with the view to exploring their applicability to the Nigerian 330KV electric power grid, towards seeking ways to enhancing the performance of the grid for better asset utilization. A typical transmission line in the grid was modeled to assess its strengths and weaknesses. Enhancement techniques were then applied to assess their impacts on the line.

Keywords: *power transmission, reactive power, active power, voltage profile, line transfer capability*

1. Introduction

Electricity markets deregulations have made electric energy more like a commodity that is bought in bulk from one market location and sold in another and from one country to another. These deregulations, coupled with increase in energy demand due to increasing population and economic activities, have in some cases resulted in over utilization of the transmission system as some of the transmission lines are stressed beyond their design limits.

The increase in demand on the transmission system has made some of them to be loaded more than their power transfer capacities and this has in some places reduced the quality of power delivered and in some other cases, resulted in cascading overloads with consequent power outages [1]. Because of these, researchers are working round the clock to devise means of extending the power transfer capacities of the transmission lines while maintaining the quality of power delivered through them. Technically, the limitation on power transfer capacity on a transmission line can always be removed by addition of new transmission capacity, but the economic, political and environmental considerations in building of new transmission facilities have made this option not always desirable [2]. Meanwhile, an important aspect of the contract terms for these inter regional and international power purchases is that the transmission system be adequate for the power evacuation [3,4].

Currently, in Nigeria, the previously government-owned power generation and distribution companies have been privatized. It is expected that this privatization will result in better management of these utility companies resulting in increased generation and more efficient power distribution and revenue collection [5]. It is a fact that the success of the privatized system still depends on the effectiveness of the transmission system.

This work seeks to evaluate the strength of the Nigerian electric power grid as well as ways to enhance the performances of the transmission lines in the grid with the aim of achieving better utilization of the available transmission assets.

The rest of the paper is organized as follows. Section II outlines the major concepts in transmission line performances. Section III presents Nigerian electric power grid and the methodology. Results and observations are presented in section IV, while conclusions are made in section V.

2. Transmission Line Performance

2.1. Line Parameters

All transmission lines in a power system are made to transport energy from power generation stations to distribution stations. Hence, they exhibit the electrical properties of resistance, inductance, capacitance, and conductance. The resistance is due to the nature of the conductor while inductance and capacitance are due to the effects of magnetic and electric field around the conductor. These parameters form the basis for the development of transmission line models used in power system studies. The shunt conductance accounts for leakage current which flows across insulators and ionized pathways in the air.

The leakage currents are negligible compared to the current flowing in the transmission lines and may be neglected [6].

2.2. Power Flow through Lines

The aim of power lines is to effect power transmission from one bus to another or from one grid to another. Transmission lines are built either in single circuit or double circuit lines (for higher capacities). Power flow through a line is determined by the voltage magnitudes of the two buses connected by the line, the line impedance and the difference in angles of the voltages of the two buses. When double circuit (parallel line) is involved, the fact that current flows through the line of least impedance path can also bring about an imbalance in the flow of power and this can cause an overloading of one line of the circuit in some cases as the other line operates quite below its power rating [7].

For a simple 2-bus transmission system shown in Figure 1, the power flow through the line is represented by equations 1-2, where $\delta = \delta_1 - \delta_2$.

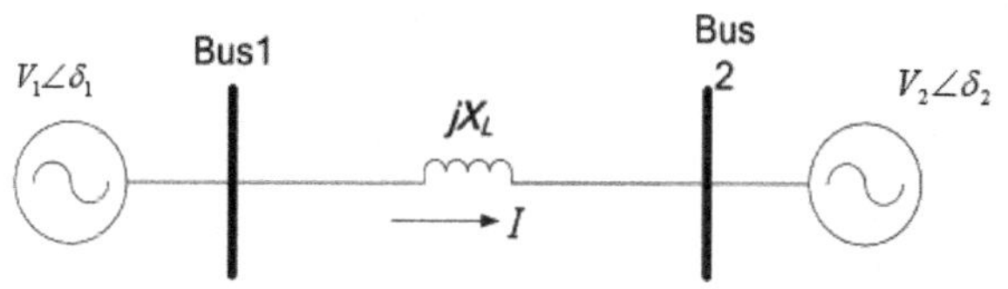

Figure 1. Simplified model of a power transmission system

$$P_2 = \frac{V_1 V_2 \sin\delta}{X_L} \tag{1}$$

$$Q_2 = \frac{V_2\left(V_2 - V_1\cos\delta\right)}{X_L} \tag{2}$$

$$V = f(P,Q) \tag{3}$$

Equations above indicate that the active and reactive power/current flow can be regulated by controlling the voltages, phase angles and line impedance of the transmission system. Real power transfer can be increased by increasing the line voltage, reducing the line reactance or by increasing δ. From (1), the active power flow will reach the maximum when the difference in phase angles δ is 90° but in practice, a small angle difference, between 35° to 45°, is used to keep the system stable from the transient and dynamic oscillations [8]. Also, from equation (2), reactive power at any bus is dependent on the voltage drop along the line, such that reactive power increases with increase in voltage difference between the two buses. As a result of (1) and (2), the voltage at the receiving–end of the line becomes a function of the active and reactive power flow through the line as given in (3). For effective operation of a transmission line, these transmission parameters are kept under control. Power flow balance is achieved in a parallel transmission line using series compensation or phase shifting transformers [9].

2.3. Surge Impedance Loading

The surge impedance loading (SIL) of a transmission line is the MW loading of the transmission line at which a natural reactive power balance occurs. It is simply the MW loading at a unity power factor at which the lines MVar usage is equal the line's MVar production, causing reactive powers to cancel out. SIL is given in equation form as:

$$\text{SIL} = \frac{(V_{Lrated})^2}{Z_c} \tag{4}$$

Where Z_C is the characteristic impedance or surge impedance of the line.

The concept of SIL is important in transmission line studies because it sets the theoretical limit for stable operation (power delivery) for very long lines and indicates where the reactive requirements of the line are small.

2.4. Line Loadability

There is always an inherent trade-off between increasing utilization of the grid and security of grid's operation. Loadability of a transmission line is defined as the optimum power transfer capability of a transmission line under a specified set of operating criteria. Many research works have been carried out on the improvement of transmission transfer capability with a variety of operation constraints, such as stability, voltage security and thermal limits [10]. St. Clair curve [11] provides a simple means for estimating power transfer capabilities of transmission lines. It concerns three limiting factors: thermal limit, voltage drop limit and angular stability limit which usually affects short, medium and long transmission lines respectively. Voltage limit and stability limit fall below thermal limit. Compensation can be used to modify the natural parameters of transmission lines and to increase loadability of long lines toward their thermal limit.

2.5. Transmission Systems Enhancement Strategies

Several methods which could be used to enhance the performances of transmission lines have been reported in [9,12,13], which include:

1. **Installation of New Transmission Line:** This is usually the first option that comes to mind whenever a transmission line is limited in the amount of power it can transmit, so as to alleviate overloading by providing additional paths for power flow. It is beneficial by increasing the reliability of the transmission system. However, it has to pass through economic, political and environmental hurdles.

2. **Reconductoring Transmission Line/Terminal Equipment Replacements:** A line can be reconductored with a larger conductor with more power-carrying capability if the original transmission line conductor is inadequate to carry expected power flows, provided that the transmission line towers do not need to be significantly altered to support the heavier conductor. In addition, some terminal equipment may need to be upgraded to match the desired rating.

3. **Conversion from single circuit to double circuit:** This involves making necessary modifications to the existing transmission towers and adding a second transmission line to the structure. This option extensively reduces the line impedance and increases current-carrying

capacity of the line, thereby increasing power transmission capacity of the line.

4. **Voltage Upgrade:** Another option is to increase the operating voltage of the transmission line, such as upgrading the voltage from 132 kV to 330 kV. In this instance, for example, the nominal rating of the line maybe drastically increased while using the same conductor. This type of improvement may require upgrading the transmission towers to meet National Electric Safety Code (NESC) clearance levels. In addition, the switching stations and substations must also be upgraded with higher voltage circuit breakers, switches, transformers, and other related equipment.

5. **Reactive Power Compensation**: Addition of reactive power compensation devices (especially FACTS devices) in the form of series compensation, shunt compensation or the combination of the two, (depending on the nature of the line and its identified deficiency and need) to the transmission line is also another means of enhancing the performances of a transmission line. The addition of compensation modifies the electrical impedance of the line and therefore increases the power flows across the line. This can be an effective and economical means of increasing the transmission capability as a whole, by taking advantage of transmission lines that are not loaded to their thermal limits.

6. **Phase Shifting:** Inequality in line impedances of a double circuit transmission line can bring about imbalance in load sharing between the circuits of the line: a situation where one circuit suffers overload while the other is under loaded. This is because electric current naturally flows through the path of least impedance. Addition of phase shifting device in the line helps to regulate and to better control the power flows on the system to optimize the existing transmission capability by distributing the power flows across the transmission lines.

3. Nigerian 330kv Power Grid; A Test Case

The Nigerian 330kv power transmission grid is made up of 28 buses connected by a total of 5523.8km length of transmission lines. The grid, shown in Figure 2, is characterized by poor voltage profile in most parts of the network, inadequate dispatch and control infrastructure, radial and fragile grid network, inadequate available generation capacity, and inadequate or total absence of spinning reserve [14]. The grid is marred by incessant voltage and frequency instability issues [15], leading to load curtailments. These instabilities have often resulted in total system collapse such that the grid experienced an average of thirty-one (31) system collapse per year for the period between years 2000-2014 [16,17].

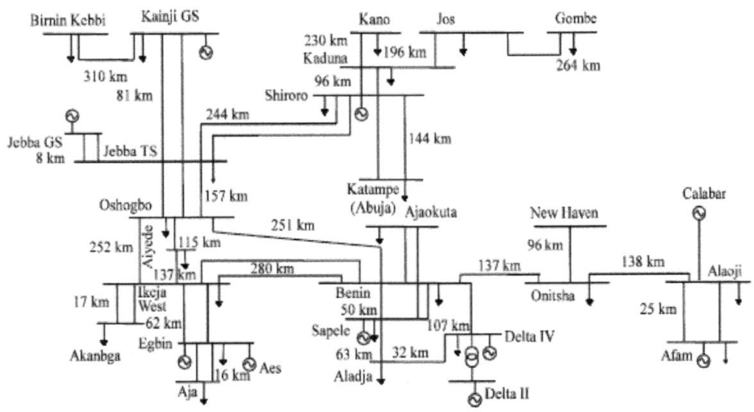

Figure 2. One-line diagram of Nigerian 330kv grid

Table 1. Different Generation levels and violations observed [18]

S/N	GENERATION LEVELS	GENERATION		DEMAND		LOSSES		No. of Voltage Violations (-/+5%)	Thermal Rating Violation (>760MVA)
		PG (MW)	QG (MVAR)	PD (MW)	QD (MVAR)	PL (MW)	QL (MVAR)		
1	Base Case at 2284MW Generation	2284.50	-549.26	2230.60	1394.10	54.04	-1989.58	1	No violation
2	Generation Increase to 3063.71MW	3036.71	92.16	3000.00	1875.00	63.50	-1782.86	1	No violation
3	Generation Increase to 4114.38MW	4114.38	974.10	4054.80	2534.50	59.75	-1530.36	1	No violation
4	Generation Increase to 4500MW	4505.00	1867.03	4300.00	2687.50	22.14	-766.11	13	1 violation
5	Generation Increase to 4711.77MW	4711.77	2051.14	4500.00	2812.50	225.60	-713.20	13	1 violation
6	Generation Increase to 7290.09MW	7290.09	4719.21	6768.40	4230.25	650.12	1948.19	12	5 violations
7	Generation Increase to 10024.51MW	10024.51	7829.00	9000.00	5625.00	1173.62	3902.43	16	7 violations

The power grid delivers low power quality while transmitting the present available generating capacity of 4949MW [17], and should deliver even lower power quality while transmitting the current peak demand forecast of 12800MW. Analysis [18] carried out on the grid (Table 1) shows that the grid, at its present state, cannot stably transmit the current power demand of the grid and will be much less capable if the generation becomes sufficient, which formed the motivation for this present work.

This study was carried out using the line, which connects Onitsha and New Haven substations. The line was constructed and commissioned in the year 1982. It has a total length of 96km. The line is composed of two bundled ACSR 350 mm^2, 54/7 BISON conductors per phase with horizontal configuration. The parameters of the line are given in Table 2 below. The performance of the line was evaluated to deduce ways to enhance it for better utilization.

Table 2. Parameters of the transmission line

Parameter	Value
Line Voltage, V	330KV
Line Length, ℓ	96km
Line Inductance, L	1.06302mH/km/phase
Line Capacitance, C	0.01076μF/km/phase
Line Resistance, R	0.033698Ω/km/phase
Line Conductance, G	0.00 Siemens/km/phase (assumed)
Current Capacity, I	1360A/phase
Line Power Rating	760MVA

Two-port network (Figure 3) model of transmission lines relates the sending-end quantities with the receiving-end quantities using ABCD constant as shown in (5)

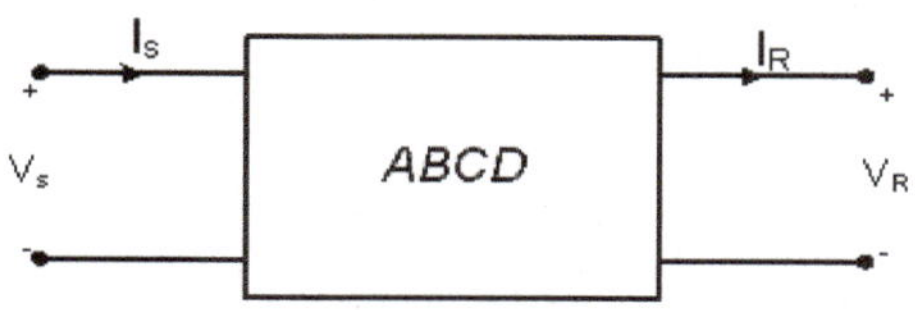

Figure 3. Two-port representation of a transmission line

$$\begin{bmatrix} V_S \\ I_S \end{bmatrix} = \begin{bmatrix} A & B \\ C & D \end{bmatrix} \begin{bmatrix} V_R \\ I_R \end{bmatrix}. \tag{5}$$

The transmission line was modeled and simulated at different loading levels. The simulation was carried out in MATLAB platform using power system analysis toolbox provided in [6]. The sending-end to receiving-end relationships were observed at the different loading levels to determine the voltage drop characteristics of the line. St. Clair's curve was also generated for the line, which indicated the thermal and the stability limits of the line. Different enhancement strategies were then applied to find out whether they could affect the line performances.

4. Observations and Discussion

4.1. Open-Line Characteristics

Figure 4 shows the voltage profile of the line while open-circuited at the receiving end (i.e. at no load when the line is being charged). The figure shows that there is a rise in the voltage of the line at the receiving end. Though it is still within tolerance limit (330kv±5%) for the line under study, it can rise above the limit for a longer line as depicted in Figure 3b where the same line design was used but the length was extended to 500km. In the case of a longer line, (Figure 4(b)), the receiving end voltage has even risen to 383KV. It also shows that the voltage level of Figure 4(a) was brought back to 330KV after the line has been compensated with a shunt reactor of reactance XLsh = 6206.21Ω which consumed 19.46Mvar from the line, while that of 4(b) was brought back to 330KV after it was compensated with a shunt reactor of reactance XLsh=1164.66Ω which consumed 93.72Mvar from the line.

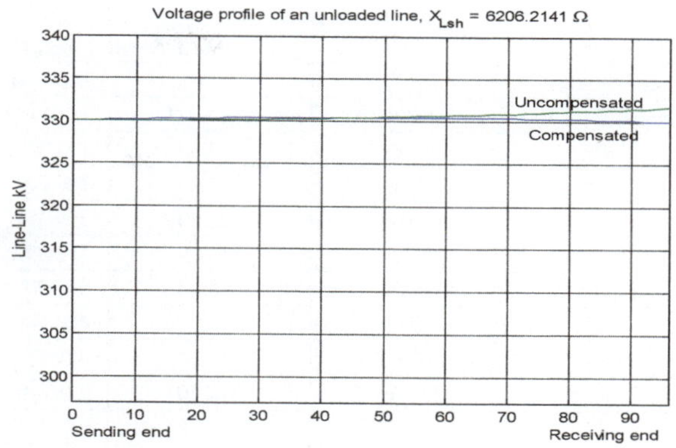

Figure 4. (a) open-line voltage profile (normal length)

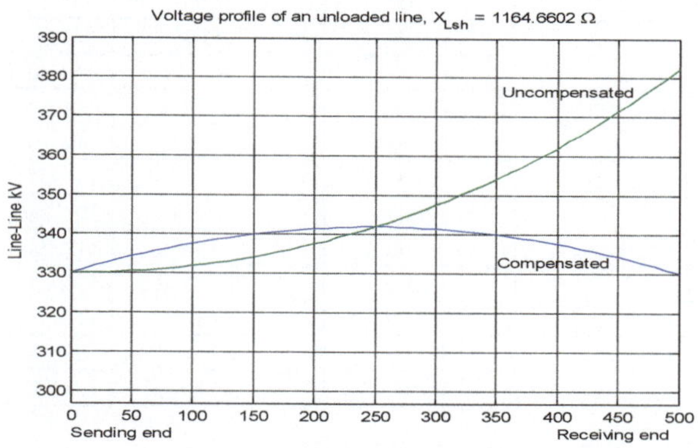

Figure 4. (b) open-line voltage profile (500km)

4.2. Line's Voltage Profile

It was observed, after subjecting the line to varying sending-end power levels, that the voltage at the receiving end kept decreasing as the sending-end power increased until at about 185MW when voltage violation occurred. 185MW at 0.8 p.f. is equivalent to 231MVA, which is far below the line rating of 760MVA. It was also observed in Table 3 that as the load increases, current through the line increases with a consequent increase in power (I^2R) loss on the line. Percentage voltage regulation also increases with increase in power transmitted, and the efficiency of the line decreases. Figure 5 presents the voltage profiles of the line at different loading levels. It shows that for lines of about 80km and above, the receiving-end voltage becomes totally unusable if the line is loaded to its rated capacity. This means that the lines in the system cannot be utilized to their rated capacities especially considering the radial nature of the grid network.

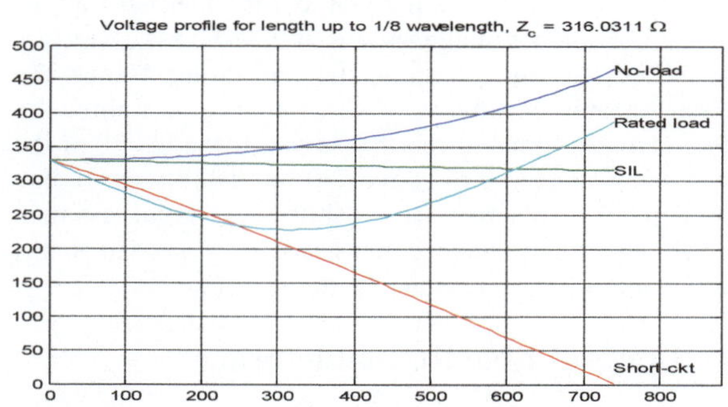

Figure 5. The line's voltage profiles

Table 3. Sending-end to receiving-end characteristics

Sending-end		Recieiving-end				Losses		V. Reg. %	Eff. %
Voltage VS(KV)	Power Ps(MW)	voltage VR (KV)	IR (A)	PR (MW)	QR (MVAR)	PL (MW)	QL (MVAR)		
330	50.00	324.80	153.90	49.83	70.62	0.17	-33.12	2.24	99.67
330	75.00	322.20	206.20	74.67	87.53	0.33	-31.28	2.97	99.56
330	100.00	320.20	259.30	99.45	103.87	0.55	-28.87	3.61	99.45
330	125.00	318.20	312.90	124.17	119.63	0.83	-25.88	4.26	99.33
330	150.00	316.20	366.70	148.83	134.83	1.17	-22.33	4.91	99.22
330	175.00	314.20	420.70	173.43	149.45	1.57	-18.20	5.57	99.11
330	200.00	312.30	474.70	197.98	163.50	2.02	-13.50	6.22	98.99
330	225.00	310.40	528.80	222.47	176.99	2.53	-8.24	6.88	98.88
330	250.00	308.50	583.00	246.90	189.90	3.10	-2.40	7.54	98.76
330	275.00	306.60	637.20	271.27	202.23	3.73	4.02	8.20	98.65
330	300.00	304.70	691.50	295.59	214.00	4.41	11.00	8.87	99.52
330	325.00	302.90	745.70	319.85	225.20	5.15	18.55	9.53	98.42
330	350.00	301.03	800.00	344.05	235.83	5.95	26.67	10.20	98.30
330	375.00	299.20	854.30	368.19	245.88	6.81	35.37	10.86	98.18
330	400.00	297.40	908.60	392.27	255.36	7.73	44.63	11.53	98.07
330	425.00	295.70	962.90	416.30	264.28	8.70	54.47	12.20	97.95
330	450.00	293.90	1017.20	440.27	272.63	9.73	64.88	12.86	97.84
330	475.00	292.20	1071.60	464.18	280.40	10.82	75.85	13.53	97.72
330	500.00	290.50	1125.90	488.03	287.60	11.97	87.40	14.20	97.61
330	525.00	288.80	1180.20	511.83	294.22	13.17	99.52	14.86	97.49
330	550.00	287.10	1234.60	535.56	300.29	14.44	112.22	15.53	97.38
330	575.00	285.50	1288.90	559.24	305.77	15.76	125.48	16.19	97.26
330	600.00	283.89	1343.30	582.87	310.69	17.14	139.31	16.85	97.14

4.3. Line transfer capability

St. Clair curve on Figure 6 has in display, the theoretical stability limit, practical stability limit and thermal limit of the test case transmission line. It can be seen that the power transfer by the line is not affected by any of the stability limits being that the line is not long enough.

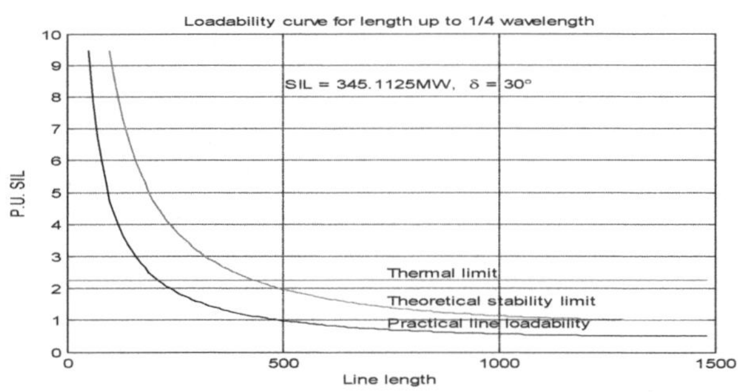

Figure 6. The line's loadability curve at 330KV operating voltage

Worthy of note also is the fact that the thermal limit of the line at 330KV voltage level is at 2.3P.U. SIL, which is at 794MW, far above the line standard rating. Hence, the power transfer capability of the line is not limited by either stability limits or thermal limits. If the length of the same line is increased to above 500km, stability of the line will limit its power transfer capability to a point less than its SIL, and then, enhancements would be needed to get the line to transmit a reasonable amount of power.

4.4. Effect of Line Reconductoring

A case of changing the sizes of the conductors of the transmission line was examined to determine its effects. The conductors of the line were replaced with the same conductor size but three conductors bundle. Increasing the bundle size increases the geometric size of the conductor thereby affecting both the resistive and the magnetic characteristics of the conductor. It was observed that reconductoring the line did not show any significant effect on the voltage profile and the angular stability of the line as shown in Figure 7. The effect observed was in the thermal characteristics, where the SIL became 388MW and the thermal limit rose to 3p.u.SIL which is 1164MW.

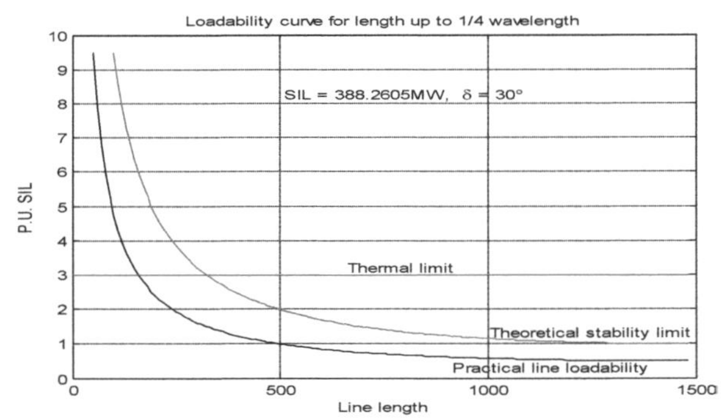

Figure 7. The line's loadability curve for reconductored line at 330KV

4.5. Effect of Increase in Voltage Level

Figure 8(a) and Figure 8(b) show the effects on the line when the line voltage was increased to 450KV and 500KV respectively. It can be seen that at 450, the SIL has increased to 641.738MW and the thermal limit is at about 1.7 P.U. SIL (about 1090MW), and at 500KV, the SIL has increased to 792.269MW and thermal limit is at 1.5 P.U. SIL (about 1188.4MW). It is also worth noting that the line is still able to transfer reasonable amount of power at 500km length even with the effects of stability limits.

These show that upgrading the voltage level can increase line loadability and increase asset utilization.

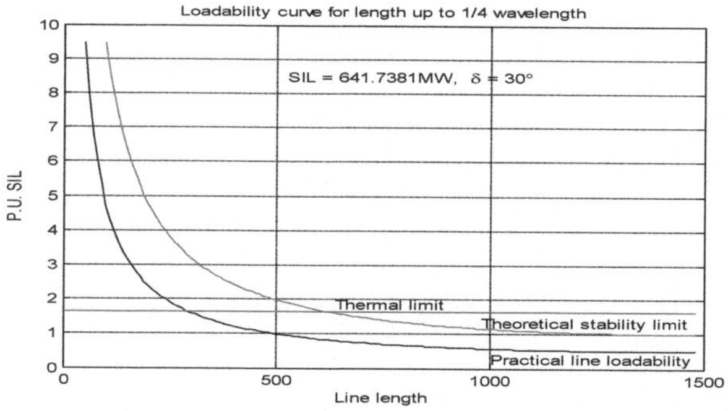

Figure 8a. Loadability at 450kv voltage level

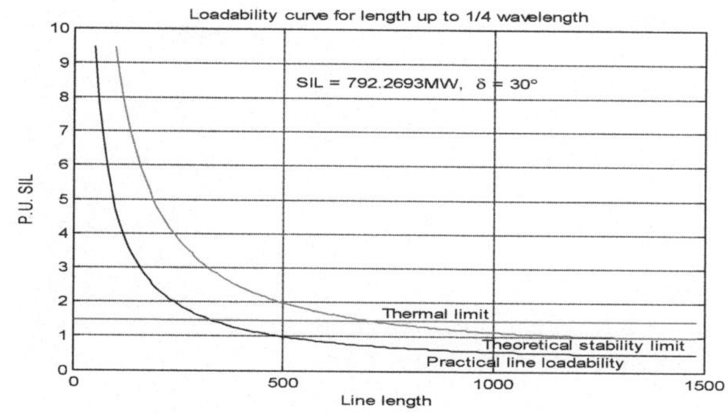

Figure 8b. Loadability at 500kv voltage level

4.6. Effect of Compensation

Table 4. Effects of compensation on the transmission line

VS/VR	Recieiving-end			Losses		V. Reg. (%)	Eff. (%)	Reguired Capacitor Spec.		
	IR (A)	PR (MW)	QR (MVAR)	PL (MW)	QL (MVAR)			Ohm	µ-Farad	MVAR
330.00	89.79	49.83	12.29	0.08	-34.63	0.52	99.85	1867.05	1.42	58.33
330.00	131.65	74.67	9.32	0.17	-33.71	0.52	99.77	1392.50	1.91	78.20
330.00	174.33	99.45	6.18	0.30	-32.43	0.52	99.70	1114.75	2.38	97.69
330.00	217.30	124.17	2.86	0.47	-30.77	0.52	99.63	932.57	2.84	116.77
330.00	260.39	148.83	-0.64	0.67	-28.76	0.52	99.55	803.84	3.30	135.47
330.00	303.52	173.43	-4.32	0.91	-26.39	0.52	99.48	708.20	3.75	153.77
330.00	346.67	197.98	-8.18	1.18	-23.65	0.52	99.41	634.34	4.18	171.68
330.00	389.81	222.47	-12.20	1.50	-20.56	0.52	99.33	575.60	4.61	189.19
330.00	432.92	246.90	-16.41	1.84	-17.11	0.52	99.26	527.86	5.03	206.31
330.00	475.99	271.27	-20.78	2.23	-13.31	0.52	99.19	488.31	5.43	223.01
330.00	519.04	295.59	-25.33	2.64	-9.15	0.52	99.11	455.01	5.83	239.33
330.00	562.06	319.85	-30.06	3.10	-4.63	0.52	99.04	426.63	6.22	255.26
330.00	605.03	344.05	-34.95	3.59	0.23	0.52	98.97	402.17	6.60	270.78
330.00	647.96	368.19	-40.02	4.12	5.45	0.52	98.90	380.90	6.96	285.90
330.00	690.85	392.27	-45.26	4.68	11.02	0.52	98.82	402.41	6.59	270.62
330.00	733.71	416.30	-50.67	5.23	16.94	0.52	98.75	345.77	7.67	314.95
330.00	776.53	440.27	-56.25	5.90	23.20	0.52	98.68	331.13	8.01	328.88
330.00	819.32	464.18	-62.00	6.57	29.82	0.52	98.61	318.05	8.34	344.40
330.00	862.06	488.03	-67.92	7.27	36.78	0.52	98.53	306.32	8.66	355.52
330.00	904.78	511.83	-74.01	8.00	44.10	0.52	98.46	295.74	8.97	368.23
330.00	947.45	535.56	-80.26	8.78	51.75	0.52	98.39	286.16	9.27	380.55
330.00	990.10	559.24	-86.69	9.58	59.75	0.52	98.32	277.48	9.56	392.46
330.00	1032.93	582.87	-93.28	10.43	68.10	0.52	98.24	269.57	9.84	403.97

Deploying shunt reactors to the unloaded line was able to restore the receiving-end voltage to an acceptable level, while shunt capacitor deployed at the receiving-end of the loaded line and varied according to the reactive power demand was able to maintain the receiving-end voltage at an acceptable level. Table 4 summarizes the effects of compensation on the transmission line. It shows the specifications of shunt capacitors required to restore the receiving-end voltage shown in Table 3 back to the 330KV. It can also be observed that the line losses reduced and the voltage regulation and efficiency of the line improved as a result of the compensation. This shows that by placing FACTS devices at the receiving-end of the line, the desired compensation at any load level can automatically applied and keep the voltage at a desired level.

5. Conclusion

The existing Nigerian national power grid is being underutilized due mainly to the radial nature of the network and lack of voltage control devices in the network. The lines in the 330KV transmission network are not limited by stability limits (the longest being the Kainji-Brinin kebbi line of 310km), and they are not limited by thermal limit but their limitations are their voltage drop levels, which cause load curtailment and limit the lines at load levels far below their thermal limits. There is need for investment into expanding and strengthening the power grid.

Reconductoring, as a power grid strengthening strategy, may not yield a reasonable improvement on the grid in question. Though transmission at Ultra High Voltage (UHV) levels can create rooms for more power evacuation, it requires much investment in changing terminal and substation equipment, and amendments of transmission towers. Hence, this may not be a feasible option at the moment but having UHV transmission backbone is still desirable for effective power evacuation.

The power grid should be fortified with voltage control equipment especially in the form of FACTS devices located at strategic points on the grid to control voltage sags and to increase the power handling capacity of the grid thereby enhancing utilization of the grid.

References

[1] US-Canada-Power-System_Outage-Task-Force, "Final Report on the August 14, 2003 Blackout in the United States and Canada: Causes and Reecommendations," *2004. [Online] Available: www.energy.gov/sites/prod/files/oeprod/DocumentsandMedia/BlackoutFinal-Web.pdf.*

[2] L. Furby, P. Slovic, B. Fischhoff, and R. Gregory, "Public perceptions of electric power transmission lines," *Journal of Environmental Psychology,* vol. 8, pp. 19-43, 1988.

[3] UNITED-NATIONS, "Multi Dimensional Issues in International Electric Power Grid Interconnections" *United Nation Department of Economic and Social Affiars, 2006, [Online]. Available: http://www.un.org/esa/sustdev/publications/energy/interconnections.pdf.*

[4] NERC, "NERC Policy 2: Transmission," *[Online] Available: http://www.nerc.com/pa/Stand/Version%200%20Relaibility%20StandardsRD/Operating%20Policies%20Markup.pdf.*

[5] B. E. Dikki, "Update-Privatization Issues," ed: A presentation at the 1st National Council on Power Conference (NACOP) Abuja Nigeria, August 11, 2014.

[6] H. Saadat, "Power System Analysis,(2nd)," ed: McGraw-Hill Higher Education, 2009.

[7] J. Verboomen, D. Van Hertem, P. H. Schavemaker, W. L. Kling, and R. Belmans, "Phase shifting transformers: principles and applications," in *Proceedings of the IEEE 2005 International Conference on Future Power Systems*, 2005, pp. 6 pp.-6.

[8] L. Gyugyi, "Application characteristics of converter-based FACTS controllers," in *Proceedings of the IEEE International Conference on Power System Technology, PowerCon.* , 2000, pp. 391-396.

[9] E. Wirth and A. Kara, "Innovative power flow management and voltage control technologies," *Power Engineering Journal,* vol. 14, pp. 129-140, 2000.

[10] H. Jin and X. Wilsun, "Extended transmission line loadability curve by including voltage stability constrains," in *Proceedings of the IEEE 2008 Electric Power Conference, EPEC, Canada*, 2008, pp. 1-5.

[11] R. Gutman, P. Marchenko, and R. Dunlop, "Analytical development of loadability characteristics for EHV and UHV transmission lines," *IEEE Transactions on Power Apparatus and Systems,* pp. 606-617, 1979.

[12] WESTERN-AREA-POWER-ADMINISTRATION, "Transmission Enhancement Technology Report "*July 2002 [Online]. Available: www.pserc.wisc.edu/documents/.../wapa_technology_report.pdf.*

[13] EEI, "Transmission Projects: At A Glance," *Edison Electric Institute, 2015. [Online] Available: www.eei.org/issuesandpolicy/documents/trans_project_lowres_bookmarked.pdf.*

[14] H. S. Labo, "Current status and future outlook of the transmission network," in *Investors' Forum for the Privatization of PHCN Successor Companies, Abuja Nigeria, January 18*, 2010.

[15] J. U. Ndiagwalukwe, "Frequency Control and Grid Stability," *National Power Summit, Abuja Nigeria, December,* 2012.

[16] I. Samuel, J. Katende, S. A. Daramola, and A. Awelewa, "Review of System Collapse Incidences on the 330-kV Nigerian National Grid," *International Journal of Engineering Science Invention,* vol. 3, pp. 55-59, 2014.

[17] *Presidential Task Force on Power, "2014 Year In Review", Annual Report of the Presidential Task Force on Power (PTFP),* 2015.

[18] C. C. Okoro and K. C. Achugbu, "Performance Analysis and Indices for the Existing Nigerian 330KV National Power Grid," in *Proceedings of International Conference and Exhibition on Power and Telecommunication, ICEPT*, Abuja Nigeria, 2008.

Optimal Control for the Voltage Regulation of a Power Line Including an Interconnection with a Wind Park and FACTS

N. Schinas*, A. Moullas

Department of Electrical Engineering, Technological Educational Institute of Western Greece, Patras, Greece
*Corresponding author: nickschinas@gnail.com

Abstract In this paper the voltage regulation of a grid bus is studied. The bus accepts real power from a wind park through a medium voltage cable. The changes of the wind speed result in large alterations in the exchange of real and reactive power between the cable and the grid. As a result, the voltage of the bus is not stable. The installation of a Static Var Compensator(SVC) is a rational choice for the stabilization of the voltage. The SVC needs a controller which according to the difference of the measured voltage from its nominal value, it defines the equivalent susceptance of the SVC. So, the right amount of reactive power between the grid and the SVC is achieved. Two design techniques for the controller are proposed, both based on the theory of optimal control. The full theory of the optimal control used in the design procedure, is given clearly. In the first place, the nonlinear model is linearized in two ways and then the mathematical principles of optimal control are applied. The second input of the linearized model, regarded as disturbance, has been taken into account in the design procedure. The proposed application of the optimal control theory has resulted in controllers that can be regarded approximately as PI controllers andso, their implementation is very easy. The whole nonlinear system has been simulated in SIMULINK so as the response of the proposed system to be judged. The results are satisfactory. Despite the large variations in the exchange of real and reactive power between the wind park and the grid, the variation of the grid voltage is under 1.5%.

Keywords: *optimal control, voltage regulation, wind turbine, FACTS*

1. Introduction

Many scientific papers regarding voltage control in power systems can be found in the literature. For instance, there are many new control schemes that deal with voltage control on large - scale power systems ([1]) or distribution power systems ([2,3]). The integration of renewable energy systems into the main grid makes the voltage even more unstable according to the fluctuations of the produced energy and the load. One solution to the stabilization of the voltage is the installation of FACTS devices in power systems with wind generation ([4,5,6]). Many schemes are presented in these papers for the control of various FACTS devices and the right performance of the grid voltage, although it may be difficult for some of them to be implemented due to their complex form.

In this paper, a specific interconnection between a wind power station and the main grid is studied and simple PI controllers for the stabilization of the voltage are designed, according to the optimal control theory.

Many applications of wind parks require off - shore installations that means interconnection with the main grid through submarine cables. As a result, the grid voltage is susceptible to alterations from the rated value because of the significant changes of the real and reactive power being interchanged between the grid and the cable.

In this paper a Static Var Compensator (SVC) scheme equipped with a specific controller is proposed which is able to maintain the grid voltage constant despite the large alterations of the wind speed .The design of the controller is based on optimal control theory and finally a simple PI controller can be applied.

The system under study is shown in Fig. 1. It can be seen the bus of the grid to which the cable of the wind park is connected. The SVC is connected to the same bus. The goal of the controller of the SVC is to keep the voltage of the bus as constant as possible under large variations of real and reactive power from the cable into the grid. The rated voltage of the cable is 49 kV so it is connected to the main grid through a transformer 150/49 kV. The wind park (WP) consists of 26 doubly fed wind turbines (WT) using asynchronous generators. The rated power of each WT is 1.5 MVA and the rated voltage 575 V. For simplicity reasons there is a central transformer 49 kV/ 0.575 kV for the interconnection of the WP to the cable. In the Appendix, there are the values of all the parameters of the system. The whole nonlinear system has simulated in the SIMULINK software using its libraries for the main components such as the WT, the cable, the SVC etc.

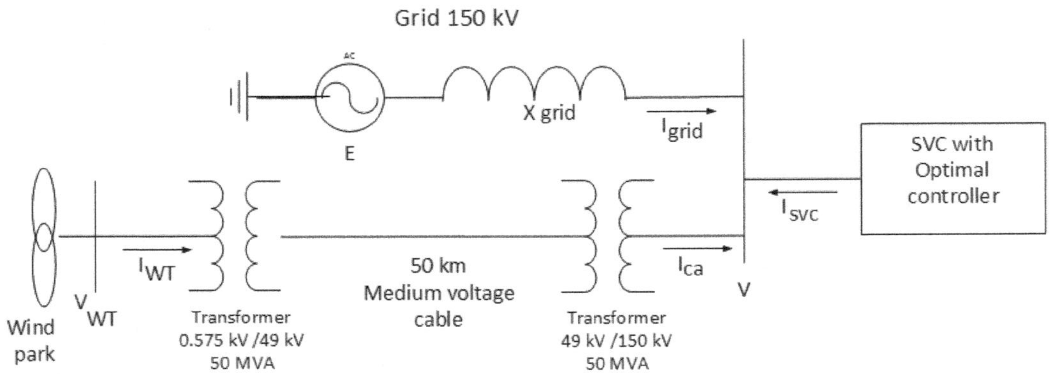

Figure 1. System under study

2. Description and Modeling of the System

In this paragraph, we develop the original nonlinear mathematical equations of the various components of the system and then we linearize them around the operating point. The operating point has been chosen to be the steady state of the system when the WP produces its maximum real power S_0 = 40 MW under rated voltages. For the whole system it has been decided to be S_b = 100 MVA. So in p.u. it is: S_0 = 0.4. The voltage of the bus under study is Vand the voltage of the bus at the output of the wind park is $V_{WT.}$ For the operating point it is $V_0 = V_{0WT} = 1$ in p.u.

For short lines of length 100 km or less, the total shunt admittance is very small and can be neglected. So the equivalent electric circuit of the line is the circuit given in Figure 2a. The relationship between sending-end voltage (V_s) and receiving-end voltage (V_r) can be written as:

$$\begin{bmatrix} V_S \\ I_S \end{bmatrix} = \begin{bmatrix} 1 & Z \\ 0 & 1 \end{bmatrix} \begin{bmatrix} V_r \\ I_r \end{bmatrix},$$

where I_s, I_r are the sending-end and receiving-end currents respectively.

For, lines more than 100 km long, or cables, charging currents due to the shunt capacitance can not be neglected. The capacitance is uniformly distributed throughout the length of the cable, but for cables not too long, the nominal – πrepresentation can be used. In this method, the total line capacitance Y is divided into two equal parts which are lumped at the sending and receiving ends of the cable. The equivalent circuit is given in Figure 2b. From Figure 2b, it is:

$$I_S = I_r + \frac{1}{2}V_rY + \frac{1}{2}V_SY,\ V_S = V_r + (I_r + \frac{1}{2}V_rY)Z.$$

Finally, we have in matrix form:

$$\begin{bmatrix} V_S \\ I_S \end{bmatrix} = \begin{bmatrix} (1+\frac{1}{2}YZ) & Z \\ Y(1+\frac{1}{4}YZ) & (1+\frac{1}{2}YZ) \end{bmatrix} \begin{bmatrix} V_r \\ I_r \end{bmatrix} = \begin{bmatrix} A & B \\ C & D \end{bmatrix} \begin{bmatrix} V_r \\ I_r \end{bmatrix}.$$

In this form, the constants A,B,C,D of the cable have been included.

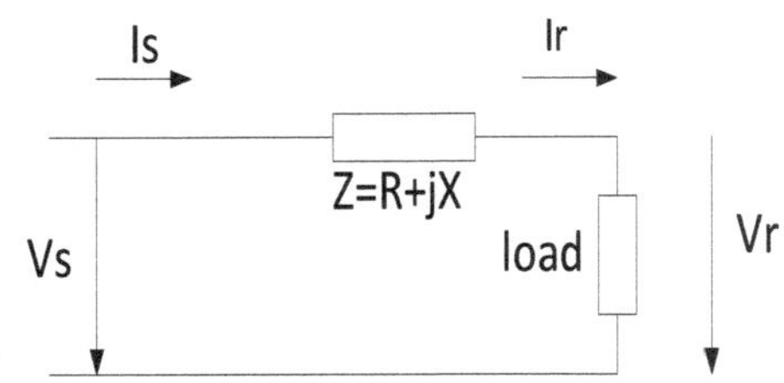

Figure 2a. Equivalent circuit of a short line

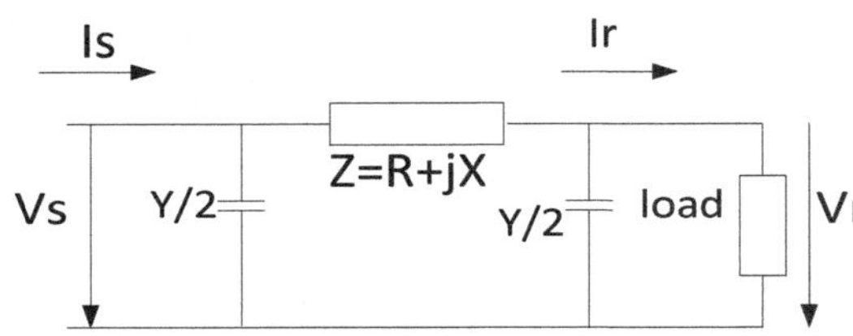

Figure 2b. Cable representation

The cable carries a current I_{ca} (= I_r) towards its connection to the grid bus and accepts a current I_{WT} (= I_s) from the wind park. From the values of the Table A2 in the Appendix, the model of the cable in the s-plane it is:

$$Y = C*s = 0.00002s, Z = 4.55 + 0.055s = B,$$

$$A = D = 1 + 0.5*Y*Z = 1 + 0.000022s + 0.0000055s^2,$$
$$C = Y(1 + 0.25*Y*Z)$$
$$= 0.00002s + 4.55*10^{-10}s + 5.5*10^{-12}s \approx 0.00002s$$

$$I_{WT} = C*V + A*I_{ca} = 0.00002s*V + \begin{pmatrix} 1+0.000045s \\ +0.0000055s^2 \end{pmatrix} * I_{ca}$$

$$\Rightarrow I_{ca} = \frac{181818}{s^2+8.18s+181818}I_{WT} - \frac{3.64s}{s^2+8.18s+181818}V$$

For V_b = 49 kV it is I_b = 1178 A. So the previous equation in pu is written in linearized form as:

$$\Delta I_{ca} = \frac{154.35}{s^2+8.18s+181818}\Delta I_{WT} - \frac{0.003s}{s^2+8.18s+181818}\Delta V \quad (1)$$

The grid is simulated through a voltage source E and an inductor equal to X_{grid} = 0.18 H. This value corresponds to a short-circuit capacity of 400 MVA which is ten times the rated power of the wind park. We suppose a 'weak' node of the system because in this occasion, the voltage is more sensitive to the alterations of I_{ca}. The internal voltage E is not a parameter in the linearized model, and if we notice that I_b = 385 A in 150 kV it finally is in pu:

$$I_{grid} = \frac{E-V}{sL} = \frac{E}{0.095s} - \frac{V}{0.095s}$$
$$\Rightarrow \Delta I_{grid} = -\frac{0.014}{s}\Delta V \qquad (2)$$

The bus voltage regulation at a bus conventionally is achieved through static shunt reactive power compensationwhich means the installation of a combination of capacitors at the bus. Modern SVC schemes use power electronics technology and so fast and smooth adjustment of the voltage is achieved. They may use either thyristor controlled reactors (TCRs) or thyristor switch capacitors (TSCs). In any case, the SVC can be viewed as a reactive power source from the grid side. With the right pulsation procedure, the equivalent susceptance B_{SVC} of the SVC is adjusted in order the right amount of reactive power to be interchanged between the SVC and the grid for the successful regulation of the bus voltage. So, it isgenerally:

$$Q_{SVC} = \frac{V^2}{X_{SVC}}$$

and in steady state operation it will be in p.u.:

$$Q_{SVC0} = \frac{V_0^2}{X_{SVC0}} = \frac{1}{X_{SVC0}} = B_{SVC0}.$$

The steady state power flow model of the cable is:

$$P_{WT} = \left|\frac{D}{B}\right|\left|V_{WT}\right|^2\cos(b-a) - \frac{\left|V_{WT}\right|\left|V\right|}{\left|B\right|}\cos(b+\delta),$$

where δ is the angle between the voltages V_{WT} and V, P_{WT} is the real power of the WP and the parameters $A = |A|\sphericalangle a, B = |B|\sphericalangle b$ are the A,B,C,D parameters from the equivalent Πcircuit of the cable. From the rated parameters of the cable it is:

$$A = D = 1 + \frac{1}{2}YZ = 0.923\sphericalangle 1.044^0 \ p.u.,$$

$$B = 0.89\sphericalangle 77.7^0, C = Y(1+\frac{1}{4}YZ) = 0.17\sphericalangle 90.5^0.$$

For maximum real power production from the WP (P = 40 MW or 0.4 p.u.) and rated voltages (all equal to 1 p.u.) we have from the above equation that δ_0 = 20.5^0. The steady state reactive power Q_{ca0} from the cable into the grid is given by:

$$Q_{ca} = -\left|\frac{D}{B}\right|\left|V\right|^2\sin(b-a) + \frac{\left|V_{WT}\right|\left|V\right|}{\left|B\right|}\sin(b-\delta)$$

and by substitution Q_{ca0} = -0.066 p.u., which means that the cable absorbs reactive power from the grid.

The real power coming from the WP inserts into the grid, so from the equations

$$P = \frac{EV}{X}\sin\delta, Q = \frac{V(E\cos\delta - V)}{X}$$

we can compute that the reactive power from the grid is $Q_{grid\,0} = 0.02\ p.u.$ So, in steady state operation, the equation of the reactive power flows at the bus gives:

$$Q_{grid0} + Q_{ca0} + Q_{SVC0} = 0$$
$$\Rightarrow Q_{SVC0} = 0.046 p.u. = B_{SVC0}.$$

The current from the SVC generally is equal to $I_{SVC} = V * B_{SVC}$, ([7]), so the linearization gives: $\Delta I_{SVC} = B_{SVC0} * \Delta V + V_0 * \Delta B_{SVC}$. In order a simple delay in the pulsation of the SVC to be taken into consideration, the previous equation will become:

$$\Delta I_{SVC} = B_{SVC0} * \Delta V + V_0 * \frac{1}{1+T_d s} * \Delta B_{SVC}.$$

The time delay T_d is taken equal to 3 ms, so the linearized form of the alteration of the SVC current, finally becomes

$$\Delta I_{SVC} = 0.046 * \Delta V + \frac{1}{1+0.003s} * \Delta B_{SVC}. \qquad (3)$$

Applying Kirchhoff's law regarding the bus currents we have:

$$\Delta I_{ca} + \Delta I_{grid} + \Delta I_{SVC} = 0.$$

Substituting from (1), (2), (3) and after some algebra, we finally result in the following basic equation that gives the linearized model of the bus voltage:

$$\Delta V = -0.32\frac{s^2+22227s}{s^2+329s-100}\Delta B_{SVC}$$
$$-\frac{3355.43s}{s^3+7.82s^2+181782s-55326}\Delta I_{WT} \qquad (4)$$
$$= h_1(s) * \Delta B_{SVC} + h_2(s) * \Delta I_{WT}.$$

This equation describes the dynamic behavior of the bus voltage (output in our model) according to the susceptance of the SVC (input) and the current from the WP (disturbance). Our control system will be designed according to this model. Besides, the reactive power from the SVC will be:

$$Q_{SVC} = B_{SVC}V^2 \Rightarrow \Delta Q = 2B_{SVC0}V_0\Delta V + V_0^2\Delta B_{SVC}$$
$$\Rightarrow \Delta Q = 0.092\Delta V + \Delta B_{SVC}. \qquad (5)$$

3. Optimal Control Design

3.1. SimpleNormalized Model of the Systemand Controller Design

In order a fast response to be achieved, the previous linear model of the system is adjusted so that it contains the normalized values of the corresponding parameters. The normalized value of ΔV is assigned as $\overline{\Delta V}$ and is equal to: $\overline{\Delta V} = \frac{\Delta V}{V_0}$. Similarly the normalized values of the rest of the parameters have been defined as: $\overline{\Delta B_{SVC}} = \frac{\Delta B_{SVC}}{B_{SVC0}}, \overline{\Delta I_{WT}} = \frac{\Delta I_{WT}}{I_{WT0}}$. From (4) and neglecting the third order of s we have:

$$\frac{\Delta V}{V_0} = -0.32\frac{s^2+22227s}{s^2+329s-100}\frac{\Delta B_{SVC}}{B_{SVC0}} - \frac{3355.43s}{s^3+7.82s^2+181782s-55326}\frac{\Delta I_{WT}}{I_{WT0}}$$

$$\overline{\Delta V} = \frac{-0.0145s^2-322.3s}{s^2+329s-100}\overline{\Delta B_{SVC}} - \frac{154.45s}{s^2+23246s-7075}\overline{\Delta I_{WT}} = h'_1(s)*\overline{\Delta B_{SVC}} + h'_2(s)*\overline{\Delta I_{WT}}. \quad (6)$$

The normalized equation of the SVC reactive power is derived from (5) as:

$$\frac{\Delta Q_{SVC}}{Q_{SVC0}} = 0.092\frac{\Delta V}{V_0} + \frac{\Delta B_{SVC}}{B_{SVC0}} \quad (7)$$

$$\Rightarrow \overline{\Delta Q_{SVC}} = 2\overline{\Delta V} + \overline{\Delta B_{SVC}}.$$

The first simple design procedure is based on the above model and only on the input-output transfer function h'₁ (s), so the disturbance parameter has not been taken into consideration.

In state space equations, the transfer function h_1' is transformed into the matrices:

$$A = \begin{bmatrix} -329 & 100 \\ 1 & 0 \end{bmatrix}, B = \begin{bmatrix} 1 \\ 0 \end{bmatrix}, \quad (8)$$

$$C = [-317.53 \ -1.45], D = -0.0145.$$

The main element used in the optimization criterion is the reactive power ΔQ_{SVC} from the SVC and so the optimization criterion is defined as:

$$I = \int_0^\infty [a\,\overline{\Delta Q_{SVC}}^2(t) + \overline{\Delta B_{SVC}}^2(t)]dt = \int_0^\infty [a\,\overline{\Delta Q_{SVC}}^2(t) + u^2(t)]dt$$

since the input u to the control system is the susceptance of the SVC. From (6) and (7) it is:

$$I = \int_0^\infty \begin{bmatrix} 4a\,\overline{\Delta V}^2(t) + 4a\,\overline{\Delta V}(t)\overline{\Delta B}(t) \\ +(a+1)\,\overline{\Delta B_{SVC}}^2(t) \end{bmatrix} dt. \quad (9)$$

From the (8) it actually is: $\overline{\Delta V} = -317.53X_1(t) - 1.45X_2(t)$, where $x_1(t)$, $x_2(t)$ are the state variables of the system. By substitution into (9), we have:

$$I = \int_0^\infty \begin{bmatrix} 403300aX_1^2(t) + 3684aX_1(t)X_2(t) \\ +8.4aX_2^2(t) + 2(-635aX_1(t) \\ -2.9aX_2(t)) + (a+1)u^2(t) \end{bmatrix} dt$$

which is of the form

$$I = \int_0^\infty [X^T(t)Q_1X(t) + 2X^T(t)Q_{12}u(t) + u^T(t)Q_2u(t)]dt$$

([8,9]). Apparently, the weighting matrices are:

$$Q_1 = \begin{bmatrix} 403300a & 1842a \\ 1842a & 8.4a \end{bmatrix}, Q_{12} = \begin{bmatrix} -635a \\ -2.9a \end{bmatrix}, Q_2 = a+1.$$

The problem is solved through solving the relevant Riccati equation:

$$A^TS + SA - (SB + Q_{12})Q_2^{-1}(B^TS + Q_{12}^T) + Q_1.$$

The state feedback L is then:

$$L = Q_2^{-1}(B^TS + Q_{12}^T).$$

The L has been calculated through MATLAB software for various values of the parameter a. The control law is u = -Lx. Then an equivalent series controller has been calculated in order the final control system is of the form depicted in Fig.3. For instance, for a=1, the equivalent series controller which was derived from the optimal state feedback law, is:

$$G_C(s) = \frac{0.95s^2+312.3s+201.4}{s^2+324.5s-2*10^{-16}} \cong \frac{0.95s^2+312.3s+201.4}{s^2+324.5s}.$$

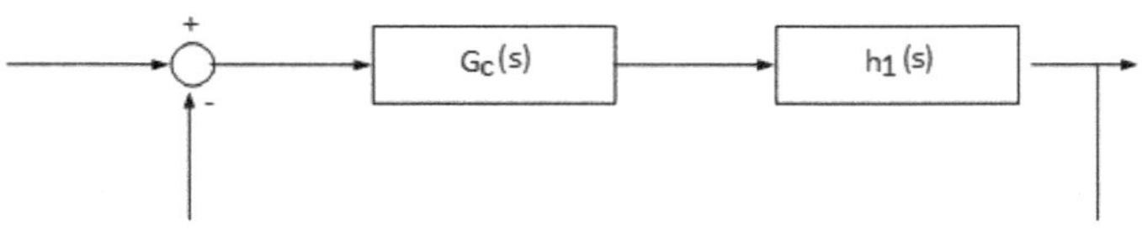

Figure 3. Simple control loop for the design of the SVC controller

We can use this controller, but if we neglect the second order of s, then we conclude to a simple PI controller with parameters $K_p = 0.96$ and $K_I = 0.62$ which is very easy to be implemented in a microprocessor. So, we see that through the optimal theory we reached in a simple PI controller for our application. Other values for ‘a’ give not very different values for the parameters of the PI controller, for example, if a=1000 then $K_p = 1.6$ and $K_I = 0.62$. As we will see in the next paragraph, the simulation of the nonlinear system showed that there are no significant differences in the response of the system for various values of the parameter ‘a’.

3.2. Augmented Normalized Model and controller Design

In the previous design technique, the controller decides about the proper susceptance of the SVC having taken into consideration only the difference between the rated voltage and the actual one (transfer function $h_1'(s)$).

In order the dynamics of the presence of the disturbance $\overline{\Delta I_{WT}}$ to be taken into consideration, an augmented system is proposed for the design of the optimal controller. The state space model of the system given by $h'_2(s)$ is:

$$A=\begin{bmatrix}-23246 & 7075\\ 1 & 0\end{bmatrix}, B=\begin{bmatrix}1\\0\end{bmatrix}, C=[-154.45 \quad 0], D=0.$$

The augmented system in state space will be:

$$A=\begin{bmatrix}-329 & 100 & 0 & 0\\ 1 & 0 & 0 & 0\\ 0 & 0 & -23246 & 7075\\ 0 & 0 & 1 & 0\end{bmatrix}, B=\begin{bmatrix}1 & 0\\ 0 & 0\\ 0 & 1\\ 0 & 0\end{bmatrix},$$
$$C=[-317.53 \quad -1.45 \quad -154.45 \quad 0], \qquad (10)$$
$$D=[-0.0145 \quad 0].$$

The state-variables matrix x(t) contains the four state variables of the system $X(t)=[X_1(t)\ X_2(t)\ X_3(t)\ X_4(t)]^T$ and the vector u for the input of the system is $u=[\overline{\Delta B_{SVC}}\ \overline{\Delta I_{WT}}]^T$. The optimization criterion now is the same as before, but the input now will be named as u_1 as only the parameter $\overline{\Delta B_{SVC}}$ is the control signal. So:

$$J=\int_0^{\infty}[a\,\overline{\Delta Q_{SVC}}^2(t)+\overline{\Delta B_{SVC}}^2(t)]dt$$
$$=\int_0^{\infty}[a\,\overline{\Delta Q_{SVC}}^2(t)+u_1^2(t)]dt$$
$$=\int_0^{\infty}[4a\overline{\Delta V}^2(t)+4a\,\overline{\Delta V}(t)u_1(t)+(a+1)u_1^2(t)]dt.$$

From (10) it is:

$$\overline{\Delta V}(t)=-317.53X_1(t)-1.45X_2(t)-154.45X_3(t).$$

By substitution and after some algebra, we finally result in:

$$J=\int_0^{\infty}\begin{bmatrix}a(403300X_1^2(t)+8.4X_2^2(t)+95420X_3^2(t)\\ +3684X_1(t)X_2(t)+1792X_2(t)X_3(t)+\\ +392340X_1(t)X_3(t))+2a(-635X_1(t)\\ -3X_2(t)-309X_3(t))u_1+(a+1)u_1^2(t)\end{bmatrix}dt.$$

The weighting matrices now are:

$$Q_1=a\begin{bmatrix}403300 & 1842 & 196170 & 0\\ 1842 & 8.4 & 896 & 0\\ 196170 & 896 & 95420 & 0\\ 0 & 0 & 0 & 0\end{bmatrix},$$

$$Q_{12}=a\begin{bmatrix}-635 & 0\\ -3 & 0\\ -309 & 0\\ 0 & 0\end{bmatrix}, Q_2=\begin{bmatrix}a+1 & 0\\ 0 & 0\end{bmatrix}.$$

The problem is solved through solving the relevant Riccati equation for the calculation of the state feedback gain L. The L has been calculated through MATLAB software for various values of the parameter 'a'. As before, an equivalent series controller has been calculated in order the final control system is of the form depicted in Fig.3. For instance, for a=1, the equivalent series controller which was derived from the optimal state feedback law, if we neglect high orders of s, is:

$$G_C(s)=\frac{21.96s-3.317}{5.306s-3.4*10^{-18}}\cong\frac{21.96s-3.317}{5.306s}.$$

Apparently, this can be implemented again easily as a PI controller with K_p = -4.14 and K_I =0.62. One can notice that, the design procedures result in controllers of the same form and the K_I parameter is the same in both of the design schemes with normalized values for the model parameters and for the same value of 'a'.

4. Nonlinear System Simulation Results

The system under study as is depicted in Fig. 1 has been simulated in the SIMULINK software. The nonlinear models for the DFIM and SVC have been taken from the software library. The rated values for the machines are listed in the Appendix.

In the first case, we test the response of the system under severe changes of the wind speed which bring about huge alterations of real and reactive power from the cable into the grid. As a result, the voltage of the grid changes. The proposed optimal controllers of the SVC take on to adjust the grid voltage to its nominal value. In Fig. 4 the wind speed is depicted. The wind speed increases rapidly from 8 m/s to 14 m/s at t = 80 sec and then decreases to 8 m/s at t = 150 sec.

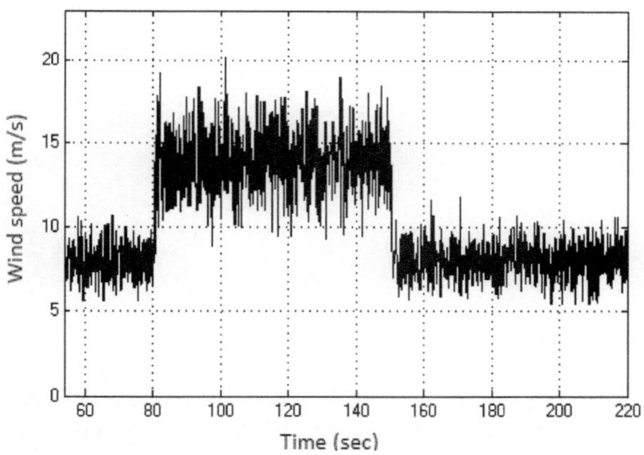

Figure 4. Wind speed waveform.

In Figure 5 the real power coming from the wind park is shown which alters according to the wind speed(p.u. refers to the rated power of the wind park).

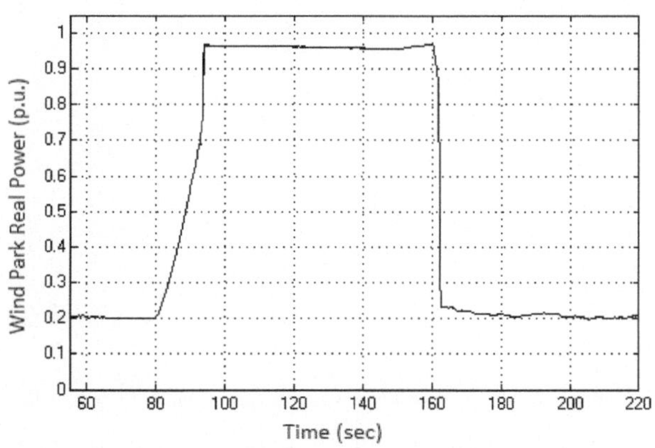

Figure 5. Real Power of the Wind Park

In Figure 6 (a) we see the response of the grid voltage with the optimal controller designed according to the simple normalized model for a =100, while in Figure 6 (b) the response with the optimal controller according to the augmented normalized model.

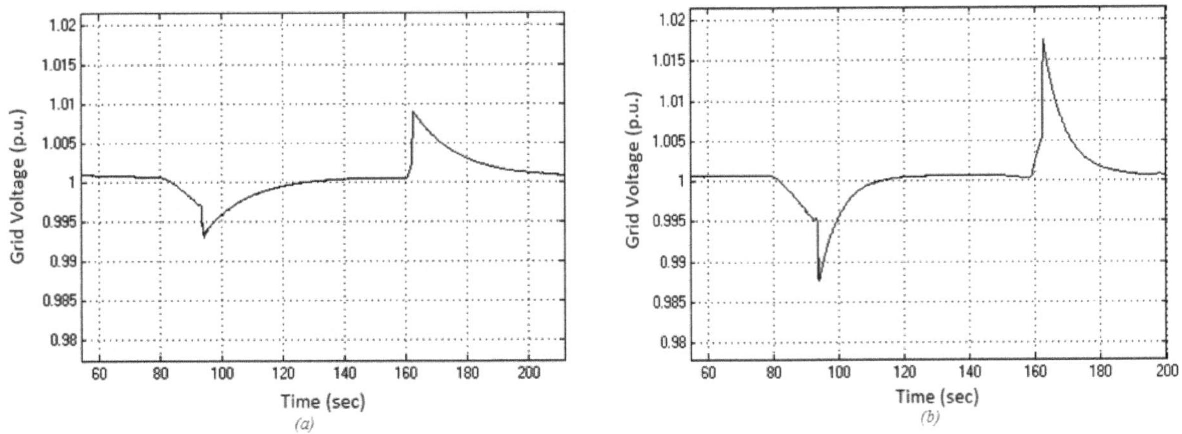

Figure 6. Grid voltage response under the two optimal controllers\

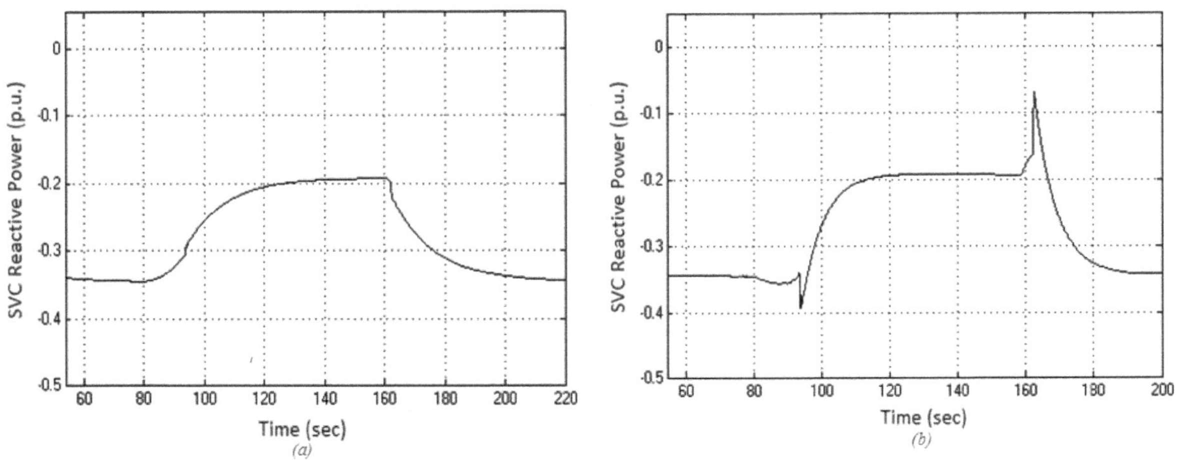

Figure 7. Reactive power of SVC under the two optimal controllers

In Figure 7 (a) we see the change in the reactive power of the SVC due to the first controller and in Figure 7 (b) due to the second controller.

Generally, the response of the system is satisfactory in both situations and with both of the control schemes.

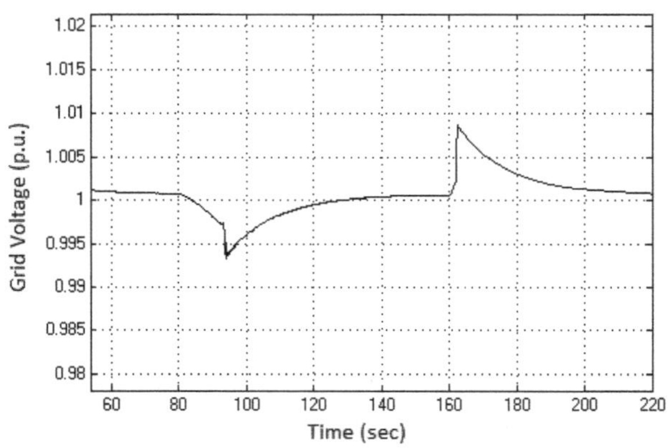

Figure 8. Grid voltage response with the simple model for a = 1000

We can see that the simple normalized model results in a controller that has small overshoots but with slow response. Under a severe wind speed change, the variation of the bus voltage is hardly 1.5%. On the contrary, the augmented normalized method results in a controller with larger overshoots but much faster than the first one (about half the time of the previous controller). The final choice depends on the demands.

Finally, in Figure 8 there is the response of the grid voltage with the simple normalized model for a = 1000.

We can see that there is no significant change in comparison with the Figure 6a. The only subtle difference is that the peaks of the transient periods are smaller in this case.

5.Conclusions

A high voltage grid bus is considered in this paper to which a wind park is connected through a medium voltage cable. The severe changes of the real and reactive power coming from the cable into the grid make the voltage unstable. The installation of an SVC improves the voltage response. The SVC is controlled through two different proposed controllers which have been designed according to the theory of optimal control. The theory results in simple PI controllers which can be implemented easily. The response of the system is satisfactory and the voltage variation due to the changes of the power from the wind park can be confined between +/- 1.5% in a time-limited transient period.

6. Appendix

Table A1. WT PARAMETERS ([10,11])

Low speed inertia J_t	800 kgm^2
Coupling stiffness K_{tg}	12500 Nm/rad
Coupling damping D_{tg}	130 Nms/rad
High speed inertia J_g	90 kgm^2
Nominal active power P_N	1.5 MW
Nominal electrical torque T_{elN}or T_g	9555 Nm
Stator voltage V_{SN}	700 V
Nominal generator speed η_{go}	1500 rpm
Speed range of generator	900-2000 rpm
Pole pairs	2
Blades diameter d	60 m
Nominal wind speed V_{wN}	13 m/sec
Maximum power coefficient C_p	0.44
Air density	1.125 kg/m^3
Nominal turbine speed η_{to}	18 rpm
Speed range of turbine speed	9-18 rpm

Table A2. MEDIUM VOLTAGE CABLE ([12])

Rated voltage V_N	49 kV
Inductive reactance X_o	0.417 Ω/mi
Resistance	0.091 Ω/mi
Capacitive Reactance	6700 Ω*mi
Total Length	50 mi

References

[1] Vu, H., Pruvot, P., Launay, C., and Harmand, Y. (2002), *An improved voltage control on large – scale power system*, IEEE Transactions on Power Systems, 11(3), pp. 1295-1303.

[2] Senjyu, T., Miyazato, Y., Yona, A., and Urasaki, N. (2008), *Optimal distribution voltage control and coordination with distribution generation*, IEEE Transactions on Power Delivery, 23(2), pp:1236-1242.

[3] Novalio, D., Biswarup, D., and Jaydev, S. (2015), *Robust voltage regulation in unbalanced radial distribution system under uncertainty of distribution generation and loads*, International Journal of electrical Power & Energy Systems, 73, pp:516-527.

[4] Dawood, A. (2016), *DSTATCOM modelling for voltage stability with fuzzy logic PI controller*, International Journal of electrical Power & Energy Systems, 76, pp: 129-135.

[5] Elmitwally, A., and Eladl, A. (2016), *Planning of multi-type FACTS devices in restructured power systems with wind generation*, International Journal of electrical Power & Energy Systems, 77, pp:33-42.

[6] Khezri, R., and Bevrani, H. (2015), *Voltage performance enhancement of DGIG-based wind farms integrated in large-scale power systems: Coordinated AVR and PSS*, International Journal of electrical Power & Energy Systems, 73, pp:400-410.

[7] Wang, X.F., Song, Y., and Irving, M.(2008), *Modern Power System Analysis*, Springer, New York.

[8] Astrom, K.J., and Wittenmark, B.(2013),*Computer Controlled Systems*, Dover, New York.

[9] Dorf, R.C., and Bishop, R.H. (2011), *Modern Control Systems*, Prentice-Hall, London.

[10] Vittal, V., and Ayyanar, R. (2013), *Grid Integration and Dynamic Impact of Wind Energy*, Springer, New York.

[11] Abad, G., Lopez, J., Rodriguez, M.A., Marroyo, L., and Iwanski, G. (2011), *Doubly Fed Induction Machine*, IEEE Press, New Jersey.

[12] Gonen, T. (2014), *Electrical Power Transmission System Engineering, Analysis and Design*, CRC Press, New York.

Electrodynamics Settings of the Communicator

V.G. Kudrya*

Odesa National Polytechnic University, Ukraine
*Corresponding author: vcudryagoogl@gmail.com

Abstract Studied electromagnetic properties of connections between component. The analysis of electromagnetic field components parallel radial highway lanes of the printed and volumetric mounting electronic devices. Found conversion algorithms mutual capacitive, inductive, dielectric and semiconductor parameters of the symmetric and regular communication structures.

Keywords: *between component connections, electromagnetic interference, printed tracks, communicator, electrodynamic parameters, MEMS, NEMS*

1. Introduction

The theme of this study were between the component connection of computer and other information and communication tools. The current level of technology design and manufacturing of micro and nanotechnology integrated circuits characterized by the use of a variety of physical, chemical and the biological processes. Synergetic approach the study of phenomena and processes of such devices must, first of all, based on the theory and methods of mathematical physics, including electrodynamics [1-8].

Nanotechnology allows the manufacture of monolithic integrated circuits with dividing resolution of 20 nanometers, thereby dramatically raise the frequency of signal processing and increase the integration of elementary components per unit volume.

At the same time, technology design of integrated structures virtually abode at chain theory of electrical circuits, as evidenced by a number of applications of computer simulation: MicroCap, OrCAD, MultiSim more. Some attempts to correct the adequacy models by requiring minimize the total length of highway lanes, such as Altium Designer, do not solve the problem and shaken methods modeling of electrical circuits that are based solely on the classical Kirchhoff's law. However, in [5,6,7,8] expressed reservations that modern monolithic integrated circuit components and means of the relationship between them have acquired properties that can not be accepted mathematical models that incorporated in the software architecture of the above software packages.

2. The Formulation of the Problem

For the first time on the study of electrical circuits as object electrodynamics [1,2], led by Professor O.V. Tozoni started at the Institute of Cybernetics of AS in Ukraine almost half a century ago. A mathematical model was designed for powerful computing systems - computer type BESM-6. In many cases, individual work brought to the numerical results as of Nikolsky V.V. [3], who studied the chain of high frequency directing devices were only illustrative and had a rather scientific, rather than practical. Lack query, explained the complexity of algorithms, insufficient big at the time integration circuits and weak capacity of computers. A step forward was research of professor A.I. Knyazya in Odessa Electrotechnical Institute of Communications [1,4]. The proposed methods of calculation fields in the quasi-stationary approximation, was based on the analogy of different orthogonal coordinate systems, allowing to simplify the complexity of numerical algorithms. However, a number of unresolved issues, including the problem of space-time distribution of sources of flow vectors induction and selection of idealization the communicator structure of and a number of other poorly formalized tasks are not allowed to bring results to the finished application. To solve the problems set out accounting electromagnetic interference between component connections leading scientists have joined efforts within research areas, which are called micro- and nanoelectronic mechanical systems (MEMS, NEMS) [6]. However, the search for simple algorithmic mapping procedures morphology monolithic integrated circuit design is extremely urgent problem

3. Statement of the Tasks

In generalized models of analysis and synthesis high frequencyof electrical [1-9] circles one of the main problems is the calculation of the flow of the electric and magnetic induction in different environments, conductor, dielectric, magnetic, characterized by the relevant material parameters γ, μ, ε. The source of these streams are currents and voltages trees communicator. Under the communicator is to be understood compositional element of the circuit, that transfers energy from one component to another component. The form and value of the energy in the idealized communicator remains unchanged. In other words, if from the electronic device to remove all are concentrated components, that make information and changing energy, then we get the communicator. First of all - a single-layer or multi-layer boards, for example, Figure 1.

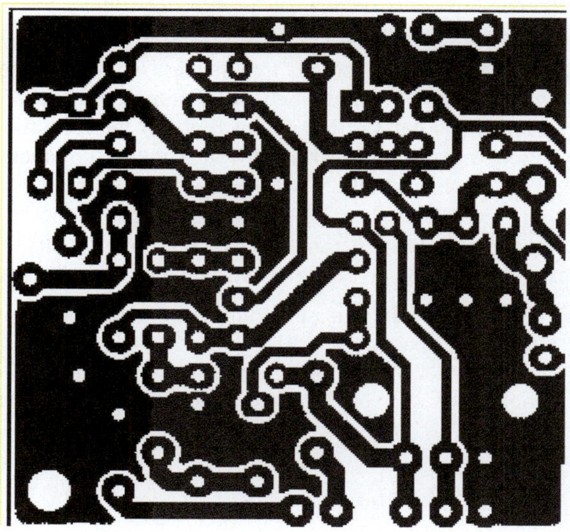

Figure 1. Communicator - a piece of PCB

The source of these streams are currents and voltages trees the communicator. The surfaces through which flows calculated asked morphology device. Depending on the need to identify those or other electromagnetic parameters such surfaces are: lateral surface highway lanes; closed circuit formed highway lanes, or surface tree (node communication highway lanes), as part of the device. Under the trees means a series of parts of the communicator on which it crumbles when removing from the construction the lumped components and their replacement by point equivalents insulated from each other all poles (their geometrical extensions) at this point.

Model the communicator as compositional elements of the electronic device may be formed in a matrix equations that reflect the relationship descriptors (sets of actions and reactions) at its poles. In general, impacts and reactions can be any physical quantities: current, voltage, the flow of the electric and magnetic induction, charges, power and so on. For example, under the influence can be regarded as a set of arc voltage components and reactions - currents poles, or vice versa.

In general, using a variety of natural phenomena and processes in the transfer of energy and information the communicator complexity of calculating matrix elements determined the communicator varying degrees of detail physical, electromagnetic, chemical and biological processes that contribute to its functioning.

Analysis of structures the communicator type electrodynamic restrict the following tasks: preparation system matrix equations that establish the relationship between currents and voltages pole arcs the communicator; formulation of the problem of calculating the parameters of the communicator in the quasi-stationary approximation; determine the potential of the electromagnetic field; calculating the flow of magnetic and electric induction for most common morphological structures the communicator; determine the relationship between the magnetic and electrostatic settings of communicator; performed numerical calculations to confirm the results.

4. The Main Part

Consider a system of matrix equations, which connects the voltage and current curves pole device. Let the electric circuit consists of k components, the communicator has n polesthat combine these components in the electrical circuit. Not narrowing generalization assume that the number of branches of the communicator equals the number of its poles, and the total number of trees - (r + 1). Within quasistationary approach can assume that every tree, like the "topological" node of the circuit has one and the same potential. Thus, the number of independent trees (nodes) will be equal to r, that is one less of the total, and the number - independent paths defined cycloramic number of the graph corresponding electrical circuit

$$q = n - r + 1 - k, \tag{1}$$

4.1. Resistive Communicator Parameters

To form ohmic losses conductors such as contour ξ can be written

$$\mathbf{U}_{\xi}^{Rd} = \mathbf{R}_{\xi,1}^{d} * \mathbf{I}_1 + \mathbf{R}_{\xi,2}^{d} * \mathbf{I}_2 + \ldots + \mathbf{R}_{\xi,n}^{d} * \mathbf{I}_n \tag{2}$$

where ξ - additional voltage ξ - circuit, caused by the voltage drop across all its areas of conductor, marks on uppercase Rd do not have the numbers, and indicate that the corresponding values are wood communicator (d) and belong to the ohmic losses (R); $\alpha = \overline{1,n}$ - number pole trees (number of poles trees and poles components identical); $\mathbf{R}_{\xi,\alpha}^{d}$ - resistance plot communicator "tied" to the current pole $\mathbf{I}_{\alpha}$.

Keep in mind that the current pole is not always equal currents tree branches and resistant individual conductors may differ from equivalent ohmic values of parameters. Really $\mathbf{R}_{\xi,\alpha}^{d}$ - is a common aspect ratio between voltage and current at the poles, rather than any conductor within the device.

Defining communicator, fixing ohmic losses $\mathbf{R}_{\xi,\alpha}^{d}$ are problems of electrodynamics, formulation which will be executed later.

Number of independent equations of type (2) in accordance with the second Kirchhoff's Law defined by the formula (1).The matrix form of writing is:

$$\left[\dot{U}^{R}\right]_{q,1} = \left[R^{d}\right]_{q,n}\left[\dot{I}\right]_{n,1} \tag{3}$$

Model calculation ohmic losses in conductors of communicator represented in the form (3), called current, that is such, a source of additional voltages in which is the current poles components.

4.2. Inductive Communicator Parameters

Consider another type of current models that allows accounting the magnetic flux through the contours of the communicator.

According to Faraday's law of independent circuits are shown in additional voltage due to currents create magnetic flux tree branches of communicator also can be written as (3)

$$\left[\dot{U}^{\Phi}\right]_{q,1} = \left[Z^{d}\right]_{q,n}\left[\dot{I}\right]_{n,1}, \quad (4)$$

where a separate line presented as

$$\mathbf{U}_{\xi}^{d\Phi} = Z_{\xi,1}^{d} * \mathbf{I}_1 + Z_{\xi,2}^{d} * \mathbf{I}_2 + \ldots + Z_{\xi,n}^{d} * \mathbf{I}_n, \quad (5)$$

and every element of the matrix $\left[Z^{d}\right]_{q,n}$, which is at the intersection ξ- row and α- column

$$Z_{\xi,\alpha}^{d} = j\omega L_{\xi,\alpha} \quad (6)$$

Note that the vectors described voltages and currents the poles of trees (4) are in complex form for subsequent formation of a closed system of equations in symbolic analysis method of complex amplitudes. Substituting (6) into (5) and dividing the result by $j\omega$ available

$$\Phi_{\xi}^{d} = L_{\xi,1}^{d} * \mathbf{I}_1 + L_{\xi,2}^{d} * \mathbf{I}_2 + \ldots + L_{\xi,n}^{d} * \mathbf{I}_n, \quad (7)$$

indicating that each ξ- circuit $\left(\xi = \overline{1,q}\right)$ currents that flow from the poles generate magnetic fluxes mediated by. The immediate source of these flows are real leaders the currents trees communicator. Equation (7) in a body can also be present in the matrix form

$$\left[\Phi\right]_{q,1} = \left[L^{d}\right]_{q,n}\left[\dot{I}\right]_{n,1}, \quad (8)$$

allowing you to add comprehensive induced voltage circuits (4) as

$$\left[\dot{U}^{\Phi}\right]_{q,1} = j\omega\left[\Phi\right]_{q,1} = j\omega\left[L^{d}\right]_{q,n}\left[\dot{I}\right]_{n,1}. \quad (9)$$

It should also be borne in mind that it is current only at the point of connection of communicator and component, not the real currents that generate real magnetic fluxes.

The value of induced voltage in the ξ- circuit (ξ = 1…q) depends on the currents of all n of poles components. Thus, the voltage that occurs in ξ-th independent circuit is modeled as a dependent voltage source, the value of which is determined by a linear combination of currents poles (5).

4.3. Conductor Device Settings

The loss of communicator due to substandard insulation properties between the trees we model based on equations that establishes the relationship between the leakage current through the insulation of ξ- tree

$$I_{\xi}^{Gd} = G_{\xi,1}^{d} * U_1 + G_{\xi,2}^{d} * U_2 + \ldots + G_{\xi,n}^{d} * U_n, \quad (10)$$

where I_{ξ}^{Gd} - additional current ξ- tree which caused partial currents $G_{\xi,\alpha}^{d} * U_{\alpha}$ from ξ- tree, $\xi = \overline{1,r}$, under voltage U_{α} arc α of communicator for the remainder of the trees through certain areas of the insulation environment that came under voltages.

Indications on uppercase Gd does not have the numbers, and indicate that the corresponding values are tree communicator (d) and relate to energy losses in insulator (G); $\alpha = \overline{1,n}$ - number of arcs communicator (number of arcs communicator components and the same); $G_{\xi,\alpha}^{d}$ - equivalent the conductivity of volume of communicator which is tied to α- tree arc voltage U_{α}, $\alpha = \overline{1,n}$.

It should again be noted, that the arc voltage is not always equal to the voltage between different branches of trees, and some parts of the conductivity of the insulation space may not coincide with equivalent values conductor parameters $G_{\xi,\alpha}^{d}$.

Defining the parameters of communicator, that fix the energy loss in the insulation environment $G_{\xi,\alpha}^{d}$ of is the task of electrodynamics in spreading current conduction in the conductive space.

Number of independent equations of the first Kirchhoff's law such as (10) is equal to r. The matrix form of writing is:

$$\left[\dot{I}^{G}\right]_{r,1} = \left[G^{d}\right]_{r,n}\left[\dot{U}\right]_{n,1} \quad (11)$$

Model calculation of energy losses in the conductor's communicator represented in the form (11), called voltage's, that is such a source of additional current which has a voltage arcs device.

Capacitive options communicators. Consider a model that allows accounting of the electric flow from trees communicator, characterized in that form displacement currents between the trees, i.e. currents, the nature of which is not related to polarization or charge transport in space.

The current displacement creates a variable flow of the electric induction, like current between the capacitor plates in a vacuum (with a perfect insulator) between them.

Displacement currents of trees communicator in matrix form is written as

$$\left[\dot{I}^{N}\right]_{r,1} = \left[Y^{d}\right]_{r,n}\left[\dot{U}\right]_{n,1}, \quad (12)$$

which is presented as a separate line

$$I_{\xi}^{dN} = Y_{\xi,1}^{d} * U_1 + Y_{\xi,2}^{d} * U_2 + \ldots + Y_{\xi,n}^{d} * U_n, \quad (13)$$

and every element of the matrix $\left[Y^{d}\right]_{r,n}$, which is at the intersection ξ- row and α- column

$$Y_{\xi,\alpha}^{d} = j\omega C_{\xi,\alpha}. \quad (14)$$

Vectors induced currents and voltages of arcs trees (12) are in complex form for subsequent formation of a closed system of equations in symbolic form in accordance with the method of complex amplitudes. Substituting (14) into (13) and dividing the result by $j\omega$ can be obtained

$$N_{\xi}^{d} = C_{\xi,1}^{d} * U_1 + C_{\xi,2}^{d} * U_2 + \ldots + C_{\xi,n}^{d} * U_n, \quad (15)$$

indicating that from each of ξ- tree $\left(\xi = \overline{1,r}\right)$ currents flowing from the poles indirectly generate electrical flow. The immediate source of electrical flow is a real voltage between of the trees communicator. Equation (15) can be represented in matrix form

$$\left[\dot{N}\right]_{r,1} = \left[C^d\right]_{q,n}\left[\dot{U}\right]_{n,1}, \quad (16)$$

that allows us to formulate complex matrix form a plurality of displacement currents of trees (12) in the form

$$\left[\dot{I}^N\right]_{r,1} = j\omega\left[N\right]_{r,1} = j\omega\left[C^d\right]_{r,n}\left[\dot{U}\right]_{n,1}. \quad (17)$$

It should also be borne in mind that U_α this is a voltage only on the surface of energy component verge of contact arc and arc the communicator, not the real voltage that generates electrical flow within the device. The value of the bias current ξ- tree (ξ = 1…r) depends on the voltage of arcs all n components. Thus, the current shift from ξ-independent tree is modeled as dependent energy sources, the value of which can be formed by (13).

Thus, analyzing the results it should be noted, that electrodynamics problem statement, you should first find the real sources of space-time, which create a stream of electrical and magnetic induction. Guided by the values of these indirect sources, ie descriptors model (topological currents of poles and voltage of arcs components), real space-time voltage and current can be determined only by the terms of morphology analysis device. This poorly formalized problem solved only by creating libraries of composite components communicator (of trees and contours).

Second, the need to develop a library of electrodynamic parameters for the standard, the most common form of highway lanes canonical conductor structures. Thirdly, it is necessary to determine how structural disintegration of complex shapes communicator with a view to their possible reproduction of canonical structures based on superposition. These canonical structures should first consider the components of the communicator in a system and parallel or radial conductors.

5. Vector Potential of the System of Parallel Conductors with Currents

The conductors of communicator usually are laid on the a parallel track and on the track with radial structure in the places of their joining to the electrical components. Field of (n + 1) parallel conductor simply calculated, for example, is known to vector potential in the space between parallel conductors, that directed along the axis z, is independent of the coordinate z and has only one non-zero component. Under these conditions both in Cartesian (x, y, z), and in polar coordinates it satisfies the Laplace equation

$$\frac{\partial^2 A(x,y)}{\partial x^2} + \frac{\partial^2 A(x,y)}{\partial y^2} = 0, \quad (18)$$

solution which is a function that in polar coordinates

$$\rho = \sqrt{x^2 + y^2},\ \alpha = arctg\frac{y}{x}. \quad (19)$$

takes the form

$$A(\rho,\alpha) = \frac{\mu\sum_{k=1}^{n+1} I_k \ln\sqrt{\rho_k^2 + \rho^2 - 2\rho\rho_k\cos(\alpha-\alpha_k)}}{-2\pi} \quad (20)$$

where ρ_k, φ_k - polar coordinate axis of k-th conductor, μ - magnetic permeability of environment in which there are conductors, I_k - current of k-th conductor positive direction which coincides with the direction of the vector potential.

By placing observation point on the surface of the m-th wire by laying

$$\rho = \rho_m \pm a_m \qquad \alpha = \alpha_m \quad (21)$$

where m - number a_m - radius m-th conductor, that is, after substituting (21) into (20) we obtain the vector potential of the surface of the m-th wire

$$A_m = \mu\sum_{k=1}^{n+1} I_k P_{mk} \quad (22)$$

where

$$P_{m,k} = \frac{\ln(\rho_k^2 + (\rho_m \pm a_m)^2 - 2(\rho_m \pm a_m)\rho_k\cos(\alpha_m - \alpha_k))}{-4\pi} \quad (23)$$

If the result of calculations by (23) does not depend on what the sign (+ or -) stayed in (21), (23), it allows to justify the assumption coincidence of fields generated strings with fields thin wires. According to the first Kirchhoff's equation total current autonomous system of conductors is zero.

$$I_{n+1} = -\sum_{k=1}^{n} I_k \quad (24)$$

After substituting (24) in (22) can be folded n equations, each of which determines the value of the vector potential at the m-th wire

$$A_m = \mu\sum_{k=1}^{n} I_k(P_{mk} - P_{m,n+1}) \quad (25)$$

The magnetic flux Φ_{mk} through the area between the wires m and q with length l, is:

$$\Phi_{mq} = l\left(A_m - A_q\right) = l\mu\sum_{k=1}^{n} I_k\begin{pmatrix} P_{mk} - P_{m,n+1} \\ -P_{qk} + P_{q,n+1}\end{pmatrix}. \quad (26)$$

Volume flow in wires neglected. As the only independent threads n, then without limiting generalization, organize sequence independent review areas in (26) by changing the index $q = m+1$; $m = 1,2,\ldots n$. Subject to then (26) can be rewritten as:

$$\Phi_{m,m+1} = l\mu\sum_{k=1}^{n} I_k B_{mk} \quad (27)$$

where

$$B_{m,k} = P_{m,k} - P_{m+1,k} - P_{m,n+1} + P_{m+1,n+1} \quad (28)$$

In (27) double indexes flow of magnetic induction through area (m, m + 1) length l can also be replaced by a single index m without reducing the information content and submit (27) in matrix form

$$\begin{bmatrix} \Phi_1 \\ \dots \\ \Phi_m \\ \dots \\ \Phi_n \end{bmatrix} = l\mu \begin{bmatrix} B_{11} & \dots & B_{1k} & \dots & B_{1n} \\ \dots & \dots & \dots & \dots & \dots \\ B_{m1} & \dots & B_{mk} & \dots & B_{mn} \\ \dots & \dots & \dots & \dots & \dots \\ B_{n1} & \dots & B_{nk} & \dots & B_{nn} \end{bmatrix} \begin{bmatrix} I_1 \\ \dots \\ I_k \\ \dots \\ I_n \end{bmatrix} \quad (29)$$

The form (29) corresponds to the matrix equation (8), in accordance with which the matrix of magnetic parameters of the parallel wires can be represented as

$$[L]_{n,n} = l\mu [B]_{n,n}, \quad (30)$$

where

$$[B]_{n,n} = \begin{bmatrix} B_{11} & \dots & B_{1k} & \dots & B_{1n} \\ \dots & \dots & \dots & \dots & \dots \\ B_{m1} & \dots & B_{mk} & \dots & B_{mn} \\ \dots & \dots & \dots & \dots & \dots \\ B_{n1} & \dots & B_{nk} & \dots & B_{nn} \end{bmatrix}, \quad (31)$$

Algorithm for computing, for example, Lmk each element of this matrix is calculated based on the geometric characteristics of the sequence using formulas (21) and (23), (28) in the cyclic procedures. The number of cycles in each procedure is determined by the number of conductors.

Field current round conductors using conformal mappings can be counted in the field almost parallel conductors of arbitrary cross section, and therefore the results can be generalized to more complex forms of intersections of parallel conductors.

When placing the conductors in the same plane ($y_m = y_k = 0$), or on the surface of a circular cylinder ($\rho = R$) Pmk coefficients calculated by the simplified equations, respectively:

$$P_{mk} = \frac{-1}{2\pi} \ln | x_m - x_k + a_m | \quad (32)$$

$$P_{mk} = \frac{\ln[R^2 + (R + a_m)^2 - 2R(R + a_m)\cos(\alpha_{mk})]}{-4\pi} \quad (33)$$

6. Vector Potential System of Radial Conductors

To determine the flows in the system wires crossed appropriate to apply the theory of spherical parallel fields. In spherical coordinates (r, θ, α)) coordinate surfaces are: r = C - concentric spheres, θ = C - circular cones, α= C - half-plane. Instead θ, we will consider the variable β:

$$\beta = \ln \gamma = \lg tg \frac{\theta}{2} = \lg \frac{\sqrt{x^2 + y^2}}{r + z}, \quad (34)$$

which together with the coordinates

$$\alpha = arctg \frac{y}{x}, \qquad r = \sqrt{x^2 + y^2 + z^2} \quad (35)$$

specifies the location of points in space.

Spherical parallel magnetic field is called the field vector potential which is independent of the coordinates r:

$$\bar{A}_r = A_r \bar{r}^0 = A\bar{r}^0, \qquad \frac{\partial A_r}{\partial r} = \frac{\partial A}{\partial r} 0, \quad (36)$$

An important property of spherical parallel field is that imposed coordinates β, α (34), (35) vector potential (36), two-coordinate satisfies the Laplace equation similar to (18):

$$\frac{\partial^2 A(\alpha, \beta)}{\partial \alpha^2} + \frac{\partial^2 A(\alpha, \beta)}{\partial \beta^2} = 0. \quad (37)$$

This means that the field system of parallel filaments with the currents corresponding to spherical field parallel thread-rays system of with current. Potential $A(\gamma, \alpha)$ fields filaments-rays system of with the currents invention similar to (20):

$$A(\gamma, \alpha) = = \frac{-\mu}{2\pi} \sum_{k=1}^{n+1} I_k \ln \sqrt{\gamma_k^2 + \gamma^2 - 2\gamma\gamma_k \cos(\alpha - \alpha_k)}. \quad (38)$$

With the potential of spherical- parallel magnetic field can find flows without conducting operations integration. We illustrate this for the flow of magnetic induction. Consider a filled circular sector 1234, Figure 2.

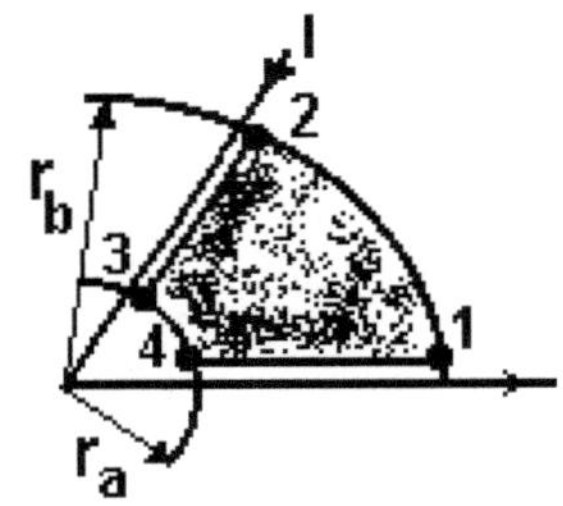

Figure 2. Definition of magnetic flux ring sector

The flow through the sector is

$$\int_S \bar{B} d\bar{S} = \oint_l \bar{A} d\bar{l} = \int_{23} A dl + \int_{41} A dl = = -\int_{r_a}^{r_b} A_2 dr + \int_{r_a}^{r_b} A_1 dr = (r_b - r_a)(A_1 - A_2) \quad (39)$$

When taking the integral (39) took into account equality

$$\int_{12} \bar{A} d\bar{l} = 0, \qquad \int_{34} \bar{A} d\bar{l} = 0 \quad (40)$$

performed because of orthogonality radial vector (36) and by integrating passing through the area. At intervals - rays 23 and 14 values of the vector potential A is a constant, so he carried under the integral sign. Thus, the flux of magnetic induction, which accounts for unit length in the direction of the radius $\bar{r}$ is equal to the potential difference in the points through which the rays that create the boundaries of the annular sector.

In determining the flow field plane parallel threads used by potential field and limit the flow calculation was selected at a short distance from the a thread (21). This distance equal to the radius of the wire. The same procedure for the sphere parallel field consists in choosing the boundary beam - limits conductive sector, which is located near the beam-a thread with the current. To determine the potential value of A on the surface of the conical conductor Figure 3a, it is necessary to put in (38)

$$\gamma = \gamma_1 \pm \gamma_0, \, a = a_0, \tag{41}$$

where

$$\gamma_1 = tg\frac{\theta_1}{2}, \qquad \gamma_0 = tg\frac{\theta_0}{2} \tag{42}$$

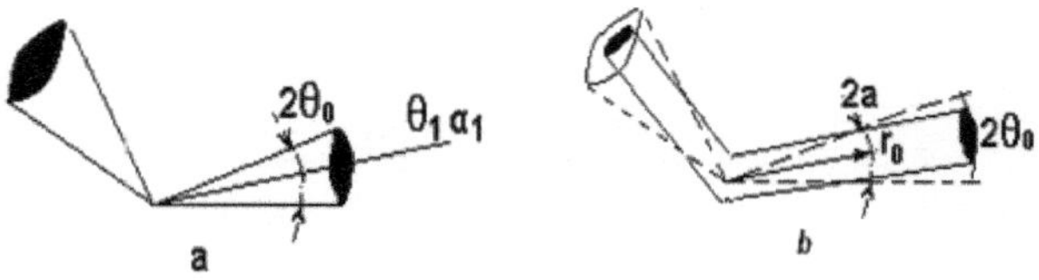

Figure 3. Replacement conical – a cylindrical – b conductor wire

To practice case where the wire has two sharp cones, Figure 3a, less interesting, than the design of wires as thin curved cylinder, Figure 3b. The flow between the cone surface of the conductor, Figure 3a, there will be little different from the stream you want to find, for curved cylinder, Figure 3b, at small angles θ_0, when executed ratio

$$\theta_0 \approx \frac{a(r_0)}{r_0}, \qquad \gamma_0 = tg\frac{\theta_0}{2} \approx \frac{\theta_0}{2} \cong \frac{a}{2r_0} = \frac{a}{l}. \tag{43}$$

where, for construction, Figure 3b, a cylindrical wire length
$l = 2r_0$, $a(r_0)$ - radius conical wire at a distance r_0. Choice of size r_0 at a given radius of a cylindrical wire "a" is related to issues such replacement cylinder cone to flow in finding the place would not have a significant error. Under these conditions, the coordinates of the observation point on the surface of any m-th cone m = 1 ... n according to (41) can set as

$$\gamma = \gamma_m \pm \gamma_{0m}, \, \alpha = \alpha_m, \tag{44}$$

Thus, the calculation of magnetic flux in the system of cylindrical radial wires with a common beginning can also be calculated by formulas (27) and (28), provided that the coefficients (23) are calculated by the formula

$$P_{mk} = \\ = \frac{\ln[\gamma_k^2 + (\gamma_m + \gamma_{0m})^2 - 2(\gamma_m + \gamma_{0m})\gamma_k \cos(\alpha_m - \alpha_k)]}{-4\pi} \tag{45}$$

It should also be recalled that the results should not depend on the choice of sign in (44), in other words the length of conductors must exceed in order of their radius.

The practical interest is the location of all rays incident on the plane z = 0. In this case $\theta_k = \pi/2$ and

$$\gamma_k = \gamma_m = tg\frac{\pi}{4} = 1, \tag{46}$$

which simplifies the expression (45):

$$P_{mk} = \frac{-\ln[1 + (1+\gamma_{0m})^2 - 2(1+\gamma_{0m})\cos(\alpha_m - \alpha_k)]}{4\pi} \tag{47}$$

7. Electrostatic Settings of the Communicator

Electrostatic field system of parallel and radial conductors as described by scalar potential $\varphi(\rho, \alpha)$ and $\varphi(\gamma, \alpha)$. Analytical expressions for them to within a constant factor coincide with (20) and (38) respectively. As algorithmic of procedure consistent with the above-mentioned, give only the final results.

Streams vector electrostatic induction between n + 1 parallel or radial wires characterized by autonomy determined through electrostatic voltage between the wires as

$$[N]_{n,1} = [C]_{n,n}[U]_{n,1}, \tag{48}$$

where the matrix of proportionality

$$[C]_{n,n} = \varepsilon l[B]^{-1}_{n,n} \tag{49}$$

between the flow of the electric induction and arc voltage is the matrix of electrostatic the communicator settings and ε - permittivity of environment. The elements of the matrix [B], as such that are formed geometry device also calculated by formulas (28), (23) and (45).

Analyzing the electrostatic matrix (49) and magnetic (30) parameters should be noted the close relationship between their computation algorithms. Doing multiple reciprocal matrix of can get

$$[L]_{n,n}[C]_{n,n} = l^2\varepsilon\mu[E]_{n,n} \tag{50}$$

where [E] - identity matrix. Equation (50) allows to obtain a matrix of electrostatic parameters through a matrix of magnetic parameters and vice versa.

$$[L]_{n,n} = l^2\varepsilon\mu[C]^{-1}_{n,n}; \qquad [C]_{n,n} = l^2\varepsilon\mu[L]^{-1}_{n,n} \tag{51}$$

Given the analogy of electrostatic fields and field conduction currents between the wires can be calculated in the same way conductor matrix of parameters (10), (11).

To do this in the formula for the scalar potential of electrostatic task replace the value ε in conductivity environment σ, or use the known results for the conversion formula [[10], (3.27)], which is easy to be represented as a matrix equality

$$[G]_{n,n} = \frac{\sigma l^2}{\varepsilon}[C]_{n,n} \tag{52}$$

where σ - conductivity conductor. As for the calculation resisting parameters of, (2), (3), their calculation can be performed similarly to the calculation parameters of conductor (52), which is equivalent to solving problems of electrodynamics in spreading current of conduction environment. But the easiest way to do it knowing the resistivity of the conductive tracks their square, autopsy and length using the formula

$$R = \frac{l}{\sigma S}$$

where S - cross-sectional area of the conductor.

Examples calculation of the parameters communicator. The algorithms were tested in the software environment MathCad and compared to the simplest forms of communicator with other algorithms [11,12]. Give the results of solving specific problems for the elements of communicator, Figure 1.

Variant 1. Given a 2 wire one millimeter in diameter, length and distance between the axes - 10mm. The results of calculating C = 0.1158 nF, and in [11] - 0.1206 nF. The difference is 2%.

Variant 2: Two wires with a diameter of 1 mm and a length of 1m located at a distance of 10 mm. Inductance L = 2,031 was uH. Moreover 100% match results are in three algorithms, proposed in this article (30), the algorithm in [[2] p. 139 (3.28)], and by recalculation the result of some other communicator's parameters (51) according to [10].

8. Conclusion

In this paper made a systematic study between componental connections (communicator) for high-frequency, including nanotechnology, electronics. Synergetic effect of creating modern monolithic integrated schemes with the dividing ability to 20nm requires the construction of mathematical models that take into account the interaction of all composite components that form their characteristics. Such models must take into account not only the classic circuit engineering as well design and, in particular electromagnetic interaction and compatibility between componental connections.

Components and communicator have different descriptor. Concentrated components described by time descriptors, and for the communicator - space-time. The relationship between them is represented as a linear combination of (2), (7), (10), (15). They allow to formulate the task of electrodynamics to determine the influence electromagnetic field of communicator. Such influence is modeled in the form of magnetic, electrostatic, resisting and conductor parameters between which, as shown in the numerical examples, there is a close relationship (51), (52).

Determining the flow of magnetic and electric induction through certain fragments surfaces make possible algorithms for calculating matrix elements electrodynamic parameters (30), (49), (52) for composite components of communicator, presented in a system of parallel and radial wires (highway tracks) that come together to focused component.

Thus, our study allows to raise adequate modeling of electronic products.

References

[1] Tozoni O.V., Knyaz A.I. Electrodynamic iteration in the design of computers. - "E-Mod". Kiev: Naukova Dumka, 1979, №2, p. 57 - 63.(Rus).

[2] Tozoni O.V. Method of secondary sources in electrical engineering. M .: Energia, 1975, 296 p. (Rus).

[3] Nikolskiy V. and others. Computer-aided design UHF devices. M .: Radio and Communications, 1982. 272 p.(Rus).

[4] Knyaz A.I. Electrodynamics of information systems. - M .: Radio and Communications, 1994. - 392 p. (Rus)

[5] Butkovsky A.G. Characteristics of systems with distributed parameters (handbook). - M .: Nauka, 1979, 224 p.(Rus).

[6] Nazmul Islam. Microelectromechanical Systems and Devices. [Text] [Electron resource] / Dr.Nazmul Islam (Ed.), InTech, 2012, ISBN: 978-953-51-0306-6. Chapter 8: Petrenko A.I. "Macromodels of Micro-Electro-Mechanical Systems (MEMS) / (Eng).

[7] Kudrya V.G. Modeling nanotechnology electronic means. Monograph [Text, Electronic resource] // Odessa National Academy of Food Technologies - Kherson. Oldie-Plus, 2013. 780 p.

[8] Kudrya V.G. Simulation of high-speed electronic devices. - Proceedings of the National Technical University of Ukraine "KPI" Series - Radio engineering. Radioaparatobuduvannya. 2013.-№54. P. 151-159.

[9] Kudrya V.G. Simulation of electromagnetic interference communicator of electronic equipment frequency range 0.03 ... 3 GHz. / In.: Proceedings of the Third International Scientific and Practical Conference "Modern information and electronic technologies" - Odessa: SIET-2002, 2002. P. 52 (Ukr).

[10] Goldshtein L.D., Zernov N.V. Electromagnetic fields and waves. M .: Sov. radio. 1971. 664 p.(Rus).

[11] Iossel Y.A., Kochanov E.S., Strunsky M.G. Calculation of Capacitance - 2nd ed. Revised and enlarged. L .: Energoizdat. 1981, 288 p.(Rus).

[12] Kalantarov P.L., Tsejtlin L.A. Calculation of Inductance: Reference book. - 3rd ed. Revised and enlarged. L .: Energoatomizdat. 1986. 488 p.(Rus).

15

Internet of Things (IoT) – A Technological Analysis and Survey on Vision, Concepts, Challenges, Innovation Directions, Technologies, and Applications (*An Upcoming or Future Generation Computer Communication System Technology*)

Gourav Misra[1,*], Vivek Kumar[1], Arun Agarwal[1], Kabita Agarwal[2]

[1]Department of Electronics & Communication Engineering, Institute of Technical Education & Research, Faculty of Engineering & Technology, Siksha 'O' Anusandhan University, Khandagiri Square, Bhubaneswar-751030, Odisha, India
[2]Department of Electronics and Telecommunication Engineering, CV Raman College of Engineering, Bhubaneswar-751054, Odisha, India
*Corresponding author: gourav.misra.ima@gmail.com

Abstract The Internet of Things (IoT) has been inscription in this review paper. Internet of Things is a keyword to cover various challenges related to internet and the web to the real physical world. We know that, today internet has already taken an important part of everyday life and it has also dramatically changed the lives of human being. The most important factor of this invention is, integration or combination of several technologies with the communication system solutions. The most applicable factors of IoT is the identification and tracking various factors for smart objects. The universal sensing networks is enabled by Wireless Sensing Networks (WSN) and these technologies cuts across many areas of modern day living. The escalation of these devices in a communicating and actuating network will create the Internet of Things (IoT). Here the sensors and actuators combine easily with the environment around us and the information is shared across various platforms in order to develop a common operating picture (COP). Internet of Things predicts the future that, the advance digital world and the physical world will get linked by means of proper information and wireless communication system technologies. In this survey paper we have mentioned the visions, concepts, technologies, various challenges, some innovation directions, and various applications of Internet of Things (IoT).

Keywords: *IoT, internet, identifications and tracking, Wireless Sensing Networks (WSN), sensors, actuators, Common Operating Picture (COP)*

1. Introduction

A group of two or more than two computers or mobile systems linked together is called as a network. The Internet is a global system of interconnected computer networks that use the standard Internet protocol suite (TCP/IP) to link several devices worldwide. The Internet is basically carrying a high range of data or information and services and applications to support email, and telephony. Internet of things (IoT), it is nothing but the combination of network of various physical objects with electronics, software and network connectivity, which enables these physical objects to collect and exchange data between various souses to destinations. Nowadays, people use the Internet for browsing the Web, sending and receiving e-mails, accessing various multimedia content and services, playing games, using various types of social networking applications and many other tasks. One of the most important thing is that, the Internet will be existing as a base line of classic networks and networked objects.

In [1], the authors have mentioned that, The Internet of Things (IoT) is a novel design which is rapidly gaining the importance in modern wireless communication systems. The basic or fundamental idea behind this concept is the pervasive presence around us of a variety of things or objects – such as Radio-Frequency Identification (RFID) tags, sensors, actuators, mobile telephones, etc. are able to interact with each other and cooperate with their neighbors to reach common goals. One thing is IoT will have a very high impact on everyday life. From the private user's point of view, one of the most prominent effects of the IoT will be visible in both working and domestic fields. In the United States of America, the National Intelligence Council (NIC) predicts that by 2025 Internet nodes may occupy in everyday things. Many developed countries like United states of America, Europe and some countries of Asia like India, china and japan, they are now a days considering internet of things (IoT) as an area of innovation and growth. So to develop this system, an extensive research is going on in various universities and many research organization across the globe.

In [2], it is mentioned that, the revolution will be empowered by the embedding of electronics into everyday physical objects, making them ''smart'' and letting them seamlessly integrate within the physical world and this will be giving rise to new opportunities for the Information and Communication Technologies (ICT) sector, paving the way to new services and applications able to leverage the interconnection of physical and virtual realms. In [3,5], it is mentioned that the word or expression "Internet-of-Things'' (IoT) is broadly used to refer the resulting global network interconnecting smart objects by means of advanced Internet technologies. Devices like Radio Frequency Identification (RFID), sensors, actuators, machines etc. are necessary to support the applications related to IoT and also the whole applications, services to provide such kind of technologies to open new business and market opportunities. In this review paper we are providing a brief interpretation or standpoint on the development and fundamental concept of Internet of Things (IoT), including various kind of research challenges, technologies etc. The IoT is facing many problems like various network challenges and also the "things" formulating the IoT will be specified by low resources in terms of both computations and energy capacities [3]. Several industrial and research bodies are currently working in the activity of research and development of solutions to satisfied the pinnacled technological requirements. This review paper also gives a picture or brief idea of the current state of the creative activity on the Internet of Things (IoT). The further development and potential of Internet of Things has been estimated by using some technologies like cloud computing, Big Data, networking, robotics etc. and these concepts are overlapping in some part of the development process. Many big industries like Cisco, Ericsson, they have identified the development or evolution of internet of things (IoT) embedded with smart environment and smart platforms. In the below section some fundamental concepts and visions have been described.

2. Vision and Fundamental Concept Behind IoT

Internet of Things (IoT) is a concept that considers extensive presence in the environment of a variety of smart objects that through wireless and wired connections. The unique addressing schemes are able to interact with each other and cooperate with other smart objects or we can say smart things to create many new applications or services and also to achieve a common goal.

In [3,5,7,8] the authors have mentioned that, the Internet of Thongs leads to the technological development in the field of ICT. From a conceptual point of view, the IoT has three supports, something related to the capability of a smart object to be notable, to communicate and to interact with each other, and designing of networks of various interconnected objects. IoT will be based on the idea of "smart objects" or "things", in which a final touch to the existing entities will be given by them in the internet [9]. Smart objects or "Things" should have some physical symbol and physical features like size, shape etc. It should have a proper communication functionalities like- it should be able to accept the incoming messages and reply to them. The entities of smart objects or "Things" should be associated with one name and one address. The most important thing is, the name should be human-readable description of the object. The address should be machine-readable string and it can be used to communicate to various objects. The smart objects should be able to perform some basic fundamental computations.

The Internet-of-Things can be viewed as a highly dynamic network system, which is consisting of a very large number of smart objects or "Things" producing and consuming information from a system-level view or observation and the capacity to interface with the physical area or zone is achieved through the presence of devices able to sense physical phenomena and translate them into a stream of information or data as well as through the presence of devices able to activate actions having an impact or collision on the physical zone or area. In [45], it has been mentioned that, Internet of Things (IoT) is a new revolution of the Internet. Here in this technology various physical smart objects or things make themselves auto recognizable and intelligence has been obtained by themselves only by enabling context related decisions. They can access information that has been aggregated by other things, or they can be components of complex services.

We can say from a service-level point of view, the main issue associate to how to integrate or unite the functionalities and/ or resources provided by smart objects into services [10,11]. The definition of architectures and methods for ''virtualizing'' objects by creating a standardized or systemized representation of smart objects or "Things" in the digital domain or field and methods for seamlessly combining and composing the resources or services of smart objects or "Things" for end users.

Generally a wide range of opportunities are provided by Internet of Things (IoT) to the manufacturers, companies and various industries. In many productive industries (such as: environmental monitoring, inventory, product management etc.) IoT technology will find much applicability.

IoT will be categorized by a large heterogeneity in terms of devices taking part in the technology system and these are expected to present very different capabilities from the computational and communication viewpoints. In IoT technology scalability level arises in different levels. This includes data communication and networking (or computer networking) - because of very high number of interconnections among a very large number of entities or organizations.

In Internet of Things (IoT), wireless communication technology or systems will play a vital role and they are enabling the smart objects or "Things" to be networked. The universal assumption of wireless communication systems for exchanging data or information will create various issues in terms of availability of frequency spectrum and it will impel towards the assumption of Cognitive radio Systems (smart radio system) [12]. The abilities of wireless communication system can provide and identify the operation of Internet of Things (IoT) and also it will be possible to locate the smart objects or "Things" in the physical area or zone. It is important that the various operations of IoT the energy which will be spent for wireless communication and computing purpose should be very less. The most important thing is techniques or related to energy gathering i.e. the energy should be handled with proper care while dealing with the devices which are used in IoT technology.

The IoT technology is nothing but, it is about the interchanging of information from one device to another device and also analyzing the huge or massive amount of data. So in order to turn them in to very useful information and also to ensure the interoperability between two or more applications, it is most important to provide sufficient data with some specified formats and well defined languages. The security for IoT system or technology should be a key property and also it should be taken in to account while the architectural design.

In Figure 1 (which is mentioned below), the technologies, concepts, and standards of IoT systems are classified and highlighted.

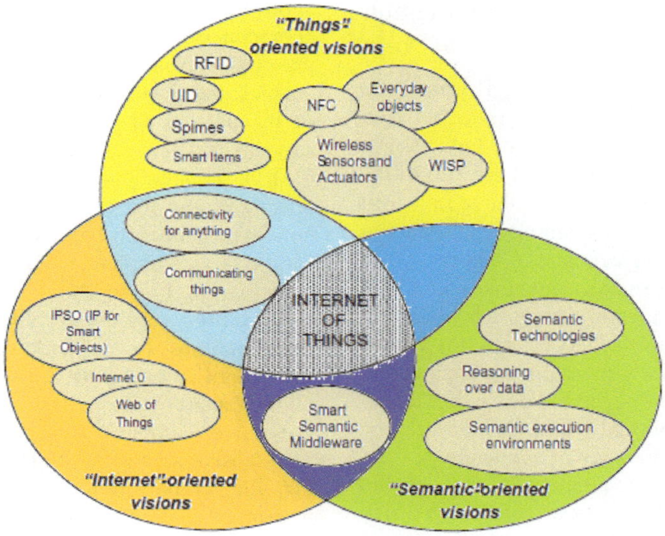

Figure 1. Architectural diagram [3]

In the above diagram the very first definition "Things oriented visions" considers some simple items like Radio Frequency Identifications (RFIDs). In [13], it is mentioned that, the Internet of Things assigns to Auto –ID lab, which is a worldwide network of academic research laboratories in the field of networked RFIDs and some kind of emerging sensing technologies. In this field the prime focus is on the development of Electronic product Code (EPC) to support the spread use of Radio Frequency Identification (RFID) in worldwide trading centers and to improve the object visibility these standards are designed.

In [14], it is mentioned that, in technology field RFID stands in a leading position and also a wide range of portfolio of networks devices and technologies will develop the IoT. In this scenario there are some basic or fundamental components (such as: Near Field Communications known as NFC, Wireless Sensors and Actuator Networks together with RFID) which will link the real world with the modern digital world. The definition of IoT finds some kind of real world implementations is so called smart items or smart objects and also these are the auto sensors not only equipped or furnished with normal wireless communication systems or technologies, memory, and capabilities, but also with new kind of capabilities.

In [17], it is mentioned that, an IoT imagination statement, which goes just out of boundary ''RFID centric" approach and this has been given by the consortium CASAGRAS. The community CASAGRAS consortium proposes a basic imagination of IoT as a global infrastructure or design which makes a connection to both virtual and various physical collective objects and spotlight the importance of including existing or developing Internet and developments of network in this valuable imagination. In this sensation, IoT will become the natural enabling architectural design for the deployment of various integrated services and applications.

There is a statement in [20] is that, a similar approach of reducing the complexity of the IP abundance to achieve a protocol designed to route ''IP over anything" is followed by internet oriented visions and also in some of the meetings this has been mentioned that, the easiest way to move from the Internet of Devices to the Internet of Things. According to Internet oriented vision approach, the IoT will be implemented by means of simplification of Internet Protocol to change it to any object and also make those objects addressable and reachable from any kind of location.

In [21,23], the authors have mentioned that, the basic and fundamental idea behind the "semantic oriented" vision is that the number of items or components in the future internet system will be very high. Some of the important things like how to gather information, interconnection, search and communication by IoT will become a challenging job. So in this case semantic oriented imagination could play a very important role in the Internet of Things systems.

So this all about the vision (or imagination) and fundamental concepts behind Internet of things (IoT).

3. Validating Technologies and Challenges in IoT

In this section we have discussed the most admissible ones. It is not our purpose to provide a survey of each and every technology. Our main goal is to provide a pictorial representation of the role they would play in the IoT. The fundamental research is required to achieve Internet of Things (IoT) and some of the technologies and challenges in the areas like massive scaling, creating knowledge and big data, robustness openness have been mentioned.

3.1. Identification, Sensing and Communication System Technologies

The wireless communication systems as taken an important role today and also today the ratio between radios and human is nearly equal 1 to 1. The reduction in terms of size, weight, energy consumption, and cost of the radio can take us to a new span where the above ratio will increase of orders of magnitude and this will allow us to combine radios in almost all objects and to add the world ''anything" to the above valuable imagination, which leads to the basic and fundamental concepts behind IoT. One thing has been found out that, RFID systems can be used to control various objects in real-time application, without the necessity of line-of-sight and this will allow to map the real world into the virtual world. So, they can be used in an extraordinarily wide range of applications.

In [25], the authors have mentioned that, Radio Frequency Identification (RFID) is small microchip which will be attached to an antenna (Microstip patch antenna) which is just similar to paste or gum and In this case dimensions would be very low (Hitachi has developed the dimensions of 0.4mm × 0.4mm × 0.15mm). Here signal will generate an alternating current into the tag antenna by induction process and the current will be utilized to supply the microchip which will transmit the tag ID. Usually, the gain of such kind of systems are very low. However, we are very thankful to the highly directive antennas utilized by the readers by which tags ID can be received almost correctly within a radio several frequency bands covering from low frequencies at 124– 135 kHz up to ultra-high frequencies at 860 – 960 MHz that will have the longest range.

Sensor networks will be playing an important role in the IoT technology systems. They can also cooperate with RFID systems to better track the status of things or smart objects (such as: their location, temperature, movements, etc.). Sensor networks basically consist of a certain number of special nodes (or junction points) called sinks and also a very large scientific theory has already been produced on sensor networks in the recent past years. The basic design objectives of the proposed solutions are energy efficient, scalability (i.e. the number of nodes can be very high), reliability (i.e. the network may be used to report some urgent events), and robustness. The main advantage of integrating sensing networks with the RFID tag is that, it will enable a lot of new featured applications to the IoT context. For this several kind of solutions has already been proposed. For example we can say WISP (it stands for wireless identification and sensing platform) ad this project is taken by Intel labs.

3.2. Middleware

Middleware is simply a software layer which is placed between the technology and the application. The middleware is gaining now a days so much of importance in the last years due to its major role play in simplifying or rationalizing the development of new services and the combination of upshot technologies into new the new technologies.

The middleware architectures proposed in the last years for the IoT and it frequently follows the Service Oriented Architecture (SOA) approach. The assumption of the SOA principles allows for decomposing complex systems into applications consisting of simpler and well-defined components. The use of common interfaces and standard protocols gives a view of an enterprise system. SOA enables, the development of business, which is the result of the process of designing workflows of synchronized services and these are associated with objects actions. In [28], it is mentioned that, the hardware and software reusing is allowed by the SOA approach because it does not enforce any kind of scientific approach or technology for the service implementation.

Below in Figure 2 we have drawn the SOA-based architecture for IoT middleware systems. System functionalities to the end user and this layer is not considered to be a part of the middleware but it exploits all the functionalities of the middleware layer.

Service Composition layer is the second layer of the SOA-based architectural design. Various functionalities for the composition of single services offered by networked objects to build specific applications is provided by this layer. The only advantages are services and here there is no any kind of idea of devices on this layer. The fundamental concept or logic behind the creation and the management of complex services could be explained in terms of workflows of business processes by using workflow languages and in [31], it is mentioned that, the business processes that interact with external entities through Web Service operations, defined by using the Web Service Definition Language (WSDL).

The main functions that are expected to be available for each object and that allow for their management in the IoT scenario is basically provided by service management layer. So this layer plays a vital role in this architecture.

Applications Layer
Service Composition Layer
Service Management Layer
Object Abstraction Layer
Objects

Figure 2. SOA-based architecture for IoT middleware systems

Application layer is the top of the architecture. The IoT depends on a large set of objects and each of the objects provide some kind of specific functions. So here there is a need of an abstraction layer, which will be capable of coordinating the access to the different devices with a common language and procedure or protocol. It consist of two sub layers known as interface sub layer and communication sub layer. The interface sub layer provides a web interface introducing the methodologies available in a standard web service interface and also it is responsible for the management of all the incoming or out coming of messages involved in the communication with the external world. The Communication sub-layer implements the basic logic behind the web service methodologies or functions. After that it translates these methods or functions into a set of device-specific commands to communicate with the real-world objects. So this all about the validating technologies that are used for IoT technology.

3.3. Massive Scaling

The route of the numbers of smart devices being deployed, it implies that billions and trillions of things will be on the Internet in near future. Since many of the things on the Internet will require their own energy sources, will energy searching and exceptionally low power circuits eliminate the need for batteries? How will the massive amounts of data be used, and stored? What kind of longitudinal studies will be performed? And also how the real-time and reliability aspects will be supported? How will be the mobile devices discovered? It is unlikely that any solution immediately becomes the norm.

3.4. Creating Knowledge and Big Data

In IoT, there will be existing a vast or big amount of raw or fundamental data are continuously collected. Generally it is very necessary to develop any techniques which will convert this raw or fundamental data into usable knowledge. For example, in medical science field, raw streams or data of sensor values should be converted into some meaningful activities performed by a person such as eating, exhibiting signs of depression etc. The main challenges for data interpretation and the formation of knowledge include addressing noisy, physical world data and developing some new inference techniques that do not suffer the limitations of Bayesian schemes.

The amount of collected data will be exceptional. It is also expected that a large number of real-time sensor data streams will exist and it will be common for a given flow of data which will be used in many different ways. Another well-known technique i.e. Data mining techniques. These Data mining techniques are very much expected to provide the creation of most valuable knowledge from all this data. It will enable the data streams to act as primitives for unexpected future inferences which is really an interesting research problem now a days. Apart from this, trust is one of the important aspect of the usefulness of big data and essential elements of trust are security and privacy. However, as a basis for trust it is also necessary to develop something new in-field sensor calibration techniques. So without these basic system-level capabilities, the further inference might be operating with wrong data. One approach is to ensure that, all the information is accompanied by a confidence level in the form of a probability that the information is correct or incorrect.

3.5. Robustness

Internet of Things (IoT) applications are generally based on deploying sensing, actuation as well as communication platform and also in these deployments it is common for the devices to know their respective locations, know their neighbor devices when collaborating, appropriate power levels for communication, and pair-wise security keys. However, over time these conditions can deteriorate and the most common example of this deterioration problem is with clock synchronization. For example, some nodes unexpectedly may be physically moved here and there and it is quite possible that more and more nodes may become out of place and to make again system-wide node locations coherent, node re-localization needs to occur. We can also consider this issue as a form of entropy where a system will decline unless energy in the form of re-running protocols and other self-healing mechanisms is applied.

Another problem hardly addressed is that in some Internet of Things (IoT) applications, especially for safety purpose, run time assurances must be given to authorities. Now let's consider a fire fighting system deployed in a sky scraper office building to detect fires, alert fire station and aid in ejection. Periodically, it is very necessary to demonstrate to various certification authorities and such kind of IoT applications will need services that can support run time certification.

4. Innovation Directions of IoT

The strategic development of enabling new technologies (such as nano-electronics, communication systems, sensors, embedded systems technologies, cloud computing, networking, and software etc.) will be essential to provide capability all the time everywhere. This will lead to support important future IoT product innovations affecting many different industrial sectors. The gap between cyber space and the physical world of real things has been connected by embedded systems.

In [47], the authors have mentioned that, the final report of the Key Enabling Technologies (KET), of the High-Level Expert Group has identified the enabling technologies. It is very critical to many of the existing chains of the European economy such as Micro and Nano electronics, Photonics, Biotechnology, and Advanced Manufacturing Systems. From a technology point of view, the rapid grow in the integration density proposed by Dr. Gordan Moore made possible by a dimensional scaling. The reduction in the critical dimensions while keeping the electrical field constant, and at the same time a user obtained at a higher speed with reduction in power consumption of a digital MOS circuits. So now a days these two parameters became driving or key forces of the microelectronic industry along with the integration density. And also now a days the International Technology Roadmap for Semiconductors has highlighted in its early editions and its associated benefits in terms of performances, the traditional parameters in Moore's Law.

In [50], it is written that, the second trend has been categorized by functional diversification of semiconductor-based devices. The non-digital functionalities contribute to the miniaturization of electronic systems. Consequently, in view of added various functionality, this trend has already been designated as "More-than-Moore". From a technological survey, it has been estimated that, Mobile data traffic is double each year between now and 2015. So mobile operators are facing problem to provide the required bandwidth to the client and customers. In many countries there is no any kind of extra spectrum that can be assigned. So proposed solutions are the seamless integration of existing Wi-Fi networks into the mobile ecosystem and also this will have a very big impact on Internet of Things (IoT) ecosystems. The chips designed to fulfil this integration process are known as "multicom" chips. In a single silicon package, it is expected to cover Wi-Fi and baseband communications. The architecture of mobile devices is likely to change as well as the baseband chip will be taking control of the routing process, so the connectivity components are connected to the baseband or integrated. So as a result of this change in architectural design, an increasing share of the integration work is likely done by baseband manufacturers.

Today many European projects address Internet of Things technologies, knowledge and also it has been mentioned that these topics can be heterogeneous and specialized, also there is a strong need for integration of the individual results. In this context, the integration of knowledge has been conceptualized as the process through which some specialized cognizance situated in multiple projects across Europe is applied and assimilated. The Agenda of Strategic Research and Innovation has been developed with the proper support of a European-led community of interrelated projects and their stakeholders with dedication to the innovation, creation, development and use of the Internet of Things (IoT) technology.

After release of the first edition of Strategic research and Innovation agenda, many active research on several IoT topics have been witnessed. Now a days researches are going on in some specialized areas such as: cloud computing, cyber-physical systems, autonomic computing, and social networks. Because of these advanced research, the scope of IoT convergence have been successfully modified. In [45], [49], and [51] it is mentioned that, the previous versions of this Strategic Research and Innovation Agenda has been incrementally progressed because of the updated release and the main research topics that are incorporated with the development of some IoT infrastructures, key enabling technologies, and applications with an outlook towards 2020 has been highlighted [53].

5. IoT and Related Future Internet Technologies.

5.1. Cloud Computing

The most important feature of the future internet technology and new technology enabler have been encouraged at different levels and allowed the various paradigms known as "Applications as a Service", "Platforms as a Service". In [54] it has been noted that, Internet of Things (IoT) applications (Sensor-based services) will be on demand through a cloud based environment only. If we will apply Internet of Things (IoT) systems in large scale then in that case security would be critical barrier and also apart from that it will never be sufficient, if we will apply the extending existing IT security architecture to the IoT. In order to protect these smart objects (or smart things) over a long period of time then in that case cloud based security services with efficient resources is very much essential.

It is observed that, with the rapid growth of Internet of Things (IoT), we are just moving towards a cyber-physical paradigm, where generally we are integrating the computing technology and communication technology with the connected smart objects (or smart things) with the ability to control their basic operations. In below Figure 3, the interchanging of cyber and physical world have been mentioned.

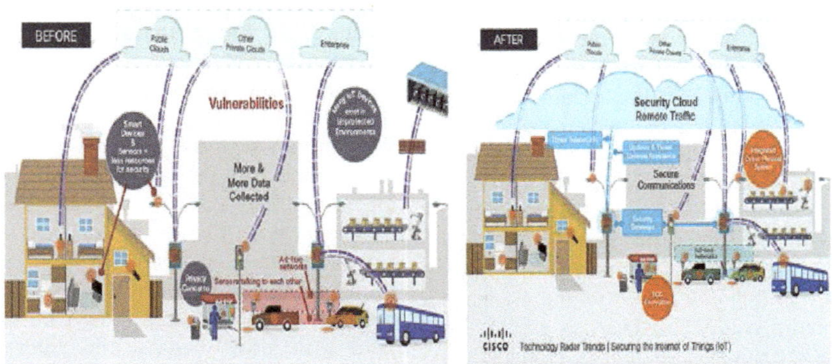

Figure 3. Securely Integrating the Cyber and Physical Worlds [58]

Here in that case many security threats are coming from the interaction between the cyber and physical world. So the improvement in security threat is very much essential. In variations to today's networks, where assets under protection are typically inside firewalls (a network security systems, that controls the incoming and outgoing n/w traffic based on set of rules) and protected with access control devices, many things in the IoT arena will operate in unprotected or highly vulnerable environments. Protecting such things poses additional challenges beyond enterprise networks [55].

Mobility (or potency) support and geo-distribution are required by many Internet of Things applications in addition to the awareness of location and low latency (the amount of time a message takes to traverse a system.), while the data is to be processed in "real-time" in micro clouds or fog. Many new applications are enabled by Micro cloud and Fog computing and services applies a different data management system, which extends the Cloud Computing paradigm (a certain pattern)/ model to the boundary of the network. Similar to Cloud, Micro Cloud/Fog provides data, compute, storage, and application services to end-users. In below Figure 4, the Fog computing paradigm have been shown.

Figure 4. Fog Computing Paradigm [59]

The Micro cloud or Fog needs to have some features (such as: Low latency and location awareness, Wide-spread geographical distribution, Mobility/ potency, huge number of nodes, Predominant role of wireless access, Strong presence of streaming and real time applications, Heterogeneity) in order to efficiently implement the required IoT applications. Moreover, if we generalize the serving scope of an Internet-connected object or things beyond the "sensing service", it is not hard to imagine virtual objects (or things) that will be integrated into the fabric of future IoT services and reused in many different contexts.

5.2. IoT and Semantic Technologies

The IERC SRIAs have identified the importance of semantic technologies towards discovering devices, as well as towards a goal to achieve semantic interoperability. Upcoming research on IoT is likely to embrace the concept of Linked Open Data. This could build on the earlier integration of sensor technologies into IoT infrastructures and applications.

Semantic technologies will be playing a vital role also in enabling (or validating) sharing and re-use of virtual objects as a service through the cloud. The semantic enrichment of virtual object descriptions will realise for Internet of Things, what semantic interpretation of web pages has enabled in the Semantic Web. Internet of Things (IoT) users will be assisted by the associated semantic-based reasoning to find more independently the relevant proven virtual objects to improve the performance.

5.3. Networking Technology

Today Mobile traffic has been driven by a predictable activities such as making calls, receiving email, and watching videos. Over the next 5 to 10 years, billions of Internet of Things (IoT) devices with less traffic patterns will join the network, including vehicles, machine-to-machine (M2M) modules, that requires all the time bandwidth, or different types of sensors that send out tiny bits of data each day. It has been observed that, the rise of cloud computing requires new network strategies for the evolution of fifth generation mobile cellular networks (or 5G), and a good convergence of network access technologies has been represented very clearly. The architecture of such network has to integrate the needs for IoT applications. Fast, and high-capacity network is needed to make the IoT and M2M communication possible.

5G networks will be providing 1,000 to 5,000 times more capacity than 3rd Generation (3G) and 4th Generation (4G) networks and will be made up of cells that provides peak rates of between 10 and 100Gbps. It will be taking data 1–10 milliseconds to get from one designated point to another. Another focus is to separate communications infrastructure and the mobile users will be allowed to move very smoothly between 5G, 4G, and Wi-Fi, which will be fully integrated with the cellular network. Networks will be programmable, and the operators will be allowed to make any kind of changes to the network without touching the physical infrastructure/ model. These features are very much important for IoT applications. The evolution of present communication technologies has the potential to grow to extraordinary levels in the near future by including the developing Internet of Things.

6. Applications of IoT

Internet of Things has many applications in various fields. So we have categorized the applications in to four different fields and they are such as: Home, Enterprise, Utilities, and Mobile telephone systems. Figure 5 which is mentioned below represents the applications of IoT in these four fields.

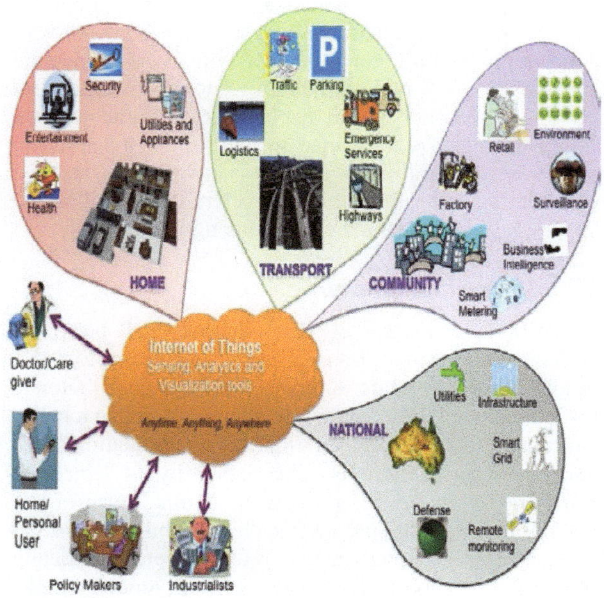

Figure 5. Applications of IoT based on data [6]

5.4. Home

The sensor information is used only by those individuals who directly own the network and basically Wi-Fi is used as the backbone which, enables the higher bandwidth data transfer as well as higher sampling rates. In home also a Smartphone can be used for communication along with several interfaces like Bluetooth for interfacing sensors measuring physiological parameters, and there are several applications available for Apple iOS, Google Android and Windows Phone operating systems that measure various parameters. In [32,33], the authors have mentioned that, for elderly care the expansion of computer networks is creating a home monitoring systems, which will help the doctors to help their patients and aged people thereby reducing hospitalization cost.

5.5. Enterprize

Sensors have always been a most important part of any kind of industry setup for security, automation, and control etc. This can be replaced by a wireless communication system which will be giving the flexibility to make any kind of changes to the setup whenever it will be required and this is nothing but called an IoT subnet dedicated to factory or industry maintenance. Basically we are referring to the 'Network of Things' within a work environment as an enterprise based application and here the Information that has been collected from such kind of networks are generally used by the owners or users and the data may be released selectively. Here in this case Environmental monitoring is the prime application which is generally implemented to keep track of the number of participants and manage the utilities within the building.

5.6. Utilities

Another critical application that is being addressed using IoT is water network monitoring and quality assurance of drinking water. Here in this case sensors are measuring critical water parameters and these parameters are installed at important locations in order to ensure very high supply quality. So that, this will avoid accidental contamination or pollution among storm water drains, drinking water, and sewage disposal. So the same network can be extended or expanded to control the irrigation in various agricultural land and the network can also be extended for controlling soil parameters which will allow informed decision making concerning agriculture.

5.7. Mobile Telephones

Due to the nature of data sharing and backbone implementation smart transportation and smart logistics are placed in a separate domain. The main contributor to traffic noise pollution is urban traffic and it is also a major contributor to urban air quality degradation and greenhouse gas emissions. The popularity of Bluetooth technology (BT) devices reflects the current entry of IoT in a number of digital products such as mobile phones, car hands-free sets, navigation systems, etc. Basically Bluetooth devices emit signals with a unique Media Access Identification (MAC-ID) number that can be read by any kind of Bluetooth sensors within the coverage area and also the readers placed at different locations can be used to identify the movement of the Bluetooth devices. So these are the applications of Internet of Things (IoT) in various fields.

7. Conclusion

The Internet has changed drastically the way we are maintaining our lifestyle. For example such as moving interactions between people at a virtual level in several surrounds covers from the professional life to the social relationship and also the IoT has the potential to add a new dimension to this process by enabling the wireless communication systems with and among smart objects or "Things". One of the most interesting model or pattern which appears in the Internet of the future or upcoming generation is the so called "Web Squared". The most important thing behind this concept that, it will be obtained by taking into account the information about the user background collected by the sensors.

The Internet of Things will grow to 26 billion units and it will be installed in 2020 representing an almost 30- fold increase from 0.9 billion in 2009. The revenue will be incremented up to $300 billion by the IoT product and service suppliers in 2020. It is expected that "ghost" devices with unused connectivity will be common due to the low cost of adding the capability of IoT to the consumer products and there will be a combination of products that have the built in capability but require software to "activate" it. Various enterprises will make a very large-scale use of IoT technology, and there will be a wide range of products (such as: advanced medical devices, factory automation sensors and applications in industrial robotics) sold into various markets.

By the year 2020, the cost of all the component will have come down to a certain point/ level that, the connectivity will become a standard feature, even for processors costing less than $1. This leads to the possibility of connecting almost anything, from the very simple to the very complex to offer remote control access, monitoring and sensing. The IoT incorporates/ comprises of sensors, actuators, electronic processors, microcontrollers, and information services associated with the things. The IoT applications are still implemented by the different industries with a very high importance in manufacturing, healthcare sciences and home/ buildings.

So in this review paper we have surveyed the most basic, important and fundamental aspects of Internet of Things technology which will be done in near future and also some the aspects has already been done and now also still extensive research is going on this IoT technology. So we conclude that, the great invention of this IoT technology will lead the computer communication systems to a different level in near future.

References

[1] D. Giusto, A. Iera, G. Morabito, L. Atzori (Eds.), "The Internet of Things", Springer, 2010.

[2] Daniele Miorandi, Sabrina Sicari et.al., "Internet ofThings-Vision, applications and research challenges",Ad Hoc Networks, vol. 10, 2012, pp.1497-1516.

[3] L. Atzori, A. Iera, and G. Morabito, "The internet of things: A survey," Comput. Netw., vol. 54, no. 15, pp. 2787-2805, 2010.

[4] R.V. Kulkarni, A. Förster, G.K. Venayagamoorthy, "Computational intelligence in wireless sensor networks: a survey", IEEE Communications Surveys & Tutorials, vol. 13, 2011, pp. 68-96.

[5] The Internet of Things, ITU Internet Reports, 2005. <http://www.itu.int/internetofthings/>.

[6] Jayavardhana Gubbi, Rajkumar Buyya et.al., "Internet of Things (IoT):A vision, architectural elements, and future directions", Future Generation Computer Systems, vol. 29, 2013, pp. 1645-1660.

[7] J. Buckley, "From RFID to the Internet of things: pervasive networked systems", Final Report on the Conference organised by DG Information Society and Media, Networks and Communication Technologies Directorate, March 2006. <ftp://ftp.cordis.europa.eu/ pub/ist/docs/ka4/au_conf670306_buckley_en.pdf>.

[8] "Internet of Things: Strategic Research Agenda", September 2009. <http://ec.europa.eu/information_society/policy/rfid/documents/in_cerp.pdf>.

[9] G. Kortuem, F. Kawsar, V. Sundramoorthy, D. Fitton, " Smart objects as building blocks for the internet of things", IEEE Internet Comput., vol. 14, 2010, pp. 44-51.

[10] D. Guinard, V. Trifa, F. Mattern, E. Wilde, "From the Internet of Things to the Web of Things: Resource Oriented Architecture and Best Practices", Springer, New York, Dordrecht, Heidelberg, London, 2011 (Chapter 5).

[11] L. Chen, M. Tseng, X. Lian, "Development of foundation models for Internet of Things", Front. Comput. Sci. China, vol. 4, 2010, pp. 376-385.

[12] S. Haykin, "Cognitive radio: brain-empowered wireless communications", IEEE J. Sel. Areas Commun. Vol. 23, 2005, pp. 201-220.

[13] Auto-Id Labs, <http://www.autoidlabs.org/>.

[14] M. Presser, A. Gluhak, "The Internet of Things: Connecting the Real World with the Digital World", EURESCOM mess@ge – The Magazine for Telecom Insiders, vol. 2, 2009, <http://www.eurescom.eu/ message>.

[15] N. Bressan, L. Bazzaco, N. Bui, P. Casari, L. Vangelista, and M. Zorzi, "The deployment of a smart monitoring system using wireless sensor and actuator networks", in Proc. IEEE Smart Grid. Comm., Gaithersburg, MD, 2010, pp. 49-54.

[16] P. Casari et al., "The WIreless SEnsor networks for city-Wide Ambient Intelligence (WISE-WAI) project," MDPI J. Sensors, vol. 9, no. 6, pp. 4056-4082, Jun. 2009.

[17] A. Dunkels, J.P. Vasseur, "IP for Smart Objects, Internet Protocol for Smart Objects (IPSO) Alliance", White Paper #1, September 2008, <http://www.ipso-alliance.org>.

[18] P. Bellavista, G. Cardone, A. Corradi, and L. Foschini, "Convergence of MANET and WSN in IoT urban scenarios," IEEE Sens. J., vol. 13, no. 10, pp. 3558-3567, Oct. 2013.

[19] D. Cuff, M. Hansen, and J. Kang, "Urban sensing: Out of the woods," Commun. ACM, vol. 51, no. 3, pp. 24-33, Mar. 2008.

[20] N. Gershenfeld, R. Krikorian, D. Cohen, "The internet of things", Scientific American, vol. 291, no. 4, pp. 76-81, 2004.

[21] I. Toma, E. Simperl, Graham Hench, "A joint roadmap for semantic technologies and the internet of things", in: Proceedings of the Third STI Roadmapping Workshop, Crete, Greece, June 2009.

[22] A. P. Castellani, M. Dissegna, N. Bui, and M. Zorzi, "WebIoT: A web application framework for the internet of things," in Proc. IEEE Wireless Commun. Netw. Conf. Workshops, Paris, France, 2012.

[23] I. Vázquez, "Social Devices: Semantic Technology for the Internet of Things", Week@ESI, Zamudio, Spain, June 2009.

[24] R. Roman, J. Lopez, and P. Najera, "A Cross-layer Approach for Integrating Security Mechanisms in Sensor Networks Architectures," Wireless Communications and Mobile Computing, vol. 11, no. 2, Feb. 2011, pp. 267-276.

[25] A. Jules, "RFID security and privacy: a research survey", IEEE Journal on Selected Areas in Communications, vol. 24, no. 2, pp. 381-394, 2006.

[26] N. Maisonneuve, M. Stevens, M. E. Niessen, P. Hanappe, and L. Steels, "Citizen noise pollution monitoring," in Proc. 10th Annu. Int. Conf. Digital Gov. Res.: Soc. Netw.: Making Connec. Between Citizens, Data Gov., 2009, pp. 96-103.

[27] S. Lee, D. Yoon, and A. Ghosh, "Intelligent parking lot application using wireless sensor networks," in Proc. Int. Symp. Collab. Technol. Syst., Chicago, May 19-23, 2008, pp. 48-57.

[28] J. Pasley, "How BPEL and SOA are changing web services development", IEEE Internet Computing, vol. 9, no. 3, pp. 60-67, 2005.

[29] X. Li, W. Shu, M. Li, H.-Y. Huang, P.-E. Luo, and M.-Y.Wu, "Performance evaluation of vehicle-based mobile sensor networks for traffic monitoring," IEEE Trans. Veh. Technol., vol. 58, no. 4, pp. 1647-1653, May 2009.

[30] R. Bonetto, N. Bui, V. Lakkundi, A. Olivereau, A. Serbanati, and M. Rossi, "Secure communication for smart IoT Objects: Protocol stacks, use cases and practical examples," in Proc. IEEE IoT-SoS, San Francisco, CA, USA, 2012, pp. 1-7.

[31] OASIS, Web Services Business Process Execution Language Version 2.0, Working Draft, <http://docs.oasis-open.org/wsbpel/2.0/ wsbpelspecificationdraft.pdf>.

[32] L. Haiyan, C. Song, W. Dalei, N. Stergiou, S. Ka-Chun, "A remote markerless human gait tracking for e-healthcare based on content-aware wireless multimedia communications", IEEE Wireless Communications, vol. 17, 2010, pp. 44-50.

[33] G. Nussbaum, "People with disabilities: assistive homes and environments", in: Computers Helping People with Special Needs, 2006.

[34] A. Sarma and J. Girão, "Identities in the Future Internet of Things," Wireless Personal Comm., Mar. 2009, pp. 353-363.

[35] T. V Lakshman, U. Madhow, "The performance of TCP/IP for networks with high bandwidth-delay products and random loss", IEEE/ACM Transactions on Networking, vol. 5, no. 3, 1997, 336-350.

[36] http://www.cen.eu.

[37] <http://www.iso.org>.

[38] <http://www.ist-e-sense.org>.

[39] <http://www.ist-ubisecsens.org>.

[40] <http://www.etsi.org>.

[41] L. Ren, F. Tian, X. Zhang, L. Zhang, DaisyViz: a model-based user interface toolkit for interactive information visualization systems, Journal of Visual Languages and Computing 21 (2010) 209-229.

[42] G.P. Bonneau, G.M. Nielson, F. Post (Eds.), Data Visualization: The State of the Art, Kluwer Academic, London, 2003.

[43] R.N. Murty, G. Mainland, I. Rose, A.R. Chowdhury, A. Gosain, J. Bers, et al., "CitySense: an urban-scale wireless sensor network and testbed", 2008, pp. 583-588.

[44] P. Kumar, S. Ranganath, W. Huang, K. Sengupta, "Framework for real-time behavior interpretation from traffic video", IEEE Transactions on Intelligent Transportation Systems 6 (2005) 43-53.

[45] O. Vermesan, P. Friess, P. Guillemin, H. Sundmaeker, et al., "Internet of Things Strategic Research and Innovation Agenda", Chapter 2 in Internet of Things – Converging Technologies for Smart Environments and Integrated Ecosystems, River Publishers, 2013.

[46] R.K. Rana, C.T. Chou, S.S. Kanhere, N. Bulusu, W. Hu, Ear-phone: an end-toend participatory urban noise mapping system, in: ACM Request Permissions, 2010.

[47] "Key Enabling Technologies", Final Report of the HLG-KET, June 2011.

[48] W. Arden, M. Brillouët, P. Cogez, M. Graef, et al., "More than Moore" White Paper, online at http://www.itrs.net/Links/2010ITRS/IRC-ITRSMtM- v2%203.pdf.

[49] O. Vermesan, P. Friess, P. Guillemin, S. Gusmeroli, et al., "Internet of Things Strategic Research Agenda", Chapter 2 in Internet of Things - Global Technological and Societal Trends, River Publishers, 2011.

[50] D. Donoho, Compressed sensing, IEEE Transactions on Information Theory 52 (2006) 1289-1306.

[51] IERC – European Research Cluster on the Internet of Things, "Internet of Things - Pan European Research and Innovation Vision", October, 2011.

[52] Kjær, K., E.: A Survey of Context-Aware Middleware. In: 25th conference on IASTED International Multi-Conference: Software Engineering, pp. 148-155. ACTA Press (2007).

[53] EU Research & Innovation, "Horizon 2020", The Framework Programme for Research and Innovation, online at http://ec.europa.eu/ research/horizon2020/index en.cfm.

[54] M. Yuriyama and T. Kushida, "Sensor-Cloud Infrastructure – Physical Sensor Management with Virtualized Sensors on Cloud Computing", NBiS 2010: 1-8.

[55] Cisco, "Securely Integrating the Cyber and Physical Worlds", online at http://www.cisco.com/web/solutions/trends/tech-radar/securing-theiot. Html

[56] Arun Agarwal, Gourav Misra, and Kabita Agarwal. "The 5th Generation Mobile Wireless Networks- Key Concepts, Network Architecture and Challenges." *American Journal of Electrical and Electronic Engineering* 3.2 (2015): 22-28.

[57] Aberer, K., Hauswirth, M., Salehi, A.: Middleware Support for the Internet of Things. In: 5th GI/ITG KuVS Fachgespr"ach "Drahtlose Sensornetze", pp. 15-21. (2006).

[58] https://techradar.cisco.com/theme/www.ciscotechradar.com/img/trends/Securing-the-Internet-of-Things/securing-internet-things-comp.png

[59] https://www.google.co.in/url?sa=i&rct=j&q=&esrc=s&source=images&cd=&cad=rja&uact=8&ved=0ahUKEwjbj-

h9KbJAhVYco4KHQDRBd8QjRwIBw&url=http%3A%2F%2Fdistressnet.nfshost.com%2Fabout&psig=AFQjCNGY_JS1JPzK7B6MReE39GU-VEE45A&ust=1448380518679181.

[60] https://www.google.co.in/url?sa=i&rct=j&q=&esrc=s&source=images&cd=&cad=rja&uact=8&ved=0ahUKEwj_0Zne-abJAhVSCY4KHWw4AkwQjRwIBw&url=http%3A%2F%2Fwww.theregister.co.uk%2F2014%2F05%2F07%2Ffreescale_internet_of_things%2F&psig=AFQjCNE05sJjWHQHCEpi0EjRSMhi2zTvow&ust=1448381820961573.

Reduction of Damping Oscillations of Power Systems with Combinational Operation of LFC & AVR in Single Area

Jyothilal Nayak Bharothu*, G Venkatesh

Associate professor of Electrical & Electronics Engineering, Sri Vasavi Institute of Engineering & Technology, Nandamuru, India
*Corresponding author: nayakeee@gmail.com

Abstract The main aim of this paper is to reduce the damping oscillations. In this paper a new model is proposed to show the interaction between load frequency oscillations and automatic voltage regulator loops and to achieve more accuracy in dynamic and steady state response. The major problem arising from large scale electric power system interconnection is the low frequency oscillation of interconnected system. In order to maintain all the scheduled power transactions to the constant value, automatic generation control of a power system with closed loop feedback is used. By using the data acquisition system which conducts the actual operating conditions of the systems and the errors are computed. In ordered to maintain the level of voltage and frequency automatic generation control method through LFC loops is used.

Keywords: *damping oscillations, LFC, AVR, MAT LAB, Simulink*

1. Introduction

One of the most important stability problems arising from large-scale electric power system interconnections is the low frequency oscillation of interconnected systems. The low-frequency oscillations are attributed to the oscillations of the mechanical mode of the system and can be approximately analyzed with a linear one-machine infinite bus model [1]. A complete system model for low-frequency oscillation studies must be included of mechanical and electrical loops. It has been recognized that these oscillations can be controlled by adjusting exciter and speed-governor control parameters. Furthermore, it has been shown that the Load-voltage characteristic of the power system has a significant effect on its dynamic responses, and suggestions have been made for the proper representation of these characteristics in simulation studies [2,3]. In practice, Automatic Generation Control of a power system is a set of equipment and computer programs that applies closed loop feedback control to regulate the power system frequency to a scheduled value, to maintain all scheduled power transactions to the contract value as well as the net power interchange at the value required by the interchange contracts, and to maintain each generation unit's operation at the most economic value (Economic Dispatch).

Modern automatic generation control functions are performed at a central location, the Energy Management System (EMS) or the Control Center. The AGC is driven by the scheduled frequency and the scheduled net interchange of the controlled entity. The data(errors)are computed [4,5,6]. The errors are used to determine the control signals to the individual generators in such a way that the frequency returns to the scheduled value and the power transactions are regulated to the scheduled value. A lot of studies have been made about LFC over the last decades but these researches don't have any attention to AVR effects on the results [7,8]. The two main control loops of a generation care Load Frequency Controller (LFC) and Automatic Voltage Regulator (AVR) as seen in Figure 1.1.

The turbine fed by controllable rate of steam and the Automatic Generation Control method deals with frequency through the LFC loop and with voltage through with the AVR loop, where the main purposes of these controllers are to maintain levels of voltage end frequency at the acceptable values. However, these studies are based on the assumption that there is no interaction between the power/frequency and the reactive power/voltage control loops. This assumption is permissible acquisition system collects the actual operating conditions of the system and the deviations only when the speed of the excitation system is much faster than that of the LFC system; in the practical systems, during dynamic perturbations, exists some interaction between these two control channels due to the weak coupling relationship between the AVR and LFC, the voltage and frequency are regulated separately.

The study of coupling effects of the AVR and LFC loops can be found in [6] and [9]. But in all of these researches there are sketchy attentions to these mutual effects. Furthermore there is no any attention to the turbine output power in the steady state. Also, by neglecting the effect of voltage deviation on load demand, an important interaction in LFC systems is ignored. In order to prevent this problem and improve the accuracy of responses in this paper, a new combined classical model is proposed. A complete system model for low.

Frequency oscillation studies must be included of mechanical and electrical loops. So single machine - infinite bus model of a power system is combined to load frequency control system.

This proposed combined model is tested on one-area power system. Simulation results are shown that there are mutual effects between LFC and excitation system. Furthermore the results are shown that by using this model for low-frequency oscillation study of power systems, more accuracy will be reachable in dynamic and steady state responses.

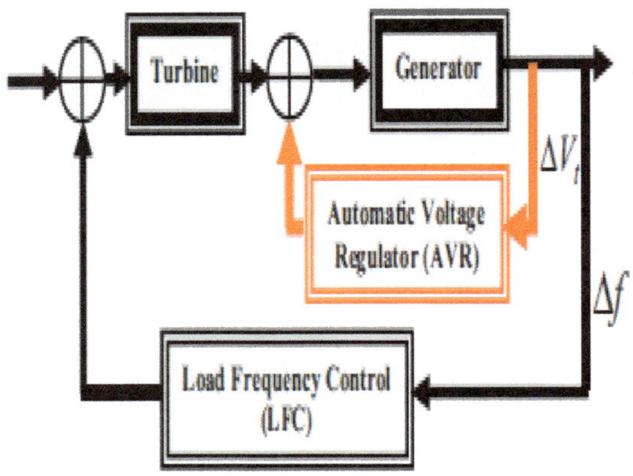

Figure 1.1. Automatic generation control with LFC and AVR loops

2. Load Frequency Controller

2.1. Introduction

The generation of power units and consumption of loads connected to the UCTE network needs to be controlled and monitored for secure and high-quality operation of the synchronous areas. The load frequency control, the technical reserves and the corresponding control performances are essential to allow TSOs to perform daily operational business. Within the UCTE synchronous area, the control actions and the reserves are organized in a hierarchical structure with control areas, control blocks and the synchronous area with two co-ordination centers. UCTE Pyramid in Policy Co-ordination Center (UCTE South) Control Block Control Block Control Area Control Area Control Area Co-ordination Center (UCTE North) Control Block Control Block Control Area Control Area Control Area Figure 1.1 Hierarchical control structure of UCTE synchronous area composed of control areas (ca), control blocks (cb) and co-ordination centers (cc) Control actions are performed in different successive steps, each with different characteristics and qualities.

Control scheme and actions starting with the system frequency primary control starts within seconds as a joint action of all parties involved. Secondary control replaces primary control over minutes and is put into action by the responsible parties / TSOs only. Tertiary control partially complements and finally replaces secondary control by re-scheduling generation and is put into action by the responsible parties / TSOs. Time control corrects global time deviations of the synchronous time in the long term as a joint action of all parties / TSOs. On the time axis, the different control reserves cover different time frames the principles, how in case of an incident with a large frequency drop (the dotted Line beginning before activation of Primary Control shows the principle plot of the frequency Deviation) the activation of primary control reserve (activated within seconds) is followed up by secondary control reserve (activated within minutes) and secondary control reserve is supported and followed up by tertiary control reserve. UCTE OH – Policy 1: Load-Frequency Control - Final Version (approved by SC on 19 March 2009) Time directly activated Tertiary Control Reserve Schedule activated Tertiary Control Reserve Activated Secondary Control Reserve UCTE-wide activated Primary Control Reserve.

Principle frequency deviation and subsequent activation of reserves The "TSO-Forum" (organized by the UCTE WG "Operations & Security") serves as the Common body in the UCTE of all TSOs for all operational and organizational items in the Framework of Load-Frequency Control and Performance. Please refer to the glossary of terms of the UCTE Operation Handbook for detailed definitions of terms used within this policy and to Appendix 1 for basics and principles of load-frequency control and performance.

This version of the document (version 3.0, level C, dated 12.03.2009) has "rev15" status. This document and other chapters of the UCTE Operation Handbook as well as excerpts from it may not be published, redistributed or modified in any technical means or used for any other purpose outside of UCTE without written permission in advance. UCTE OH – Policy 1: Load-Frequency Control - Final Version (approved by SC on 19 March 2009)

2.2. Primary Control

The objective of primary control is to maintain a balance between generation and consumption (demand) within the synchronous area. By the joint action of all interconnected parties / TSOs, primary control aims at the operational reliability of the power system of the synchronous area and stabilizes the system frequency at a stationary value after a disturbance or incident in the time-frame of seconds, but without restoring the system frequency and the power exchanges to their reference values. Adequate primary control depends on generation or load resources made available to the TSOs basics and principles of primary control. Nominal Frequency.

The nominal frequency value in the synchronous area is 50.000 Hz. The set-point frequency (or scheduled frequency) f0 defines the target value of the system frequency f for system operation. Outside periods for the correction of synchronous time the scheduled frequency is the nominal frequency. Frequency Levels and Frequency Deviations.

A frequency deviation away from the nominal frequency results from an imbalance between generation and demand that occurs continually during normal system operation or after an incident like a loss of generation. a frequency deviation results from an unwanted imbalance between generation and demand. Different criteria are used to distinguish the size of this deviation: A-D2.1. Activation of primary control.

Primary control activation is triggered before the frequency deviation towards the nominal frequency exceeds20 MHZ (the sum of the accuracy of the local frequency measurement and the insensitivity of the controller maximum permissible quasi-steady-state frequency deviation after reference incident. a quasi-steady-state frequency deviation of 180 MHZ away from the nominal frequency is permitted as a maximum value in the UCTE synchronous area after occurrence of a reference UCTE oh – policy 1: load-frequency control - final version incident after a period of initially undisturbed operation. when assuming that the effect of self-regulation of the load is absent, the maximum permissible quasi-steady-state deviation would be 200 MHZ this deviation causes full activation of primary control within the UCTE synchronous area and (passive) self-regulation of load.

Minimum instantaneous frequency after loss of generation. the minimum instantaneous frequency is defined to be 49.2 Hz (that corresponds to -800 MHZ as maximum permissible dynamic frequency deviation from the nominal frequency in response to a shortfall in generation capacity equal to or less than the reference incident according to. maximum instantaneous frequency after loss of load. the maximum instantaneous frequency is defined to be 50.8 hz (that corresponds to +800 MHZ as maximum permissible dynamic frequency deviation from the nominal frequency p1-A-D1) in response to a loss of load or interruption of power exchanges equal to or less than the reference incident according to p1-A-D3.1. A-D2.5. full activation of primary control reserves. in case of a quasi- steady state deviation of the system frequency of 200 MHZ from the nominal frequency, all available primary control reserves are expected to be fully activated. A-D3. reference incident.

The maximum instantaneous power deviation between generation and demand in the un-split synchronous area (by the sudden loss of generation capacity or load-shedding / loss of load) to be handled by primary control starting from undisturbed operation depends on the size of the synchronous area and of the largest generation unit or generation capacity connected to a single bus bar. A-D3.1. reference incident for the UCTE synchronous area for the UCTE synchronous area the maximum instantaneous power deviation is defined to be 3000 MW, based on operational characteristics concerning system reliability and size of loads and generation units 2.

Observation Incident Size for System Response Analysis. Large incidents, such as the sudden loss of generation or load, that exceed 600 mw (first level) respectively 1000 mw (second level) in the UCTE synchronous area, trigger a UCTE system response analysis procedure. A-D4. primary control characteristics.

The following key values of the primary control UCTE oh – policy 1: load-frequency control - final version p1–6 A-D4.1. Self-regulation of load. the self-regulation of the load in the UCTE synchronous area is assumed to be 1 /hz, that means a load decrease of 1 % occurs in case of a frequency drop of 1 hz. see a1-e for the calculated network power frequency characteristic of self regulation for the UCTE synchronous area. A-D4.2. quasi steady-state security margin. for frequency control, the quasi steady-state security margin is defined to be 20 MHZ A-D4.3. Minimum network power frequency characteristic of primary control. the minimum network power frequency characteristic of primary control for the UCTE synchronous area is calculated out of p1-A-D3.1 and (including the security margin p1-A-D4.2) to 15000 MW/HZ. A-D4.4. average network power frequency characteristic of primary control. on average, the network power frequency characteristic of primary control is experienced to be 30 % higher than the minimum network power frequency characteristic of primary control, that results on average to 19500 MW/HZ. A-D4.5. surplus-control of generation. The surplus-control of generation is an experienced linear responses of approximately 50 % of all generation units reacting to frequency deviations. this results in an additional value of self-control of generation, that is calculated based on the mean generation power in the system (see a1-e for the calculated values). This surplus control of generation exists in addition to the minimum network power.

2.3. Characteristic of Primary Control

Overall network power frequency characteristic. The overall network power frequency characteristic for the UCTE synchronous area is the sum of contributions from primary control of 19500 MW/HZ, the surplus-control of generation and self regulation of load for the up-to-date value of the overall network power frequency characteristic for the UCTE.

2.3.1. Synchronous Area

Overall primary control reserve. With respect to the size of the reference incident of 3000 mw the overall primary control reserve for the UCTE synchronous area is agreed to be 3000 mw. Load-frequency control - final version. Primary control reliability and target. in case of a first contingency or incident according to, such as the loss of generation or load or interruption of power exchanges in an undisturbed situation, primary control must maintain reliable system operation. starting from undisturbed operation a reference incident must be handled by primary control alone, without the need for under-frequency automatic load-shedding or disconnection of generation in response to a frequency deviation. Primary control organization. an organizational procedure to cover requirements and obligations for primary control actions and reserves performed by third parties in the control area including a monitoring procedure must be in place (e.g. grid code, regulation, association agreement or contract).

Primary control action by generators or loads. the action of the generators or loads performing primary control must have the following characteristics, to be ensured by all tsos:a-s2.1. Accuracy of frequency measurements. for primary control, the accuracy of local frequency measurements used in the primary controllers must be better than or equal to 10 MHZ. Adjustment of power and insensitivity of controllers.

Power under primary control must be proportionally adjusted to follow changes of system frequency. The insensitivity range of primary controllers should not exceed ± 10 MHZ where dead bands exist in specific controllers, these must be offset within the control area / block concerned.. physical deployment times. the time for starting the action of primary control is a few seconds after the incident, the deployment time for 50 % or less of the total primary control reserve is at most 15 seconds and from 50 % to 100 % the maximum deployment time rises linearly to 30 seconds. Each tso must check the deployment times within his control area / block on a regular basis. by this, the total primary control within the entire synchronous area (as well as within each control area /block) follows the same deployment times. Duration of delivery. Primary control power must be delivered until the power deviation is completely offset by the secondary / tertiary control reserve of the control area / block in which the power deviation has occurred UCTE oh – policy 1: load-frequency control - final version. Joint action for primary control. Primary control is based on the principle of joint action to ensure system reliability and interconnected operation.

This includes an overall distribution of reserves and control actions, as determined and decided by the "tso-forum" on an annual basis for the next calendar year contribution to primary control reserve. The total primary control reserve (in mw) required for operation of the UCTE synchronous area is of the same size as the reference incident. Each control area / block must contribute to the primary control reserve proportionally, so that the sum of all shares amounts to the total required primary control reserve. the respective shares (mandatory primary control reserves) are defined by multiplying the overall primary control reserve for the entire synchronous area and the contribution coefficients of the various control areas / blocks (see. any control area can increase its primary control reserve by 30 % by offering to cover (part of) the obligations of other control areas. however, every control area is allowed to increase its primary control reserve by 90 mw to cover the obligations of other control areas (which is approx. 3 % of total UCTE primary control reserve and corresponds to the limitation for loss of primary a-s3.2. contribution to primary control action.

Each control area / block must contribute to the correction of a disturbance in accordance with its respective contribution coefficient for primary control. Limitation for loss of primary control reserve. if a generator, on which primary control reserve is allocated, trips, the immediate loss of primary control reserve must be limited to 90 mw or 3 % of the reference incident size (by this the decline of the network power FREQUENCY characteristic will be limited).

Responsibilities in the process. each reserve connecting to is responsible for the primary control contribution of the control area, including cross-border primary control contributions for other control areas. This includes the contribution, the monitoring, the general procedure and the information exchange for primary control reserve. Information exchange between tsos the reserve receiving tso communicates to the reserve connecting tso by which amount the value of its k-factor must be increased or decreased before the beginning or termination of any exchanging transaction of power for primary control. In the case of transactions within a control block with pluralistic control mode these k-factor values must be communicated to the control block UCTE oh – policy 1: load-frequency control - final version (approved by sc on 19 march 2009) operator.

The control block operator will then verify whether the total sum of the new k-factors kri involved in the transaction is equal to the total sum of the original k-factor values. a-s4. Primary control reserve characteristics. The primary control reserve needs to have certain characteristics to be usable for primary control. availability of reserves. as a minimum in an undisturbed situation, the mandatory primary control reserve for each control area must be available continuously without interruption, not depending on the unit commitment in detail. primary and secondary control reserves must be available for activation independently. a-s4.2. operational usability of reserves.

The mandatory primary control reserve for each control area must be fully activated in response to a quasi-steady-state frequency deviation of 200 MHZ or more border-crossing primary control reserve. Exchange of reserves for primary control crossing the border of the control area can only be allowed if the concerned tsos have previously confirmed this exchange and if the reserve is exchanged directly between adjacent control areas or remains inside the same control block. Additional power flows induced by cross-border primary control reserve activation have to be considered in the determination of the capacity reliability margin. each tso has to consider worst case conditions for the primary control power flow contribution of primary control reserve to one control area. Individual generation units or individual loads may only deliver primary control reserve / power to one control area at any time, primary control may not be split for different control areas.

A generation unit or load can only have obligations to one reserve receiving tso at any time to assure transparent verification of primary control contributions from a generation unit or load to the single reserve receiving tso. a-s4.5 minimum amount of primary control reserves within control areas / blocks. each tso must declare to the "tso-forum" on annual basis the individual minimum amount of the primary control reserves that needs to be kept within the control area / block due to security needs (as a share a-s4.6. maximum amount of primary control reserves transferred from control areas / blocks. each tso must declare to the "tso-forum" on annual basis the individual maximum amount of the primary control reserves that can be transferred safely to other control areas out of the UCTE oh – policy 1: load-frequency control - final version own control area. monitoring of redistribution of primary control reserves. since the redistribution of primary control reserve is a global topic that affects all tsos, any change in the distribution of primary control reserves outside of the control area and control block needs to be declared to the "tsoforum". a-s5. monitoring and observation. a-s5.1. observation of outages. outages in production or consumption exceeding the size of the observation incident are recorded for analysis by the tso. the required information about location, time, size and type of the disturbance / incident is recorded and made available to the members of the association via the "tso-forum". for second level incidents all tsos are obliged to submit information on their primary contribution to the "tso forum".a-s5.2. cross-border reserves.

Monitoring must be executed by the reserve connecting TSO. Monitoring is based on the frequency measurement and the generation unit or load active power measurements, located at the network injection point in the network of the reserve connecting tso.a-s5.3. Measurement cycle for frequency observation. the cycle for measurements of the system frequency for control area observation must be in the range of 1 second (strongly recommended) to at most 10 seconds. the values of the incident sizes are determined and proposed by the "tso-forum" and need approval by all UCTE members.

Determination of overall network power frequency characteristic. the final value of the overall network power frequency characteristic is determined by the "tso-forum" on a regular basis, contribution coefficients determination. the contribution coefficients are determined and published annually for each control area / block the contribution coefficients are binding for the corresponding interconnection partner / tso for one calendar year. They are based on the share of the energy generated within one year in proportion to the entire synchronous area the sum of all contributions coefficients must amount to 1 the "tso-forum" determines and decides about the contribution coefficients of each control area / block for the load-frequency control area on an annual basis and sets these values into operation on the 1st of January of the next year. a-g4 constant network power frequency characteristic. in order to ensure that the principle of joint action is observed, the network power frequency characteristics of the various control areas is taken to remain as constant as possible.

This applies particularly to small frequency deviations, where the "dead bands" of generators may have an unacceptable influence upon the supply of a-g5. Contracting primary reserve in a neighboring control area or inside a control block: a tso may contract primary reserve in a neighboring control area or inside a control block.

Contracts for providing this service may be signed as tso-tso or as tso-gen co. this implies that an agreement between the concerned tsos has been signed, e.g. concerning the setting of the k-factors on the secondary controller the "tso-forum" will perform simulations to check the consequences for the overall system security on an annual basis. as a result, ultimate limits can be given to individual tsos. Contracting primary reserve in a neighboring control area / block. control performance measurement. the network power frequency characteristic is calculated in response to a disturbance (such as an observation incident), based on measurements of the system frequency and other key values and on a statistical analysis. results of the control performance analysis on a regular basis in the "regular report of the performance of the primary and secondary load–frequency control", prepared by the "tso-forum". UCTE oh – policy 1: load-frequency control.

3. Automatic Voltage Controller

3.1. Introduction

A voltage regulator is an electrical regulator designed to automatically maintain a constant voltage level. A voltage regulator may be a simple "feed-forward" design or may include negative feedback control loops. It may use an electromechanical mechanism, or electronic components. Depending on the design, it may be used to regulate one or more AC or DC voltages.

Electronic voltage regulators are found in devices such as computer power supplies where they stabilize the DC voltages used by the processor and other elements. In automobile alternators and central power station generator plants, voltage regulators control the output of the plant. In anelectric power distribution system, voltage regulators may be installed at a substation or along distribution lines so that all customers receive steady voltage independent of how much power is drawn from the line.

The output voltage can only be held roughly constant; the regulation is specified by two measurements: load regulation is the change in output voltage for a given change in load current (for example: "typically 15mV, maximum 100mV for load currents between 5mA and 1.4A, at some specified temperature and input voltage").line regulation or input regulation is the degree to which output voltage changes with input (supply) voltage changes - as a ratio of output to input change (for example "typically 13mV/V"), or the output voltage change over the entire specified input voltage range (for example "plus or minus 2% for input voltages between 90V and 260V, 50-60Hz").

Other important parameters are: Temperature coefficient of the output voltage is the change in output voltage with temperature (perhaps averaged over a given temperature range), while.

Initial accuracy of a voltage regulator (or simply "the voltage accuracy") reflects the error in output voltage for a fixed regulator without taking into account temperature or aging effects on output accuracy.

Dropout voltage is the minimum difference between input voltage and output voltage for which the regulator can still supply the specified current. A Low Drop-Out (LDO) regulator is designed to work well even with an input supply only a Volt or so above the output voltage. The input-output differential at which the voltage regulator will no longer maintain regulation. Further reduction in input voltage will result in reduced output voltage. This value is dependent on load current and junction temperature.

Absolute maximum ratings are defined for regulator components, specifying the continuous and peak output currents that may be used (sometimes internally limited), the maximum input voltage, maximum power dissipation at a given temperature, etc. Output noise (thermal white noise) and output dynamic impedance may be specified as graphs versus frequency, while output ripple noise (mains "hum" or switch-mode "hash" noise) may be given as peak-to-peak or RMS voltages, or in terms of their spectra.

Quiescent current in a regulator circuit is the current drawn internally, not available to the load, normally measured as the input current while no load is connected (and hence a source of inefficiency; some linear regulators are, surprisingly, more efficient at very low current loads than switch-mode designs because of this).

Response is reaction of a regulator when a (sudden) change of the load current (called the load transient) or input voltage (called the line transient) occurs. Some regulators will tend to oscillate or have a slow response time which in some cases might lead to undesired results. This value is different from the regulation parameters, as that is the stable situation definition. The transient response shows the behaviour of the regulator on a change. This data is usually provided in the technical documentation of a regulator and is also dependent on output capacitance.

3.2. Electronic Voltage Regulators

Simple voltage regulator can be made from a resistor in series with a diode (or series of diodes). Due to the logarithmic shape of diode V-I curves, the voltage across the diode changes only slightly due to changes in current drawn. When precise voltage control is not important, this design may work fine.

Voltage regulators operate by comparing the actual output voltage to some fixed reference voltage. Any difference is amplified and used to control the regulation element in such a way as to reduce the voltage error. This forms a negative feedback control loop; increasing the open-loop gaintends to increase regulation accuracy but reduce stability (avoidance of oscillation, or ringing during step changes). There will also be a trade-off between stability and the speed of the response to changes. If the output voltage is too low (perhaps due to input voltage reducing or load current increasing), the regulation element is commanded, up to a point, to produce a higher output voltage–by dropping less of the input voltage (for linear series regulators and buck switching regulators), or to draw input current for longer periods (boost-type switching regulators);

If the output voltage is too high, the regulation element will normally be commanded to produce a lower voltage. However, many regulators have over-current protection, so that they will entirely stop sourcing current (or limit the current in some way) if the output current is too high, and some regulators may also shut down if the input voltage is outside a given range (see also: crowbar circuits

3.3. Electromechanical Regulators

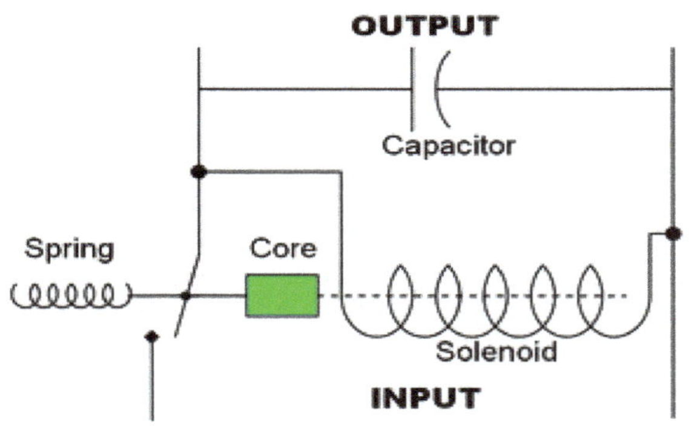

Figure 3.1. Circuit design for a simple electromechanical voltage regulator

In electromechanical regulators, voltage regulation is easily accomplished by coiling the sensing wire to make an electromagnet. The magnetic field produced by the current attracts a moving ferrous core held back under spring tension or gravitational pull. As voltage increases, so does the current, strengthening the magnetic field produced by the coil and pulling the core towards the field. The magnet is physically connected to a mechanical power switch, which opens as the magnet moves into the field. As voltage decreases, so does the current, releasing spring tension or the weight of the core and causing it to retract. This closes the switch and allows the power to flow once more.

If the mechanical regulator design is sensitive to small voltage fluctuations, the motion of the solenoid core can be used to move a selector switch across a range of resistances or transformer windings to gradually step the output voltage up or down, or to rotate the position of a moving-coil AC regulator.

Figure 3.2. a voltage stabilizer using electromechanical relays for switching

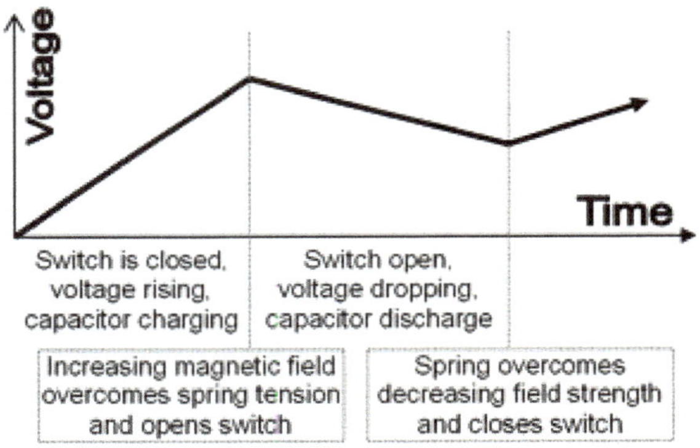

Figure 3.3. Graph of voltage output on a time scale

Early automobile generators and alternators had a mechanical voltage regulator using one, two, or three relays and various resistors to stabilize the generator's output at slightly more than 6 or 12 V, independent of the engine's rpm or the varying load on the vehicle's electrical system. Essentially, the relay(s) employed pulse width modulation to regulate the output of the generator, controlling the field current reaching the generator (or alternator) and in this way controlling the output voltage produced.

The regulators used for DC generators (but not alternators) also disconnect the generator when it was not producing electricity, thereby preventing the battery from discharging back into the generator and attempting to run it as a motor. The rectifier diodes in an alternator automatically perform this function so that a specific relay is not required; this appreciably simplified the regulator design.

More modern designs now use solid state technology (transistors) to perform the same function that the relays perform in electromechanical regulators.

3.4. Coil-rotation AC Voltage Regulator

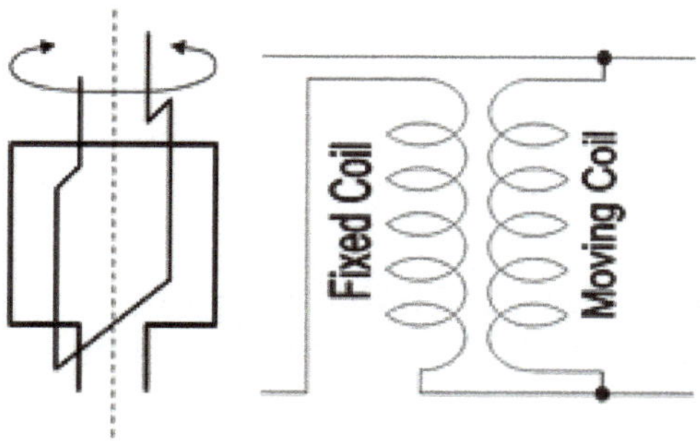

Figure 3.4. Basic design principle and circuit diagram for the rotating-coil AC voltage regulator

This is an older type of regulator used in the 1920s that uses the principle of a fixed-position field coil and a second field coil that can be rotated on an axis in parallel with the fixed coil.

When the movable coil is positioned perpendicular to the fixed coil, the magnetic forces acting on the movable coil balance each other out and voltage output is unchanged. Rotating the coil in one direction or the other away from the center position will increase or decrease voltage in the secondary movable coil.

This type of regulator can be automated via a servo control mechanism to advance the movable coil position in order to provide voltage increase or decrease. A braking mechanism or high ratio gearing is used to hold the rotating coil in place against the powerful magnetic forces acting on the moving coil.

3.5. AC Voltage Stabilizers

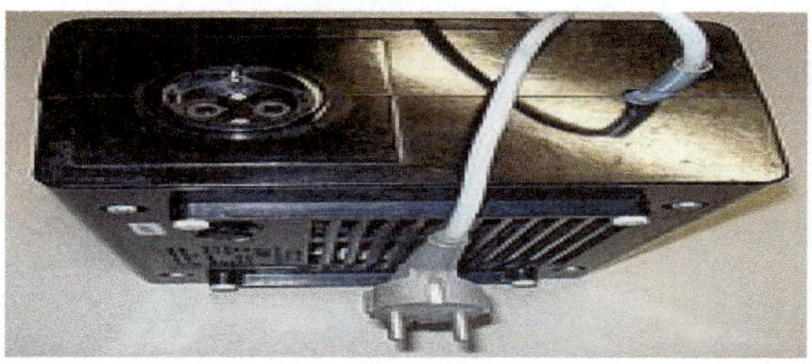

Figure 3.5. Magnetic mains regulator Electromechanical

Electromechanical regulators, usually called voltage stabilizers, have also been used to regulate the voltage on AC power distribution lines. These regulators operate by using a servomechanism to select the appropriate tap on an autotransformer with multiple taps, or by moving the wiper on a continuously variable autotransformer. If the output voltage is not in the acceptable range, the servomechanism switches connections or moves the wiper to adjust the voltage into the acceptable region. The controls provide a deadband wherein the controller will not act, preventing the controller from constantly adjusting the voltage ("hunting") as it varies by an acceptably small amount.

The ferroresonant transformer, ferroresonant regulator or constant-voltage transformer is a type of saturating transformer used as a voltage regulator. These transformers use a tank circuit composed of a high-voltage resonant winding and a capacitor to produce a nearly constant average output voltage with a varying input current or varying load. The circuit has a primary on one side of a magnet shunt and the tuned circuit coil and secondary on the other side. The regulation is due to magnetic saturation in the section around the secondary.

The ferroresonant approach is attractive due to its lack of active components, relying on the square loop saturation characteristics of the tank circuit to absorb variations in average input voltage. Saturating transformers provide a simple rugged method to stabilize an AC power supply.

Older designs of ferroresonant transformers had an output with high harmonic content, leading to a distorted output waveform. Modern devices are used to construct a perfect sine wave. The ferroresonant action is a flux limiter rather than a voltage regulator, but with a fixed supply frequency it can maintain an almost constant average output voltage even as the input voltage varies widely.

The ferroresonant transformers, which are also known as Constant Voltage Transformers (CVTs) or ferros, are also good surge suppressors, as they provide high isolation and inherent short-circuit protection.

A ferroresonant transformer can operate with an input voltage range ±40% or more of the nominal voltage.

Output power factor remains in the range of 0.96 or higher from half to full load.

Because it regenerates an output voltage waveform, output distortion, which is typically less than 4%, is independent of any input voltage distortion, including notching.

Efficiency at full load is typically in the range of 89% to 93%. However, at low loads, efficiency can drop below 60% and no-load losses can be as high as 20%. The current-limiting capability also becomes a handicap when a CVT is used in an application with moderate to high inrush current like motors, transformers or magnets. In this case, the CVT has to be sized to accommodate the peak current, thus forcing it to run at low loads and poor efficiency.

Minimum maintenance is required, as transformers and capacitors can be very reliable. Some units have included redundant capacitors to allow several capacitors to fail between inspections without any noticeable effect on the device's performance.

Output voltage varies about 1.2% for every 1% change in supply frequency. For example, a 2 Hz change in generator frequency, which is very large, results in an output voltage change of only 4%, which has little effect for most loads.

It accepts 100% single-phase switch-mode power supply loading without any requirement for derating, including all neutral components.

Input current distortion remains less than 8% THD even when supplying nonlinear loads with more than 100% current THD.

Drawbacks of CVTs are their larger size, audible humming sound, and the high heat generation caused by saturation. Also, the regulation is not as good as solid state devices, so these units are obsolete for most purposes.

3.6. DC Voltage Stabilizers

Many simple DC power supplies regulate the voltage using a shunt regulator such as a zener diode, avalanche breakdown diode, or voltage regulator tube. Each of these devices begins conducting at a specified voltage and will conduct as much current as required to hold its terminal voltage to that specified voltage. The power supply is designed to only supply a maximum amount of current that is within the safe operating capability of the shunt regulating device (commonly, by using a series resistor).

If the stabilizer must provide more power, the shunt regulator output is only used to provide the standard voltage reference for the electronic device, known as the voltage stabilizer. The voltage stabilizer is the electronic device, able to deliver much larger currents on demand.

4. Linear Model of Low-Frequency Oscillation Studies

4.1. Introduction

In this study, the simulation of low-frequency oscillations is based on a single machine connected to infinite bus system. This one machine-infinite bus system with a local load is

One machine-infinite bus power system.

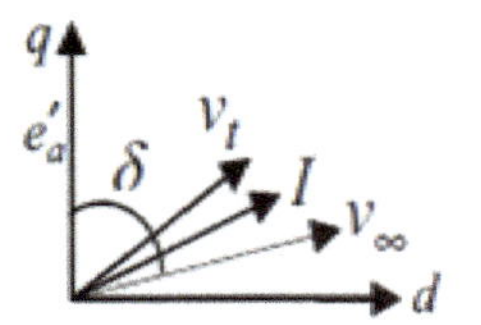

Current and voltage phasors.

Figure 4.1. one machine infinite bus system

Where, Z: series impedance of transmission line and Y: shunt admittance as a local load. As shown in, the current and voltage phasor is are drawn, and a torque angle defined as the angle between the infinite bus voltage and the q axis. For convenience, the following parameters are introduced:

The magnitude of the generator terminal voltage can be expressed in terms of its d and q components as:

$$\Delta v_t^2 = v_d^2 + v_q^2 \tag{4.1}$$

And the deviation as:

$$\Delta v_t = \left(v_d^0 / v_t^0\right)\Delta v_d + \left(v_q^0 / v_t^0\right)\Delta v_q \tag{4.2}$$

Substituting (4.2) into the linearized results of (4.1) and Δvd and Δvq thus obtained into gives:

$$\Delta V_t = \frac{\partial V_t}{\partial \delta}\Delta\delta + \frac{\partial V_t}{\partial\left|E_a'\right|}\partial\left|E_a'\right| = K_5\Delta\delta + K_6\partial\left|E_a'\right| \tag{4.3}$$

In this section, the initial values of currents, voltages and their component are all identified by superscript 0, e.g., 0 *id*. These initial values in the steady state are usually found from a load flow study. For a one machine-infinite bus system, these values are calculated is case of the electric power, reactive power and the machine terminal voltage are given.

5. Combined Model for Lfc and Avr Loops in One-Area Power System

5.1. Introduction

The rotational inertia equations describe the effect of unbalance between electromagnetic torque and mechanical torque of individual machines in a LFC system. By having small perturbation and small deviation in speed, the complete swing equation becomes.

$$\Delta f = \frac{K_p}{1 + sT_p}\left[\Delta P_m - \left(\Delta P_L + \Delta P_e\right)\right] \tag{5.1}$$

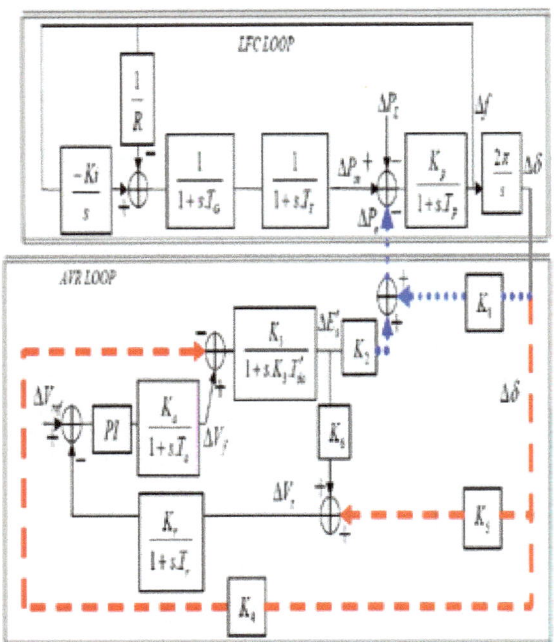

Figure 5.1. proposed combined model for LFC AVR loops

Where, Pe is the deviation of internal electrical power that is sensitive to load characteristics. In this section with using (22), (25), (30) and one-area load frequency control system [6], and by substituting Pe from(22) into (32), a combined model including load frequency control and AVR system is proposed. This model can be used to show the mutual effect between LFC and AVR loops and depict the slight change in response of turbine output power in the steady state. This proposed model is shown in Figure 4.1. Note that the dashed and dotted lines show the connection between LFC and AVR loops.

6. Results

6.1. Introduction

In the simulation study, the combined proposed model is applied for one area power system. To illustrate the performance of this model, simulations are performed for two scenarios of possible operating conditions of a power system. In this simulation the performance of the proposed combined model is compared with a classic model of a load frequency control system ignoring AVR loop and excitation system. Note that the simulations are done using platform. Results are shown in following graphs.

6.2. Scenario

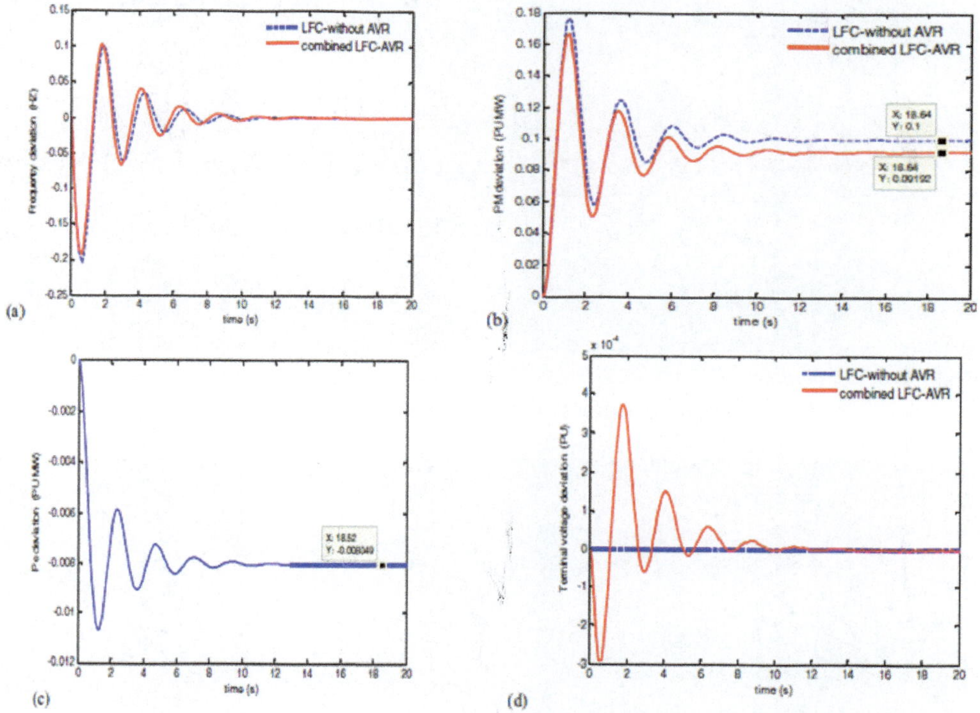

Figure 6.0. (a) Frequency deviation (rad/s), (b) Turbine output power (pu MW), (c)Terminal voltage deviation (pu), (d) Internal electrical power deviation (pu) (Results for this operating condition: P=0.9 pu, Q=0.5 pu, Vt=1 pu)

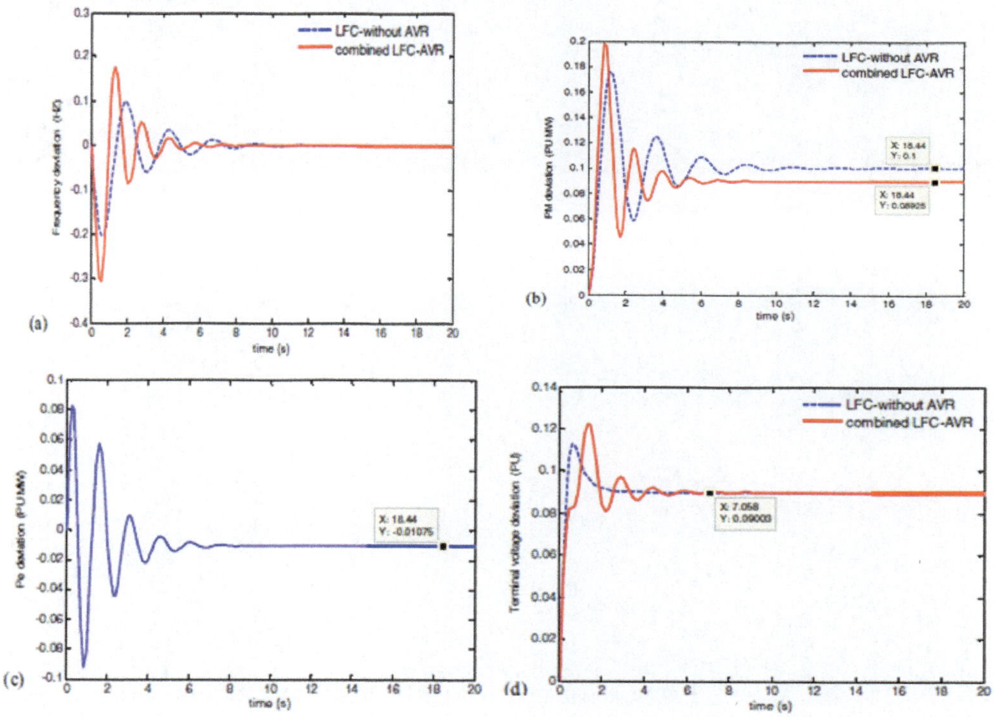

Figure 6.1. (a) Frequency deviation (rad/s), (b) Turbine output power (pu MW), (c)Terminal voltage deviation (pu), (d) Internal electrical power deviation (pu) (Results for this operating condition: P=0.9 pu, Q=0.61 pu, Vt=0.91 pu)

In this scenario, it is assumed that the electric power is 0.9 pu, reactive power is 0.5 pu and the machine terminal voltage is equal to 1 pu. For the LFC system, the load change in real power is set at 10%.Values of the parameters required for this model can be found. The frequency deviation, plant power change, deviation of internal electrical power and deviation of terminal voltage.

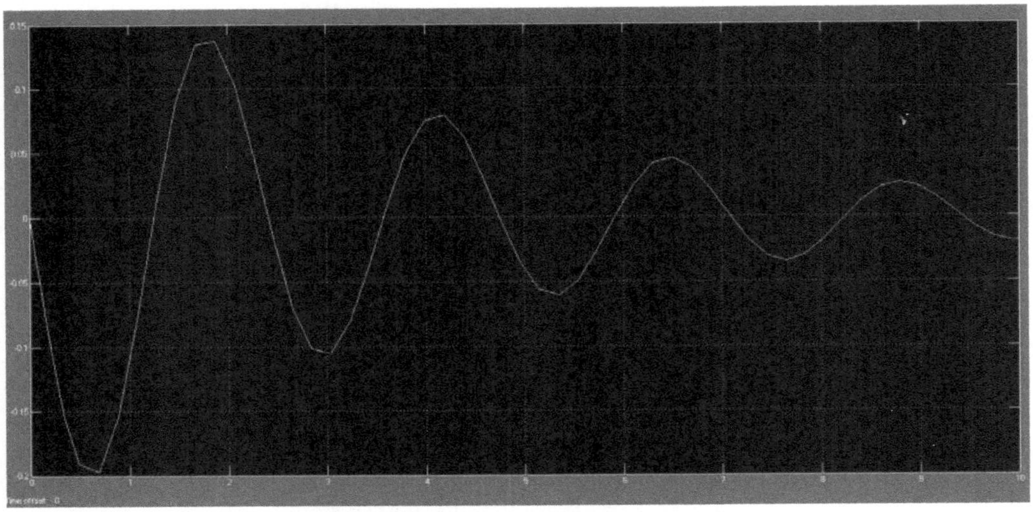

Figure 6.2. FREQUENCY DEVIATION WITHOUT AVR

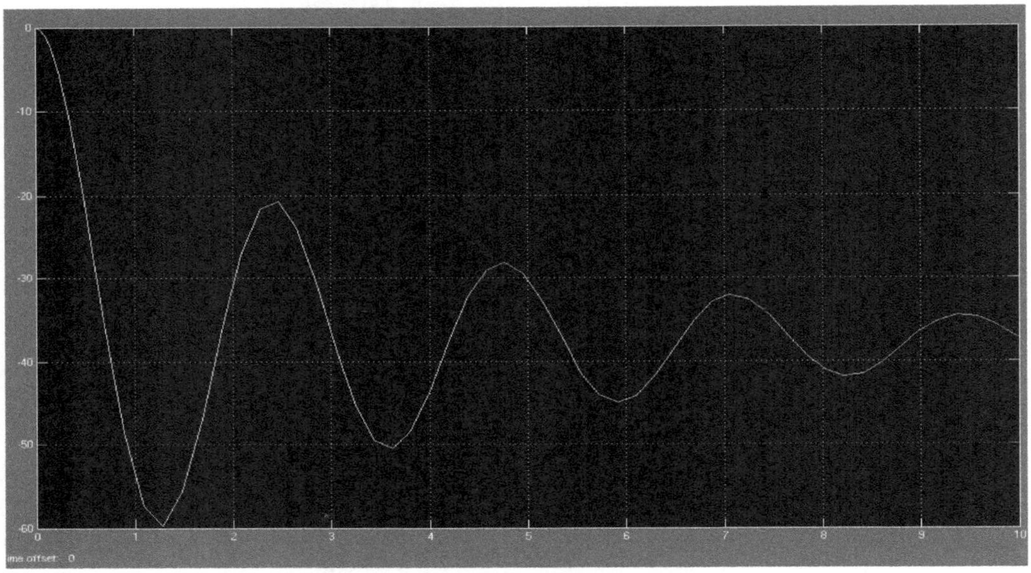

Figure 6.3. PM DEVIATION WITHOUT AVR

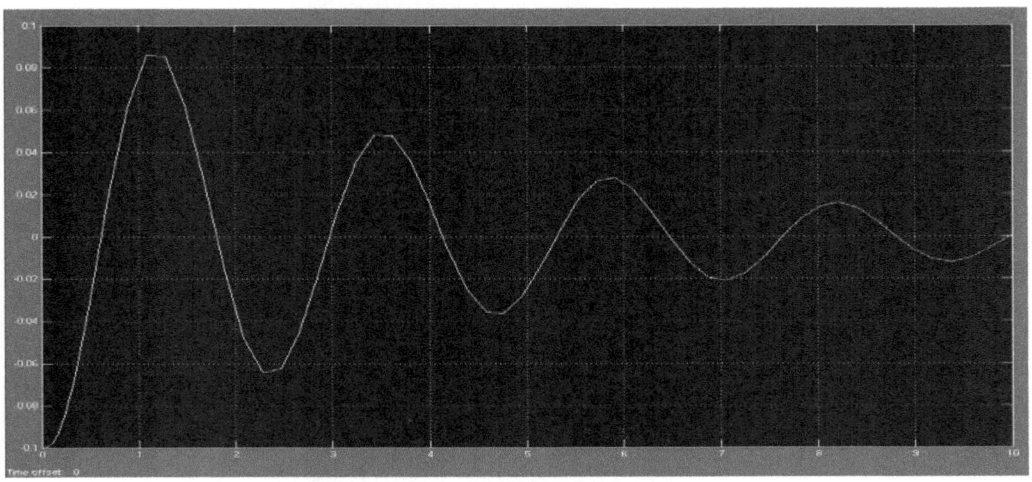

Figure 6.4. PE DEVIATION WITHOUT AVR

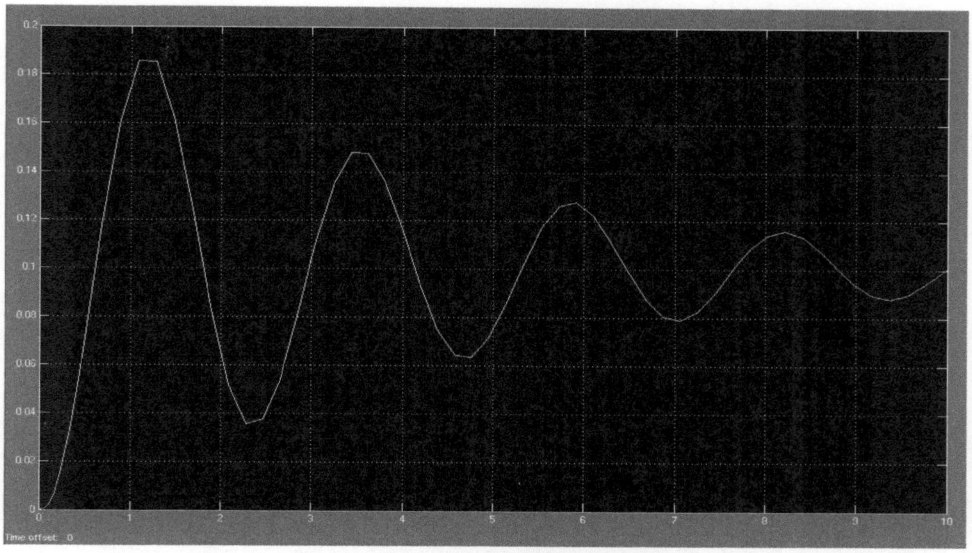

Figure 6.5. TERMINAL VOLTAGE DEVIATION WITHOUT AVR

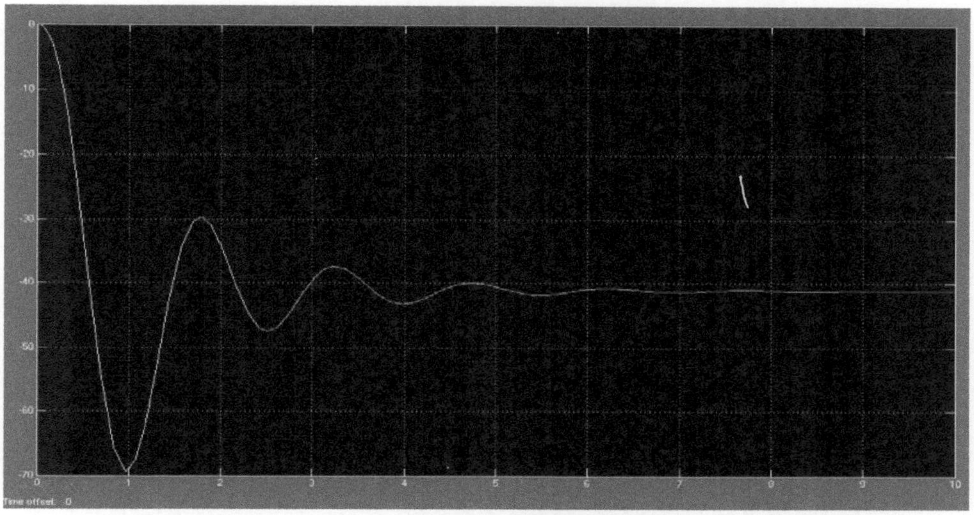

Figure 6.6. FREQUENCY DEVIATIONS COMBINED WITH LFC –AVR LOOPS

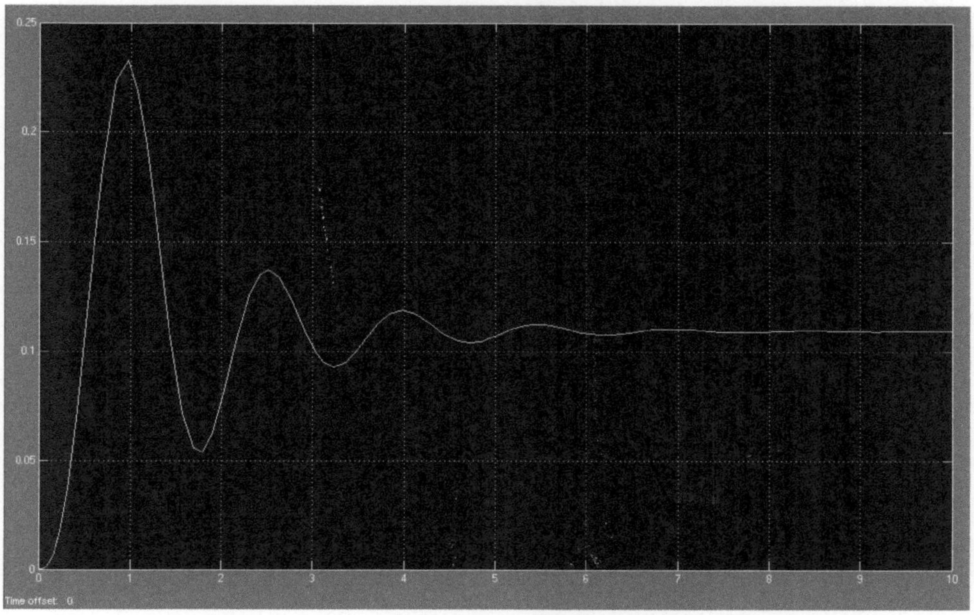

Figure 6.7. PM DEVIATION COMBINED WITH LFC-AVR LOOPS

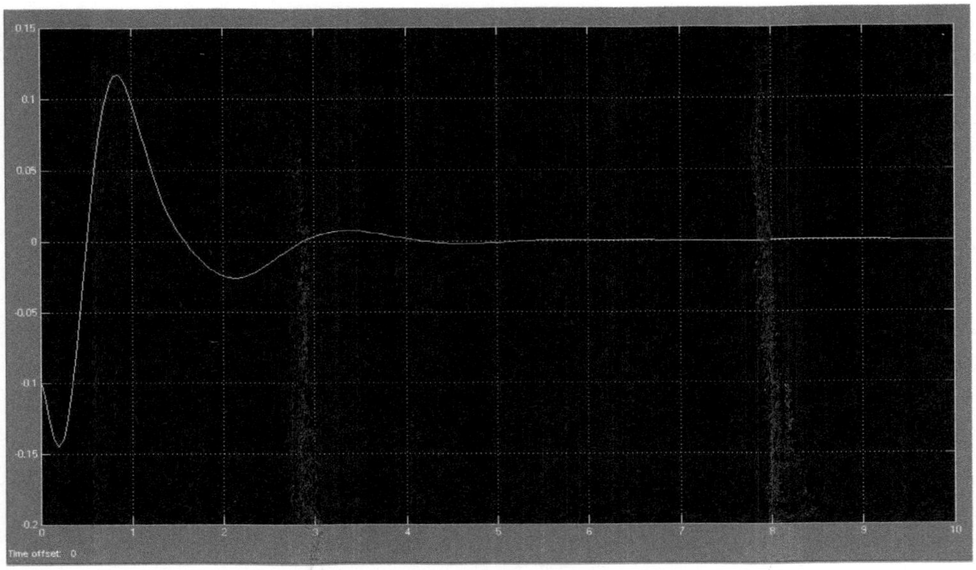

Figure 6.8. PE DEVIATION COMBINED WITH LFC-AVR LOOPS

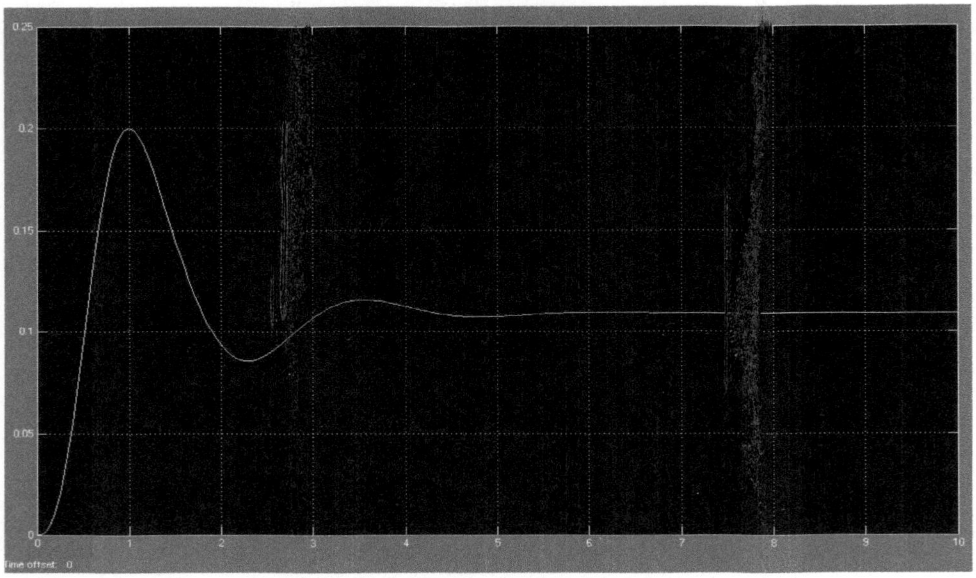

Figure 6.9. TERMINAL VOLTAGE COMBINED WITH LFC-AVR LOOPS

7. Conclusion

In this project, by extending the liberalized AGC system to include the excitation system with using one-machine infinite bus model and classic LFC system, a new combined model is proposed. Based on simulations this model shows interaction between the LFC and the AVR loops. This proposed combined model is tested on one-area power system. The results are shown that by using this model more accuracy will be reachable in dynamic and in the steady state responses. The speed governor and the exciter control loops should be considered as interacting loops when a complete study is done. Note that the choice of values for the governor regulation parameter R and the exciter speed stabilizer gain has a significant effect on the damping of intersystem oscillations as well as area frequencies.

Acknowledgements

I thank to our Institute Executive directors Mr. T Sai kumar & Mr. D Baba for providing creative environment for this work. Also I am very much thankful to our Institute principal Dr. K Ramesh for his kind permission and encouragement to write research paper. I would like to extend my heartfelt thanks to my colleagues. And finally I am very much obliged to my respected parents who inspiring me around the clock.

References

[1] Yau-NaN yu, Electric Power System Dynamics, London, AcademicPress, 1983.

[2] S. C. Tripathy, N. D. Rao, and L. Roy, "Optimization of exciter andspeed governor control parameters in stabilizing intersystem

oscillationswith voltage dependent load Characteristics," *Electric Power and EnergySystems,* vol. 3, pp. 127-133, Jul. 1981.

[3] K Yamashita and H. Miyagi, "Multivariable self-tuning regulator forload frequency control system with interaction of voltage on loaddemand," in Proc. 1991 IEE Control Theory and Applications Conf.,vol.138, pp. 177-183.

[4] N. Jaleeli, L.S. VanSlyck, D.N. Ewart, L.H. Fink and A.G. Hoffmann,"Understanding automatic generation control," IEEE Trans. on Power Systems, vol. 7, no. 3, pp. 1106-1112, Aug. 1992.

[5] Ibraheem, P. Kumar and D.P. Kothari, " Recent Philosophies of Automatic Generation Control Strategies in Power Systems," Power IEEE Trans. on Systems, vol. 20, no. 1, pp. 346-357, Feb. 2005.

[6] H. Saadat, *Power system analysis,* USA: McGraw-Hill; 1999.

[7] J. Sadeh and E. Rakhshani, "Multi-area load frequency control in a deregulated power system using optimal output feedback method," 5th International Conference on European Electricity Market, May. 2008.

[8] E. Rakhshani and J. Sadeh, "Simulation of two-area AGC system in a competitive environment using reduced-order observer method," 5th International Conference on European Electricity Market, May. 2008.

[9] P. Kundur, Power System Stability and Control, McGraw-Hill Inc., 199.

[10] O.I.Elgerd and C.E.Fosha presents the development of state variable model of the megawatt frequency control problem of multi-area electric energy systems and applications of theorems of modern optimal control theory.

[11] Ibrahim and P Kumar paper presents a comprehensive study on dynamic performance of two area power system when subjected to parameter uncertainties.

[12] M.M. Adibi, J.N. Borkoski, R.J. Kkafka, and T.L. Volkman present the frequency response rates and reserve rates developed for typical combustion turbines, steam electric units and hydro electric units.

[13] N.N.Benjamin and W.C.Chan presents a two level optimal control for the load frequency control of inter connected power system using optimal control and multi level system techniques

[14] H. Shayeghi, H. A. Shayanfar, "PSO based neuro-fuzzy controller for LFC design including communication time delays", International Journal on TPE (Technical and Physical Problems of Engineering) IJTPE, Vol. 1, No. 2, pp. 28-36, June 2010.

[15] A. Soundarrajan, Dr. S. Sumathi, C. Sundar, "Particle swarm optimization based LFC and AVR of autonomous power generating system", IAENG International Journal of Computer Science, Vol. 37, No. 1, 2010.

[16] D. Karaboga, B. Basturk, "A powerful and efficient algorithm for numerical function optimization: Artificial Bee Colony (ABC) algorithm", Journal of Global Optimization, Springer Netherlands, Vol. 39, No.3, pp.459-471, 2007.

17

Efficient Baseband Digital Predistorter Using Lut for Power Amplifier (PA) with Memory Effect

B. Feuvrie[1,2], M. Diop[3,*], Y. Wang[1]

[1]IREENA Ecole polytechnique – de l'université de Nantes, Nantes, France
[2]IUT GEII Nantes - Site Fleuriaye, Av. du Prof J. Rouxel, Carquefou, France
[3]Ecole Supérieure Polytechnique, UCAD, BP 5085, Dakar-Fann, Sénégal
*Corresponding author: magdiop2002@yahoo.fr; magaye@ucad.sn

Abstract Modern communications systems will have to integrate multiple radio technologies to support a large number of wireless standards (WLAN, WPAN...). This system should operate with third generation (3G) air interface standards, where non constant envelope modulation formats (MQAM, OFDM...) are required. The amplification of a modulated signal with large variation of envelope introduces distortions when the input-output characteristic is nonlinear. These distortions are particularly important when the power amplifier (PA) has non-linear memory effects with a wideband signal. The technique proposed in this paper aims to fight against this phenomenon by optimizing the method developed in the laboratory [7,20] for a real-time implementation. This technique is able to provide exactly the reverse characteristics of a PA modeled by a Hammerstein model in base band. Its principle is to find the largest real root of a polynomial related to the analytical model of the PA and to the input signal to be amplified to calculate a pre-distorted signal. To reduce significantly the computation time, an implementation by the LUT method is proposed. This method uses two pre-distortion principles and can be implemented on a FPGA board with table bases method known for simplicity including in the same way the correction of the memory effect.

Keywords: *nonlinear distortion with memory, digital pre-distortion, power amplifier (PA), Hammerstein model, Look Up Table (LUT)*

1. Introduction

Nonlinear distortions in the transmitter are typically introduced by the RF Power Amplifier (PAs). High efficiency wide-band transmitter design for modern high speed wireless communication systems, such as worldwide interoperability for microwave access (Wimax), third generation and beyond systems, is a complex task. Simultaneously accomplishing high linearity and high power efficiency is a great challenge.

Ideally, the output signal of a PA is proportional to its input signal. But in practice, the output signal is compressed and comes to a saturation zone when the power of the input signal is too high. In this case, the output signal is subjected to the distortions of the amplitude and phase. If the input signal is a modulated signal, two undesirable phenomena can be observed at the output. The first is a distortion of the input signal's constellation, leading to an increase of the Bit Error Ratio (BER) of the system. The second is the phenomenon of spectral regrowth or the appearance of new spectral components in the adjacent channels. In digital communication systems, adjacent channels may be allocated to the other transmissions and then the interference created (ACI: Adjacent Channel Interference) by nonlinearity becomes unacceptable. The linearity is thus a fundamental characteristic for the design of power amplifiers [1,2]. One way of suppressing the nonlinear effects is to operate a non linear PA away from its saturation region by adding sufficient back-off to the input signal. However, such amplification schemes possess low-power efficiency since maximum power efficiency is only attained when the PA is operated near its saturation zone. But a good power efficiency is a highly desirable property of a power amplifier, because, it allows to minimize the heat dissipation to reduce the size and the weight of the dissipating elements and thus to increase the autonomy of the terminal. To operate the PAs with high power efficiency, it is necessary to choose an operating point near the compression zone. Under these conditions, we have to correct these undesirable effects by associating, for example, a further PA correction circuit.

Among all linearization techniques, the baseband digital predistortion (DPD) is one of the most cost effective. It is basically relying on the equivalent baseband modeling of the PA and/or its inverse. For digital signal processing convenience, it is very desirable to implement the PD in baseband.

The pre-distortion (PD) is an effective technique which is especially suitable for broadband applications. In order to obtain an output signal proportional to the input signal (Figure 1), the PD technique consists to design a circuit whose characteristics are the opposite to that of PA.

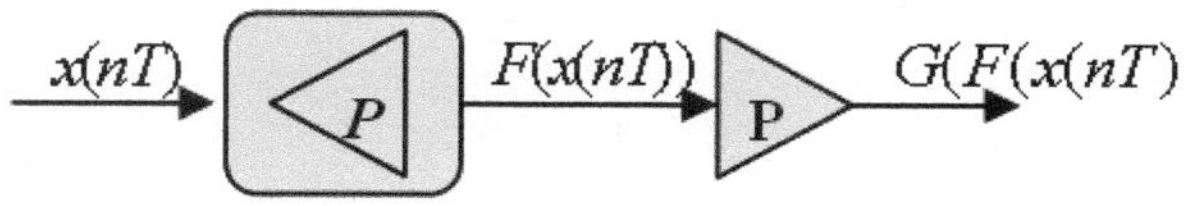

Figure 1. Principle of the pre-distortion

In the Figure 1, the function G represents the nonlinear characteristics of the PA and F those of the PD, which should be the opposite of those the PA. The output $y(nT)$ of the cascade of the two blocks, PD and AP, is written as:

$$y(nT) = G(F(x(nT))) = G_d x(nT) \tag{1}$$

where G_d is a positive constant representing the gain of the linearized amplifier and $x(nT)$ is the input signal. The module and phase of this function must be constant for the whole amplitude range up to saturation.

Different techniques, based on analogue, digital or hybrid circuits, could be adopted while designing the PD. In this paper, we present a technique for baseband adaptive digital pre-distortion (DPD). Its principle is shown in Figure 2. The modulated baseband signal is distorted in DPD, then converted into analog signal, and finally modulated onto RF signal. The pre-distorted signal will be amplified by PA with a linearized output signal.

The baseband DPD is suitable for real-time implementation. In fact, digital circuits working on baseband allow the system reconfiguration, the baseband processing and reduce the sampling constraint which is very important for a modulated signal (often a few GHz). Excellent results on linearity performances can be achieved by the baseband DPD with good power efficiency and flexibility, low costs, and reduced complexity compared to the analogue techniques. The DPD can be implemented on an ASIC (Application-Specific Integrated Circuit) [3], an FPGA (Field Programmable Gate Array) [4] or a DSP (Digital Signal Processor) [5].

In modern SDR (Software-Defined Radio) transmitters, most components must be reconfigurable in order to switch, ideally in the air, from one standard to another. According to these requirements, the DPD technique seems to be an applicable linearization technique. The only limit for the applications of DPD in baseband is the bandwidth of the signal which can be treated by the processor. This limit can be an inconvenient factor for the wide band applications.

In this paper, we present a technique to optimize the technique proposed in [7]. The characteristics of the PA are modeled by a Hammerstein model. This model is a memory less non linearity followed by a linear time invariant (LTI) system.

This model is able to take into account the nonlinearity and also the memory effects of the PA. This method allows to inverse exactly the characteristics of the PA by a direct identification of the behavioral model.

Once the model is identified, the technique corrects the signal perfectly. This method needs to search for the largest real root of a polynomial, which requires too much computation time for a real-time application [17,20]. In this paper, we propose a technique based on the principle of LUT which will significantly decrease the computation time of the original root-finding based technique.

The paper is organized as follows. The principle of the proposed method is described in Section 2. Section 3 presents the new LUT based technique. The results of the original method and the new method are compared in Section 4. The experimental results are given in Section 5. Finally, a conclusion on the proposed new technique compared to the other existing methods is presented.

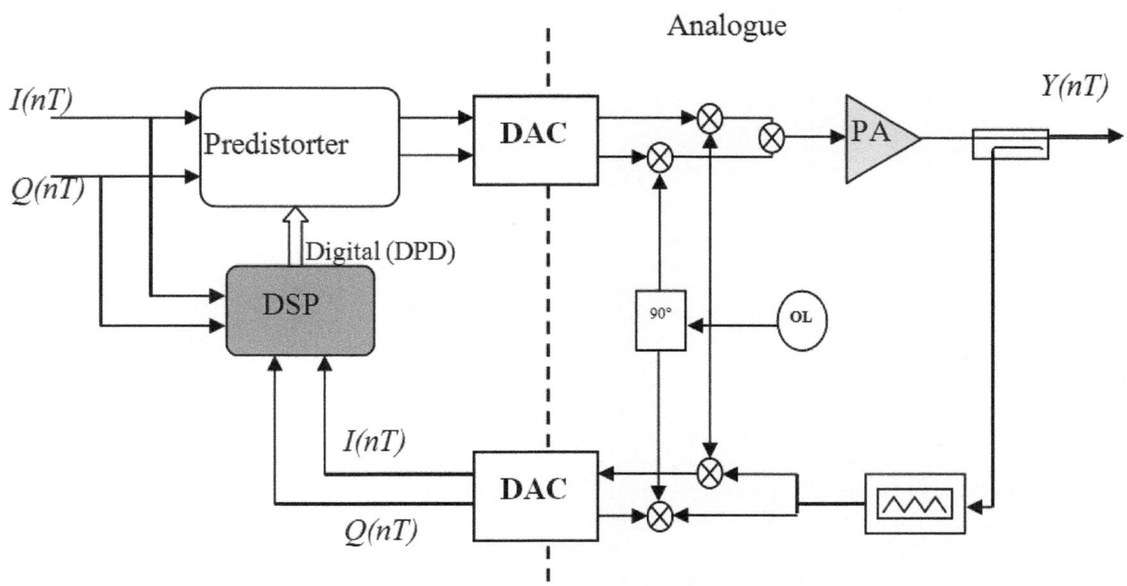

Figure 2. Baseband Digital Pre-Distortion Principle

2. Direct Inversion of the Transfer Function

2.1. Choice of the Method of PD

There are essentially two types of architecture of baseband DPD techniques: the direct and the indirect architectures [13].

The first type identifies the amplifier model firstly and then reverses it to realize a pre-compensator [14,15,16]. From the power amplifier model, it is possible to deduce the value of the output signal of the pre-compensator. The block diagram is presented in Figure 3. This method is chosen in this work. In Section 2.2, it will be discussed in detail.

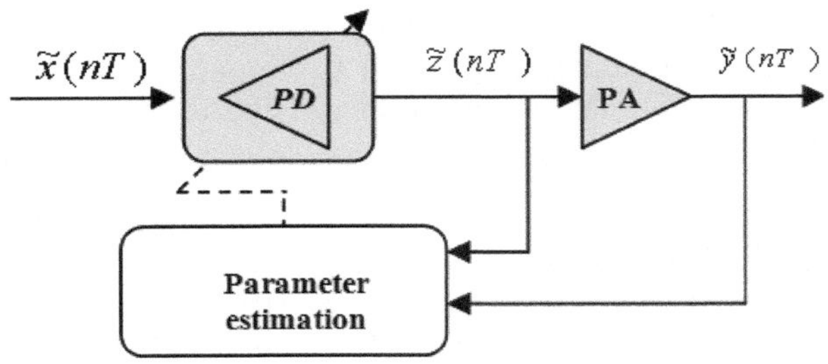

Figure 3. Linearization method based on a direct architecture

The second type is based on an indirect identification method (Figure 4) ([8,9,10,13]).

The principle is initially proposed by Gao et al. [11] to compensate the nonlinearity in the speaker, and then used by Eun et al. [12] in the DPD with memory.

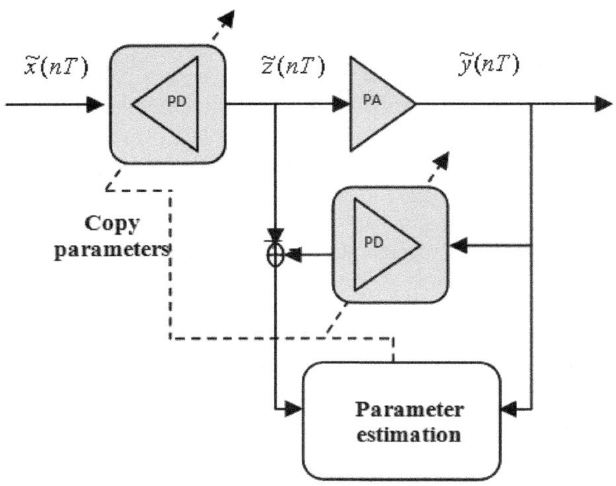

Figure 4. Linearization method based on an indirect architecture

This class of methods is called indirect method of identification or post-distortion, because the post-inverse of the amplifier is identified and placed before the amplifier [13]. The output signal samples of PA divided by the gain are the inputs of model and the input signal samples of PA are the outputs of model.

The advantage of this identification method is that it can be realized in a single step. Thus, we do not need to concern the modeling of the PA, but directly get its inverse. This greatly facilitates the linearization procedure, even with complex models taking into account the memory effects of the PAs [19]. However, this method has two disadvantages which can affect its performance. The first one is the measurement noise *b(n)*, which appears at the entrance of the identification block. The post-reverse thereby results errors in the identified model. The covariance matrix may be ill-conditioned in some cases. The second one is if the memory of the amplifier is limited, its inverse characteristics will have an infinite memory. Thus some approximations should be made.

2.2. Model of the Power Amplifier

The baseband digital pre-distortion technique with the direct architecture (Figure 3) is chosen in this study.

To model the characteristics of the PA, a Hammerstein model (Figure 5) is used. The static nonlinearity is modeled by a polynomial of odd order (coefficients a_{2i+1}), and the memory effect is represented by a FIR (finite impulse response coefficients h_k) filter.

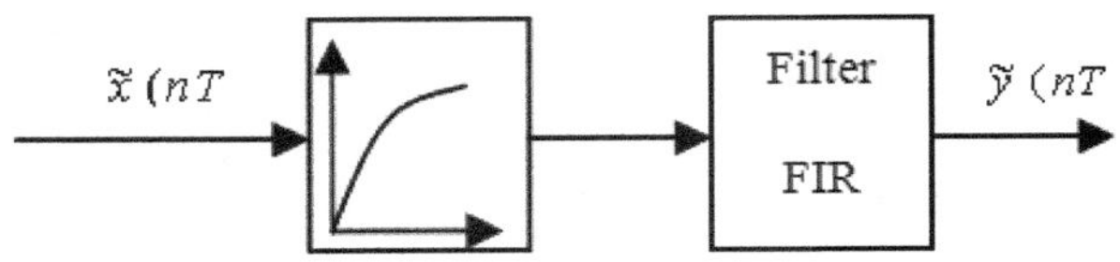

Figure 5. Hammerstein model of PA

The baseband output signal is written as:

$$\tilde{y}(nT) = \sum_{k=0}^{P-1}\sum_{i=0}^{N} h_k a_{2i+1} \left|\tilde{x}[(n-k)T]\right|^{2i} \tilde{x}[(n-k)T] \quad (2)$$

This model can be identified with the least squares method.

2.2. Principle of the Pre-compensator

From the above Hammerstein PA model, it is possible to deduce the value of the output signal of PD. The block diagram of the proposed PD is given in Figure 6:

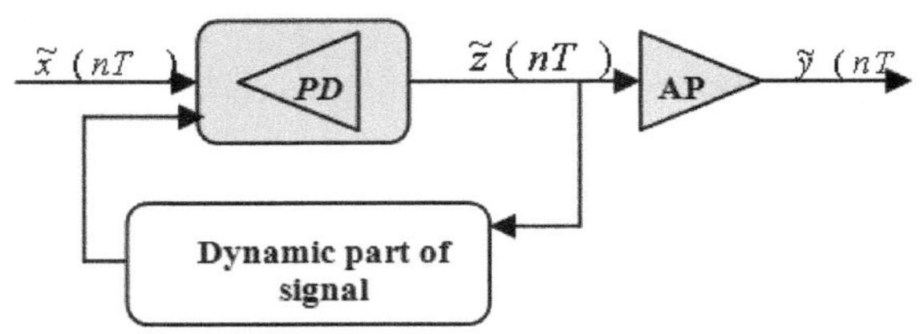

Figure 6. Principle of the adaptive PD linearization technique

The PD is composed of two blocks: the direct loop compensates the amplifier nonlinearities and the back loop compensates its memory effect.

The relation between the complex envelopes of the output signal $\tilde{y}(nT)$ and that of the input signal of the power amplifier $\tilde{z}(nT)$ is given by:

$$\begin{aligned}\tilde{y}(nT) &= \sum_{k=0}^{P-1}\sum_{i=0}^{N} h_k a_{2i+1} \left|\tilde{z}[(n-k)T]\right|^{2i} \tilde{z}[(n-k)T] \\ &= G_d \tilde{x}(nT)\end{aligned} \quad (3)$$

where $\tilde{z}(nT)$ is also the pre-distorted signal (output of DPD), G_d is the desired linear gain of the amplifier and $\tilde{x}(nT)$ is the input signal of DPD representing the signal to be amplified.

Separating the components of the instant and those associated with the previous in formula (3), we get the static part and dynamic part. The static part can be expressed as follows:

$$h_0 \sum_{i=0}^{N} a_{2i+1} \left|\tilde{z}(nT)\right|^{2i} \tilde{z}(nT) \quad (4)$$

and the dynamic part representing the memory effect is given by:

$$\eta(nT) = \sum_{k=1}^{P-1} \sum_{i=0}^{N} h_k a_{2i+1} \left|\tilde{z}[(n-k)T]\right|^{2i} \tilde{z}[(n-k)T \quad (5)$$

Then, equation (3) can be rewritten as:

$$\sum_{i=0}^{N} a_{2i+1} \left|\tilde{z}(nT)\right|^{2i} \tilde{z}(nT) = \frac{G_d \tilde{x}(nT) - \tilde{\eta}(nT)}{h_0} \quad (6)$$

The aim of the system is to find the values of $\tilde{z}(nT)$ at the input of the PA to compensate the nonlinearities and the memory effect of the PA. We should solve equation (6) to find $\tilde{z}(nT)$. The dynamic part related to the previous values, and the input signal of the system are known at time nT, the amplifier coefficients are identified during the initialization procedure.

The algorithm of the pre-distortion can then be written as:

Table 1. PD algorithm

Initialization phase: $\tilde{\eta}(1) = 0$
$n = 1$
Loop
{
- Solve equation (6) to find $\tilde{z}(n)$
- Calculate:

$$\tilde{\eta}((n+1)T)) = \sum_{k=1}^{P-1} \sum_{i=0}^{N} h_k a_{2i+1} \left|\tilde{z}[((n+1)-k)T]\right|^{2i} \tilde{z}[((n+1)-k)T]$$

- Increment the loop (n=n+1)
}

Solving equation (6) can be divided into two parts. Firstly, we should determine the module $Z(nT) = |\tilde{z}(nT)|$, which is the largest real and positive root of the following polynomial:

$$\left|\sum_{i=0}^{N} a_{2i+1} \left|\tilde{z}(nT)\right|^{2i+1}\right| = \left|\sum_{i=0}^{N} a_{2i+1} Z(nT)^{2i+1}\right| = \left|\frac{G_d \tilde{x}(nT) - \tilde{\eta}(nT)}{h_0}\right| \quad (7)$$

Secondly, we should calculate the associated phase $\varphi_z = \arg(\tilde{z}(nT))$ given by:

$$\varphi_z(nT) = Arg\left[\frac{G_d \tilde{x}(nT) - \tilde{\eta}(nT)}{h_0 \sum_{i=0}^{N} a_{2i+1} Z(nT)^{2i+1}}\right] \quad (8)$$

It takes a lot of computer resources and time to calculate the largest real and positive root (theoretically, it is proven that this root always exists [7,17]).

3. Optimization of the above Pre-compensator by Using a LUT

The previous method presents excellent results if the PA model is close to the behavior of the PA. The calculation time is the main problem. Indeed, solving the polynomial equation (7) to determine the root is time-consuming, whichever method (Newton ...) is chosen.

One technique to overcome this handicap is to introduce a table which determines directly the relationship between the input signal $\tilde{x}(nT)$ and the pre-distorted signal $\tilde{z}(nT)$ .This technique has the implementation efficiency of the conventional LUT techniques. In addition, it is able to take into account the memory effect without significantly increasing the complexity.

A careful examination of equation (7) shows that the coefficients of the polynomial are constant, only the right member changes for each new sample. This observation allows us to implement a procedure to avoid the time-consuming root-finding procedure. In order to avoid the root searching, we decompose the maximum dynamic rang of the signal $\tilde{z}(nT)$ by several equal length intervals. Associated with these intervals, the values of the amplitude of the polynomial are calculated by:

$$\left|\sum_{i=0}^{N} a_{2i+1} Z(nT)^{2i+1}\right| \quad (9)$$

This table provides directly the corresponding the module of the root for each value of the right member of the polynomial (7).

Therefore, for each input $\tilde{x}(nT)$, we compute $\left|\frac{G_d \tilde{x}(nT) - \tilde{\eta}(nT)}{h_0}\right|$ to find the corresponding amplitude $Z(nT)$ and angle $\varphi_z(nT)$ given in the table. The algorithm is greatly simplified since we only need to calculate $\varphi_z(nT)$ by equation (8) and $\tilde{\eta}((n+1)T)$ by equation (5). Finally, the new algorithm can be written as given in Table 2.

Table 2. PD algorithm based on LUT

Initialization: $\tilde{\eta}(1) = 0$
$n = 1$
Loop
{- Compute $\left|\frac{G_d \tilde{x}(nT) - \tilde{\eta}(nT)}{h_0}\right|$ and find the corresponding $Z(nT)$ in the table
- Compute $\varphi_z(nT)$ by equation (8)
- Compute

$$\tilde{\eta}((n+1)T)) = \sum_{k=1}^{P-1} \sum_{i=0}^{N} h_k a_{2i+1} \left|\tilde{z}[((n+1)-k)T]\right|^{2i} \tilde{z}[((n+1)-k)T]$$

- Increment the loop (n=n+1)
}

This new technique achieves almost the same performance as the initial technique described in Table 1. Due to the significant decrease in computation time, the implementation becomes much more feasible. Figure 7 illustrates the main procedures of algorithm.

Nowadays, there are a number of FPGA modules which can produce the DPD by look up tables (they are mostly designed for linearizing memory less nonlinear PAs). The output characteristics of PA are linearized up to the saturation zone. These techniques involve calculating the pre-distorted signal and the corresponding module of the input signal by inversing the input-output characteristic of PA. The values of the input signal are quantified.

In comparison, the proposed method is comprised of a phase calculation process by equation (8) and an operation involving multiplications, divisions and additions by equation (5). These different operations can be easily implemented on FPGAs. The most important point of this technique is that it allows resolving the nonlinearities with memory effect.

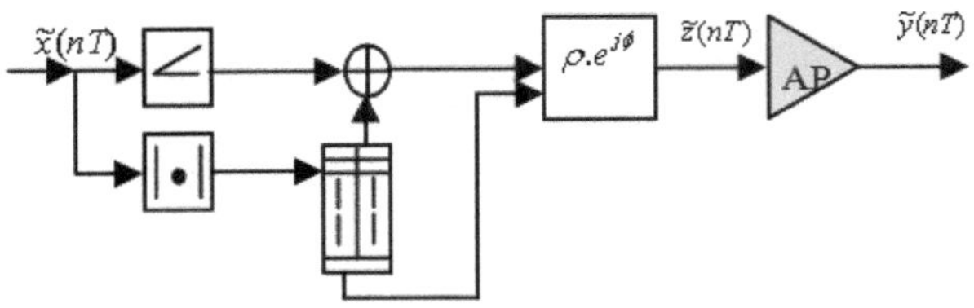

Figure 7. Principle of implementation by table

4. Simulation

4.1. Power Amplifier Model

The PA model aims to simulate the behavior of a PA. The static nonlinearity is modeled by the Saleh model [21] with coefficients given in Table 3.

Table 3. Coefficients of Saleh model

Amplitude	α_a	2
	β_a	2,2
Phase	α_p	2
	β_p	1

The Input-output relation is given by:

$$\tilde{y}(nT) = F_a\left(\left|\tilde{x}(nT)\right|\right)e^{j\left[\arg((\tilde{x}(nT)+F_p\left(\left|\tilde{x}(nT)\right|\right))\right]} \quad (10)$$

$$F_a\left(\left|\tilde{x}(nT)\right|\right) = \frac{\alpha_a\left|\tilde{x}(nT)\right|}{1+\beta_a\left|\tilde{x}(nT)\right|^2} \quad (11)$$

$$F_p\left(\left|\tilde{x}(nT)\right|\right) = \frac{\alpha_p\left|\tilde{x}(nT)\right|^2}{1+\beta_p\left|\tilde{x}(nT)\right|^2} \quad (12)$$

where F_a and F_p are the transfer functions of AM/AM and AM/PM respectively.

AM/AM and AM/PM characteristics are shown in Figure 8. 1 dB compression appears at the average input power of 0.6dBm.

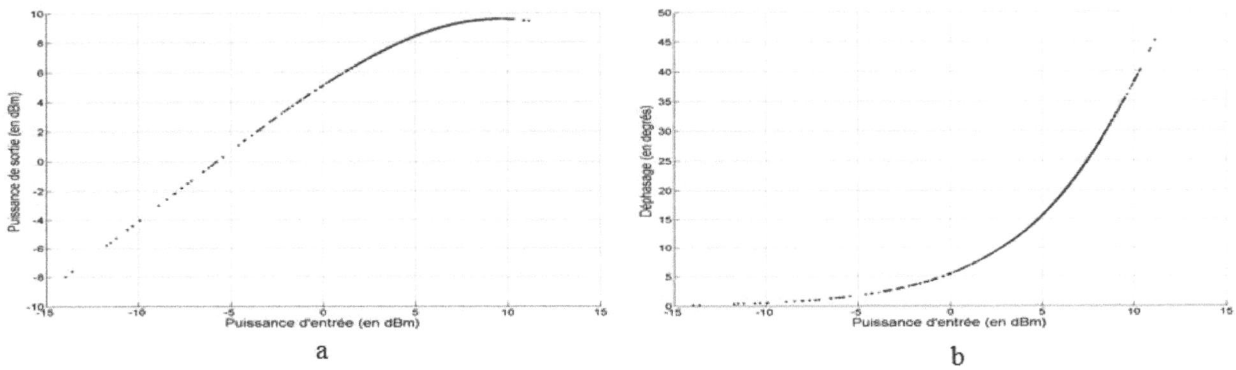

Figure 8. AM/AM Saleh model

A FIR filter follows the Saleh model to take the memory effect into account. The input-output relationship is given by:

$$s(nT) = x(nT) + (0.5+0.1j)x\left[(n-1)T\right] \quad (13)$$

4.2. Comparison of the Two Methods

The above model is used to simulate the behavior of the power amplifier.

The Saleh model with different groups of parameters and M-QAM modulation (M=16) with a raised-cosine filter for pulse shaping (roll-off=0.35) and an up-sampling factor 8 have been adopted to assess the performance.

The symbol rate is 4M symbols/s and 2000 symbols are taken. Initially, the size of the table N is 4096. The output saturation power of the model is approximately 9.75dBm.

The PAPR of a 16QAM signal filtered by a raised cosine filter being about 7.25 dB, the lowest value of output back-off of the linearized PA (LPA) for a possible linearization is $OBO_{min} \approx 7.25$ dB. We can deduce the maximum average output power $P_{linmax} \approx 2.5$dBm corresponding to the input power -3dBm. The values of the EVM and ACPR for various levels of the output power are shown in Figure 9.

The power spectral density (PSD) of input/output signals of the PA is estimated using the Welch method in Matlab. The length of each segment is 300 samples, with 50% overlap between each segment (one hundred segments totally). A window of Blackman is used. The Matlab function which performs this calculation is the function "psd (•)". The use of the Blackman window minimizes the side lobe level due to the finite length of data segments. This is very important for precisely estimating the power spectral density (PSD) of the linearized and non linearized spectrums which have an important dynamic. Lets note that other types of window can be used with results comparable to those obtained with the Blackman window (Han, Chebyshev, Parzen, Bartlett, Hanning, etc...). This method of estimating the PSD of the signals is also adopted in real measurement, presented later.

We can see that with an output power of 2.5dBm, the performance, in terms of ACPR, degrades rapidly. The speed of degradation depends on the proportion of samples which enter the saturation zone.

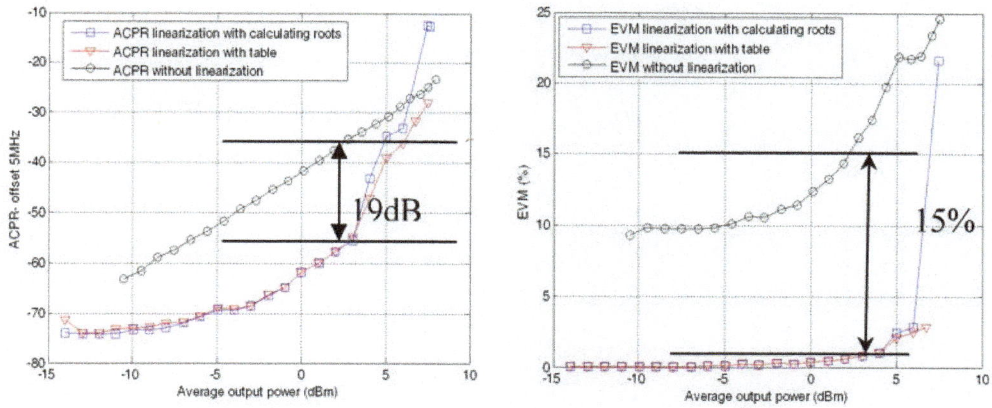

a b

Figure 9. a- ACPR, b- EVM

A significant improvement of ACPR has been observed when the average output power does not exceed 2.5dBm in Figure 9 (a) with a maximum correction of 19dB.

The memory effect is well corrected in the operating zone (2,5dBm). Without pre-distortion, in the linear zone (with low input power), we can observe a residual EVM due to the memory effect from Figure 9 (b). We can observe also that for an average output power around 3dBm and 5dBm, the pre-distortion becomes less effective for ACPR and EVM respectively.

The performances deteriorate rapidly with the method of direct polynomial root searching, because the identified model becomes less precise for the samples of high power. But with the LUT based method, the degradation is slower. As the average output power increases, the larger the samples deviate from the model, the more the errors caused by the root finding technique.

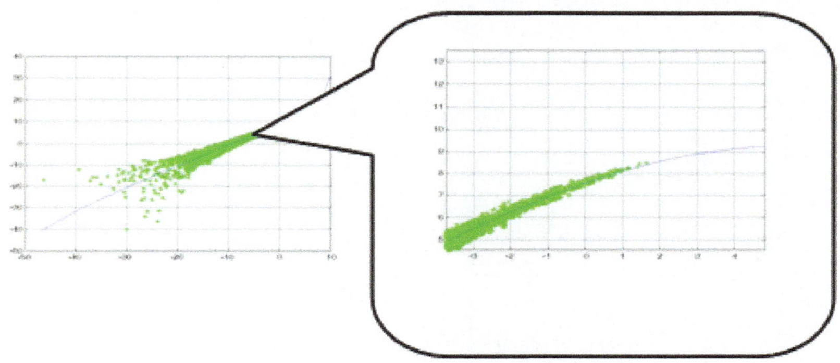

Figure 10. Polynomial identified model, Average power input -5dB

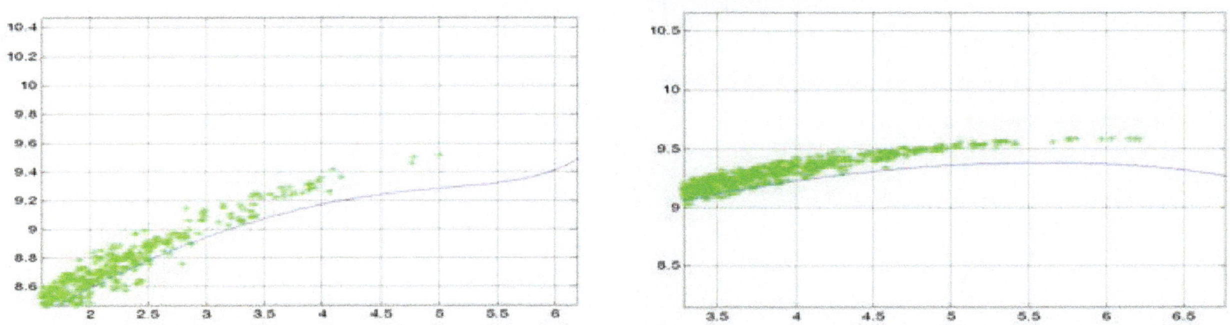

a b

Figure 11. a- Polynomial identified model (solid line: identified model; discrete lines: samples of PA, Average power input -2dBm); b- Polynomial identified model (solid line: identified model; discrete lines : samples of PA, Average power input 0dBm)

Figure 10 and Figure 11 show that when the power of input signal increases, the gap between the samples of PA output signal and the samples of the identified model increases.

In the following, to analyze the influences of different parameters, we set the input power equal to -5dBm. The identification of the Hammerstein model is achieved by the least squares method. The simulation results are presented as follows with the table size N of 4096.

The eye diagram of the nonlinear output signal in Figure 12 (b) is strongly distorted compared with the eye diagram of the input signal Figure 12(a).

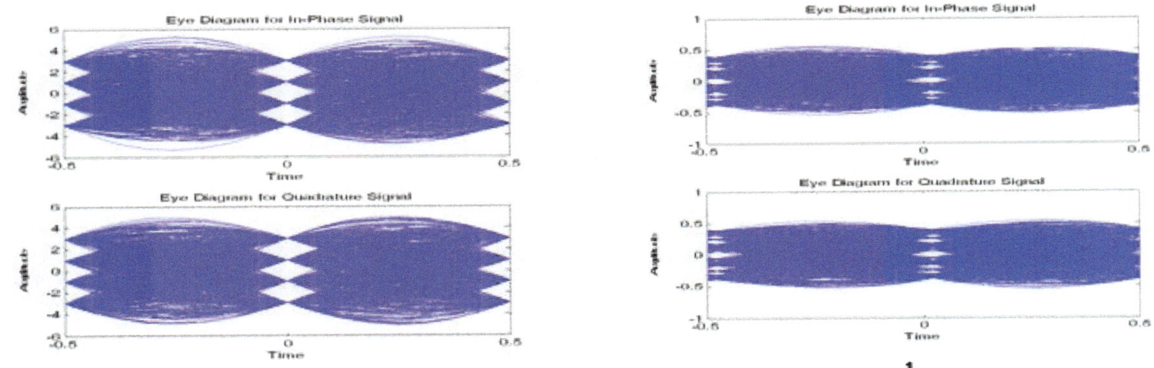

a b

Figure 12. a-Eye diagram, input signal; b- Eye diagram, output signal without linearization

Constellations of input signal and the output signal without linearization are presented in Figure 13. Rotations and dispersions of constellations are due to the nonlinearity of PA. The results with linearization are shown in Figure 14. Perfect reshaping symbols are achieved.

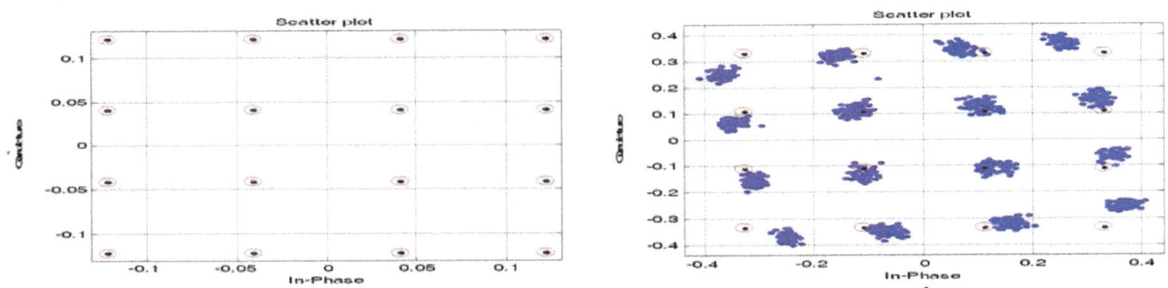

a b

Figure 13. a- Constellation of the input signal, (EVM=0%); b- Constellation of output signal without linearization (EVM=17.53%)

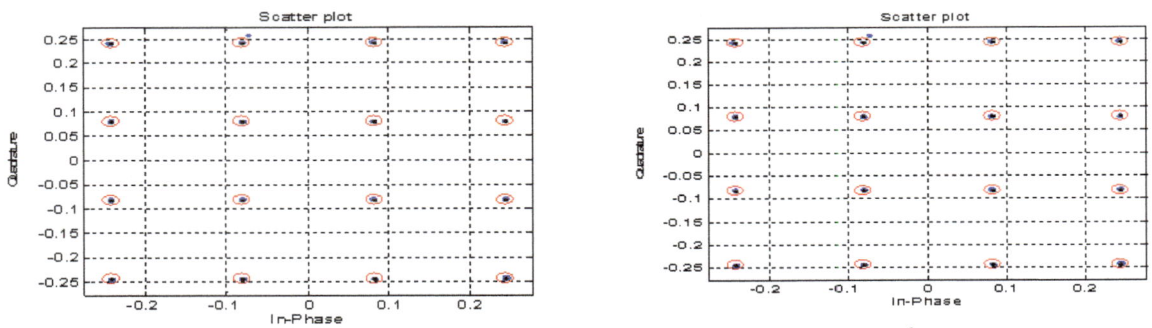

a b

Figure 14. a- Constellation with linearization, Method with root calculation (EVM=0.44%); b- Constellation with linearization, Method with LUT (EVM=0.44%)

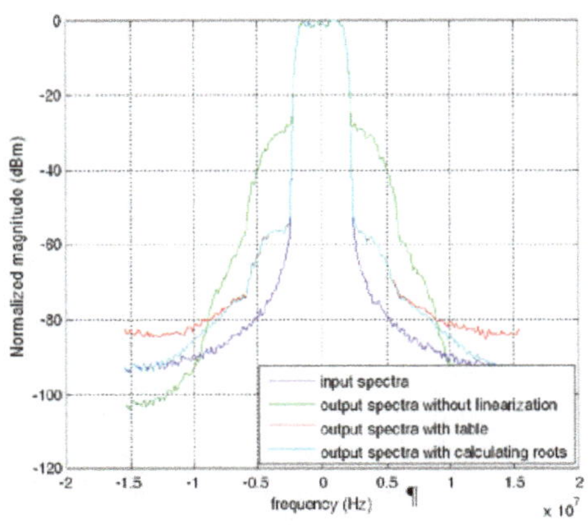

Figure 15. Spectrum after linearization (N= 4096)

In Figure 15, a significant improvement on the regrowth spectrum is observed by the proposed LUT based method which is almost identical to the method based on direct root finding of a polynomial.

The ACPR is resume in table

Table 4. ACPR after linearization

ACPR	
Input	-74dB
Output without linerization	-33.8dB
Linearized output Root solving	-60.2dB
Linearized output Method with table	-60.1dB

4.3. Influence of the Size of the Table

A simulation with a table size of 64 has been made and compared with the table size of 4096 Figure 16.

There is a residual distortion of the constellation after linearization with a table size N of 64. This is due to the quantization effect on the pre-distorted signal. But the EVM still has a very low value.

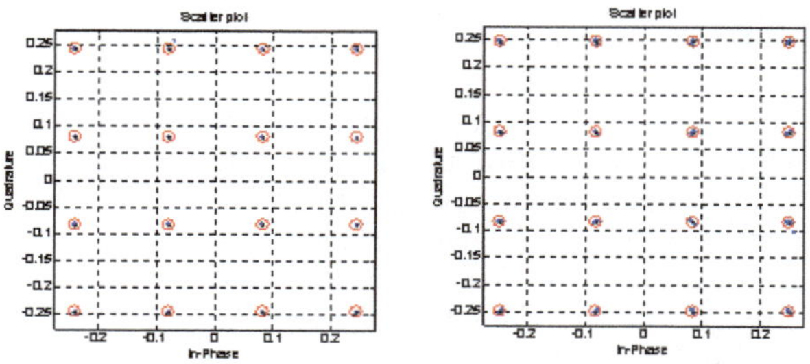

Figure 16. Linearization with table: (a) N=4096 (EVM=0.44%); (b) N=64 (EVM=2%)

Figure 17 presents a level of -52dBm in the adjacent channel for the output of the PA without linearization. Compared with the method based on the polynomial root finding, the proposed method achieves an ACPR of -49.6 dB instead of -59.6 dB. If the table size is increased to 256, the ACPR will be of -59 dB and the EVM of 0.8%.

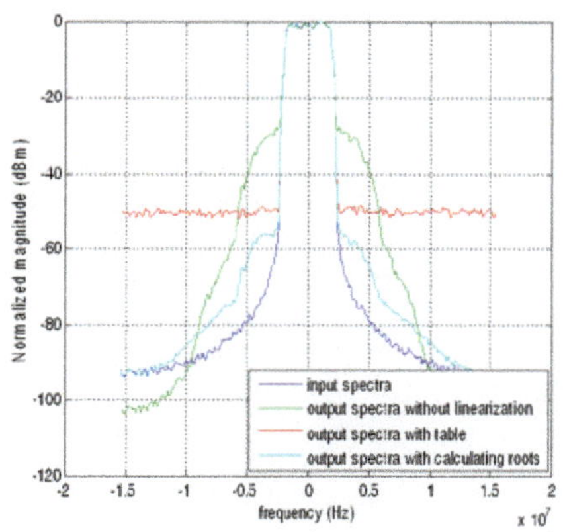

Figure 17. Spectrum before and after linearization (N=64)

The proposed method is more appropriate for a real time implementation. Of course, it has the same limitation as all the linearization techniques: the average power of input signal should not be too close to the saturation zone. In these conditions, according to the value of the PAPR, the quality of PD will decrease sharply.

5. Experimental Results

5.1. Test Bench Description

In this section, we first describe the test bench designed for the experiments. Then, we will evaluate the performance of the digital baseband pre-distortion technique, by using an amplifier from Mini-Circuits, ZFL-2500, driven by a 16-QAM modulated signal.

In order to evaluate the proposed linearization techniques, we begin by identifying a model of the PA from the input/output signals from the power amplifier under test. Then, we present the experimental results, and compare them to the theoretical ones.

The performance of the PD is evaluated by measuring two important parameters, Adjacent Channel Power Ratio (ACPR) and Error Vector Magnitude (EVM), with different values of power output signal.

The test bench [18] consists of a Vector Signal Generator (VSG), a digital oscilloscope (DO) and a personal computer (PC) (Figure 18). It is designed to be fully automatic by using Matlab toolbox. The data are generated in Matlab and then sent to the VSG. The VSG (Rhode & Schwartz SMU 200A) receives the complex envelope data via an Ethernet cable (TCP/IP) from the PC and uses a direct up-conversion from baseband to RF, produces virtually any signal within its bandwidth limits. Note that, once the data have been sent to the VSG, the latter will repeatedly send the corresponding modulated signal to the PA. A marker can be activated to trigger the DO each time when the sequence is regenerated.

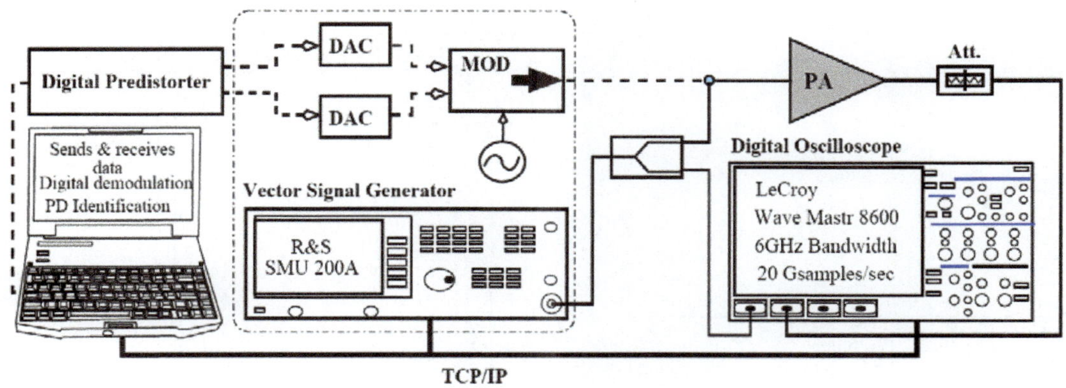

Figure 18. Test Bench

The RF input and output signals of the PA are then sampled simultaneously in the real time oscilloscope (Leroy, 4 channels Wave master 8600, 6GHz bandwidth, 20 GS/sec), and then transferred via an Ethernet cable to the PC, and finally recorded in the of workspace Matlab.

The acquisition time in the DO is fixed to be equal to the duration of the baseband signal generated by Matlab. In this way, the acquired RF signals correspond exactly to the original signal of Matlab. After that, the two sequences are digitally demodulated in Matlab, in order to identify the parameters and evaluation can be implemented in a single program running in Matlab without interruption. Note that, for signal acquisition, the spectrum analyzer "Agilent E4440A" can also be used as an alternative tool for precision, comparison and verification. In this case, the signal analysis software provided with this instrument can be used to demodulate and acquire the input and output signals separately.

The signals can be synchronized by correlating them with the original of Matlab signal.

After that, the two sequences are digitally demodulated in Matlab, in order to identify the parameters of the PD. The baseband signal is then processed by the pre-distortion function and loaded again to the VSG. Finally, the output of the linearized PA is digitized in the DO and sent back to the PC to evaluate the performance of the PD. This evaluation can be done by comparing the output spectrum (ACPR) and constellation distortion (EVM) of the PA with and without linearization at different back-off levels.

The time cost of this entire test is just several minutes since this test bench is fully automatic. In other words, the signal transmission, acquisition, identification and evaluation can be implemented in a single program running in Matlab without interruption.

Note that, for signal acquisition, the spectrum analyzer "Agilent E4440A" can also be used as an alternative tool for precision, comparison and verification. In this case, the signal analysis software provided with this instrument can be used to demodulate and acquire the input and output signals separately.

The signals can be synchronized by correlating them with the original of Matlab signal.

Measurements have been carried out on an amplifier ZFL 2500 from Mini-circuits. This wide-band (500-2500 MHz) PA is used in several types of application, typically in GPS and cellular base stations. According to its data sheet, it has a typical output power of 15dBm at a 1 dB compression point, and a small signal gain of 28 dB (±1.5).

In order to validate the study presented in Section 3, we have performed the one tone test on this PA.

The defined parameters, namely, compression, interception and saturation points are also very useful for the experimental evaluation of the DPD technique. Figure 19 shows the AM/AM characteristic of the PA under test.

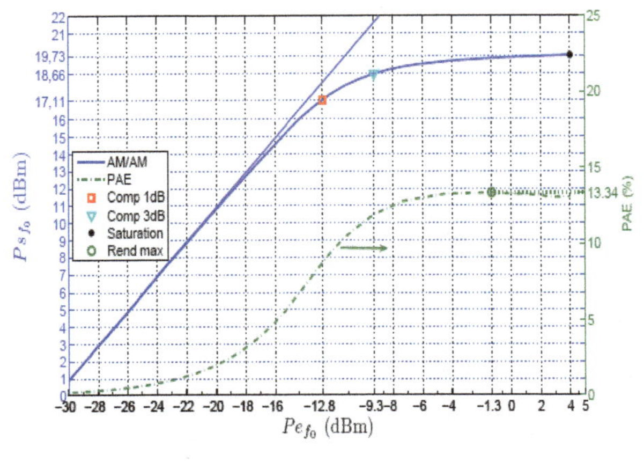

Figure 19. Compression and interception points, efficiency extracted from static power measurements (ZFL-2500 PA)

In order to compare the two methods, we use a modulated 16 QAM signal. The pulse shaping filters are raised cosine filters with a roll-off factor 0.35 extending 4 symbols on either side from the centre and 20 times oversampled. The carrier frequency is 1.8GHz and the bandwidth is 4MHz.

The parameters of the modulated signal are presented in Table 5:

For a precise comparison, we will do the same measurements to evaluate the method based on the polynomial root finding. The results presented below are obtained with the following parameters:

- 5 coefficients NL (Length of non-Linearity)
- 2 coefficients EM (Memory effect depth)
- 1024 coefficient N (Size of table)

After the identification of the Hammerstein model coefficients, the pre-distorted signal is determined by the LUT based method and then re-injected into the amplifier.

Note that the output saturation power P_{ssat} of ZFL-2500 is approximate to 19.7dBm. Knowing that the PAPR of the modulated 16QAM signal filtered by raised cosine is approximate to 7.25dB, and the lowest value of output back-off of the linearized PA (LPA) for a possible linearization is 7.25dB for OBO_{min}, with P_{linmax} of 12.45dBm. It is important to note that the probability density of the 16QAM signal amplitude and its PAPR, changing with the filter, no matter raised cosine (CS) or root raised cosine (SCR) is adopted.

In fact, the performances of PD by the two methods are very similar. The maximum correction is very close to the maximum linear power, P_{linmax}, approximately from the average output power of 12.2 dBm for ACPR and EVM to 14dBm. In Figure 20, the correction for the ACPR is just 10dB. This is due to the fact that DO does not present a sufficient quantization of the input signal (8 bits). The precision of the DSP cannot go low enough. For the EVM, the improvement is 3%. The difference between measurement and simulation is the unavoidable noise, especially the phase noise.

However, the PD by root solving and the exact optimized method by table reach nearly the same performance throughout the power range, especially near P_{linmax}.

Table 5. Parameters of modulated signal

Signal 16QAM	Amplifier
4M Symbols/s	Carrier : 1.8GHz
200 symbols	Power : variable
20 samples by a symbol	

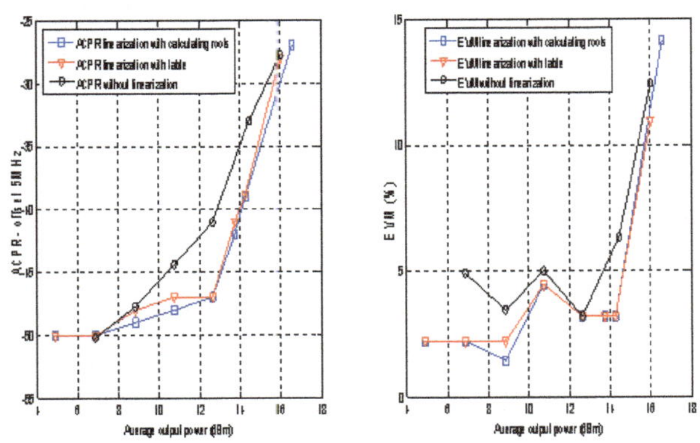

Figure 20. Performance of PD

6. Conclusion

The method proposed in this paper has two substantial advantages compared to the original method.

First, the original method proposed in [7] taking into account the memory effects becomes implementable on a FPGA. But the algorithm proposed in this paper contains only simple operation (additions, multiplications), better suited for real-time implementation. The complex step of polynomial root solving is removed.

Secondly, this method improves performances for the high power levels of input signal. In fact, the divergence of the polynomial model deteriorates the performance of PD before a certain average power in the non-linear zone, where the input signal samples enter the different zones. The implementation by table solves this problem.

As soon as the level of the input signal becomes large enough, the pre-distorted samples selected correspond to the highest level in the table. They are independent on the model coefficients if the dynamic parts of the signal for filling the table are properly chosen.

The proposed method retains the advantages of the LUT method by which correcting the memory effect becomes easier.

For a perfectly linearized amplifier, it must introduce a minor decrease (back-off) which is equal to the PAPR of the input signal. With a smaller OBO_{min}, the spectral regrowth increases very quickly.

Thus it would be important to study the existing techniques for reducing the PAPR of the modulated signals, associated with the pre-distortion technique.

References

[1] P. B. Kenington, "High Linearity RF Amplifier Design", Norwood, MA, USA: Artech House, Inc., 2000.

[2] S. C. Cripps, "RF Power Amplifiers for Wireless Communications", Second Edition (Artech House Microwave Library (Hardcover)), Norwood, MA, USA: Artech, House, Inc., 2006.

[3] L. Sundstrom, M. Faulkner, and M. Johansson, "Quantization analysis and design of a digital predistortion linearizer for RF power amplifiers", IEEE Transactions on Vehicular Technology, vol. 45, no. 4, pp. 707-719, Nov 1996.

[4] N. Safari, J. Tanem, and T. Roste, "A block-based predistortion for high poweramplifier linearization", IEEE Transactions on Microwave Theory and Techniques, vol. 54, no. 6, pp. 2813-2820, June 2006.

[5] M. Faulkner and M. Johansson, "Adaptive linearization using predistortion experimental results", IEEE Transactions on Vehicular Technology, vol. 43, no. 2, pp. 323-332, May 1994.

[6] W.-J. Kim, S. Stapleton, J. H. Kim, and C. Edelman, "Digital predistortion linearized wireless power amplifiers", IEEE Microwave Magazine, vol. 6, no. 3, pp. 54-61, Sept. 2005.

[7] Cottais, E.; Nomm, S.; Wang, Y.; Toutain, S.; Moog, C. "Baseband adaptive predistorter to linearize power amplifiers with memory", European Microwave Conference, Amsterdam, Netherlands, October 2004.

[8] D. Morgan, Z. Ma, J. Kim, M. Zierdt, and J. Pastalan, "A generalized memory polynomial model for digital predistortion of rf power amplifiers", IEEE Transactions on Signal Processing, vol. 54, no. 10, pp. 3852-3860, Oct. 2006.

[9] T. Liu, S. Boumaiza, and F. Ghannouchi, "Augmented hammerstein predistorter for linearization of broad-band wireless transmitters", IEEE Transactions on Microwave Theory and Techniques, vol. 54, no. 4, pp. 1340-1349, June 2006.

[10] L. Ding, G. Zhou, D. Morgan, Z. Ma, J. Kenney, J. Kim, and C. Giardina, "A robust digital baseband predistorter constructed using memory polynomials", IEEE Transactions on Communications, vol. 52, no. 1, pp. 159-165, Jan. 2004.

[11] F. Gao and W. Snelgrove, "Adaptive linearization of a loudspeaker", International Conference on Acoustics, Speech, and Signal Processing, 1991 (ICASSP1991), pp. 3589-3592 vol.5, Apr 1991.

[12] C. Eun and E. Powers, "A new Volterra predistorter based on the indirect learning architecture", IEEE Transactions on Signal Processing, vol. 45, no. 1, pp. 223-227, Jan 1997.

[13] Marsalek, R.; Jardin, P. & Baudoin, G., "From post-distorsion to pre-distorsion for power amplifiers linearization", IEEE Communications Letters, Vol. 7, pp. 308-310, 2003.

[14] M. Schetzen, "The Volterra and Wiener Theories of Nonlinear Systems", Melbourne, FL, USA: Krieger Publishing Co., Inc., 2006.

[15] Y. H. Lim, Y. S. Cho, I. W. Cha, and D. H. Youn, "An adaptive nonlinear prefilter for compensation of distortion in nonlinear systems", IEEE Transactions on Signal Processing, vol. 46, no. 6, pp. 1726-1730, Jun 1998.

[16] H. W. Kang, Y. S. Cho, and D. H. Youn, "On compensating nonlinear distortions of an ofdm system using an efficient adaptive predistorter", IEEE Transactions on Communications, vol. 47, no. 4, pp. 522-526, Apr 1999.

[17] E.Cottais, "Linéarisation d'amplificateurs de puissance large bande par pré-distorsion adaptative en bande de base", Ph.D.dissertation, Université de Nantes – France, 5 Dec 2005.

[18] M. Abi Hussein, "Linéarisation des amplificateurs de puissance – Pré-distorsion numérique adaptative en bande de base", Ph.D.dissertation, Université de Nantes – France, 3 Jun 2009.

[19] R.Marsalek, "Contribution à la linéarisation des amplificateurs de puissance par prédistorsion numérique adaptative en bande de base", Ph.D.dissertation, ESSIE NOISY LE GRAND– France, 12 Dec 2003.

[20] E.Cottais, Y.Wang, S.Toutain, "A new adaptive baseband digital predistortion technique", European Microwave Association, vol. 2, juin 2006, pp 154-159.

[21] A. Saleh, "Frequency-independent and frequency-dependent nonlinear models of twt amplifiers," Communications, IEEE Transactions on, vol. 29, no. 11, pp. 1715-1720, Nov 1981.

18

Analysis and Implementation of an Optimal PV Solar Generator in Tropical Zone

Andre YOUMSSI[1,2,3,*], Emmanuel Patrick FEUDJIO[3]

[1]Electric Energy Laboratory (LEE), Montreal, Canada
[2]Ecole Polytechnique de Montreal, Succursale Centre Ville, Montreal, Quebec, Canada
[3]Electromagnetic Laboratory & Process Engineering Laboratory, The University of Ngaoundere, Ngaoundere, Cameroon
*Corresponding author: andre.youmssi@yahoo.com

Abstract Many small size enterprises, other organisms and rural people mostly in underdeveloped countries do not have an easy access to the grid, and are then in the need of a decentralised electric sources which appear as crucial in some cases specially when feeding rural hospitals, or serve for pumping drinkable or irrigation water. Electric Solar Generators Technologies could then efficiently come to rescue. This article focuses on the analysis of keys factors for a design, and an implementation of such a PV Solar Generator. We have in that aim proceeded to a study of the state of art in that domain of Solar Electric Generators, which offered keys parameters of each component of a safe easy maintainable feeding solar system, and lead us to the design notably owing to the software PVSyst, to the implementation of and the experiments of a domestic flat PV solar generator. Results are constituted by the designed schematics, tables and curves of simulated and experiments electric quantities like the maximum delivered powers, and the efficiencies. The article reported an example of a complete PV system installed, and instrumented for data logging. That system included a battery and an inverter to power real loads. The system design parameters comprised the size of the PV array, the battery size, and the power rating of the inverter, as well as the power delivered to the loads. The data logged have been used to evaluate the designed system and to make assessments on how the system design could be improved so that systems of this type could serve the local needs most economically.

Keywords: *modelling, experiments, PV generator (PEF), Power Electronics*

1. Introduction

The whole planet could run out an energy crisis according to International Energy Agency which has indicated that the needs will triple by the years 2050. Some studies of Professors *Kenneth Deffeyes*, from Princeton University (2006) and *Fredrik Robelius*, from Uppsala University in Sweden (2007), show that the oil production has reached a peak value of 84.5 millions of gallons per day, and that the oil may not in finely compensates the needs rate, and this mainly because of the production goes diminishing. So many researches are carried out all over the world on the searching of new energy sources that could satisfy those needs. It could also be noticed that many people live in some geographical zones were the grid is technically impossible or really difficult to be implemented. Renewable energy sources then come to be considered with great interests among the researched new sources because they respect the environments and their present use does not fortunately affect their future use. Solar Energy seems to constitute an interesting such renewable sources that may come to rescue for challenging in one hand grid feeding in difficult geographical zones, and on the other hand guaranties the future in case of energy crisis. This solution then appears crucial, specially for people in rural zones where the grid is absent, and deeply recommended by many world development organisms like the *ASEAN Centre for Energy, Environment and Development Network for Africa,* and the *"Organizacion Latino americana de Energia"*.

A PV generator may then in some cases be preferable for many reasons: It brings autonomy for the systems that are not connected to the grid and, it could come to rescue in the case of a black out to those that are connected to the grid. The PV solutions also seem to be of interest for countries with grids fed by dams because the water could then be used with more efficiency [1-10]. A PV panel's generator equipped with batteries seems the most advantageous among the decentralised electricity production systems like generating units. The context of our study is tropical zones which could received per year an irradiance of 2.065 kWh/m^2, which means that a square meter of PV panels if used in that area could help covering during 20 months the energy needs of a family who spends a 100 kWh per month.

We will try in this study to perform a state of the art in that domain of PV solar generation and then try to reach after a deep analysis the key factors that will help designing and implementing an optimal PV panel's generator. We then consider Saheb-Koussa & Haddadi (2007) well known one diode model of a PV panel in order to express the power delivery with respect to the voltage. Our study then brings out the different technologies of components that could constitute the PV panel's generator, and the conditions of their use. Our article also addresses the energy conservation problematic and the maintenance of the installed PV generator.

We will so spend some efforts to answer the following hypothesis:

- Why it important to focus on the maximum power delivered, Pmax?
- How could we make a productive choice among all the PV systems?
- How could Pmax be influenced by the geographical position of the PV panel, by the sun irradiance, and by the temperature?
- How do the batteries behave with respect to the called current rate?
- What are the limits of the PV system?
- And the question of safety?

This article will try to answer those interrogations through three main parts: one on the literature on the domain, the second on the equipments and the methodologies of the studies, and the last one on the results and the discussions.

2. Review on the Modeling of PV Generators - Objectives of the Study

We present here well known PV cell modeling and similarly well known and well documented how variations of the position of a PV panel affect the output power and energy capture. Our main goal here is to remind how the Maximum Power Point (MPP) could be predetermined from Power-Voltage, and Current-Voltage curves that can also be predetermined for a given PV panel. We then explained how the voltage drops that lead to MPP points when loaded the PV system can be predetermined owing to the characterization and the identification of the parameters of the chosen model.

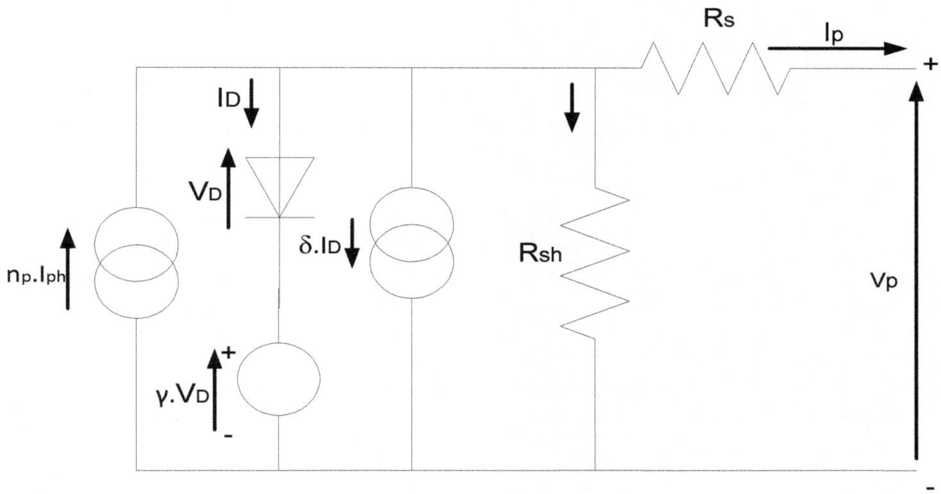

Figure 1. Complete Model of a PV Panel [1-21]

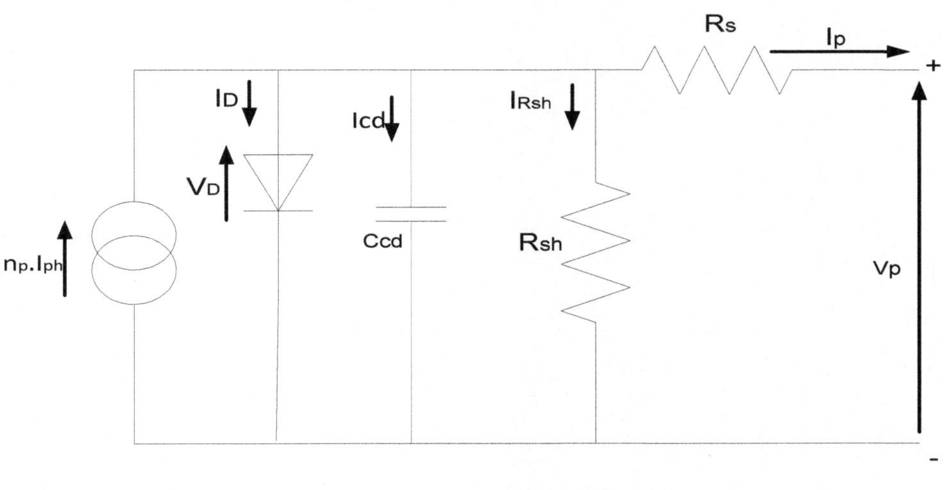

Figure 2. Model of a PV Panel with Rsh, Rs and a Capacitor [1-21]

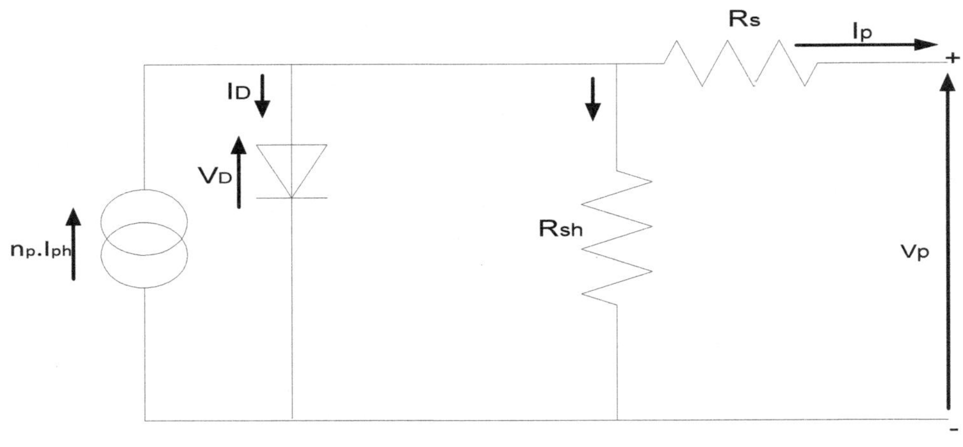

Figure 3. Model of a PV Panel with Rsh and Rs [1-21]

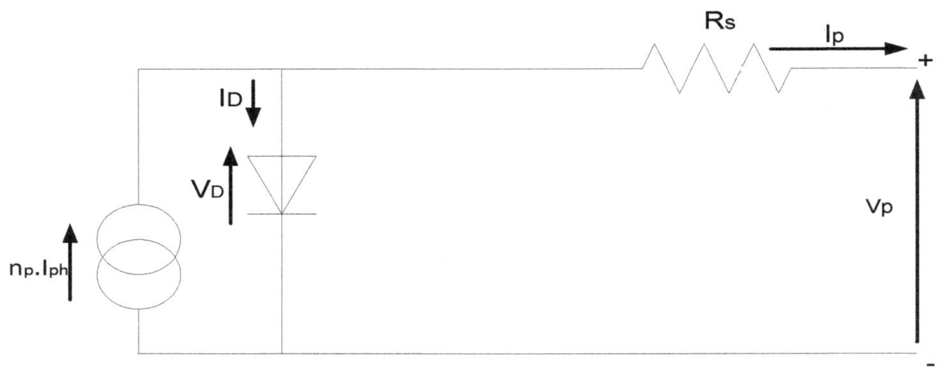

Figure 4. Simplified Model of a PV Panel [1-21]

It is now well known that PV cells behave as photo p-n junction that operates with very low efficiency. They so needed to be judiciously connected in order to obtain maximum power at the outputs [1,2]. The PV generators need to be controlled and commanded to operate at that point (Maximum Power Point, named MPP). They should then be modelled and characterised. The scientific literature on PV generators shows many types of models [1,2,3,4,5]. It so comes out that the Figure 1 to Figure4 seems to be the most used in the modeling of PV Generators (Figure 1 - Figure 6), each presenting its particular advantages.

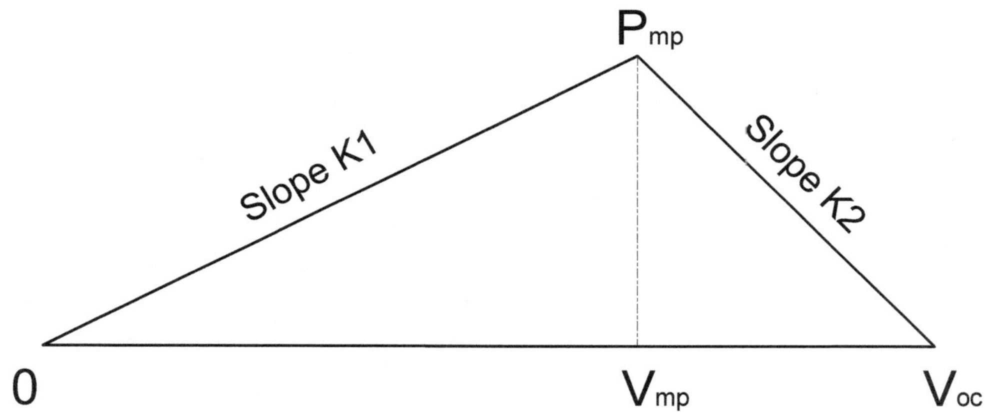

Figure 5. Deduced general modeling of the Power-Voltage Curve

All the models insist on one hand on the series output resistance Rs and on the other hand on the source current np.Iph, that comes from the direct conversion of the sun irradiance into an electric current. All those models also remind that the PV cell is a p-n junction which is here represented by a classic diode. So it can be deduced from those models that the PV-Generator will behave like a classic DC voltage source with an internal resistor which is here Rs, and which will responsible for the voltage drop when the output current will increase, the maximum current being reached at short-circuit that corresponds to zero voltage and zero power. The maximum voltage Voc will so obtained with a zero current that also corresponds to zero power. Classic values of Voc are between 0.4 V to

0.6V for silicium cells with 12 mA/cm^2 for the short-circuit currents and under standards conditions (STC). It is so easy to demonstrate that a maximum power Pmp will be reached at a couple (Vmp, Imp) that corresponds to the MPP operating point. It can so be deduced from all the models used till now (Figure 1 - Figure4), that the power-voltage characteristics of the PV-generator can then be modelled by the general graph on Figure 5. The characterisation of the PV generator could so lead to the values of the parameters that constitute the chosen models (Figure 1 - Figure 4), and then to the control of the voltage drop $\Delta V = Voc - Vmp$ on Figure 5.

The scientific literature also indicates that the generated electric current np*Iph could drastically increase with an optimal geographic position of the PV panel [2-8]. That optimal position needs to be known and set using the mechanical system that holds the panels. The angles α, β, and γ should then be implemented (Figure 6.a).

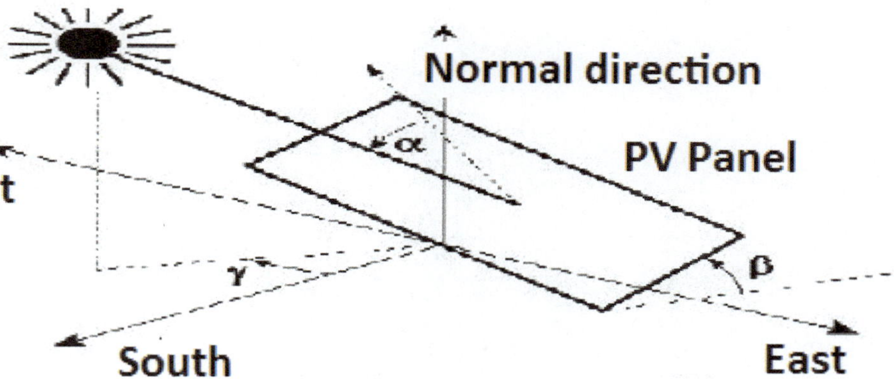

Figure 6. a: Possible Variations of the position of a PV Panel

Our study focuses on tropic regions that corresponds to the maximum W.m^{-2} in the world and where it is known that the solar irradiance over a year which is almost 1.6*10^{21} Wh may lead to satisfy if transformed the present world annual need of energy (almost 17.3*10^{15}) Wh (Figure 6.b) and as it is well explained by Jean Zin author.

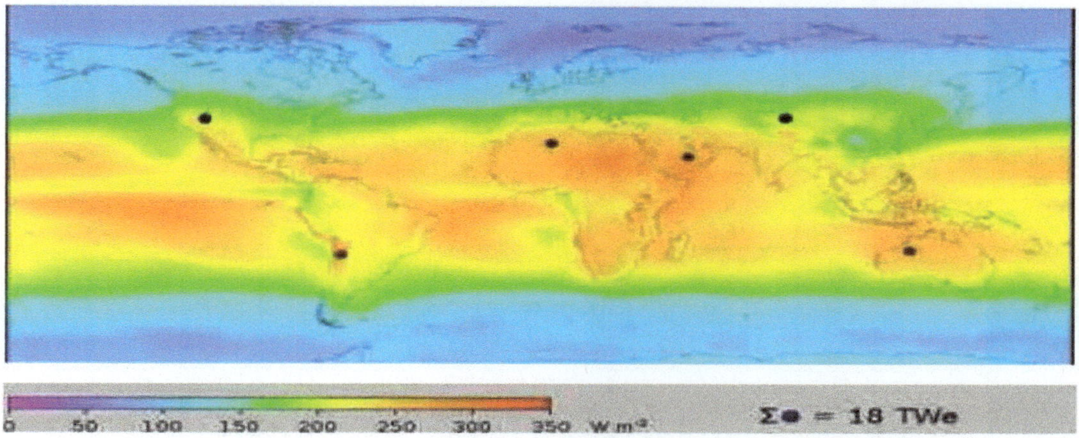

Figure 6. b: World Solar Irradiance – Solar potential of Tropical zone [6]

We would like to remind that a PV Solar Generator can be or not connected to the grid, it could run with or without a power additional income, and/or an energy storage system. The solar covering rate, also called the LOLP (*Loss of Load Probability*), as shown on Figure 6.b, constitutes an important data when dimensioning a PV Solar Generator. The value of the LOLP makes the PV Solar Generator critical or not. The PV System designed as an example for data logging was critical, and complete with batteries, inverters and controls. The designating has been in four steps:

1. The power needs per day calculated at a value of 20640 Wh/day.
2. The calculation of the solar covering rate (Figure 6b), which assures the providing of the power a year Along, the estimated minimum irradiance has been evaluated at a value of 4.66 kWh /m^2
3. The 12 V batteries with a capacity calculated at a value of 5733 Ah
4. and finally the required PV field to cover the real needs but also all the losses estimated at a value of 5537 Wc which corresponds to 93 modules of 60 wc each.

So a quick hand pre design leads to a 12 V/ 5733 Ah/ 5.54 kWc/ 61 m^2, PV Generator. We have used the software PV-Syst in order to obtain optimised parameters of that PV Generator. Such software takes into account:

- The minimum autonomy of the PV system which diminishes the batteries capacity
- The possibility of using alternatives sources when the sun is not at rendezvous
- The solar covering rate and the advantage of running the system at MPP points, which may lead to minimum number of PV panels the author Gilbert CHAMP, indicated. We then came out with an

optimal PV generator that of the following characteristics for year consumption: 12V/ 2250 Ah / 4.320 kWc / 42.3 m^2

3. Equipments and Methodologies

We carried experiments with an example of equipments constituted by: 04 PV modules of 60 Wc with thr dimensions: Lg = 1.13m; lg = 0.52 m ; useful area : 0.48 m^2, each panel built with 39 Cells. We also used a 80 Wc 36 cells module of dimensions Lg = 1.20 m; lg = 0.53 m; and a useful area of: 0.49 m^2 (Figure 7). Three junction boxes for 12V, 24V and 48V modules; Four solar batteries (Figure 8.a); three 12V, 24V, 48V/20A charges regulators (Figures 8.b); 12V – 300 W – 230V-50 Hz inverters; and many other measurements apparatus (voltmeters...etc.) have been used to instrumented to design and installed the PV system for data logging.

Figure 7. the experimented PV panels

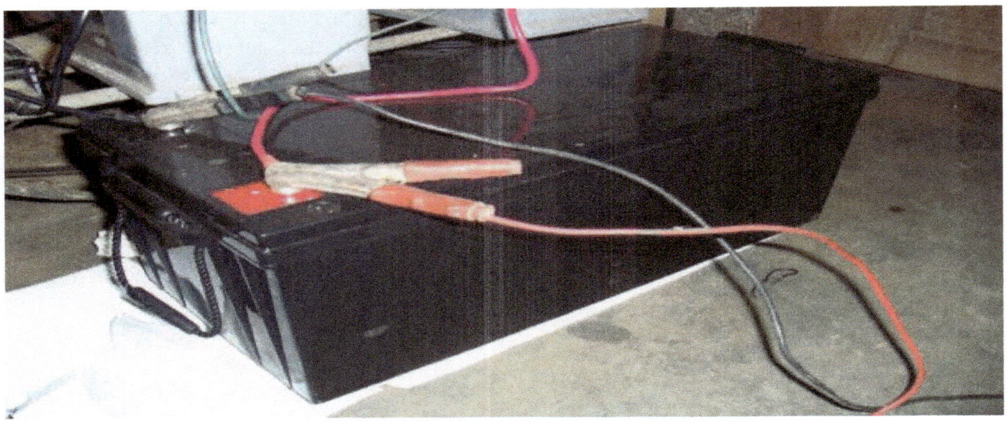

Figure 8. a: Experimented batteries

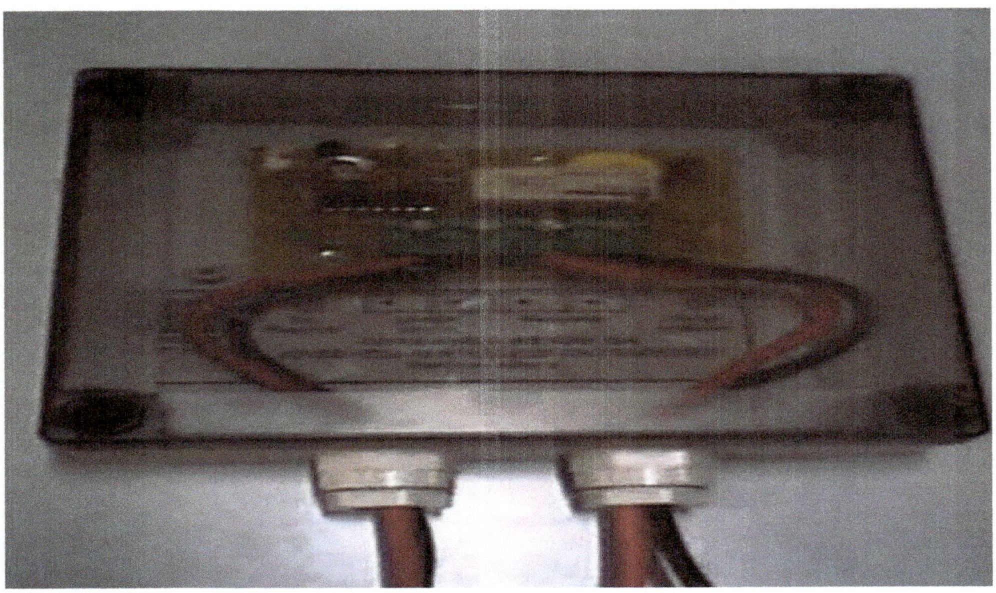

Figure 8. b: The used regulators for Batteries

Figure 8. c: Photo of the proposed Inverter

The software PVSyst has also been used to design the PV panels and other components, and also to simulate the running of the system with the possibilities of trying to know what results could be obtained when supposing the variations of different parameters. The influence of the position of the PV Panels and of the time period (the month) have so been predicted and confirmed with experiments with very good accuracy...etc. That software has then served for the simulation of each studied case. We have so come to be capable to compare the simulation and the experiments results, bringing the real system to operate at predicted desired MPP points. We have then used PVSyst software to pre design and simulate our PV system. We have then first performed a rapid pre dimensionnement of the system based on

- The wheather forcast,
- The orientation of the PV pannels
- The real electric charge including all the losses,
- The depth of the batteries discharge
- And the covering rate

We then came out with the power of the PV panel and the capacity of the batteries to be installed.

In the other hand, we have been able to make a simulation through PVsyst going from the first results and then compare with experiments. It should be indicated here that the losses have also been taken into account by the simulations.

4. Results and Discussions

We have been obliged to design and build a judicious multi positions stand which has helped to obtain the Maximum irradiance and bring the system to always operate at MPP point which varies with months (Figure 9.a).

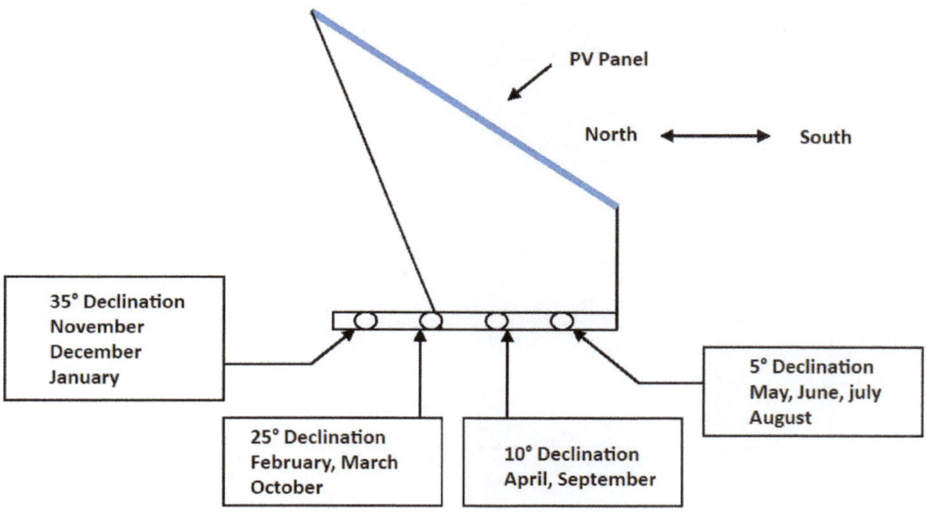

Figure 9. a: Built PV Panel Multi positions stand

We have then come out with the remarks that the characteristics (optimal positions ...etc) of the installed PV modules were the same like the ones obtained by software simulations (Figure 9.a). The whole designed and installed system can then be represented by the following schematic. On Figure 9.b. The system could then feed Ac and DC charge. The autonomy being assure through batteries (Figure 9.b).

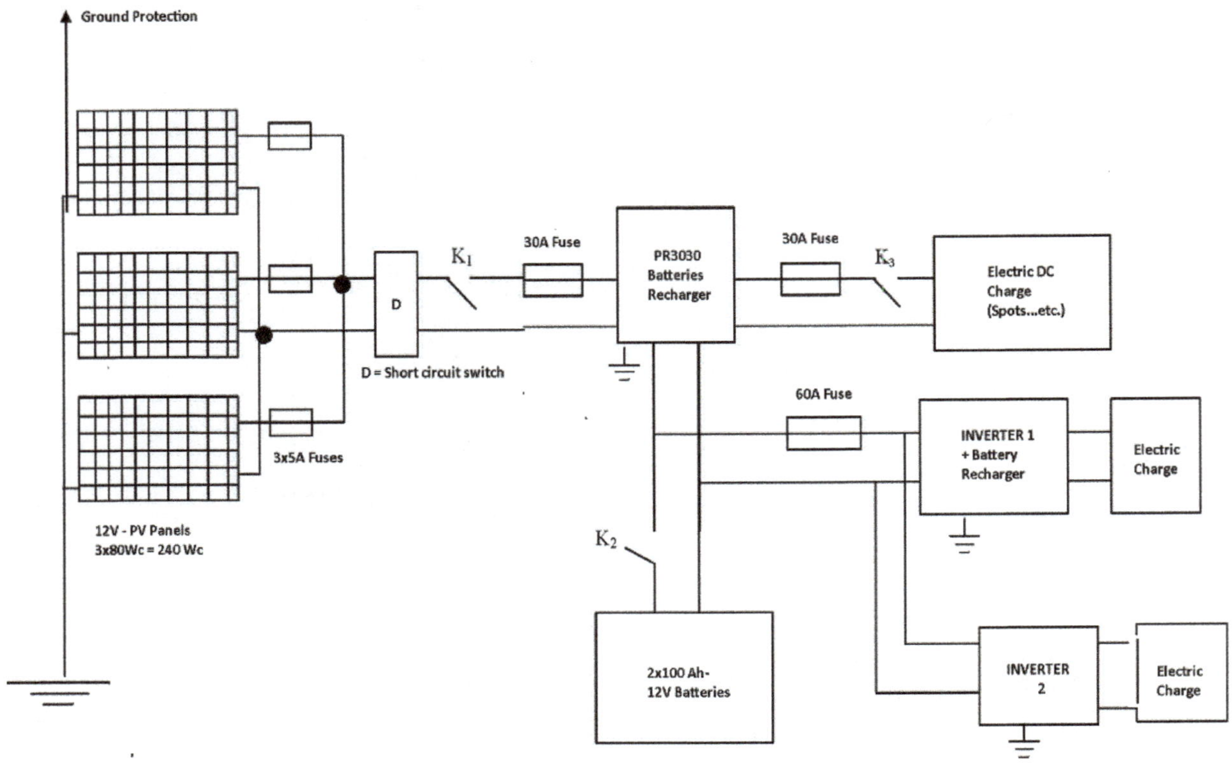

Figure 9. b: Schematic of the designed and installed PV Solar Generator

Some algorithms have been written in the solar systems software PVSyst which have helped simulating the efficiency of a module with the possibility to vary operating cell temperature in on hand, and the irradiance in the other hand, the values of the efficiencies remaining low, under 12 %. This is due to the limits encountred in all PV technologies. The silicium poly crystalline 60 Wc that we have used in experimentation generates a maximum of 125 W/m, while the mono crystalline technology 80 Wc, the other module generates 163.1 W/m (Figure 10 & Figure 11).

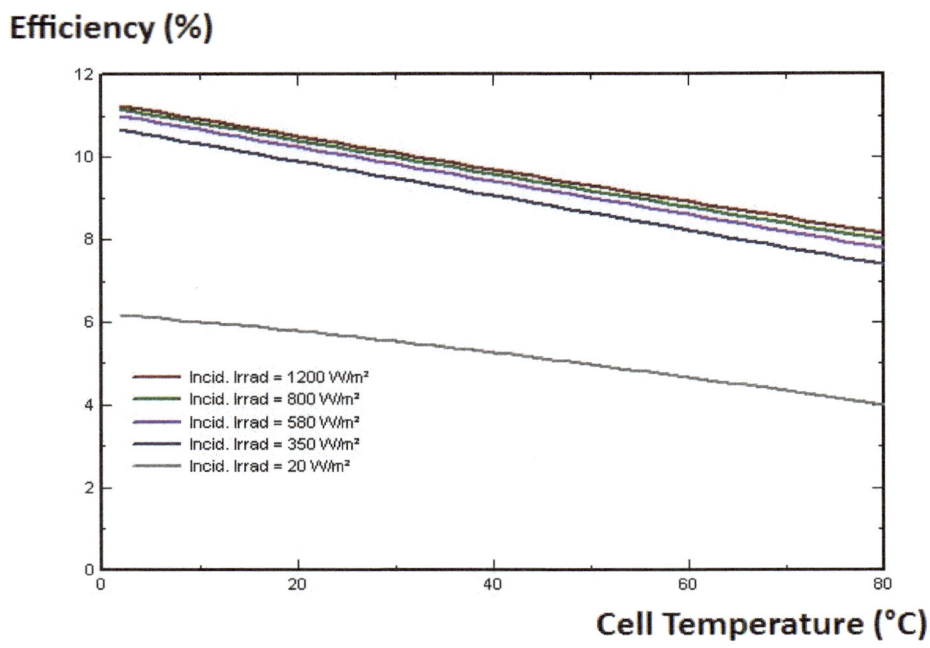

Figure 10. Efficiency with respect to the cell temperature

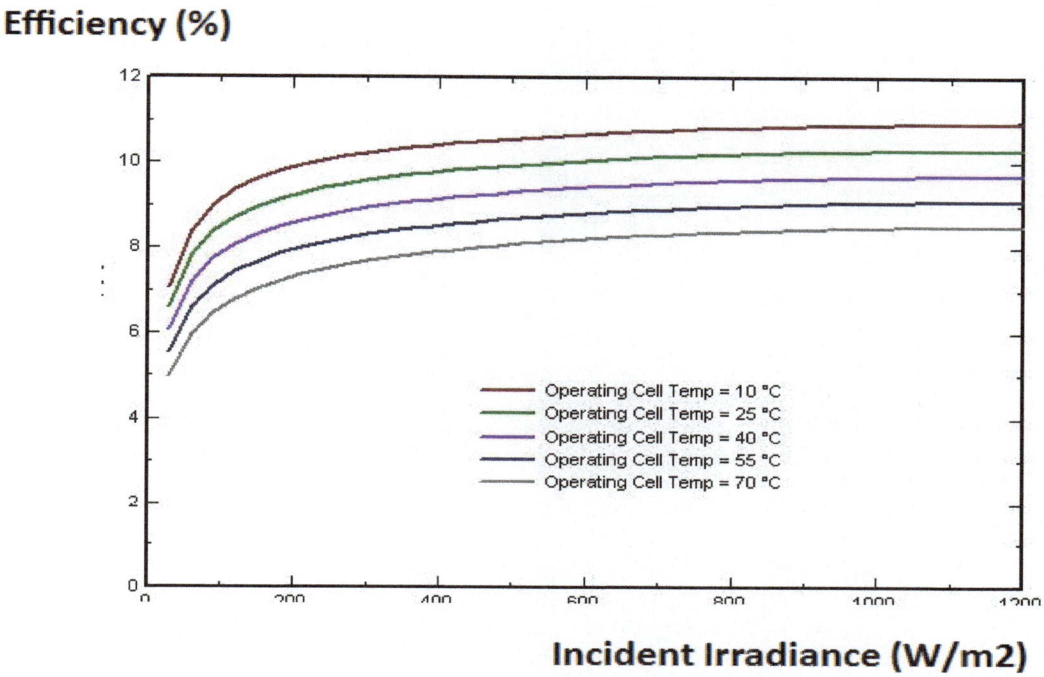

Figure 11. Efficiency with respect to the incident irradiance

Our study has been carried in a town of coordinates 7° to 8° degree North latitude Nord, and between 13° to 14° degree East longitude. So the panels have been disposed facing south according to cooper equation [6]:

$$\delta = 23,45\sin\left(2\pi\frac{284+n}{365}\right)$$

Where n is a number that represents the day (n=1 for January 1[rst], n=32 for February 1[rst]), δ the obtained declination to be applied.

We then obtained from PVSyst some useful data of some equatorial towns that might be of great help, this study being concerned as mentioned above (Table 1.a).

Table 1. a: Solar Irradiance and Temperature of some towns (Ndéré means here Ngaoundere town in Cameroon)

	Ndéré 7°N – 13,6°E		Garoua 9°N – 13,4°E		Mora 11°N – 14°E		Yaoundé 4°N – 11,5°E		Douala 4°N - 9,7°E		Tombouctou 17°N – 3°O		Moscou 56°N – 38° E	
	kWh/m²	°C	kWh/m²	°C	kWh/m²	°C	kWh/m²	°C	kWh/m²	°C	kWh/m²	°C	kWh/m²	°C
Janv	6,44	24,7	6,01	26,3	5,57	27,8	5,50	23,2	5,56	24,6	4,91	21,8	0,50	-9,7
Fev	6,82	26,2	6,48	28,1	6,41	30,0	5,62	23,7	5,59	24,8	5,98	24,5	0,94	-7,7
Mars	6,57	27,0	6,64	30,1	6,80	34,4	5,22	24,0	4,97	24,7	6,83	28,2	2,63	-2,2
Avr	5,80	25,5	6,30	28,9	6,58	35,7	4,94	24,2	4,57	24,7	7,26	32,1	3,07	5,8
Mai	5,42	24,4	5,88	27,1	6,37	33,4	4,62	24,0	4,41	24,5	7,45	34,2	4,69	13,1
Jun	4,99	23,2	5,48	25,5	5,95	30,0	4,33	22,9	4,02	23,6	7,27	33,0	5,44	16,6
Jul	4,66	22,2	5,07	24,4	5,36	26,0	4,06	22,2	3,46	22,8	6,93	29,9	5,51	18,2
Aout	4,67	22,1	4,92	24,2	5,07	26,4	3,99	22,3	3,08	22,8	6,62	28,7	4,26	16,4
Sept	4,80	22,5	5,23	24,8	5,61	27,9	4,26	22,5	3,45	23,0	6,65	30,5	2,34	11
Oct	5,22	23,1	5,74	26,1	5,85	31,3	4,10	22,7	3,69	23,4	6,25	31,2	1,08	5,1
Nov	6,11	24,7	6,12	28,0	5,72	30,9	4,57	22,9	4,33	23,7	5,49	27,6	0,56	-1,1
Dec	6,25	24,9	5,90	26,9	5,33	28,3	5,60	23,0	5,13	24,1	4,72	23,3	0,36	-6,1
Annual	**5,64**	24,2	**5,81**	26,7	**5,88**	30,2	**4,69**	23,1	**4,35**	23,9	**6,35**	28,8	**2,63**	5,0

We will now present the testing of specific PV modules and related discussions. We hope this will contribute to provide much deepest insights. We then discuss simulated versus experimental results for a module. The no load output voltage Voc on Figure 5 was 21.6 V and the short circuit current 4.01 A. The no load output voltage Voc was here 21.62 V and the short circuit current in this case was 4.9 A (Figure 5).

Simulations performed owing to PVSyst indicate that a 20° declination with 0° azimuth correspond to the maximum incident irradiance at our experiment geographic point in February. The experiment measurements lead to 25° declination with 0° azimuth. Table 1.b gives for the other months the corresponding theoretical and experiments (in blue) positions for maximum incident irradiance.

Table 1.b: Positions for Maximum incident irradiance during the year

Month	Jan	Fév.	Mars	Avr	Mai	Jun	Jul.	Août	Sept	Oct.	Nov.	Déc.
Decl Exp	35°	20° - 25°	17° - 25°	15° - 10°	4° - 5°	3° - 0°	2° - 0°	5°	7°	22°	34°	40°
Az	0°	0°	0°	0°	0°	0°	0°	0°	0°	0°	0°	0°

Table 2. indicates the theoretical and experiments (in blue) day positions along the month of February

DTps	7h	8h	9h	10h	11h	12h	13h	14h	15h	16h	17h	18h	19h
Decl	80°	70°	60°	50°	40°	20°	35°	40°	50°	60°	70°	80°	-
Exp			50°	45°	35°	25°	30°	36°	45°	65°			
Az	-75°	-60°	-45°	-30°	-15°	0°	+20°	+35°	+45°	+50°	+60°	+75°	-
Exp			-50°	-35°	-10°	0°	+25°	+30°	+50°	+50°			

Exp: Experiments; Decl: Declination: Az: Azimuth: DTps: Day time

We have then proceed to the characterization of four diffrerent modules through experiments that bring out the Voltage-Current, and Power-Voltage characteristics. (Figure 12, Figure 13, Figure 14, and Figure 15). All those figures present predicted shapes. We can then noticed the MPP (Imp, Vmp) point for each module as well as the short circuit current.We also have access to the no load voltage Voc, and the slope that leads to the voltage drop (Voc-Vmp).

Module A: Experiments at 10h20 at 30°, 1st/02/2013 - 60 ohms Load – 1.215 W.

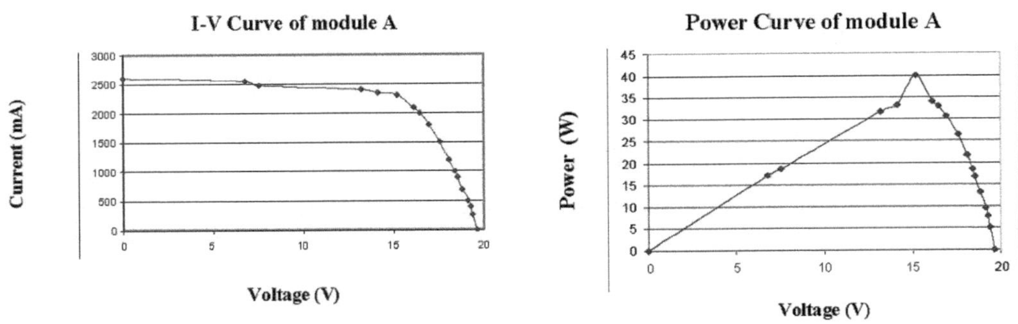

Figure 12. Module A I-V and P-V characteristics

Module B: Experiments at 11h00 at 34°, 1st/02/2013 - 60 ohms Load – 1.215 W.

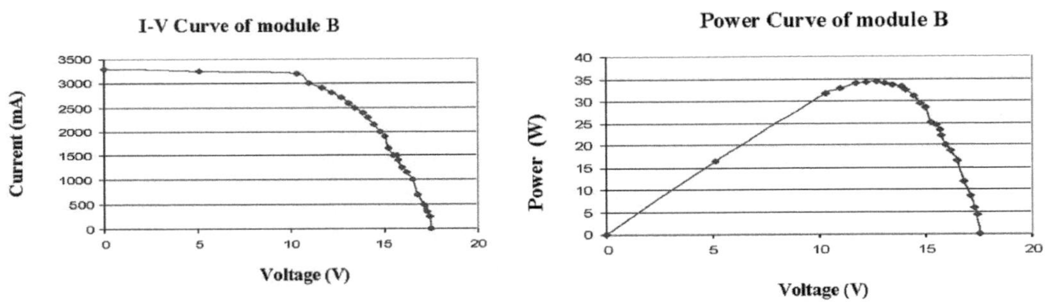

Figure 13. Module B I-V and P-V characteristics

Module C: Experiments at 12h32 at 38°, 1er/02/2013-60 ohms Load – 1.215 W.

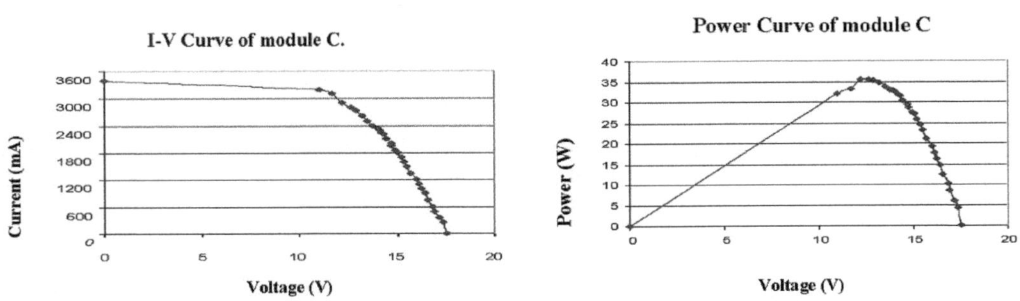

Figure 14. Module C I-V and P-V characteristics

Module D: Experiments at 12h12 at 35°-1st/02/2013 - 60 ohms Load – 1.215 W.

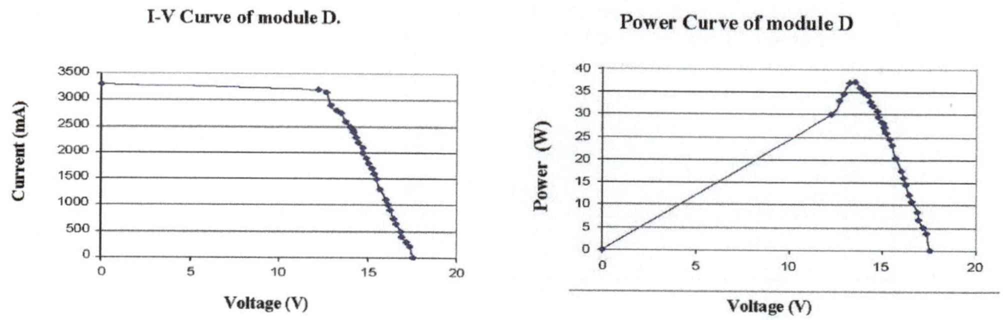

Figure 15. Module D I-V and P-V characteristics

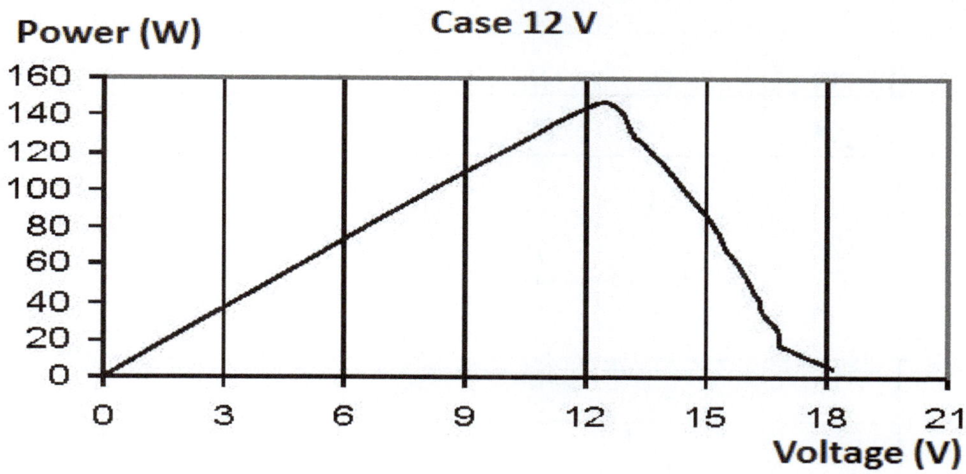

Figure 16. Power Curve of a 12 V PV Panel

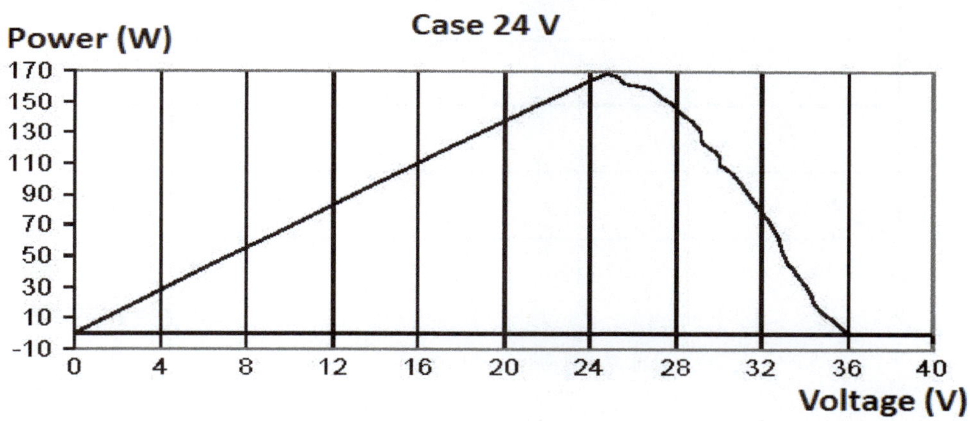

Figure 17. Power Curve of a 24 V PV Panel

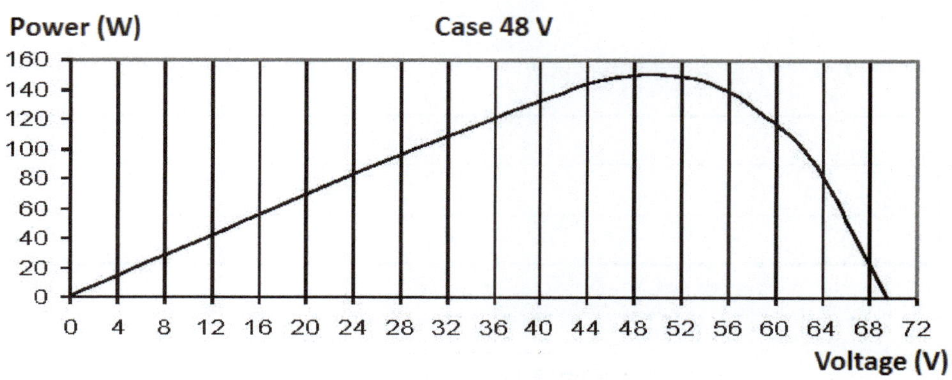

Figure 18. Power Curve of a 48 V PV Panel

Some other measurements have been carried out on March 25 between one o'clock PM to one thirty the day temperature was 42°C, the obtained experiments characteristics came to correspond to the expected ones, the maximum power of the 12 V PV panel was the lowest one 145 W, meanwhile the 24 V PV panel shown the greatest value of Pmp, 168W. The maximum value of the power delivered by 48V PV154W, it so came out that a PV panel system should be characterised before it current used (Figure 16, Figure 17, & Figure 18).

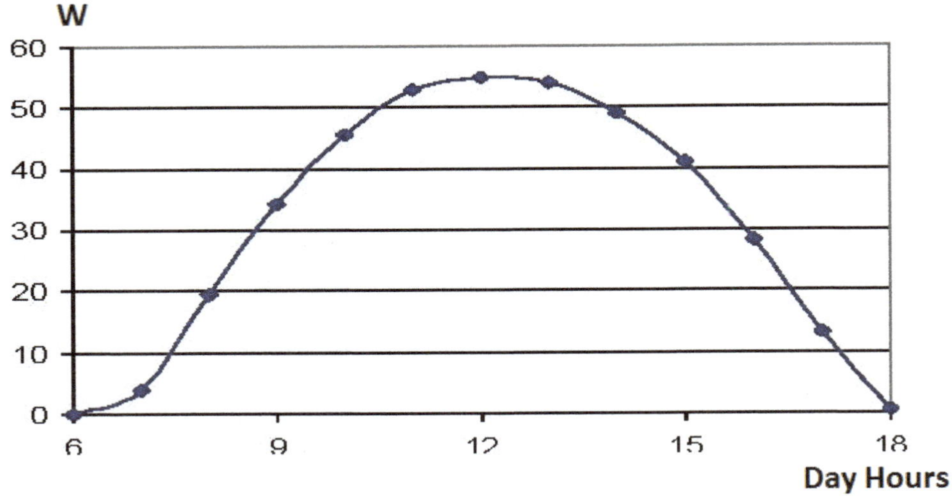

Figure 19. Simulated Power of an horizontal module

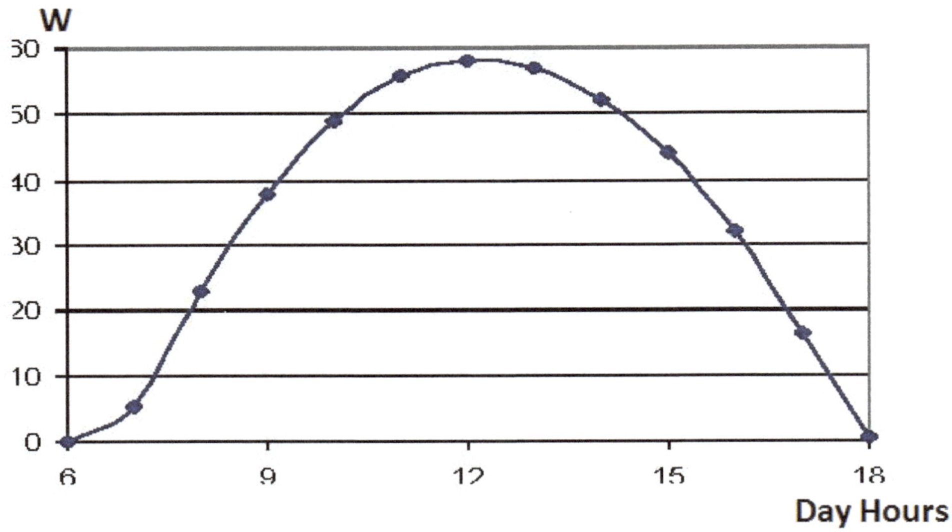

Figure 20. Simulated Power of a 25° declined module

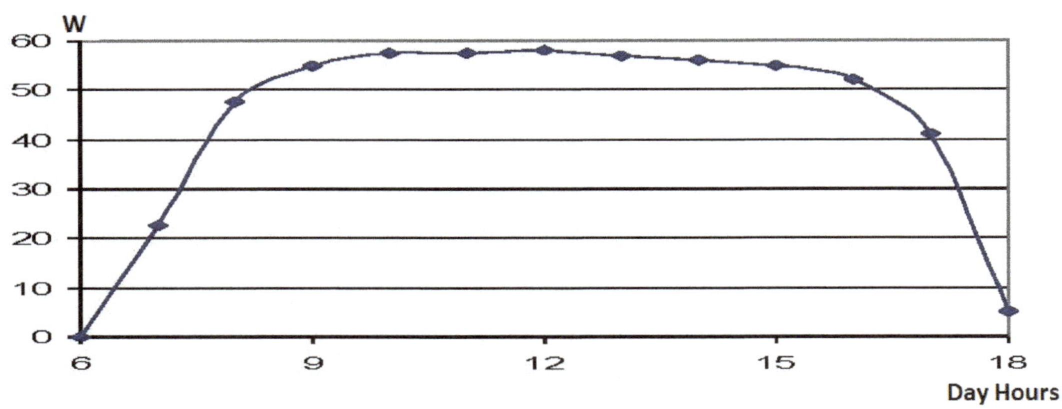

Figure 21. Simulated Power of a module that tracks the sun

A judicious use of PVSyst based on the parameters of our systems as inputs of the software lead to a quantity of data results among which the solar fluxes, the cell temperatures, the MPP points, and the voltages, results from which we have deduced the Expected behavior [9,10,11,12] of the power the day along and this according to the geographic position (Figure 19, Figure 20, Figure 21).

We then proceed to experiments, measuring the power with the same conditions. The estimated curves of power correspond to the measured ones [13-20]. The MPP points are almost the same (Figure 22, & Figure 23).

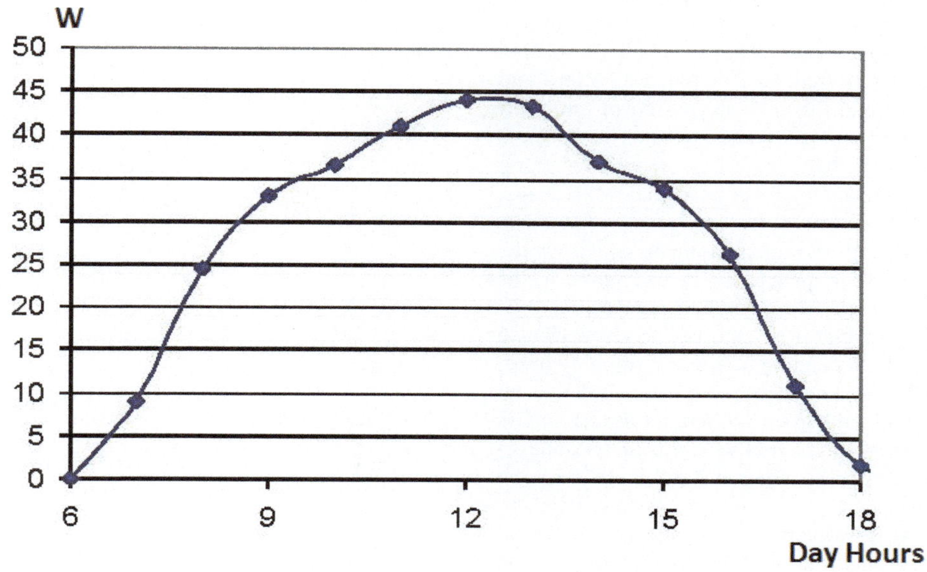

Figure 22. Measured Power of a 10° declined module

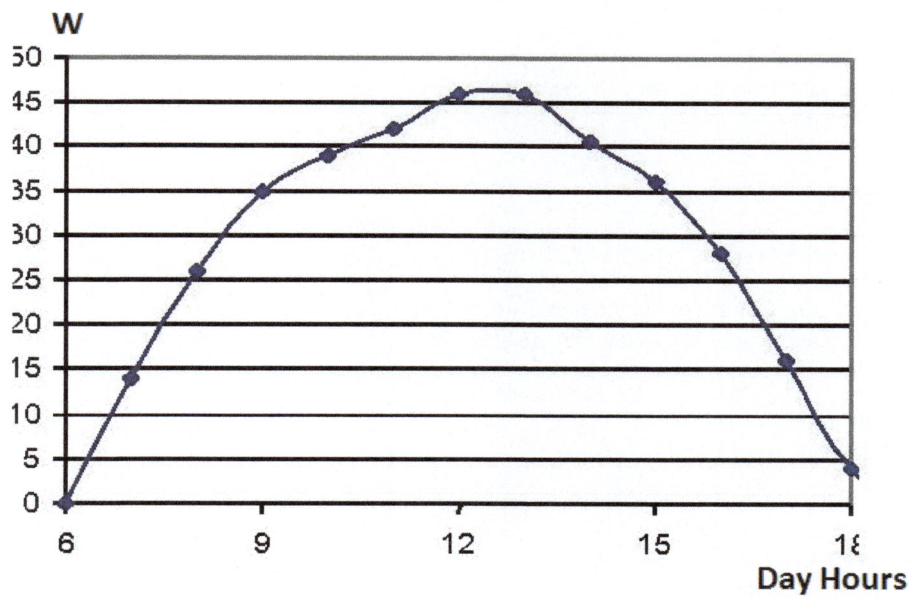

Figure 23. Measured Power of a 25° declined module

It finally comes out that the best choice of a PV Panel should take into account: The maximum power (Wc), the cell temperature (NOCT), and the certifications IEC 61215 for crystalline modules or IEC 61646 for thin modules. Class protection II (SKL II) should also be respected for safety reasons.

In the study context, economic results shows that the installed PV system will generates 20 % maintenance more that if connected to the grid while the investment is seven times the price for a direct connection to the grid. So we can conclude that PV systems will be more appreciated even in town when the different components will be at accessible prices [6].

5. Conclusion

This manuscript discusses several issues related to analysis of PV systems bringing out some insights related to practical deployment in tropic regions. We hope our work constitutes new contributions beyond what is well known from textbooks and scientific literature. We have also put on the spot what is readily provided by tools such as PVSyst which has lead to the capability of designing and implementing such PV generators for domestic uses in tropical zone.

We have reached to interesting results related to simulated versus experimental results obtained for a module, and have found be of interest to extend this work further and report the results. For example, have a complete PV system has been installed, and instrumented for data logging. This system had included a battery and an inverter to power real loads over a period of one year. The system design parameters had taken into account the size of the PV array, the battery size, and the power rating of the inverter, as well as the power delivered to the loads. The data logged have so been used to evaluate the designed system and to make assessments on how the system design could be improved so that systems of this type would serve the local needs most economically.

We can here remind that it is important to know how to orientate a PV system depending on the geographic points, the month, and the moment during the day. A PV Panel should also be electrically characterised, its model should so be known and used for predictive needs. The Voltage-Current and the Power-Voltage curves could then be known leading to use the system with good efficiency. The handling of software such as PVSyst might be also of great helps since it could allow some simulations, and the variations of parameters. It could serve as a guide and may permit comparisons studies. Theoretical and experiments results may then be beneficially confronted. We would like to indicate here that PVsyst Software provides detailed outputs, and useful data bases of sensors, batteries, and inverters, despite the fact that this is not sufficient. Its interface with users seems practice and of easy to use.

Acknowledgements

Authors would like to acknowledge respectively "École Polytechnique de Montreal", in Canada, through its Department of Electrical and Computer Sciences and its Laboratory of Electrical Energy, and to "Ecole Nationale Supérieure d'Electricité et de Mécanique de Nancy", in France, through its Laboratory "Groupe de Recherches en Electrotechnique et Electronique de Nancy", G.R.E.E.N, both for visitor ships and facilities to Andre YOUMSSI.

Nomenclature

MPP: Maximum Power Operating Point

AM: Air Mass is an Optical air mass that expresses the mass of the air in atmosphere that the sun radiations go through

Azimuth: orientation with respect to the south counted negatively from the east.

NOCT: Nominal Operating Cell Temperature which is the real operating temperature of the cells of a PV module that operates under 20°C, with an irradiance of 800W/m^2, a 1.5 AM spectrum, and a wind velocity of 1m/s. Normal values of NOCT are between 45° and 60°.

PV: Photovoltaic.

LOLP: Loss of Load Probability

IAM: Incidence Angle Modifier.

Icc: Short circuit current of a photovoltaic module

STC: Standard Test Conditions, for an irradiance of 1.000 W/m^2, a 25 °C junction cell temperature and a 1.5 AM spectrum

Vco: photovoltaic module voltage at no load.

W, kW, MW: Watt, Kilowatt (103), Mégawatt (106).

Wh, kWh: Watt-hour, KiloWatt-hour.

Wc, Wp: Watt peak value, very used for PV cells, Wp is obtained under STC.

References

[1] Bhattacharjee S & Bhakta S. "Analysis of system performance indices of PV generator in a cloudburst precinct", Sustainable Energy Technologies and Assessments, vol 4, pp. 62-71, Dec 2013.

[2] K.A kim et al. "Photovoltaic Converter Module Configurations for maximum power point operation". Proc IEEE Power Energy Conference, pp. 77-82, 2010.

[3] Jih-Sheng Jason Lai et al. "Design of a Photovoltaic simulator with a novel reference signal generator and two Stage LC Output Filter", IEEE Transactions on Power Electronics, Vol 25, N°5, pp. 1331-1338, May 2010.

[4] Teuvo Suntio et al. "Photovoltaic Generators as an input source for Power electronic Converters". IEEE Transactions on Power Electronics, Vol 28, N°6, pp. 3028-3038, June 2013.

[5] M-F Almi et al. "Protection of hybrid (Wind/Photovoltaic) Generator connected to Mono- phase Electrical Network Supply". Proc the Second International Conference on Informatics Engineering & Information Science - pp 74-82, 2013.

[6] Jean Zin, l'avenir du solaire, http://perso.orange.fr/marxiens/sciences/solaire.htm, janvier 2008.

[7] Yun Tiam Tan et al. "A model of PV generation suitable for stability analysis", IEEE Transactions on Energy Conversion, vol 19, Issue 4, pp. 748-755, 2004.

[8] Yang Chen et al. "A cost-effective single-stage inverter with maximum power point tracking", IEEE Transactions on Power Electronics, vol 19, Issue 5, pp. 1289-1294, 2004.

[9] Tina, G.M et al. "Probabilistic energy performance comparison among flat and tracking PV systems", Power Electronics Electrical Drives Automation and Motion (SPEEDAM), 2010 International Symposium on, pp. 169-174, 2010.

[10] Vitorino, M.A. et al. "Using the model of the solar cell Using the model of the solar cell for determining the Maximum power point of photovoltaic systems", 2007 European Conference on Power Electronics and Applications, pp. 1-10, 2010.

[11] Skoplaki E. & Palyvos J.A. "On the temperature dependance of photovoltaic module Electrical Performance: A review of efficiency /Power correlations", Solar Energy, N°83, pp. 614-624, 2009.

[12] Tahri K & Benyoucef B. "Modelisation Studies of a Photovoltaic Generator", Journal of Scientific Research, N°0, vol 1, pp. 192-197, 2010.

[13] Hove T; Manyumbu E; Rukweza G. "Developing an improved global solar radiation map for Zimbabwe through correlating long-term ground- and satellite-based monthly clearness index values", Renewable Energy, vol 63, pp. 687-697, march 2014

[14] Myers D.R. "Solar Radiation Resource Assessment for Renewable Energy Conversion", Comprehensive Renewable Energy, Volume 1: Photovoltaic Solar Energy, vol 1, pp. 213-237, 2012

[15] Yeong-Chan Kuo et al. "A high-efficiency single-phase three-wire photovoltaic energy conversion system". IEEE Transactions on Industrial Electronics, vol 50, Issue 1, pp. 116-122, 2003.

[16] MacAlpine, S.M. et al. "Characterization of Power Optimizer Potential to Increase Energy Capture in Photovoltaic Systems Operating under No uniform Conditions". IEEE Transactions on Power Electronics, vol 28, Issue 6, pp. 2936-2945, 2013.

[17] Bialasiewicz, J.T. "Renewable Energy Systems With Photovoltaic Power Generators: Operation and Modeling" IEEE Transactions on Industrial Electronics, vol 55, Issue 7, pp. 2752-2758, 2008.

[18] Velasco-Quesada, G et al. "Electrical PV Array Reconfiguration Strategy for Energy Extraction Improvement in Grid-Connected PV Systems", IEEE Transactions on Industrial Electronics, vol 56, Issue 11, pp. 4319-4331, 2009.

[19] Suntio, T et al. “Issues on Solar-Generator Interfacing With Current-Fed MPP –Tracking Converters”, IEEE Transactions on Power Electronics, vol 25, Issue 9, pp. 2409-2419, 2010.

[20] Poshtkouhi, S et al. “A General Approach for Quantifying the Benefit of Distributed Power Electronics for Fine Grained MPPT in Photovoltaic Applications Using 3-D Modeling”, IEEE Transactions on Power Electronics, vol 27, Issue 11, pp. 4656-4666, 2012.

[21] Locment, F et al. “DC Load and Batteries Control DC Load and Batteries Control Limitations for Photovoltaic Systems Experimental Validation”, IEEE Transactions on Power Electronics, vol 27, Issue 9, pp. 4030-4038, 2012.

Simple Mathematical Model for Ferromagnetic Core Inductance and Experimental Validation

B. Nana[1,*], S. B. Yamgoué[1], R. Tchitnga[2], P. Woafo[3]

[1]Department of Physics, Higher Teacher Training College, University of Bamenda, PO Box 39 Bamenda, Cameroon
[2]Laboratory of Electronics and Signal Processing, Faculty of Science, University of Dschang, PO Box 67 Dschang Cameroon
[3]Laboratory of Modelling and Simulation in Engineering, Biomimetics and Prototypes, Faculty of Science, University of Yaounde I, PO Box 812 Yaounde, Cameroon
*Corresponding author: na1bo@yahoo.fr, url: http//www.lamsebp.org

Abstract In determining the properties and inductance of ferromagnetic core inductor, hysteresis modeling is of high importance. Many models are available to investigate those characteristics but they tend to be complex and difficult to implement. In this paper, we report a new mathematical model based on the experimental data of hysteresis for ferromagnetic core inductor. The proposed model can restore the hysteresis curve with a little RMS error. We used the model to determine analytically the expression of the current in a RL series circuit forced by an alternating source. A good agreement is found between our theoretical and experimental results.

Keywords: *nonlinear, inductance, ferromagnetic, core, hysteresis*

1. Introduction

Magnetic materials are used in electrical power devices such as transformers, motors and generators. They are also present in measurement and control of electrical power for actuators and sensors, contactors and relays. Another application field is found to be in the magnetic recording industry, either by analog or digital, as in audio and video recording or in the storage of data information on magnetic disks and tapes for computer applications. The purpose of a magnetic material core in an inductor is to produce by a given applied field, a higher flux compared to the flux produced in air, and to form a magnetic path for the flux.

In general, all electrical circuit elements respond in a nonlinear fashion to any form of electrical input. However, simple passive elements, such as resistors, capacitors, and air core inductors, respond to a first order approximation nearly linearly. But in devices that have ferromagnetic cores, the relationship between the flux density and the magnetic field strength in the magnetic core is nonlinear. This nonlinear relationship depends on several factors, such as the chemical constitution and structure of the magnetic material, technological process, the way the material is worked and applied. The nonlinear characteristic of the magnetic material is irreversible, hence exhibits hysteresis.

The ferromagnetic cores are basic constructive elements of transformers and inductors for wide application area, ranging from analog and digital microelectronics toward power converters and power systems. It is known that the rigorous study and design optimization of such electromagnetic devices is difficult because of nonlinearity, electromagnetic inertial behavior and other related phenomena, as saturation, anisotropy, magnetic hysteresis and induced eddy currents.

Hysteresis is a wide research topic in science and engineering. Mathematicians, physicists and engineers have proposed numerous models to describe that phenomenon [1-9]. They are generally fully analytic models based on certain physical theories like Weiss, Preisach or Jiles-Atherton, often restricted to particular applications [10,11,12]. Some apparently accurate models are inadequate when used in circuit simulation because of discontinuities which can provoke severe convergence problems and therefore, alter the simulation results. For many known models used in circuit simulators [13,14], it is difficult to identify the correct correspondence between the model parameters and the real device ones. However, all indicates that this research topic is still far from being fully exploited and that it is difficult to find the ultimate solution and a universal model. This has been and seem to remain a long-term challenge for researchers [15,16].

There are basically three approaches in the modeling of inductance nonlinearity that include hysteresis. The first approach looks at the physical properties of the material: domain alignments, wall movements, spin rotations, etc [17,18]. The second approach prefers a macroscopic description of hysteresis using mathematical models to predict the B - H curve but without completely neglecting the physics of the material [19]. The third approach needs the equivalent circuit to be modeled in existing computer programs [20,21].

Thus, the merit of the model developed in this work is that it defines the complex requirements of the particular modeling approach that provides a solution to this problem with excellent accuracy.

The method of analysis used in this research is a combination of direct harmonic domain approach, the Harmonic Balance Method, orthogonalized polynomials and piecewise approximations. It employs the methodology that analyzes the influence of all significant factors in more detail and it includes the representation of hysteresis. The modeling procedure has a wide range of generality, being suitable to be extended to transformers, linear actuators and rotating motors.

The paper is organized as follows: in Section 2, the measurement method is briefly described and the mathematical models are given. Next we analyze the dynamical behavior of a circuit containing a resistor and a ferromagnetic core inductor forced by an alternating generator in Section 3. In Section 4, theoretical and experimental results are presented and the paper is concluded with some remarks in Section 5.

2. Measurement Method and Mathematical Models

2.1. Measurement Methods

The relation between magnetic induction and magnetic field in ferromagnetism can be presented with a hysteresis loop. Several methods of measurement are commonly applied for the recording of hysteresis loop of ferromagnetic materials, such as Roland's and Juing's method [22] for measuring with direct current, and oscilloscope measuring method which employs AC current.

The AC method requires extensive preparation for the measurement procedure, such as the calibration of the oscilloscope and a proper selection of the components in the secondary electric circuit. In addition, measurement results cannot be stored in an easy and simple way in the proper format. For this measurement, we will use the DC method applied to the connection scheme shown in Figure 1.

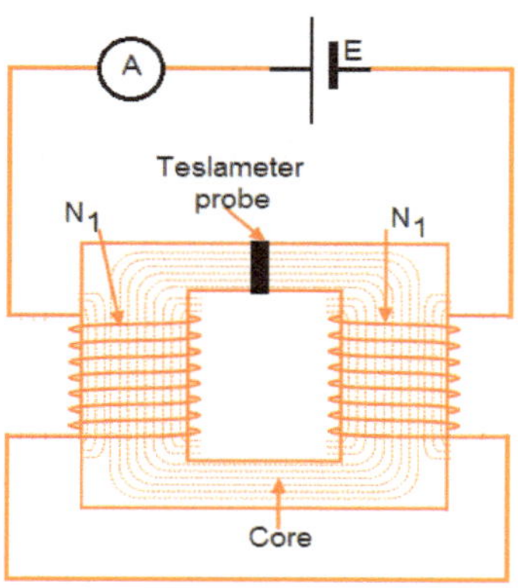

Figure 1. Connection scheme for recording hysteresis loop.

When the current I measured by the ammeter is proportional to the strength of the magnetic field H, the magnetic induction B is directly measured using a teslameter.

The hysteresis loop we obtained is shown in the graph of Figure 2. It is a three steps illustration. Step one represented by circles corresponds to the variation of the power supply from 0 V to +2 V. Without switching off the circuit, but varying back the power supply from +2 V to –2 V, then forth from –2 V to +2 V, we obtain the second step (triangles) respectively the third depicted by squares.

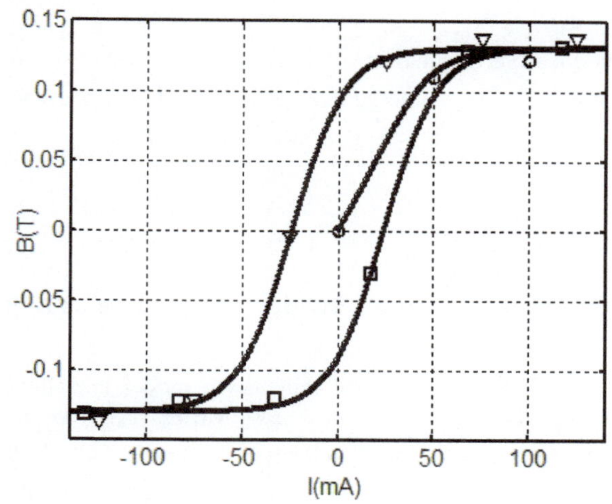

Figure 2. Experimental recorded data (circles, triangles and squares) and splitting (solid line) B - I curves

2.2. Mathematical Model of Hysteresis

Just by considering the shapes of the curves formed by the triangles and squares points, we describe the relation between the magnetic induction and the magnetic field in the following form:

$$B = \mu_0 H + B_s \tanh\left(\frac{\alpha H - \sigma}{2}\right) \text{with } \sigma = \beta sign\left(\frac{dH}{dt}\right). \quad (1)$$

B_s, α and β are parameters to be determined and the step function $sign(x)$ is defined as

$$sign(x) = \begin{cases} 1 & \text{if } x > 0 \\ 0 & \text{if } x = 0 \\ -1 & \text{if } x < 0 \end{cases} \quad (2)$$

μ_0 is the permeability of free space. One can notice that, for the part of the hysteresis loop with triangles points, H or I is decreasing, then $\sigma = -\beta$ and for the other part, H is increasing, means that $\sigma = \beta$. The relation between magnetic induction and current can be obtained by taking into account the Ampere law: $H\ell = NI$, where ℓ is the average length of the magnetic material and $N = 2N_1$ is the total number of turns. Under this relation, equation (1) becomes

$$B = \frac{\mu_0 NI}{\ell} + B_s \tanh\left(\frac{\alpha NI}{2\ell} - \frac{\sigma}{2}\right) \text{with } \sigma = \beta sign\left(\frac{dI}{dt}\right). \quad (3)$$

At this stage, the saturation flux density can be obtained from our recorded data and we found that $B_s = 130$ mT.

The values of α and β can be obtained by using the means square method. For the following parameters of our used inductor $\ell = 24$ cm, the cross sectional area $A = 176.71\,\text{mm}^2$ and $N_1 = 500$ we obtain $\alpha \simeq 88.22\cdot 10^{-4}$ m $\cdot A^{-1}$ and $\beta \simeq 88.42\cdot 10^{-2}$. Using these values, the curves we obtained are shown in the graph of Figure 2 with solid line. The results show a good agreement between the experimental values and our mathematical model.

Since ferromagnetic materials are generally characterized by three parameters: the remanence (B_r), the coercive magnetic field (H_c) and the saturation flux density B_s , we found general relations between them and the coefficients α and β in the following form:

$$\alpha = \frac{1}{H_c}\ln\left(\frac{B_s + B_r}{B_s - B_r}\cdot\frac{B_s - \mu_0 H_c}{B_s + \mu_0 H_c}\right),\ \beta = \ln\left(\frac{B_s + B_r}{B_s - B_r}\right). \quad (4)$$

2.3. Mathematical Model of Ferromagnetic Core Inductance

The ferromagnetic core inductance model is obtained through the relation $\phi = LI = BNA$ where A is the cross sectional area of the magnetic material and ϕ the magnetic flux through the magnetic material. We then obtain

$$L = \frac{BNA}{I}. \quad (5)$$

Since we have two expressions for B (see equation (3)), we will use the average of the B - I curves. Consequently, the inductance of an inductor that contains a ferromagnetic material has the following expression:

$$L = \frac{\mu_0 N^2 A}{\ell} + \frac{B_s NA}{2I}\times \left[\tanh\left(\frac{\alpha NI}{2\ell} - \frac{\beta}{2}\right) + \tanh\left(\frac{\alpha NI}{2\ell} + \frac{\beta}{2}\right)\right]. \quad (6)$$

For illustration, the experimental result with circles and dashed line and the theoretical result with solid line are plotted in Figure 3.

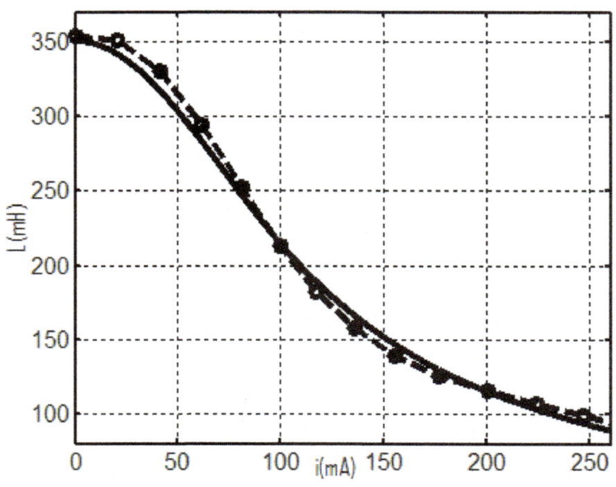

Figure 3. Experimental recorded data (dashed line) and splitting (solid line) L - I curves.

The experimental values are computed using equation (5) with the recorded values of I and B . We have increased the supply voltage from $0V$ to $2V$ but the first value of I we recorded is 1 mA. Also here, there is a good agreement between experimental and theoretical results as the experimental curve fluctuates slightly around the theoretical one.

Regarding the shape of the curve, this result is similar to those obtained experimentally by the authors of references [23,24]. One can also notice from our model that the inductance of a ferromagnetic core inductor lies between two values. The maximum value L_i and the minimum value L_f are obtained for very small values of current ($I = 0$ A) and for infinite values of current respectively. Their expressions are given as follows:

$$L_i = \frac{\mu_0 N^2 A}{\ell}\left[1 + \frac{\alpha B_s}{\mu_0(1+\cosh\beta)}\right] \text{ and } L_f = \frac{\mu_0 N^2 A}{\ell}. \quad (7)$$

From expressions of L_i and L_f , we can deduce the initial relative permeability and the final relative permeability respectively as $1 + \frac{\alpha B_s}{\mu_0(1+\cosh\beta)}$ and 1 . This is a very interesting confirmation of our model since it is known that the relative permeability of a ferromagnetic material is greater than one.

3. Dynamics of a Circuit Containing a Resistor in Series with a Ferromagnetic Core Inductor

Since the ferromagnetic core inductance is a function of a current in the circuit and the current is also a function of components in the circuit, the aim of this section is to analyze the dynamics of a circuit where a resistor is connected in series with a ferromagnetic core inductor.

3.1. Circuit and Mathematical Model

The circuit diagram is shown on Figure 4, where r represents the internal resistance of the inductor and R_0 is another resistor added, both to reduce the amplitude of current through the inductor and to measure current.

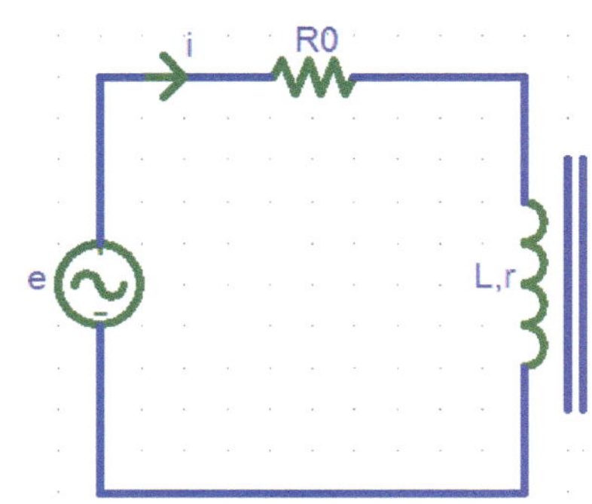

Figure 4. The circuit diagram modeled in this paper.

Since the inductance of such inductor is not constant, the application of the Kirchhoff law leads to the following equation:

$$e = \frac{d(Li)}{dt} + Ri = \frac{d(BNA)}{dt} + Ri. \quad (8)$$

$R = r + R_0$ represents the total resistance of the circuit viewed by the generator. If we replace B by the corresponding expression as in equation (3) and force the circuit with an alternating generator that has magnitude E_m and radiant frequency ω, we obtain the equation of the circuit as:

$$\left[\frac{\mu_0 N^2 A}{\ell} + \frac{B_s N^2 \alpha A}{2\ell}\left(1 - \tanh^2\left(\frac{\alpha N i}{2\ell} - \frac{\sigma}{2}\right)\right)\right] \times \frac{di}{dt} + Ri = E_m \sin(\omega t). \quad (9)$$

Here again, σ is defined as in equation (3)

$$\sigma = \beta sign\left(\frac{di}{dt}\right). \quad (10)$$

Let us define $t = \frac{\tau}{\omega_0}$ and $i = I_0 x$ where ω_0 is a constant and $I_0 = \frac{\ell}{\alpha N}$. Then, the state equation (9) can be rewritten in dimensionless form as:

$$g(x)\dot{x} = -\lambda x + E\sin(\Omega\tau) \quad \text{with} \quad g(x) = (1-\eta) + \frac{2\eta}{1+\cosh(x-\sigma)}. \quad (11)$$

The new introduced parameters are defined as follows:

$$\lambda = \frac{R}{L_0\omega_0},\ E = \frac{E_m}{L_0\omega_0 I_0},\ \Omega = \frac{\omega}{\omega_0},\quad \eta = \frac{B_s\alpha}{2\mu_0 + B_s\alpha},\ L_0 = \frac{\mu_0 N^2 A}{\ell}\left(1 + \frac{B_s\alpha}{2\mu_0}\right). \quad (12)$$

3.2. Mathematical Analysis

To derive an analytical expression of the current in the circuit, we put the function $g(x)$ in the form $g(x) = \frac{g_1(x)}{g_2(x)}$ where $g_2(x) = 1 + \cosh(x)$ and $g_1(x) = (1+\eta) + (1-\eta)\cosh(x)$. For simplification, the parameter β is neglected and equation (11) is reduced to:

$$g_1(x)\dot{x} = [-\lambda x + E\sin(\Omega\tau)]g_2(x). \quad (13)$$

Replacing expressions for g_1 and g_2 into equation (13) leads to the following equation:

$$[(1-\eta)\dot{x} + \lambda x - E\sin(\Omega\tau)]\cosh(x) = -(1+\eta)\dot{x} - \lambda x + E\sin\sin(\Omega\tau). \quad (14)$$

At this stage, some approximations are needed, we will first expand the function cosh in polynomial form as:

$$\cosh(x) = \sum_{m=0}^{M} \frac{x^{2m}}{(2m)!}. \quad (15)$$

M is the order of the expansion. Equation (14) becomes:

$$(1-\eta)\sum_{m=0}^{M} \frac{\dot{x}x^{2m}}{(2m)!} + \lambda \sum_{m=0}^{M} \frac{x^{2m+1}}{(2m)!} + (1+\eta)\dot{x} - E\sum_{m=0}^{M} \frac{x^{2m}}{(2m)!}\sin(\Omega\tau) = -\lambda x + E\sin(\Omega\tau). \quad (16)$$

For the second approximation, we find the solution in the form $x = P\cos(\Omega\tau) + Q\sin(\Omega\tau)$ where P and Q are unknown parameters to be determined and $X = \sqrt{P^2 + Q^2}$ represents the magnitude of x. We obtain after some mathematical transformations the following equation:

$$a\dot{x} + bx - c\sin(\Omega\tau) - (P^2 - Q^2)d\sin(\Omega\tau) = -2PQd\cos(\Omega\tau). \quad (17)$$

The parameters a, b, c and d are defined as:

$$a = (1+\eta) + (1-\eta)\sum_{m=0}^{M}\left[\frac{1}{m!(m+1)!}\left(\frac{X}{2}\right)^{2m}\right],$$
$$b = \lambda + \lambda\sum_{m=0}^{M}\left[\frac{2m+1}{m!(m+1)!}\left(\frac{X}{2}\right)^{2m}\right],$$
$$c = E + E\sum_{m=0}^{M}\left[\frac{1}{(m!)^2}\left(\frac{X}{2}\right)^{2m}\right],$$
$$d = E\sum_{m=1}^{M}\left[\frac{1}{(m-1)!(m+1)!}\left(\frac{X}{2}\right)^{2m-2}\right]. \quad (18)$$

Replacing expression for x into equation (17) and equating the coefficients of $\sin(\Omega\tau)$ and $\cos(\Omega\tau)$ in both sides, we obtain the following system:

$$\begin{cases} a\Omega Q + bP = -2dPQ, \\ bQ - a\Omega P = c + d(P^2 - Q^2). \end{cases} \quad (19)$$

After some mathematical manipulations, we obtain the algebraic equation that verifies the amplitude X as follows:

$$d^4X^8 + 2cd(b^2 - a^2\Omega^2 - cd)X^4 + c^4 - d^2(a^2\Omega^2 + b^2)X^6 - c^2(a^2\Omega^2 + b^2)X^2 = 0 \quad (20)$$

As the value of M increases, the power of the above equation also increases. We can derive the analytical solution for two cases $M = 0$ and $M = 1$.

3.2.1. Case $M = 0$

The corresponding equation and the solution are given in equations (21) and (22) respectively.

$$(\lambda^2 + \Omega^2)X^2 = E^2. \quad (21)$$

$$X = \frac{E}{\sqrt{\lambda^2 + \Omega^2}}. \quad (22)$$

In this case, the amplitude of the current and the magnitude of the AC generator are linearly coupled. This case corresponds to a linear inductor with inductance equals L_0 and is verified for small amplitudes of the current in the circuit. Since we know that as current increases the inductance decreases and the magnetic induction reaches the saturation point.

3.2.2. Case $M = 1$

The corresponding equation here is given by relation (23) where the transformation $Y = X^2$ is used to reduce the order of the algebraic equation.

$$a_3 Y^3 + a_2 Y^2 + a_1 Y + a_0 = 0. \tag{23}$$

The coefficients a_3, a_2, a_1 and a_0 are defined as follows:

$$a_3 = \lambda^2 + (1-\eta)^2 \Omega^2,\ a_2 = 32\lambda^2 + 32(1-\eta)\Omega^2 - E^2,$$
$$a_1 = 256\left(\lambda^2 + \Omega^2\right) - 32E^2,\ a_0 = -256E^2. \tag{24}$$

To derive solutions of equation (23), let consider the following parameters:

$$\delta = a_2^2 - 3a_3 a_1,\ q = \frac{2a_2^3 - 9a_3 a_2 a_1 + 27a_3^2 a_0}{2},\ \Delta = q^2 - \delta^3.$$

According to the values of parameters, equation (23) can have one or three real solutions and this can be decided through the sign of Δ.

If $\Delta > 0$, equation (23) has one real solution from which we can deduce the amplitude of the normalized current as:

$$X = \left[\frac{-a_2 + \sqrt[3]{q - \sqrt{\Delta}} + \sqrt[3]{q + \sqrt{\Delta}}}{3a_3}\right]^{1/2}. \tag{25}$$

If $\Delta < 0$, equation (23) has three real solutions and the corresponding expressions of X are given a follow:

$$X_1 = \left(\frac{-a_2 + 2\sqrt{\delta}\cos\left(\frac{\theta - 2\pi}{3}\right)}{3a_3}\right)^{1/2},$$
$$X_2 = \left(\frac{-a_2 + 2\sqrt{\delta}\cos\left(\frac{\theta}{3}\right)}{3a_3}\right)^{1/2}, \tag{26}$$
$$X_3 = \left(\frac{-a_2 + 2\sqrt{\delta}\cos\left(\frac{\theta + 2\pi}{3}\right)}{3a_3}\right)^{1/2}$$

where $\theta = \arctan\left(\frac{\sqrt{-\Delta}}{q}\right)$.

Even if the analytical method gives more than one solution, we think that for the stability of the circuit the appropriate solution is the one with smallest amplitude. According to this fact, we can say that the required value of X is the minimum value between X_1, X_2 and X_3.

4. Results and Comparisons

The aim of this subsection is to present our theoretical and experimental results and give comparison between them. The experimental values are $r = 4.0\,\Omega$, $R_0 = 10\,\Omega$, $\ell = 24$ cm, $A = 176.71$ mm ^{2} and $N = 1000$. E_m and $\omega = 2\pi f$ are used as control parameters with $0\,V \leqslant E_m \leqslant 30\,V$ and 10 Hz $\leqslant f \leqslant 250$ Hz. The corresponding theoretical values computed using equation (12) worth: $\lambda = 66.2\cdot 10^{-4}$, $\eta = 0.9978$, $L_0 = 423.18$ mH, $\omega_0 = 5\cdot 10^3$ rad/s and $I_0 = 27.2$ mA.

4.1. Effect of the Amplitude E_m

For our first investigation, we will maintain constant the frequency at $f = 50$ Hz ($\Omega = 0.0628$), then record the amplitude of the current as E_m is increasing. Our analytical and numerical results are shown on the graphs of Figure 5.

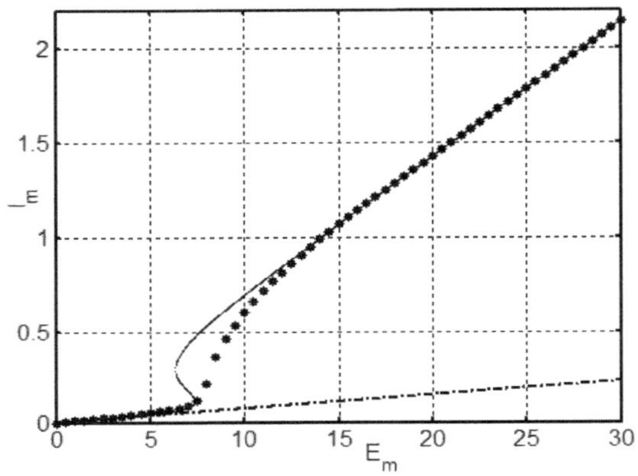

Figure 5. Amplitude of the current I_m versus the magnitude of the external generator E_m : Numerical obtained (curve with dots), first order (dashed line) and second order (solid line) analytical plotted

The curve with dashed line is obtained by using equation (22) while the curve with full line is obtained from equation (25) or (26). The curve with dots represents our numerical results obtained by solving directly the differential equation (11) with the fourth order Runge Kutta method.

As the graph reveals, the result obtained analytically with the first order approximation ($M = 0$) agrees with the numerical ones just for small values of $X_m = X \cdot I_0$. As we mentioned above, the first order approximation (linear approximation) is valid for small amplitudes of current. One can note the good agreement between results obtained analytically with the second order approximation ($M = 1$) and the numerical ones. A small disagreement is found when E_m is around 8V and we think that this can be overcome by increasing the approximation order.

Figure 6 depicts the comparison between our numerical results (curve with full line) and the experimental results (curve with dots).

Here, it can also be noticed that, the experimental results agree with the numerical ones and consequently with the analytical results obtained from the second order approximation.

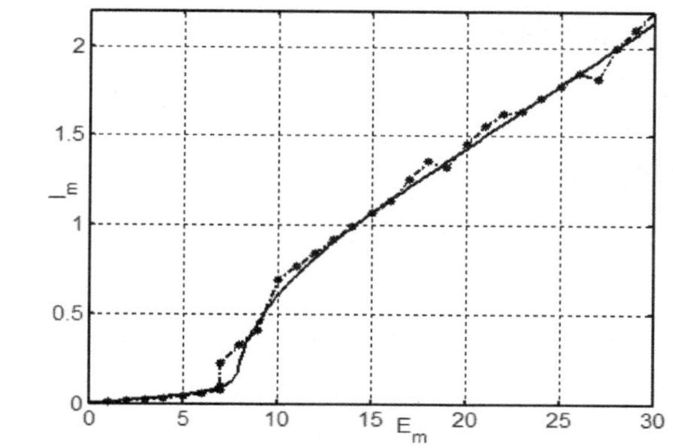

Figure 6. Amplitude of the current I_m versus the magnitude of the external generator E_m : Numerical obtained (solid line) and experimental recorded (dots and dashed lines)

Even in the case of high amplitudes of current and high amplitudes of the external generator, a good agreement remains between the numerical and analytical results as shown in the graph of Figure 7. Here, the frequency of the external generator equals 200 Hz and we plot the amplitude of the current I_m versus the magnitude of the external generator E_m. Also here, a small disagreement is found when E_m is around 35V and we think that this can be overcome by increasing the approximation order.

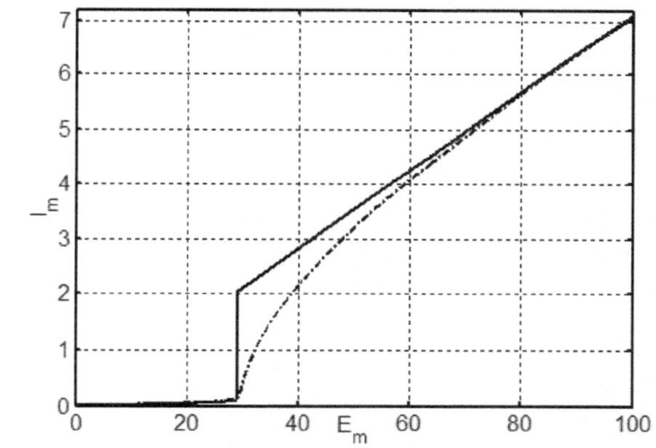

Figure 7. Amplitude of the current I_m versus the magnitude of the external generator E_m : Numerical obtained (dashed line) and analytical ploted (solid line)

4.2. Effect of the Frequency f

In this step, the magnitude of the external AC signal is kept constant and we analyze the effect of the frequency using the second order approximation for our analytical treatment. The graph of Figure 8 obtained for $E_m = 5$ V shows the comparison between the numerical results (curve with dashed line), the analytical results (curve with full line) and the experimental results (curve with squares). It can be observed that the three results are slightly identical and the shape of the curves is similar to that of the low pas filter obtained in the classical RL circuit.

Also here and as mentioned in the previous section, we found a small disagreement between analytical and numerical results when the frequency is located around 60 Hz and E_m around 8 V. For illustration, using $E_m = 10$ V we obtained the curves shown in Figure 9.

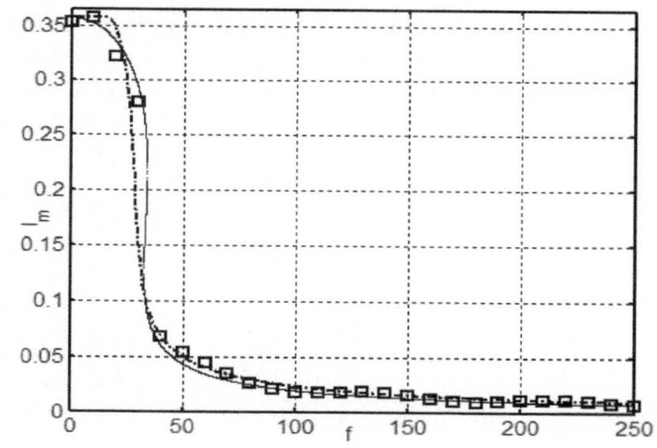

Figure 8. Amplitude of the current I_m versus the frequency of the external generator f : Comparison between the numerical (dashed line), the analytical (solid line) and the experimental results (dots) with $E_m = 5$ V.

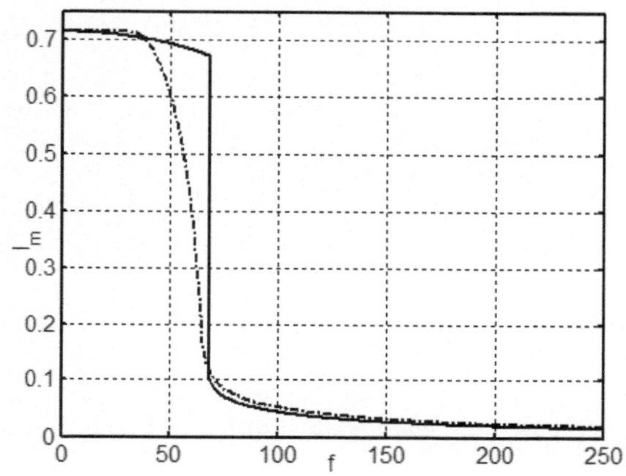

Figure 9. Amplitude of the current I_m versus the frequency of the external generator f : Comparison between the numerical (dashed line) and the analytical (solid line) results with $E_m = 10$ V.

The numerical results (dashed line) and the analytical results (full line) are slightly different in a certain frequency range.

Figure 10 highlighs the comparison between our numerical results (curve with dashed lines) and the analytical results (curve with full line) when the magnitude of the external force is increased at $E_m = 15$ V. Qualitativelly, Figure 9 and Figure 10 have the same behavior and hence the same comments.

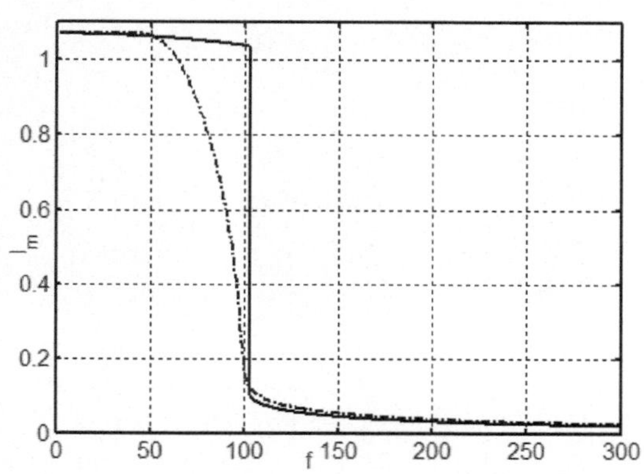

Figure 10. Amplitude of the current I_m versus the magnitude of the external generator E_m : Numerical obtained (solid line) and experimental recorded (dots and dashed lines).

The numerical simulations of equation (11) give results that agree well with those obtained experimentally (see Figure 11 and Figure 12) for $E_m = 5$ V and $E_m = 9$ V with $f = 50$ Hz

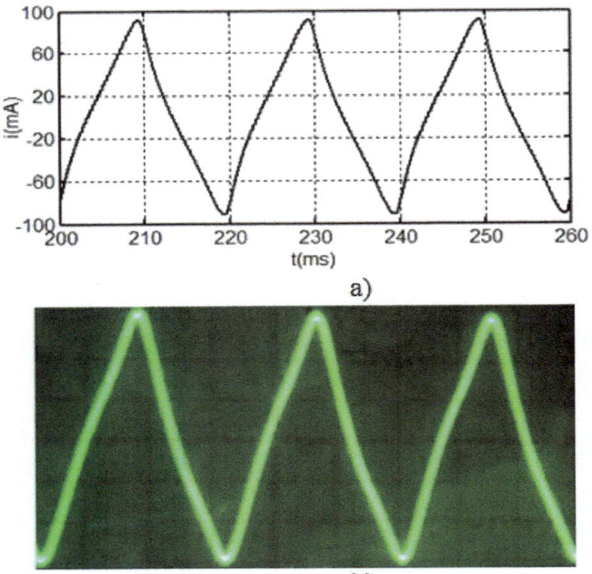

Figure 11. Time waveforms of the current obtained for $f = 50$ Hz and $E_m = 7$ V. a) Numerically plotted and b) obtained experimentally

Figure 11a) end Figure 12a) are obtained numerically while Figure 11b) end Figure 12b) are their experimental corresponding.

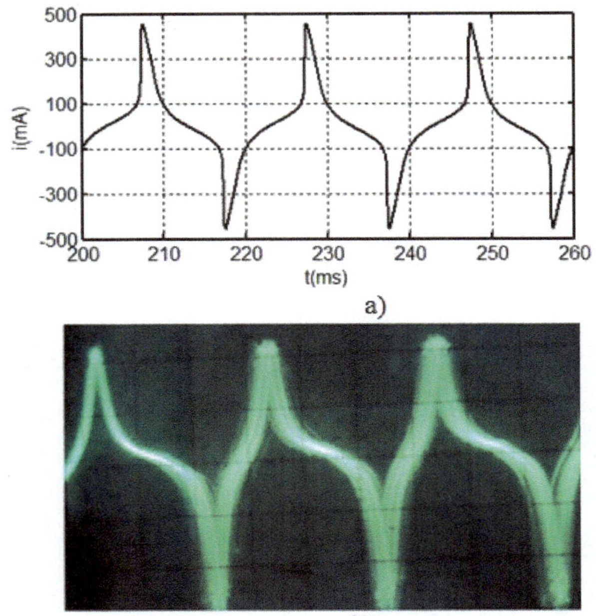

Figure 12. Time waveforms of the current obtained for $f = 50$ Hz and $E_m = 9$ V. a) Numerically plotted and b) obtained experimentally

On the other hand even if the shape of the current does not respect the sinusoidal form, the amplitudes obtained from the mathematical analysis can be compared to the experimental ones.

5. Conclusion

As far the authors are aware, the circuit of Figure 4 has not until now been modeled in a way that includes the ferromagnetic materials parameters and hysteresis. Our model has been incorporated into the differential equation that describes the circuit and has been shown to be able of displaying all the behavior observed in experiment. Another advantage of our model is that the analytical treatment can now be done to predict the amplitude of the current through a ferromagnetic core inductor. We found here a good agreement between the numerical and experimental results as well as our analytical results.

Our next investigation is to test this model on various soft and hard magnetic materials used in industrial devices. We think also that the analysis of the high frequencies effect on the accuracy of this model is another good idea. The study of electrical transformers and motors using this model is a further interesting subject.

References

[1] Visintin, A. *Differential Models of Hysteresis*, Applied Mathematical Science, Springer-Verlag, New York, 1994.

[2] Górecki, K. and Zarobski J, "Modelling of Inductors and Transformers in SPICE," *Elektronika - konstrukcje, technologie, zastosowania, Not-Sigma*, 1(2). 40-43. 2004.

[3] Górecki, K, "SPICE-aided Modelling of Coils with the Ferrite Core with Selfheating Taken into Account," *Kwartalnik Elektroniki i Telekomunikacji*, 3(5). 389-404. 2003.

[4] Chiesa, N. and Hoidalen, H.K, "Modeling of nonlinear and hysteretic iron-core inductors in ATP," in EEUGMeeting 2007, European EMTP-ATP Conference, Leon, Spania, Sep. 2007.

[5] Anderson, J.C, *Magnetism and Magnetic Materials*, Champen and Hall Ltd. London, 1968.

[6] Chua, L.O. and Stromsmoe, K.A, "Lumped-Circuit Models for Nonlinear Inductors Exhibiting Hysteresis Loops," *IEEE Transactions on Circuit Theory*, 17(4). 564-574. Nov. 1970.

[7] Carnevale, D., Nicosia, S. and Zaccarian, L, "Generalized Constructive Model of Hysteresis," *IEEE Transactions on Magnetics*, 42(12). 3809-3817. Dec. 2006.

[8] Chua, L.O. and Bass, S.C, "A Generalized Hysteresis Model," *IEEE Transactions on Circuit Theory*, 19(1). 36-48. Jan. 1972.

[9] Chiesa, N., Avendano, A.H., Hoidalen, K., Mork, B.A., D. Ishchenko, D. and Kunze, A.P, "On the ringdown transient of transformers," *in IPST'07 - International Conference on Power System Transients, Lion, France*, (229). 4-7. Jun. 2007.

[10] Jiles, D.C. and Atherton D.L, "Ferromagnetic Hysteresis," *IEEE Transactions on Magn*. 19(5). 2183-2185. Sep. 1983.

[11] Fiorillo F. and Dupré L.R, "Comprehensive Model of Magnetization Curve, Hysteresis Loops, and Losses in Any Direction in Grain-Oriented Fe-Si," *IEEE Transactions on Magn*, 38(3). 1467-1476. Jun. 2002.

[12] Carnevale D., Nicosia S. and Zaccarian L, "Generalized Constructive Model of Hysteresis," *IEEE Transactions on Magn*, 42(12). 3809-3817. Dec. 2006.

[13] Brachtendorf H.G., Eck C. and Laur R, "Macromodeling of Hysteresis Phenomena with SPICE," *IEEE Transactions on Circuits and Systems – II: Analog and Digital Signal Processing*, 44(5). 378-388. May. 1997.

[14] Chan J.H., Vladimirescu A., Gao X.C., Liebmann P. and Valainis J, "Nonlinear Transformer Model for Circuit Simulation," *IEEE Transactions on Computer-Aided Design*, 10(4) 476-482. Apr. 1991.

[15] Germay, N., Maestero, S. and Vroman, J, "Review of Ferroresonance Phenomena in High Voltage Power Systems and Presentation of a Voltage Transformer Model for Predetermining Them", *CIGRE*, 33-18.1972.

[16] Nakra, H.L. and Barton, T.H, "Three Phase Transformer Transients," *IEEE Transactions on Power Apparatus and Systems*, 93(1). 1810-1819. Nov./Dec. 1974.

[17] De Leon, F. and Semlyn, A, "A Simple Representation of Dynamic Hysteresis Losses in Power Transformer," *IEEE Transactions On Power Delivery*, 10(1). 315-321. Jan. 1995.

[18] Wright, A. and Carneiro, S, "Analysis of Circuits Containing Components with Cores of Ferromagnetic Material," *IEE Proceedings*, 121(12). 1579-1581. Dec. 1974.

[19] Greene, J.D. and Grass, A.C. "Nonlinear Modeling of Transformer," *IEEE Transactions on Industry Applications*, 24(3). 434-438. Jun. 1988.

[20] Yamada, S., Besho K. and Lu, L, "Harmonic balance finite element method applied to the nonlinear AC magnetic analysis," *IEEE Trans. On Magn.* 25(6). 2971-2973. Jul. 1989.

[21] Huang, S.R., Chung, S.C., Chen B.N. and Chen, Y.H, "A Harmonic Model for the Nonlinearities of Single-Phase Transformer with Describing Function," *IEEE Transaction On Power Delivery*, 18(3). 815-820. Jul. 2003.

[22] Radmanesh, H., Abassi, A. and Rostami, M, *Analysis of Ferroresonance Phenomena in Power Transformers Including Neutral Resistance Effect,* IEEE conference, Georgia, USA, 2009.

[23] Gianduzzo, J.C. Ygorra, S. and Lasne. L, « La ferrorésonnance, experimentation, et modélisation non linéaire, » Université de Bordeaux 1, Centre de Ressource en EEA. http://ferroresonance.free.fr.

[24] Salas, R.A. and Pleite, J, "Simple Procedure to Compute the Inductance of a Toroidal Ferrite Core from the Linear to the Saturation Regions," *Materials,* 6(3). 2452-2463. Jun. 2013.

The Next Generation Mobile Wireless Cellular Networks – 4G and Beyond

Arun Agarwal[1,*], Kabita Agarwal[2]

[1]Department of Electronics & Communication Engineering, Institute of Technical Education and Research, Siksha 'O' Anusandhan University, Khandagiri Square, Bhubaneswar, India
[2]Department Electronics & Telecommunication Engineering, CV Raman College of Engineering, Bhubaneswar, India
*Corresponding author: arunagrawal@soauniversity.ac.in

Abstract The demand of mobile user is ever increasing in this world of digital systems. Consumers demand more from their technology. Whether it is a television, cellular phone, Washing machine or refrigerator, the latest technology purchase must have new features. With the advent of the Internet, the most-wanted feature is better, faster access to information. The time not far away when access to all necessary information and the power of a personal computer, sits in the palm of one's hand. To support such a powerful system, we need pervasive, high-speed wireless connectivity. The drawbacks of the current mobile communication technology have led the researchers to come up with more advanced and efficient technologies. 4G mobile technology is the outcome in this direction. 4G is the next generation of wireless networks that will totally replace existing 3G networks. It is supposed to provide its customers with better speed and all IP based multimedia services. In the present scenario existing technologies are capable of performing functions like broadband data access, supporting voice traffic using voice over IP (VoIP), in mobile environment etc., but there is a great requirement of integrating all such technologies into a single combined system. 4G promises a solution to this problem by seamlessly integrating the terminals, networks and applications. This paper presents an overall study of the 4G systems, architecture, standard, benefits, challenges in implementation, design for the LTE and WiMAX technologies, Security issues and future scope of 4G technologies.

Keywords: *CDMA, IP, LTE, multimedia services, security, TDMA, WiMAX, 3G, and 4G wireless*

1. Introduction

The Mobile devices are getting smaller, lighter, and more powerful; they have bigger screens and longer battery life, more features and more capabilities. Bandwidth will always be the limiting factor in the development of applications and devices, be it wired, or wireless. The short fall of 3G networks is clear, it's just not fast enough, offering 384kbps doesn't meet the requirements of what the end user has come to expect these days. Some people see 3G as a stop-gap, until a fully integrated IP network is created; some countries have even chosen to bypass 3G and head straight to 4G, a method which has its advantages, and its disadvantages.

4G technology is actually an integrated, global network that will be able to provide a comprehensive IP solution where voice, data and streamed multimedia can be given to users on an "Anytime, Anywhere" basis [1]. The next generation of wireless technology (fourth generation or 4G) is intended to support broadband performance and enable voice/video multimedia applications. The enabling technologies and standards for 4G wireless communications allow for significant increases in data rates over 2G (second generation), 3G (third generation) and 3.5G wireless technologies. At the present time, LTE (Long Term Evolution) and WiMAX (Worldwide Interoperability for Microwave Access) are the two technologies considered as candidates to achieve the 4G wireless performance objectives [2].

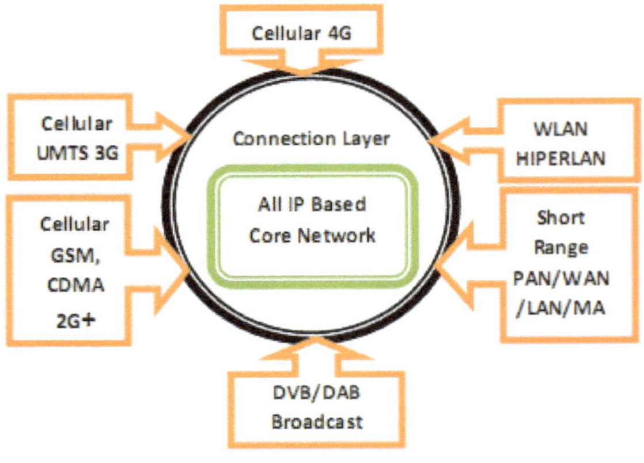

Figure 1. Seamless connections of networks [1]

These objectives include: that 4G will be a fully IP-based integrated system. 4G will be capable of providing between 100 Mbit/s and 1 Gbit/s speeds both indoors and outdoors, with premium quality and high security. The term 4G is used broadly to include several types of broadband wireless access communication systems, not only cellular telephone Systems [1]. One of the terms used to describe 4G is MAGIC—Mobile multimedia, anytime anywhere, Global mobility support, integrated wireless solution, and customized personal service. As a promise for the future, 4G systems, that is, cellular broadband wireless access systems have been attracting much interest in the mobile communication arena. The 4G systems not only will support the next generation of mobile service, but also will support the fixed wireless networks [4]. Figure 1 illustrates the seamless connectivity of the different Networks to IP.

2. Evolution of Mobile Radio

After a brief introduction to 4G, this section briefly outlines the evolution of mobile wireless technology from 1G (First Generation) to 4G.

2.1. 1G

This term refers to the first generation cellular systems that were analog telecommunications standards introduced in the 1970 s. The voice channel typically used frequency modulation, and they used FDMA techniques. Some of the standards are NMT (Nordic Mobile Telephone), AMPS (Advanced Mobile Phone System) TACS (Total Access Communications System) and C-450. The major drawbacks were Poor voice quality, Poor battery life, large phone size, No security, frequent call drops, Limited capacity and poor handoff reliability.

Table 1. Evolution of mobile radio standards [1]

Tchnology	1G	2G	2.5G	3G	4G
Year	1970	1980	1985	1990	2000
Standards	AMPS, NMT, Hicap, CDPD, TACS, ETACS	GSM, i-DEN, D-AMPS	EDGE, GPRS	CDMA 2000,WCDMA	Single Standard LTE-Advanced
Data Bandwidth	1.9 kbps	14.4 kbps	384 kbps	2 Mbps	200 Mbps
Core Network	PSTN	PSTN	PSTN packet network	packet network	INTERNET
Multiplexing	FDMA	CDMA TDMA	TDMA CDMA	CDMA	OFDMA
Service	Analog Voice	Digital Voice	Packetized Data, Higher Capacity	Higher Capacity, Broadband Data	Broadband Data With High Speed

2.2. 2G

This term refers to the second generation cellular telecommunications systems that were introduced in 1980 s. The systems were digital and were oriented to voice with only low speed data services. Systems such as GSM and US-TDMA used a mixture of TDMA and FDMA techniques. IS-95 /cdmaOne were the first CDMA system. 2G services are frequently referred as Personal Communications Services, or PCS, in the United States. 2G phones used global system for mobile communications (GSM) standard. The GSM is a circuit switched, connection oriented technology, where the end systems are dedicated for the entire call session. This causes inefficiency in usage of bandwidth and resources. The GSM-enabled systems do not support high data rates. They are unable to handle complex data such as video.

2.3. 2.5G

This term is used to describe 2G-systems that have implemented a packet switched domain in addition to the circuit switched domain. 2.5G is not an officially defined term rather it was invented for marketing purpose [1]. These systems enable high-speed data transfer over upgraded existing 2G networks. Examples: EDGE, GPRS, CDMA2000.

2.4. 3G

This refers to Third generation. The designation for systems following the 2G systems and they offer high speed data services in addition to the basic voice capability.

Typically they use CDMA techniques and include UMTS (W-CDMA), CDMA2000 1xEV-DV, CDMA2000 1xEV-DO, and TD-SCDMA. The 3G technology adds multimedia facilities to 2G phones by allowing video, audio, and graphics applications. Over 3G phones, you can watch streaming video or have video telephony. 3G promises increased bandwidth, up to 384 kbps when the device holder is walking, 128 kbps in a car, and 2 Mbps in fixed applications. However the short comings of 3G networks were High bandwidth requirement, High spectrum licensing fees, huge capital.

3. Standards and Benefits of 4G

This section outlines the details of 4G Standard and its various objectives, and comparison with existing 3G technologies. In telecommunications, 4G is the fourth generation of cellular wireless standards. It is a successor to the 3G and 2G families of standards. 4G is described as MAGIC: mobile multimedia, any-time anywhere, global mobility support, integrated wireless solution, and customized personal service. In 2009, the ITU-R organization specified the IMT-Advanced (International Mobile Telecommunications Advanced) requirements for 4G standards, setting peak speed requirements for 4G service at 100 Mbit/s for high mobility communication (such as from trains and cars) and 1 Gbit/s for low mobility communication (such as pedestrians and stationary users). Initially, candidate technologies for 4G wireless standard are: (i) HSPA+ (High Speed Packet Access) (ii) UMB (Ultra Mobile Broadband) (iii) LTE (iv) Mobile WiMAX (v) XGP(eXtended Global Platform) [2].

A 4G system is expected to provide a comprehensive and secure all-IP based mobile broadband solution to laptop computer wireless modems, smart phones, and other mobile devices. Facilities such as ultra-broadband Internet access, IP telephony, gaming services, and streamed multimedia may be provided to users. 4G technologies such as mobile WiMAX, HSPA+, and first-release Long term evolution (LTE) have been on the market since 2006, 2008, and 2009 respectively. IMT-Advanced compliant versions of LTE and WiMAX are under development and called "LTE Advanced" and "Wireless MAN-Advanced" respectively. ITU has decided that LTE Advanced and Wireless MAN-Advanced should be accorded the official designation of IMT-Advanced. On December 6, 2010, ITU recognized that current versions of LTE, WiMax and other evolved 3G technologies that do not fulfill "IMT-Advanced" requirements could nevertheless be considered "4G", provided they represent forerunners to IMT-Advanced and "a substantial level of improvement in performance and capabilities with respect to the initial third generation systems now deployed. Basically 4G standard is based on two technologies: LTE Advanced standardized by the 3GPP and 802.16 m standardized by the IEEE (i.e. Wireless MAN-Advanced).

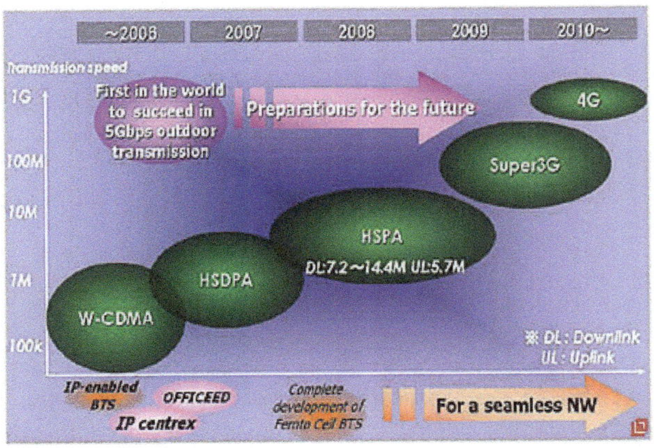

Figure 2. Evolution to 4G Technology

There are a number of advantages of 4G to answer one question "Why do we need 4G technologies?" 4G promises to deliver extremely high quality video of quality comparable to HD (high definition) TV. In addition 4G is being developed to accommodate the quality of service (QoS) and rate requirements set by further development of existing 3G applications like mobile broadband access, Multimedia Messaging Service (MMS), video chat, mobile TV, but also new services like HDTV. 4G may allow roaming with wireless local area networks, and may interact with digital video broadcasting systems.

Benefits of 4G includes: A data rate of at least 100 Mbit/s between any two points in the world. Smooth handoff across heterogeneous networks. Seamless connectivity and global roaming across a number of networks. High quality of service for next generation multimedia support (real time audio, high speed data, HDTV video content, mobile TV, etc.). Interconnection with existing wireless standards. An all IP, packet switched network. Also IP-based femtocells (home nodes connected to fixed Internet broadband infrastructure). 4G also has low cost since 4G systems can be built atop on existing networks and do not require operators to completely redesign and won't require carriers to purchase costly extra spectrum. In addition to being a lot more cost efficient, 4G is spectrally efficient, so carriers can do more with less. Table II below presents a brief difference between 4G and 3G.

Table 2. 3G vs 4G

	3G	4G
Major Requirement Driving Architecture	Predominantly voice driven-data was always add on	Converged data and voice over IP
Network architecture	Wide area cell based	Hybrid –integration of wireless LAN and wide area
Speed	384 Kbps to 2 Mbps	20 to 100 Mbps in mobile mode
Frequency Band	1800-2400 MHz	Higher frequency bands (2-8 GHz)
Switching Design Basis	Circuit and Packet	All digital with packetized voice
Access Technologies	W-CDMA,	OFDM and MC-CDMA (Multi Carrier CDMA)
Component Design	Optimized antenna design, multi-band adapters	Smarter Antennas, software multilane and wideband radios
IP	A number of air link protocols, including IP 5.0	All IP (IP6.0)

4. Hardware and Software of 4G wireless

This section presents the principal technological details of the 4G wireless system with respect to its hardware and software component of the physical layer.

4.1. OFDM (Orthogonal Frequency Division Multiplexing)

4G uses coded orthogonal frequency division multiplexing (COFDM) technology to combat the effects of Multipath fading and inter symbol interference (ISI). In addition Multi-carrier modulation (OFDM) in the downlink or single-carrier frequency-domain-equalization (SC-FDE) in the uplink is employed to exploit the frequency selective channel property without complex equalization. OFDM is derived from the fact that the high serial bit stream data is transmitted over large (parallel) number sub-carriers (obtained by dividing the available bandwidth), each of a different frequency and these carriers are orthogonal to each other.

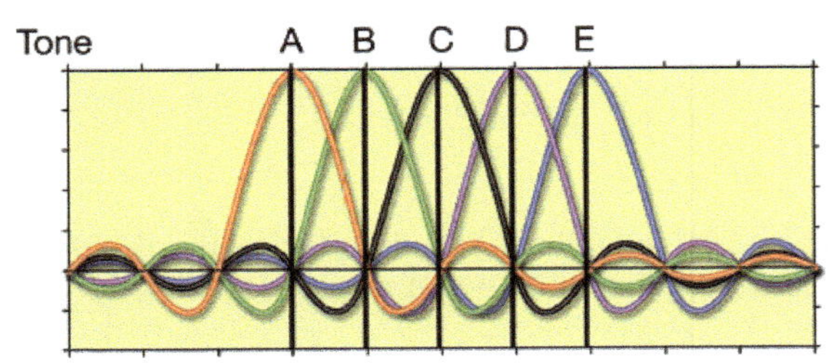

Figure 3. Orthogonal carriers in OFDM

OFDM converts frequency selective fading channel into N flat fading channels, where N is the number of sub-carriers. Othogonality (as depicted in Figure 3) is maintained by keeping the carrier spacing multiple of 1/Ts by using Fourier transform methods, where Ts is the symbol duration. Since channel coding is applied prior to OFDM symbol generation which accounts for the term 'coded' in COFDM. Turbo principle error-correcting codes is used to minimize the required SNR at the reception side. Figure 4 presents the simplified wireless OFDM block diagram.

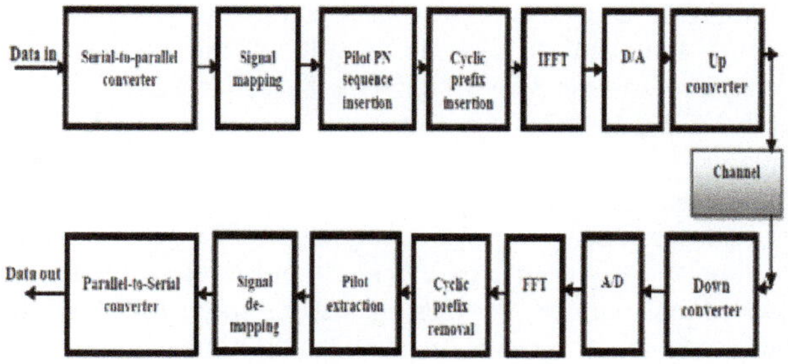

Figure 4. Wireless OFDM transceiver

4.2. UWB (Ultra Wideband Networks)

Ultra-Wideband (UWB) is a technology for transmitting message signal spread over a large bandwidth (>500 MHz) that should, in theory and under the right circumstances, be able to share radio spectrum with other users. One of the unique and valuable properties of UWB radio technology is the ability for a UWB radio system to determine "time of flight" of the direct path of the radio transmission between the transmitter and receiver at various frequencies. This helps to overcome multi path propagation, as at least some of the frequencies pass on radio line of sight. Another valuable aspect of pulse-based UWB is that the pulses are very short in space (less than 60 cm for a 500 MHz wide pulse, less than 23 cm for a 1.3 GHz bandwidth pulse), so most signal reflections do not overlap the original pulse, and thus the traditional multipath fading of narrow band signals does not exist.

The secret to UWB is that it is typically detected as noise. This highly specific kind of noise does not cause interference with current radio frequency devices, but can be decoded by another device that recognizes UWB and can reassemble it back into a signal. Since the signal is disguised as noise, it can use any part of the frequency spectrum, which means that it can use frequencies that are currently in use by other radio frequency devices. UWB provides greater bandwidth — as much as 60 megabits per second, which is 6 times faster than today's wireless networks [3]. It also uses significantly less power, since it transmits pulses instead of a continuous signal. Figure 5 shows the UWB frequency of operation between 3.1 to 10.6 GHz [4].

4.3. Advanced Antenna Systems

The performance of 4G radio communications depends on an antenna system, termed smart or intelligent antenna.

Recently, multiple antenna technologies are emerging to achieve the goal of 4G systems such as high rate, high reliability, and long range communications. One technology, spatial multiplexing, gained importance for its bandwidth conservation and power efficiency. Spatial multiplexing involves deploying multiple antennas at the transmitter and at the receiver. Independent streams can then be transmitted simultaneously from all the antennas. This technology, called MIMO (as a branch of intelligent antenna), multiplies the base data rate by (the smaller of) the number of transmit antennas or the number of receive antennas. MIMO have been used to increase the system throughput and the reception reliability. Since UWB has almost impulse-like channel response, the combination with multiple antenna techniques is preferable as well. Coupling MIMO spatial multiplexing with UWB's already high throughput gives the possibility of short-range networks with multi-gigabit rates. Apart from this, the reliability in transmitting high speed data in the fading channel can be improved by using more antennas at the transmitter or at the receiver. This is called transmit or receive diversity. Both transmit/receive diversity and transmit spatial multiplexing are categorized into the space-time coding techniques, which does not necessarily require the channel knowledge at the transmitter.

There are two types of smart antennas: Switched Beam Antennas (as seen in Figure 6) have fixed beams of transmission, and can switch from one predefined beam to another when the User with the phone moves throughout the sector.

Adaptive Array Antennas (as seen in Figure 7) represent the most Advanced smart antenna approach to date using a variety of new signal processing algorithms to locate and track the user, minimize interference, and maximize intended signal reception [4]. Smart antennas can thereby: Optimize available power, Increase base station range and coverage, Reuse available spectrum, Increase bandwidth and Lengthen battery life of wireless devices.

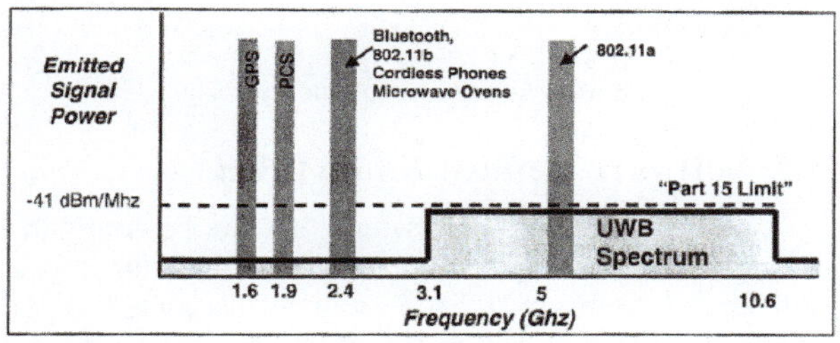

Figure 5. UWB frequency spectrum

Figure 6. Switched Beam Antenna [4]

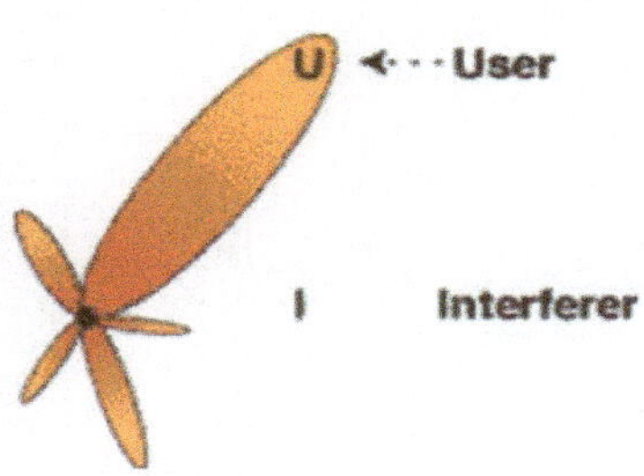

Figure 7. Adaptive Array Antennas [4]

4.4. Adaptive Modulation and Power Control

In a fading environment and for a highly loaded system there will almost exist users with good channel conditions. The modulation format for the scheduled user is selected according to the predicted signal to noise and interference ratio. By using sufficiently small time-frequency bins the channel can be made approximately constant within bins. For every timeslot, the time-frequency bins in the grid represent separate channels. For such channels the optimum rate and power allocation for maximizing the throughput can be calculated under a total average power constraint. The optimum strategy is to let one user, the one with best channel, transmit in each of the parallel channels. Figure 8 presents the adaptive modulation and coding concept.

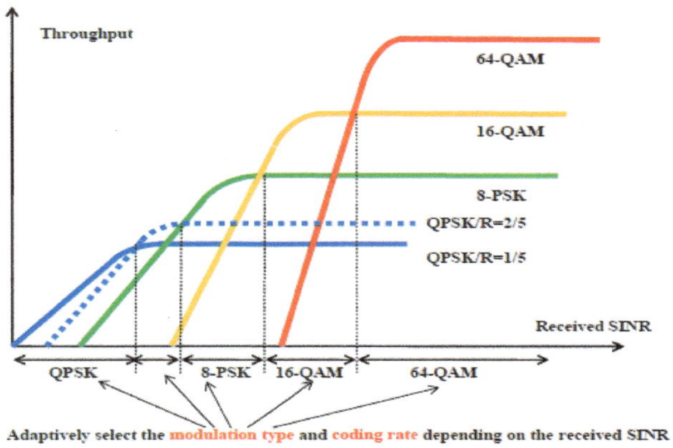

Figure 8. Adaptive modulation and coding

4.5. Software Defined Radio (SDR)

A software defined radio is one that can be configured to any radio or frequency standard through the use of software. A software defined radio in the context of 4G would be able to work on different broadband networks and would be able to transfer to another network seamlessly while traveling outside of the user's home network. A software defined radio's best advantage is its great flexibility to be programmed for emerging wireless standards. It can be dynamically updated with new software without any changes in hardware and infrastructure. Software could just download interface automatically. This is the job of the packet layer, which will split the data into small "packets." A Conceptual definition of the software defined radio (SDR) is presented in Figure 9.

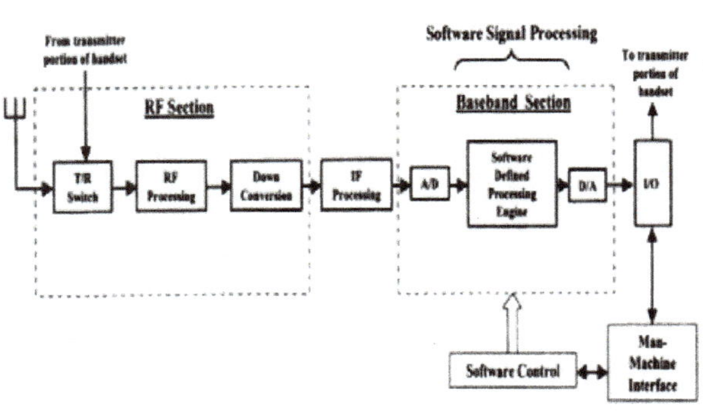

Figure 9. Conceptual definition of the software defined radio (SDR) [9]

4.6. Packet Layer IPv6

The next generation 4G addressing system uses the Internet Protocol version 6 (IPv6) to locate devices. IPv6 has a much larger address space. Its addresses take the form x: x: x: x: x: x: x: x where each x is the hexadecimal value that makes up one eighth of the address. An example of this is: FEDC: BA98: 7654: 3210: FEDC: BA98: 7654: 3210 (The Internet Engineering Task Force Network Working Group). Using this address format, there is room for approximately 3.40 * 1038 unique addresses. This is approximately 8.05*1028 times as large as the IPv4 address space and should have room for all wired and wireless devices, as well as room for all of the foreseeable expansion in several lifetimes. There are enough addresses for every phone to have a unique address. Thus, phone in the future can use VoIP over the Internet instead of continuing to use their existing network.

5. Barriers in Deployment of 4G

The security issues of 4G wireless networks and systems are a challenging one. Principal issues encountered by designers engineers of 4G wireless security include the following: (i) One of the common Security issues for 4G mobile wireless devices includes when it comes to accessing of the Internet from a fixed location with the additional requirements for flexibility and mobility (ii) Whenever additional cryptographic methods and security mechanisms will be applied to IP networks, there will be an obvious impact on the performance and traffic handling capacity of the service provider's network (iii) The Physical Layer Issues are related to both WiMAX and LTE in the sense that both are prone to Interference and Scrambling attacks but these can be minimized to some extent [2]. (iv) Integrating the IP-based systems with existing non-IP systems and thereby providing QoS (Quality of Service) guarantee for end-to-end services that involve different systems is also a big problem (v) Further Charging/ billing is a big challenge since to collect, manage and store the customers' accounts information from various service providers is difficult task [1].

6. Future Scope of 4G

There are many good reasons for 4G development and a variety of current and evolving technologies to make 4G a reality. Both service providers and users want to reduce the cost of wireless systems and the cost of wireless services. Lesser the expensive of the system, the more people who will want to own it. 4G's flexibility will allow the integration of several different LAN and WAN technologies. Finally, the 4G wireless system would truly go into a "one size fits all" category, having a feature set that meets the needs of just about everyone.

7. Conclusion

This paper briefly described the 4G wireless technology as next generation networks. It presented the 4G architecture, the technologies involved, 4G standard, benefits, challenges in deployment, design for the LTE and WiMAX technologies and security issues. It can be finally concluded that the ever increasing demands of the customers can be only fulfilled by one and only one the 4G wireless system. Today's wired society is going wireless and if it has problem, 4G is answer. 4G will be an intelligent technology that will be fast replacing 3G. But the final success of 4G mobile communication systems will depend upon when it is fully implemented and the new services and contents made available to users.

References

[1] Afaq H. Khan, Mohammad A Qadeer, Juned A Ansari, and Saria Waheed, "4G as a Next Generation Wireless Network," International Conference On Future Computer And Communication, pp. 334-338, March, 2009.

[2] N. Seddigh, B. Nandy, R. Makkar and J.F. Beaumont, "Security Advances and Challenges in 4G Wireless Networks," 2010 Eighth Annual International Conference on Privacy, Security and Trust, pp. 62-71, June, 2010.

[3] Jawwad Ahmad, Ben Garrison, Jim Gruen, Chris Kelly, and Hunter Pankey, "4G Wireless Systems," Next-Generation Wireless Working Group, May, 2003.

[4] Manu. R, "4G Wireless Systems," Technical Seminar report, Kalpataru Institute of Technology, Belgaum, Karnataka, 2009-2010.

[5] H. Harada & Ramjee Prasad, Simulation and Software Radio for mobile communications: Artech House, 2003.

[6] John. G. Proakis, "Digital Communications", 3rd edition, McGraw-Hill, 1995.

[7] Walter Tuttlebee, "Software Defined Radio", 2002 John Wiley & Sons, Ltd.

[8] "UMTS LTE Network Architecture", Technical specification TS 23.002, version 8.4.0 (Release 8), 3GPP.

[9] ITU-R M. 1645: Framework and overall objectives of the future development of IMT-2000 and systems beyond IMT-2000.

[10] http://en.wikipedia.org/wiki/4G.

Study on Anti-electricity Theft Based on Electricity Comparative Method

Gang Ma*, Guchao Xu, Linru Jiang, Tao Qian

School of Electric & Automation Engineering, Nanjing Normal University, Nanjing, China
*Corresponding author: hdmagang@163.com

Abstract Electricity stealing not only makes great economic losses, but also threatens the order and security of power supply seriously. This paper starts from electricity theft, and puts forward an anti-electricity theft design based on electricity comparative method. According to programming realization and simulation test by MATLAB, we verify the feasibility and effectiveness of the design. The anti-electricity theft design can be applied to demotic users and middle/small-sized enterprises to reduce loss from electricity theft.

Keywords: *anti-electricity theft, electricity comparative method, electric quantity gauge*

1. Introduction

Power is an important energy for human's survival and development, so good order of power supply is significant for power system operation and development. However, since the beginning of opening-up reform, in particular the establishment of economy of manufacturing, electricity stealing has become more prominent. According to statistics, due to electricity-stealing, Chinese annual economic loss is up to 20 billion Yuan [1]. With the development of science and technology, means of electricity-stealing becomes more sophisticated, and is of high technology content and hidden function, which leads to enormous difficulties to deal with various acts of electricity stealing. Electricity stealing acts will not only bring economic losses to the state and the power sector, interfering with normal supply order, but also cause fire, destroy power supply units and cause widespread power outage, threatening power grids and public security [2,3]. Therefore, it is imperative to crack down electricity stealing activities.

Although the means of electricity stealing is diverse, their basic principle is changing the metering of electric energy [4,5], which can change the metering of voltage, current and power factor. As long as any one of these is changed, it can make power meter turn slowly, stop or even reverse. Changing the structure of electric meter itself can also make a slow turn so as to achieve the goal of stealing electricity. In order to prevent electricity-stealing, power enterprise has taken variety of measures, and has made some effectiveness. But problem still exists , which mainly includes [6,7]: (1) using dedicated measuring cabinet and sealing, but general seal is easy to copy and recover after opening; (2) using anti-stolen electric power meter, it's mainly for single-phase meters .But its anti-electricity theft function is not comprehensive; (3) using high installation of power meter, which installs power meter several meters high , but it also cause troubles when reading and replacing meter; (4) using high voltage power meter, which makes theft hard to get access, but also causes inconvenience to the daily maintenance for staff.

To sum up, although there are a variety of anti-electricity theft measures, shortcomings and inconvenience, such as great limitations and high cost are still existed due to an increasingly specialized, high-tech, and hidden electricity-stealing behavior. Therefore, in view of recent various hidden electricity-stealing ways, this paper puts forward an anti-electricity theft design based on electricity comparative method. First of all, this paper introduces electricity comparative method .Then this paper makes specific anti-electricity theft policy based on the electricity comparative method .Finally, based on instance analysis, the paper verifies the rationality and effectiveness of the method so as to provide a reference for the design and application of anti-electricity theft.

2. The Basic Principle of Electricity Comparative Method

Electricity comparative method is a common way to test single-phase meter's error. It is usually used to determine whether the relative error of detection meter is qualified. Its power calculation formula is as follow:

$$\begin{cases} u = \sqrt{2}U \sin wt \\ i = \sqrt{2}I \sin(wt - \theta) \end{cases} \tag{1}$$

In the formula: U is effective value of AC voltage; I is current valid value; θ is power factor angle.

The measured power is:

$$p = ui = 2UI\sin wt \cdot \sin(wt-\theta) = UI\cos\theta\cdot(1-\cos 2wt) - UI\sin\theta\sin 2wt \quad (2)$$

Its average is:

$$p = UI\cos\theta \quad (3)$$

We can get active energy during time T:

$$W = T\cdot UI\cos\theta \quad (4)$$

Electricity comparative method is typically carried out in the following ways about single-phase active power meter.

(1)Watt second method

Under rated power, we can use standard meter to measure the time which is taken by tested meter to accumulate energy, and constant power is multiplied to get measured energy values. According to comparing with measured energy valued and power values accumulated by testing meter we can determine the relative error. Error is calculated as follows:

$$\gamma = \frac{W - P\cdot t}{P\cdot t}\cdot 100\% + \gamma_w \quad (5)$$

In the formula: γ_w is standard meter or measuring device's inherent system error (%). When there is no need to correct, $\gamma_w = 0$; P is a constant power value(W); t is standard tags measured time(s);W is the power value measured by tested meter(J).

(2)The standard method

Within a particular period of time t(s), we write down cumulative energy tested by standard meter and tested meter and the relative error of the meter can be calculated by the following formula:

$$\gamma = \frac{W' - W}{W}\cdot 100\% + \gamma_0 \quad (6)$$

In the formula: γ_0 is inherent system error of standard meter (%), When there is no need to correct, $\gamma_0 = 0$; W' is the power value displayed by tested meter(J); W is the energy value displayed by standard meter (J).

The comparative quantity of the two methods above are energy values, but one way uses time as constant value , another uses power as constant value. Their purpose is to calculate power value which is suitable for comparison. Therefore, this paper puts forward an anti-electricity theft design based on electricity comparative method.

3. Study on the Anti-electricity Theft Method Based on Electricity Comparative Method

According to the electricity comparative method, in order to identify electricity-stealing behavior, we need to compare the benchmark measurement data with the measurement data of electric quantity. In this paper, the data of relay protection equipment of line protection circuit is used to access to voltage, current and power factor directly, or we install the power factor meter between the protection circuit and the measurement circuit to collect the voltage, current and power factor, etc. directly. We calculate the standard electric quantity for a period of time and compare with the electric quantity of the measuring circuit, which can identify electricity-stealing behavior effectively.

Aiming at possible error in measurement, this paper uses electric quantity instead of time as standard, and sets the basic measurement threshold of electric quantity. We compare electric quantity when electric increment measured from protecting side reach to threshold. Using electricity as the threshold for measurement, it can not only control the error which is generated by the difference between electricity consumption in peak and valley time with a group of users, but also avoid tedious time steps and enhance the practicability of the method. Anti-electricity theft method based on electricity comparative method is shown in Figure 1.

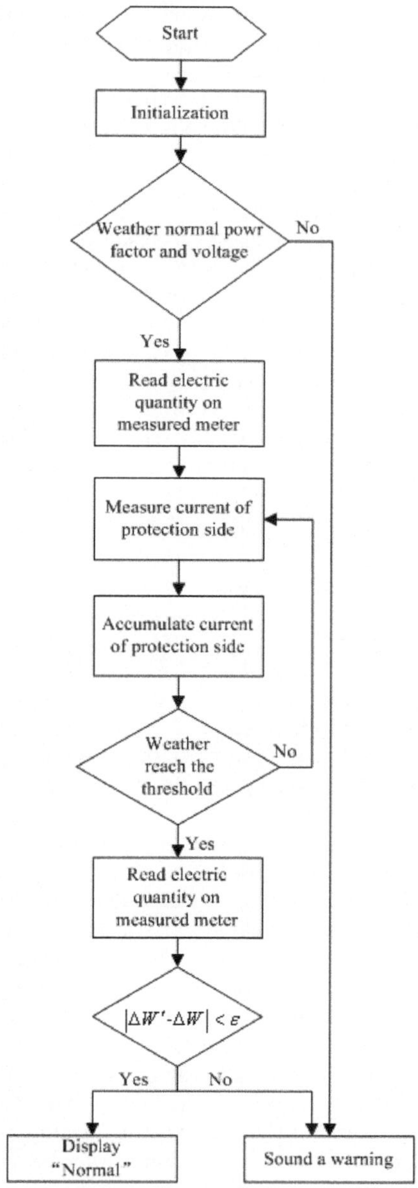

Figure 1. Anti-electricity theft method based on electricity comparative method

As shown in Figure 1. First, we should set basic measured threshold value of electric quantity ΔW as the threshold for electricity comparative method; second, we should check the operation condition of circuit, including voltage and power factor; third, we record the electric quantity of users' energy meter and calculate the sum of electric increment $\sum W$ on protection side at the same time. When $\sum W=\Delta W$, we record the electrical energy meter of the users again. The difference value between the two records is electric quantity during the period time.

Finally, we compare the power consumption of electric meter measurement on user side ($\Delta W'$) with ΔW .If $|\Delta W'-\Delta W| < \varepsilon$, The electricity-stealing could happen and the method issues warning.

Among them, the core of this method is the part of measurement and comparison, it mainly includes:

(1) Voltage comparison. During the stable operation of power system, voltage fluctuation is small, so it can be regarded as a constant value when the electric quantity is calculated. If the voltage Vp measured on protection side is normal at this time, it can be used as the input value of the voltage to calculate power. When the voltage of the protection side is less than the allowable minimum value, undervoltage warning is issued.

(2) Power factor comparison. Power system is required to maintain a high level when power system is stable operation. Therefore, it can be considered as a constant value when calculating the electric quantity. If the $\cos\varphi$ measured on protection side meet the requirement, it can be used in electric quantity calculating. When the power factor of the protection side is less than the minimum allowable value, warning is issued.

(3) Electric quantity comparison. The electric quantity comparison is the core of the electricity comparative method, and the comparison method used in this paper is as follows:

$$\frac{W_p - W}{W_p} < B \tag{7}$$

In the formula:

① $W_p = \sum W = \sum \Delta t V_p I_p \cos\varphi$ is electric quantity measured by protection side, Δt is current sampling period, V_P is protecting side voltage, I_p is real-time measured current, $\cos\varphi$ is the power factor of protection side. Protection side voltage V_P and the power factor $\cos\varphi$ are constant value, and because the user side load will change from time to time, which means I_P is real-time change and needs real-time measurement.

②W is electric quantity on measuring side. $W=W_2-W_1$, W_1, W_2 are first and second electric quantity measured by measuring side respectively. W_1, W_2 are needed to distinguish peak and valley time to prevent from misjudgment. If the first is peak time, the second is valley time, the measurement is invalid;

③B is electric quantity allowable overproof ratio, when electric quantity overproof ratio is less than B, user is judged to be normal; when the electric quantity overproof ratio greater than B, user is judged to be suspected electricity stealing. An electricity stealing warning is issued and the method increases the frequency of testing to determine the fraudulent use of electricity, the paper set B=0.05.

④The electric quantity threshold $\Delta W=W_0$, W_0 is setting electric quantity. Its value is determined by the user calculated load. According to the current measurement interval, the total measurement time is not less than 5 minutes. Under normal circumstances, in the user's daily load curve, peak load is generally no more than the computational load of 2 times [12]. In the supreme power condition, we need to ensure adequate measurement time, so the conversion ration from calculated load to setting electric quantity is 0.2. Considering the present electric energy meter accuracy is 0.01kWh, and the error of this method should not exceed 1%, therefore, the minimum electric quantity of the set is not less than 1.

In conclusion, the set value of electric quantity is:

$$W_0 = 0.2P_{ca} + 1 \tag{8}$$

In the formula: P_{ca} is calculated load of user. It determines the required total electric quantity of different users in a certain period when comparing electric quantity.

4. Test and Analysis

According to the proposed anti-electricity theft method, this paper establishes a simulation environment to verify the method, the dynamic input current of simulation system is as follows:

$$I' = 2 + 1.5\sin(0.003t + \varphi_1) + 0.1\sin(20t + \varphi_2) \tag{9}$$

In the formula: The φ_1 and φ_2 are random phases, range is from 0 to 2π; 2 is current reference value; first sine function is used to simulate the change of current in different time periods; the second sine is used to simulate current fluctuations.

Test uses the control variate method to test each function. During normal operation, the voltage is 220V, power factor is 0.9. In order to shorten the running time of the program to improve the efficiency of the test, this paper does not consider the matching of calculated load and actual power, the user calculated load is set to 0.4kW. This paper uses M file in MATLAB to realize the method.

Simulation test scheme is as follows:

(1) Turn down voltage/power factor, run program and observe the output of the voltage when voltage is too low.

①Under voltage condition

Input voltage is set to 190V, display is:

Current voltage value (V):190.00
Current power factor: 0.900
Average current (A):0.640
Peak current (A):0.727
Calculating electric quantity (KWh):1.09
Measuring electric quantity (KWh):1.08
Time used (s): 10
Electric quantity overproof ratio: 0.014
Voltage is too low!

②Low power factor condition

Input power factor is set to 0.75, display is:

Current voltage value (V):220.00
Current power factor: 0.750
Average current (A):0.899
Peak current (A):0.987
Calculating electric quantity (KWh):1.11
Measuring electric quantity (KWh):1.09
Time used (s): 7
Electric quantity overproof ratio: 0.018
Power factor is too low!

Figure 2. The simulation output of scheme (1)

It can be seen from Figure 2 that the method can detect effectively and issue a low voltage/low power factor warning when the undervoltage or low power factor condition happens.

(2) Simulate normal/electricity-stealing condition, run the program and observe the output.

①Normal condition

The input voltage is 220V, power factor is 0.9, and the program is run 5 times, and the output is as follows:

Table 1. Normal condition output display

Measurement times	1	2	3	4	5
Voltage value (V)	220.00				
Power factor	0.900				
Average current (A)	3.078	0.899	1.645	0.787	0.583
Peak current (A)	3.134	0.987	1.700	0.882	0.683
Calculating electric quantity (KWh)	1.22	1.25	1.30	1.09	1.15
Measuring electric quantity (KWh)	1.22	1.22	1.28	1.09	1.15
Time used (s)	2	7	4	7	10
Super power ratio	0.001	0.018	0.021	-0.003	0.003
Result	Normal electric quantity.				

②Simulation of electricity-stealing

When electricity-stealing happen, measuring electric quantity reduce. We assume that the electric quantity loss percentage is 30%, run the program. The results are as follows:

Table 2. The output of electricity stealing simulation

Measurement times	1	2	3	4	5
Voltage value (V)	220.00				
Power factor	0.900				
Average current (A)	1.645	3.287	0.583	2.739	0.767
Peak current (A)	1.700	3.339	0.683	2.798	0.851
Calculating electric quantity (KWh)	1.30	1.30	1.15	1.63	1.22
Measuring electric quantity (KWh)	0.89	0.90	0.80	1.12	0.83
Time used (s)	4	2	10	3	8
Super power ratio	0.315	0.307	0.302	0.309	0.319
Result	Electric quantity abnormity!				

In order to verify the results, the simulation outputs under normal condition and electricity-stealing condition are as follows:

①Normal condition:

Current voltage value (V):220.00
Current power factor: 0.900
Average current (A):1.418
Peak current (A):3.264
Calculating electric quantity (KWh):286.05
Measuring electric quantity (KWh):286.01
Time used (s): 1019
Super power ratio: 0.002
Normal electric quantity

②Electricity-stealing condition:

Current voltage value (V):220.00
Current power factor: 0.900
Average current (A):2.103
Peak current (A):3.429
Calculating electric quantity (KWh):286.46
Measuring electric quantity (KWh):200.52
Time used (s): 688
Super power ratio: 0.300
Electric quantity abnormity!

Figure 3. The output under normal and electricity stealing condition

From Figure 3, we can see, the method can make effective judgment on electricity stealing behavior and make alarm to the possible stealing behavior.

To sum up, we know that: the anti-electricity theft method can make a reasonable judgment on electricity stealing behavior, and can give alarm when the system runs in abnormal way, which increase the stability of power system operation and catch the electricity stealing behavior timely. The test result verifies the feasibility and effectiveness of anti-electricity theft method in this paper.

5. Conclusion

In this paper, anti-electricity theft based on electricity comparative method can identify the electricity stealing behavior effectively and give the alarm to ensure the stability and efficiency of power system, also the electric stealing quantity can be estimated by the output of the method. The method is concise, economic and flexible, it has good application prospect in anti-electricity theft of community, self-employment venture and middle/small-sized enterprise. At the same time, although the method is designed for single-phase electric energy meter, the same design thought with phase sequence detection can also be used in three-phase network anti-electricity theft system. However, because this method does not consider the line loss and other problems, it is not suitable for high voltage network and large power consuming unit.

References

[1] Yang, L., "Electricity Larceny Prevention in DMS," *Electric Power Automation Equipment*, 22(1). 84-85. Jan.2002.

[2] Jamil, F., "On the electricity shortage, price and electricity theft nexus," *Energy Policy*, 54. 267-272. Mar.2013.

[3] Amin, S., Schwartz, G.A., Cardenas, A.A., Sastry, S.S., "Game-Theoretic Models of Electricity Theft Detection in Smart Utility Networks," *IEEE CONTROL SYSTEMS MAGAZINE*, 35(1). 66-81. Feb.2015.

[4] Smith, T.B., "Electricity theft: a comparative analysis," *Energy Policy*, 32(18). 2067-2076. Dec.2004.

[5] Depuru, S.S.S.R., Wang, L.F., Devabhaktuni, V., "Electricity theft: Overview, issues, prevention and a smart meter based approach to control theft," *Energy Policy*, 39(2). 1007-1015. Feb.2011.

[6] Depuru, S.S.S.R., Wang, L.F., Devabhaktuni, V. High performance computing for detection of electricity theft," *INTERNATIONAL JOURNAL OF ELECTRICAL POWER & ENERGY SYSTEMS*, 47. 21-30. May.2013.

[7] Guo, L.C., Peng, Z.W., Fan, Q., "A Survey of Electric Energy Metering and Countermeasures to Electric Power Stealing," *High Voltage Apparatus*, 46(5). 86-88. May.2010.

[8] Zhu, X.H., "The Existing Misunderstandings in Watt-second Method for Single-phase Electric Meter Checking and Its Countermeasures," *Electrical Measurement & Instrumentation*, 47(7A). 98-101. Jul.2010.

[9] Cheng, H.H., Li, X.M., Huang, J.G., "Applying of Detecting The Theft of Electricity in The Power Energy Billing System," *Proceedings of the EPSA*, 13(6). 53-57. Dec.2001.

[10] Zhang, C.M., Xu, X., "Application of Preventing System from Stealing Power," *High Voltage Engineering, 31*(8). 88-89. Aug.2005.

[11] Zhao, B., Lü, Y.J., "Design of a new anti-stealing electricity device," *Power System Protection and Control*, 37(23). 116-118. Dec.2009.

[12] Sun, L.H., *Electric Engineering Foundation*, China Machine Press, Beijing, 2010, 22-38.

22

Synthetic Floating Inductors realized with only two Current Feedback Op-amps

D. R. Bhaskar[1], R. Senani[2,*]

[1]Department of Electronics and Communication Engineering, Jamia Millia Islamia, New Delhi, India
[2]Devision of Electronics and Communication Engineering, Netaji Subhas Institute of Technology, New Delhi, India
*Corresponding author: senani@ieee.org

Abstract Two floating inductance (FI) circuits are presented which employ a canonical number of passive components (namely, only two resistors and a capacitor) as well as canonical number of active elements (only two CFOAs) and realize single-resistance-tunable inductance value, without requiring any component-matching or cancellation constraints. The workability of the proposed circuits and their applications has been confirmed by hardware implementation and SPICE simulations based on AD844-type CFOAs.

Keywords: *inductance simulation, current feedback op-amps, floating inductance simulation, analog circuits*

1. Introduction

The simulation of grounded inductance (GI) and floating inductance (FI) has continued to remain an important and popular area of analog circuits research due to their applications in linear (active filters and oscillators) and nonlinear (such as chaotic oscillators) circuit designs.

Although inductance simulation circuits using a large number of active building blocks have been reported in the recent literature, those using current conveyors (CCs) and current feedback operational amplifiers (CFOAs) have found prominent place particularly due to the commercial availability of a CFOA with externally accessible z-pin such as AD844 which can be used to realize both CC-based as well as CFOA-based inductance simulators practically.

It was demonstrated in [1,2,3,4], *for the first time*, that using CCII- as building blocks, it becomes possible to realize FI simulation circuits using only three passive components *without requiring any component-matching condition(s)- a feat which was impossible to be achieved by op-amp-RC circuits prevalent in those days.* Later, when CFOA AD844 came into existence, it was demonstrated by Fabre in [5] that a grounded loss-less inductance can be simulated using two CFOAs and only three passive components (namely, two resistors and a capacitor) without any component-matching condition. In [6], it was shown that loss-less FI can be realized with only three CFOAs, three passive components and still not requiring any matching conditions. The latter works [7]-[13] have demonstrated a variety of grounded impedance simulation circuits all employing only a single CFOA with the exception of [13] in which case two CFOAs are utilized and [14] wherein three CFOAs are employed to realize a grounded-capacitor (GC) based GI[1].

Among the CFOA-based loss-less FI circuits, those in [6] and [15] (derived from the biquad of [16]) require three CFOAs, two resistors and a GC as preferred for integrated circuit implementation [17,18]. The circuits of [19,20,21] employ as many as four CFOAs although they can realize a larger variety of floating Impedances such as floating FDNR and floating FDNC also besides a FI. On the other hand, the circuit of [22] although employs a canonical number of passive components to realize loss-less FI but requires four current conveyors. Recently, the present authors reported a two CFOA-based circuit [23] which simulates a lossy/lossless floating inductance (FI) employing three resistors and two capacitors. However, the circuit of [23] uses a non-canonical number of resistors (three rather than two) and capacitors (two instead of one) and moreover, needs one cancellation constraint for realizing a lossless FI.

The purpose of this communication is to discuss two lossy FI[2] circuits which, like the circuit of [23], require only two CFOAs but by contrast, provide the following features not available in the earlier circuit of [23], namely: (i) employment of only a single capacitor along with a minimum number of (only two) resistors (ii) no requirement of any realization condition/cancellation constraint to realize the intended type of impedances and

[1]Although reference [24] shows that three passive elements and a single modified CFOA are sufficient to realize a variety of grounded impedances but the so-called modified CFOA is, in fact, actively realized from a composite connection of a CCII+ and a CCII- and therefore, would call for at least three CFOAs of the normal kind (one for CCII+ and two for the CCII-).

[2] The circuit of Figure 1(a) was briefly presented in a conference [27] and before that, had been mentioned in [33], as an *unpublished circuit.* However, the circuit of Figure 1(b) is completely new.

(iii) realizability of single-resistance–tunability of the inductance value. The workability of the described circuits has been confirmed by hardware implementation results and SPICE simulations based on AD844 type CFOAs.

2. Synthetic FI Circuits Realized with only Two CFOAs

Consider now the circuits shown in Figure 1. Assuming CFOAs to be characterized by $i_y = 0$, $v_x = v_y$, $i_z = i_x$ and $v_w = v_z$, a straight forward analysis of the proposed circuits reveals their Y-matrices to be given by:

$$[Y]=\left[\frac{1}{R_1+sC_0R_1R_2}\right]\begin{bmatrix}1 & -1\\ -1 & 1\end{bmatrix} \quad (1)$$

for the circuit of Figure 1(a) and

$$[Y]=\left[\frac{1}{R_1}+\frac{1}{sC_0R_1R_2}\right]\begin{bmatrix}1 & -1\\ -1 & 1\end{bmatrix} \quad (2)$$

for the circuit of Figure 1(b).

Thus, the circuit of Figure 1(a) simulates a floating series-RL impedance with equivalent resistance $R_{eq} = R_1$ and equivalent inductance $L_{eq} = C_0R_1R_2$, while the circuit of Figure 1(b) simulates a parallel–RL admittance with $R_{eq} = R_1$ and $L_{eq} = C_0R_1R_2$. In both the circuits, the value of L_{eq} is controllable *independently* of the associated resistive part by a single variable resistance R_2.

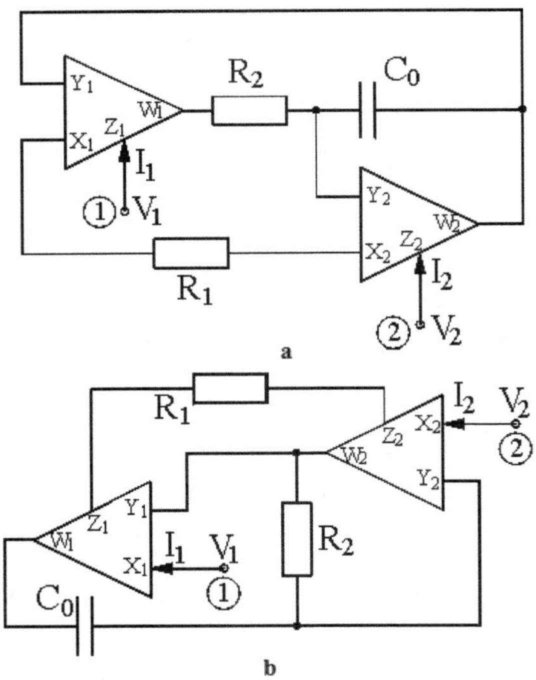

Figure 1. The canonic floating inductance simulators (a) series-RL FI simulator (b) parallel-RL FI simulator

From equations (1) and (2), it can be readily deduced that the various sensitivity coefficients of the realized equivalent inductance and resistance with respect to passive elements would be in range of

$$0 \le \left|S_{x_i}^{F}\right| \le 1 \quad (3)$$

where F represents L_{eq} or R_{eq} and x_i represents any of R_1, R_2 and C_0 and the circuits, thus, enjoy low sensitivity properties.

3. The Effect of Non-ideal Parameters of the CFOAs

Considering the various non-ideal parasitic impedances of the CFOAs, namely, the finite input impedance looking into terminal-X as R_X, the output impedance looking into terminal-Z (Z_P) consisting of a parasitic resistance R_P in parallel with a parasitic capacitance C_P and the input impedance looking into Y-terminal (Z_Y) consisting of a parasitic resistance R_Y in parallel with a parasitic capacitance C_Y, the non-ideal Y-parameters of the two circuits are found to be:

For the circuit of Figure 1(a)

$$Y'_{11}=\frac{1}{D_1(s)}=-Y'_{21} \quad (4)$$

$$Y'_{12}=\frac{-\left(1+\frac{R_2}{Z_Y}\right)}{D_1(s)}=-Y'_{22} \quad (5)$$

$$D_1(s)=(R_1+2R_X)(1+sC_0R_2+\frac{R_2}{Z_Y}) \quad (6)$$

For the circuit of Figure 1(b)

$$Y'_{11}=\frac{N_1(s)-N_2(s)}{D_2(s)} \quad (7)$$

where

$$N_1(s)=\frac{sC_0}{R_1}+\frac{1}{R_2}\left(\frac{1}{R_1}+\frac{1}{Z_p}\right) \quad (8)$$

and

$$N_2(s)=\left(\frac{R_X}{Z_P}\right)\left(sC_0+\frac{1}{R_2}+\frac{1}{Z_Y}\right)\left(\frac{1}{Z_P}+\frac{2}{R_1}\right) \quad (9)$$

$$Y'_{12}=\frac{\left(\frac{1}{R_1}+\frac{1}{Z_P}\right)\left(sC_0+\frac{1}{R_2}+\frac{1}{Z_Y}\right)}{D_2(s)} \quad (10)$$

$$Y'_{21}=\frac{\left\{\frac{1}{R_1R_2}+sC_0\left(\frac{1}{R_1}+\frac{1}{Z_P}\right)\right\}}{D_2(s)} \quad (11)$$

$$Y'_{22}=\frac{\left(\frac{1}{R_1}-\frac{R_X}{Z_P^2}-\frac{2R_X}{R_1Z_P}\right)\left(sC_0+\frac{1}{R_2}+\frac{1}{Z_Y}\right)}{D_2(s)} \quad (12)$$

where

$$D_2(s)=sC_0\left(1+\frac{R_X}{R_1}\right)+R_X\left\{\left(\frac{1}{Z_PR_2}+\frac{1}{R_1R_2}\right)+\left(sC_0+\frac{1}{R_2}+\frac{1}{Z_Y}\right)\left(\frac{1}{R_1}-\frac{R_X}{Z_P^2}-\frac{2R_X}{R_1Z_P}\right)\right\} \quad (13)$$

Note that if we take $Z_P \to \infty$, $Z_Y \to \infty$, and $R_x \to \infty$, the non-ideal Y-parameters approach their ideal values as given in equations (1) and (2). From the non-ideal Y-parameters, it is clear that like all other FI circuits, the performance of both the circuits will depart from its ideal intended one, at high frequencies.

A comparison of the various features of the proposed circuits with CFOA-based and CCII-based (realizable with CFOAs) loss-less/lossy FI circuits known earlier has been carried out in Table 1. In making this table, it is taken into account that a CCII+ is realizable with one CFOA whereas a CCII- can be realized with two CFOAs).

Table 1. Comparison with the Earlier Known Circuits

Ref.	Number and type of blocks used	Canonic in number of Passive elements?	Number of CFOAs used	Free from matching?	Can L_{eq} be tuned?
[19]	4; CFOA	No	4	Yes	Yes
[20]	4; CFOA	No	4	Yes	Yes
[4]	2; CCII-	No	4	Yes	Yes
[15]	3; CFOA	Yes	3	Yes	Yes
[6]	3; CFOA	Yes	3	Yes	Yes
[1]	1; CCII-	Yes	2	Yes	No
[2]	3/2; CCII-	Yes	4	Yes	Yes
[23]	2; CFOA	No	2	No	No
This work	**2; CFOA**	**Yes**	**2**	**Yes**	**Yes**

From Table 1, it is clear that the circuits presented in this paper are the only ones which possess the following properties simultaneously, namely, (i) employment of a canonical number of passive components (ii) employing only two CFOAs (iii) complete absence of any component-matching requirements and (iv) single-resistor-tunability of the realized FIs.

4. Experimental and Simulation Results

The validity of the proposed FI simulators has been verified by implementing them with commercially available AD844-type CFOAs and 5% tolerance RC elements, as well as by SPICE simulations based upon a macromodel of AD844.

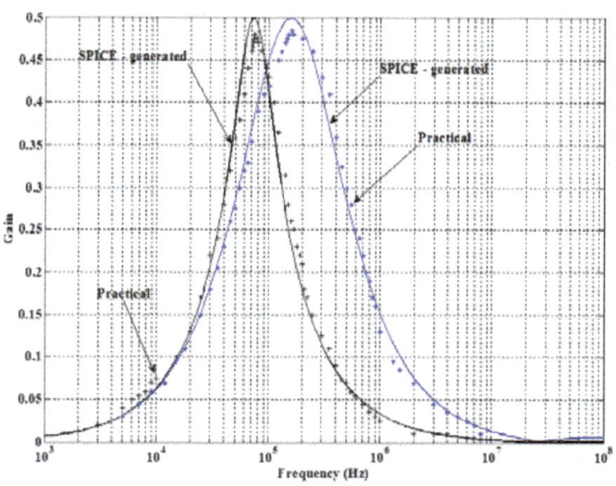

Figure 2. Frequency response of the BPF realized from the proposed FI of Fig.1 (a)

The workability of the simulated FI configuration of Figure 1(a) has been verified by employing it in the realization of a tunable band pass filter (BPF), by connecting a capacitor C_1 in series with its port-1 (or port - 2) of the simulated inductor, with a resistor R_L connected at its port 2 (or port 1) and then taking the output as the voltage across R_L. The component values chosen were: (i) Set-I: $R_1 = 1k\Omega = R_2 = R_L$, $C_1 = 1nF = C_0$ and (ii) Set-II: $R_1 = R_L = 1k\Omega$, $R_2 = 4.7k\Omega$, $C_1 = 1nF = C_0$ chosen to provide theoretical values of f_0 as 159 kHz and 73.34 kHz respectively. The experimental results of these designs have been shown in Figure 2 wherein the experimental values of f_0 have been found to be 159 kHz and 74 kHz respectively. The experimental results of Figure 2, thus, confirm the workability as well as the tunability of the inductance value (and hence, f_0) with R_2.

The workability of the parallel- RL FI of Figure 1(b) has been confirmed by using it in the design of a 4th order Butterworth filter based upon the normalized passive RLC prototype of Figure 3(a) and using transformation T-2[3] from [25,26]. This transformation scales all the impedances of the RLC prototype of Figure 3(a) by a frequency-dependent scaling factor $F(s) = 1/(1+s)$ which transforms a resistor into a parallel RC, an inductor into a parallel RL and a capacitor into a parallel combination of a capacitor and a FDNR (frequency-dependent-negative-resistance; an element having impedance of type $Z(s) = 1/Ds^2$), as shown in Figure 3(b). Note that in the transformed circuit of Figure 3(b), the floating parallel RL simulator of Figure 1(b) can be employed directly in place of both the parallel-RL branches. Furthermore, the two shunt CD-branches can also be simulated from the RC: CR transformed version of the circuit of Figure 1(b) with anyone of its two ports grounded. However, such a circuit would require two CFOAs for each grounded shunt parallel CD branch and would not be economical. Hence, to reduce the total component count, instead of using the suggested two- CFOA-based circuit, we have used a simpler one-CFOA-based circuit shown in Figure 3(c) to simulate both the shunt-CD branches encountered in the transformed prototype of Figure 3(b). The final circuit, thus, obtained has been shown in Figure 3 (d).

This circuit enables direct incorporation of the lossy FI of Figure 1(b) and grounded CD branch realized with a single CFOA circuit of Figure 3(c) into the design where the component values, as shown, have been obtained for a de-normalized cut-off frequency of $f_0 = 100$kHz.

The SPICE simulations have revealed the cut-off frequency as 98 kHz which is quite close to the theoretical value of 100 kHz. The SPICE-generated frequency response of the circuit is shown in Figure 3(e). These simulation results, thus, confirm the workability of the FI circuit of Figure 1(b).

[3] For further details of the various transformations, see [25,26].

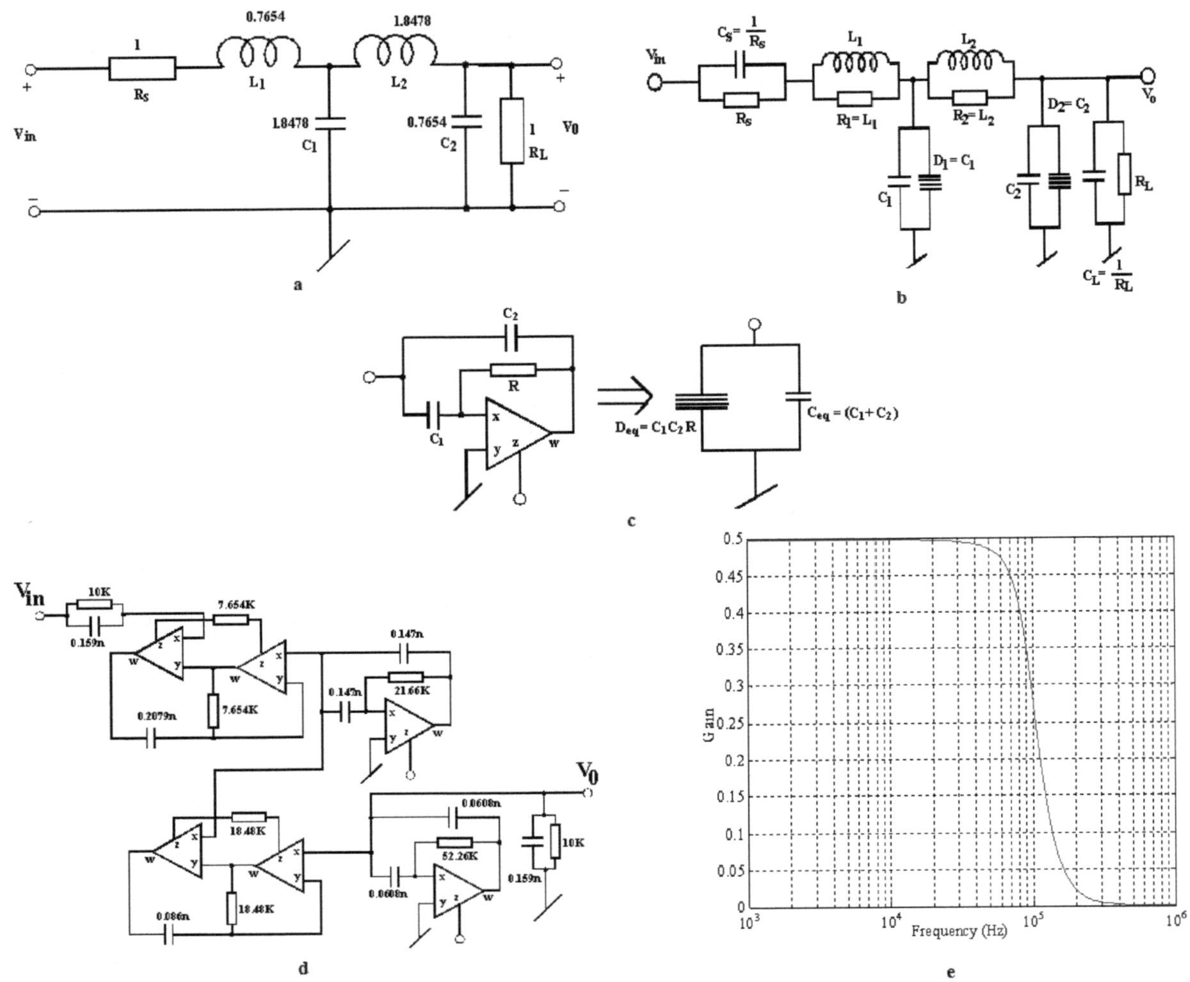

Figure 3. Application and SPICE simulation results of the FI circuit of Figure 1(b): (a) Normalized 4th-order Butterworth Low Pass Filter; (b) Filter obtained by applying Senani's transformation T-2 from [25,26] on the circuit of Figure 3(a); (c) An exemplary realization of a grounded parallel CD branch; (d) Final 4th- order Butterworth active filter obtained by replacing various RL and CD immittances of Figure 3(b) by the circuits of Figure 1(b) and Fig 3 (c); (e) SPICE generated Frequency response of the 4th-order Butterworth Low Pass Filter of Figure 3(d)

5. Discussions

Note that, in case of the BPF responses of Figure 2 while SPICE simulations take all the passive component values to be exact, in hardware implementation, RC components used were having 5% tolerances (hence, were not exact). As a consequence, the deviation of the practical responses from that exhibited by SPICE simulations is attributed to the passive component tolerances.

Further, it must be mentioned that the circuit of Figure 3(d) should not be taken as the recommended best method to design a 4th order Butterworth filter using CFOAs. This particular method has been applied here only as a vehicle to demonstrate the use of the FI of Figure 1(b) and to check its workability in higher order filter designs.

6. Concluding Remarks

Two canonic synthetic floating inductors have been discussed which, like the recently proposed FI of [23], use only two CFOAs, however, in contrast to the earlier circuit of [23], which requires two matched capacitors, three resistors (two of which are also required to be identical) and a cancellation constraint (for realizing a lossless FI), the discussed circuits provide the following advantageous features which are not available simultaneously either in the FI circuit of [28] or any other CC/CFOA based FIs known earlier: (i) use of a canonical number of passive components namely, only a single capacitor and two resistors (ii) employment of only two CFOAs for realizing an FI (iii) realization of the intended type of FIs without requiring any equality constraints or cancellation conditions, and (iv) the availability of single-resistance-tunability of the realized equivalent inductance value in both the cases.

The workability of the discussed FI circuits has been confirmed by experimental results and SPICE simulations based upon AD844-type CFOAs.

Acknowledgement

The authors wish to thank Dr. Dinesh Prasad and Dr. R. K. Sharma for their help in the preparation of this manuscript. Thanks are also due to the anonymous reviewers for their constructive suggestions and feedback.

References

[1] Senani, R., "Novel active-RC circuit for floating inductor simulation," *IEE Electron Lett*, 15(21), 679-680, Oct. 1979.

[2] Senani, R., "New tunable synthetic floating inductors," *IEE Electron Lett*, 16(10), 382-383, May 1980.

[3] Senani, R., "Floating ideal FDNR using only two current conveyors," *IEE Electron Lett*, 20(5), 205-206 March 1984.

[4] Senani, R., "On the realization of floating active elements," *IEEE Trans Circuits Syst.*, 33(3), pp. 323-324, March 1986.

[5] Fabre, A., "Gyrator implementation from commercially available transimpedance operational amplifier," *IEE Electron Lett*, 28(3), 263-264, Jan. 1992.

[6] Senani, R., "Realization of a class of analog signal processing/generation circuits: Novel configurations using Current feedback op-amps," *Frequenz*, 52(9-10), 196-206, Sept.–Oct. 1998.

[7] Liu, S.I., and Hwang, Y.S., "Realization of R-Land C-D impedances using current feedback amplifier and its applications," *IEE Electron Lett*, 30(5), 380-381, March 1994.

[8] Yuce, E., "Novel lossless and lossy grounded inductor simulators consisting of a canonical number of components," *Analog Integr Circuits Signal Process*, 59(1), 77- 82, April 2009.

[9] Abuelma'atti, M.T., "Comment on "Novel lossless and lossy grounded inductor simulator consisting of canonical number of components," *Analog Integr Circuits Signal Process*, 68(1), 139-141, July 2011.

[10] Yuce, E., "Reply to comment on "Novel lossless and lossy grounded inductor simulator consisting of canonical number of components," *Analog Integr Circuits Signal Process*, 72(2), 505-507, Aug. 2012.

[11] F. Kacar and H. Kuntman, 'CFOA-based lossless and lossy inductance simulators,' Radioengineering, vol. 20, no. 3, pp. 627-631, Sept. 2011.

[12] Abuelma'atti, M.T., "New grounded immittance function simulators using single current feedback operational amplifier," *Analog Integr Circuits Signal Process*, 71(1), 95-100, April 2012.

[13] Lahiri, A. and Gupta, M., "Realizations of grounded negative capacitance using CFOAs," *Circuits Syst Signal Process*, 30(1), 143-155, Feb. 2011.

[14] Yuce, E. and Minaei, S., "On the realization of simulated inductors with reduced parasitic impedance effects," *Circuits Syst Signal Process*, 28(3), 451-465 June 2009.

[15] Chang, C.M. and Hwang, C.S., "Comment on Voltage-mode notch, low pass and band pass filter using current-feedback amplifier," *IEE Electron. Lett*, 31(4), 246, Feb. 1995.

[16] Chang, C.M., Hwang, C.S. and Tu, S. H., "Voltage-mode notch, low pass and band pass filter using current-feedback amplifiers," *IEE Electron Lett*, 30(24), 2022-2023, Nov. 1994.

[17] Newcomb, R. W., *Active Integrated Circuit Synthesis*, Prentice-Hall, Inc. Englewood Cliffs, New Jersey, 1968, 151.

[18] Bhushan, M. and Newcomb, R. W., "Grounding of capacitors in integrated circuits," *IEE Electron. Lett*, 3(4), 148-149, April 1967.

[19] Psychalinos, C., Pal, K. and Vlassis, S., "A floating generalized impedance converter with current feedback operational amplifiers," *Int J Electron Commun (AEU)*, 62(2), 81-85, Feb. 2008.

[20] Senani, R., Bhaskar, D. R., Gupta, S. S. and Singh, V. K., "A configuration for realizing floating, linear, voltage-controlled resistance, inductance and FDNC elements," *Int J Circuit Theory Appl*, 37(5), 709-719, June 2009.

[21] Senani, R., Bhaskar, D. R., Singh, A. K. and Singh, V. K., *Current feedback operational amplifiers and their applications*, Springer, USA, 2013, Chapter 3 Simulation of Inductors and other types of Impedances using CFOAs.

[22] Senani, R., "Novel lossless synthetic floating inductor employing a grounded capacitor," *IEE Electron Lett*, 18(10), 413-414, May 1982; also see *Erratum, ibid*, August 1982 issue.

[23] Senani, R. and Bhaskar, D. R., "New loss-less/lossy Synthetic floating inductance configuration realized with only two CFOAs," *Analog Integr Circuits Signal Process*, 73(3), 981-987, Dec. 2012.

[24] Yuce, E. and Minaei, S., "A modified CFOA and its applications to simulated inductors, capacitance multipliers, and analog filters," *IEEE Trans Circuits Syst I*, 55(1), 266-275, Feb. 2008.

[25] Senani, R., "Novel higher order active filter design using Current Conveyors," *IEE Electron Lett*, 21(22), 1055-1057, Oct. 1985.

[26] Senani, R., "Network transformations for incorporating non-ideal simulated immittances in the design of active filters and oscillators," *IEE Proc. Pt. G*, 134(4), 158-166, Aug. 1987.

[27] Bhaskar, D. R. and Senani, R., "Simulation of a floating inductance: A new two-CFOA-based configuration," *Fifth Int. Conf. on Computational Intelligence, Modeling and Simulation (CIMSim), 2013*, South Korea, 24-25 Sept. 2013, 381-383, Seoul.

[28] Ferri, G. and Guerrini, N., "High-valued passive element simulation using low-voltage low-power current conveyors for fully integrated applications," *IEEE Trans Circuits Syst II*, 48(4), 405-409, April, 2001.

[29] Metin, B. and Cicekoglu, O., "A novel floating lossy inductance realization topology with NICs using current conveyors," *IEEE Trans Circuits Syst II*, 53(6), 483-486, June 2006.

[30] Maundy, B., Gift, S. and Aronhime, P., "A novel hybrid active inductor," *IEEE Trans Circuits Syst II*, 54(8), 663-667, Aug, 2001.

[31] Ferri, G., Guerrini, N., Silverii, E. and Tatone, A., "Vibration damping using CCII-based inductance simulators," *IEEE Trans Instrument Measurement*, 57(5), 907-914, May 2008.

[32] Yuce, E., "Grounded inductor simulators with improved low-frequency performance," *IEEE Trans Instrument Measurement*, 57(5), 1079-1084, May 2008.

[33] Senani, R., Bhaskar, D. R., Singh, A. K. and Singh, V. K., *Current feedback operational amplifiers and their applications*, Springer Science + Business Media, New York 2013, Ch. 3, 67.

Security Constrained Economic Environmental Dispatch through Fuzzy Based Gravitational Search Algorithm

P. K. Hota[1,*], N. C. Sahu[2]

[1]Department of Electrical Engineering Depart, Veer Surendra Sai University of Technology (VSSUT), Burla, India
[2]Department of Electrical Engineering, ITER, SOA University, Bhubaneswar, India
*Corresponding author: p_hota@rediffmail.com

Abstract A robust and reliable algorithm called as fuzzy based gravitational search algorithm has been developed for solution of the security constrained economic and environmental load dispatch (EED) problems. A fuzzy maximizing decision approach using gravitational search algorithm (GSA) is applied to a six-unit thermal power test system to obtain the best compromising solution. The usefulness and quality of the proposed algorithm is demonstrated through its application to the test system in terms of convergence speed, solution time and minimum operating cost. The obtained results from the proposed algorithm confirm the effectiveness and potential of the promising proposed algorithm compared to conventional method, real coded genetic algorithm (GA), hybrid GA and PSO methods, performed in different central load dispatch centers to solve multi-objective generation dispatch problems.

Keywords: *security constraints, economic environmental load dispatch, fuzzy decision making function, gravitational search algorithm, compromise factor*

1. Introduction

The solution of nonlinear economic environmental load dispatch is to generate optimal amount of generating power from the fossil fuel based generating units in the system by minimizing the fuel cost and emission level simultaneously by satisfying all practical constraints. But, the objective of minimum cost of generation will not provide minimum pollution level and objective of minimum emission does not provide minimum cost of generation, which make both the objectives as non-commensurable and conflicting each other. Hence, they give rise to a set of optimal solutions instead of one optimal solution. The present situation demands to handle both objectives namely, economic dispatch and environmental emission as competing objectives and therefore, are to be solved simultaneously.

Nanda *et al.* [1] have treated Economic Emission Load Dispatch as a multiple, conflicting objective problem and solved using goal-programming techniques. Palanichamy *et al.* [2] demonstrated that reduction of harmful ecological effects caused by the emission of gaseous pollutants is possible only with a noticeable increase in the operating costs of the plants by the use of the familiar co-ordination equations approach. Nanda *et al.* [3] have also solved the Economic Emission Load Dispatch problem using a classical technique based on coordination equations to minimize the total cost of generation and pollution control simultaneously with varying degrees of compromise. Venkatesh *et al.* [4] showed that GA approach is the only viable solution to solve Economic load dispatch problem for the power system networks containing combined cycle cogeneration plants that have the cost characteristics which are not differentiable as it is not possible to solve such problems by conventional techniques. Kamaraj *et al.* [5] confirmed by solving economic dispatch problem that GA approach is efficient in handling multiple constraints. Baskar *et al.* [6] presented a two-phase Hybrid Genetic Algorithm based technique to solve emission and economic dispatch problems and demonstrated that their method is valid for real-time applications. Most of the approaches adopted above fail to address the nonlinear constrained EED problems satisfactory due to the complexity and non-monotonic nature of the problems. Sudhakaran *et al.* [7] have applied a refined genetic algorithm to combined economic and emission dispatch. In another paper, Sudhakaran *et al.* have solved the EED problem [8] by particle swarm optimization (PSO) to get better nondominated solutions within less execution time. However, the quality and diversity of the non-dominated solutions presented in [8] has not been measured and assessed quantitatively. The obvious weaknesses of this approach are that it is time-consuming and tends to find weakly non-dominated solutions. The classical techniques for solving EED problem need to introduce a compromise factor in order to decide the optimal solution and these results a complicated problem formulation. Further, these methods including GA and PSO techniques are not fast enough in terms of execution time. Moreover, they do not have a mechanism to show the vague or 'fuzzy' preference of the human decision-maker in obtaining a compromise solution in presence of such conflicting objectives.

On the other-hand, fuzzy systems provide tools for representing and manipulating inexact concepts and the ambiguity prevalent in human interpretation and thought processes. Further, fuzzy sets [9] can be applied for decision making in multiple objectives involving various constraints. Many interesting applications of fuzzy sets in the power field have been reported in the literature during last decade. Hota *et al.* [10] have described a simple and efficient technique based on fuzzy set theory for the EED problem. Further, they have developed an interactive fuzzy satisfying method [11] to solve multiobjective generation dispatch problem. The major advantage of fuzzy technique applied to this kind of problems lies in having a mechanism to show the vague/fuzzy preference of the human decision-maker in obtaining a compromising solution in the presence of conflicting objectives. However, the execution time of these fuzzy techniques appears not to be very promising for real time operation where the execution time is crucial. In recent times, practical power system optimization problems are being solved by various meta-heuristic algorithms successfully with quality solutions. However, there is no specific algorithm to achieve the best solution for all optimization problems. Some algorithms give a better solution for some particular problems than others. To investigate the gravitational search algorithm (GSA) [12] in solving the EED problem is the subject of this paper. GSA has been successfully applied to economic load dispatch (ELD) problem by Swain et al. [13]. Further, Dugman et al [14] have also solved the ELD problem with valve point effects which provides the quality optimal generation dispatch solution with less execution time. However, in none of the application of GSA to EED problem, the security constraint has been considered.

In this paper, a fuzzy maximizing decision approach using gravitational search algorithm has been proposed to obtain the best compromising solution of the economic environmental dispatch problem considering security constraints. The obtained compromising solution permits the system operator to decide on different preferences for the objectives according to system operating conditions, thus resulting more flexible operation of generating units with improved man machine interface. The solutions so obtained by the proposed method show its robustness and superiority when compared with classical method, GA and PSO methods.

2. Problem Formulation

2.1. Economic Dispatch

The ELD problem may be expressed by minimizing the fuel cost of generating units under some constraints. The fuel cost curve is approximated as a quadratic function of the active power output from the generating units. Mathematically, the ELD problem can be defined as the following optimization problem.

$$\text{Minimize } F = \sum_{i=1}^{m} f_i\left(P_{gi}\right),$$

$$f_i\left(P_{gi}\right) = a_i + b_i P_{gi} + c_i P_{gi}^2 \text{ Rs/h} \quad (1)$$

where, a_i , b_i and c_i are fuel cost coefficients of ith generating unit; m is the total number of committed online generators; F is the total fuel cost; $f_i(P_{gi})$ is the fuel cost of ith generating unit which is a function of its real power generation P_{gi}.

Subject to

Power balance constraints or demand constraints: The total generation $\sum_{i=1}^{m}\left(P_{gi}\right)$ should be equal to the total system demand P_D plus the transmission network loss P_{Loss} .That is represented as,

$$\sum_{i=1}^{m}\left(P_{gi}\right) = P_D + P_{Loss} \quad (2)$$

The transmission losses must be taken into account in order to achieve true economic dispatch. To calculate the transmission losses, *B*-coefficients method and Newton-Raphson load flow solution are used in this paper. The real power loss P_{Loss} in transmission lines can be calculated from Newton-Raphson load flow solution, which has equality constraints on real and reactive power at each bus as follows:

$$P_{gi} - P_{Di} - V_i \sum\nolimits_{i=1}^{Nb} V_j \begin{bmatrix} G_{ij} \cos\cos\left(\delta_i - \delta_j\right) \\ + B_{ij}\sin(\delta_i - \delta_j) \end{bmatrix} \quad (3)$$

$$Q_{gi} - Q_{Di} - V_i \sum\nolimits_{i=1}^{Nb} V_j \begin{bmatrix} G_{ij} \sin\sin\left(\delta_i - \delta_j\right) \\ - B_{ij}\cos(\delta_i - \delta_j) \end{bmatrix} \quad (4)$$

where, i = 1, 2, ..., Nb; Nb is the number of buses; Q_{gi} is the reactive power generated at the ith bus; P_{Di} and Q_{Di} are the real and reactive load power at ith bus, respectively; G_{ij} and B_{ij} are the transfer conductance and susceptance between bus i and bus j, respectively; V_i and V_j are voltage magnitudes at bus i and bus j, respectively; δ_i and δ_j are voltage angles at bus i and bus j, respectively.

The load flow solution by Newton-Raphson method is obtained as the real power output of one generator called slack generator, is left to cover the real power loss and satisfy the equality constraint in (2). It also gives all bus voltage magnitudes and angles. Then, the real power loss in transmission line can be calculated as:

$$P_{Loss} = \sum\nolimits_{k=1}^{Nl} g_k \left[V_i^2 + V_j^2 - 2V_iV_j\cos(\delta_i - \delta_j)\right] \quad (5)$$

where, Nl is the number of transmission lines and g_k is the conductance of kth line that connects bus i to bus j.

The generator constraints: The generation real power of each unit should be between its minimum and maximum limits. That is, the following inequality constraint for each generator should be satisfied.

$$P_{gi\min} \le P_{gi} \le P_{gi\max} \quad (6)$$

where, P_{gi} is the real output of ith generator and $P_{gi\min}, P_{gi\max}$ are its minimum and maximum real power output of ith generator, respectively.

Security constraints: It is quite essential that the operator must dispatch the online generating units such that the apparent power flow through the transmission line S_i is restricted by its upper limits to ensure secured operation as described below.

$$S_{ik} \le S_{ik}^{\max}, k = 1,2,\ldots, Nl. \tag{7}$$

It is to be noted that the *k*th transmission line flow connecting bus *i* and bus *j* can be calculated as:

$$S_{ik} = (V_i \angle \delta_i) I_{ij}^*$$

where, I_{ij} is the current flow from bus *i* to bus *j* and can be calculated as:

$$I_{ij} = (V_i \angle \delta_i) \times \begin{bmatrix} (V_i \angle \delta_i - V_j \angle \delta_j) \times y_{ij} \\ +(V_i \angle \delta_i) \times j\frac{y}{2} \end{bmatrix} \tag{8}$$

where, y_{ij} is the line admittance and y is the shunt susceptance of the line.

2.2. Environmental Dispatch

The objective of emission dispatch is to minimize the total environmental degradation or the total pollutant emission due to the burning of fuels for production of power to meet the load demand. In the present work, however, SO_x, NO_x, CO_2 and particulates, are treated together as a single pollution criterion. The amount of pollutants from a fossil based generating units depend on the amount of real power generated by that unit. Mathematically, the emission dispatch problem can be defined as the following optimization problem.

$$\text{Minimize } E = \sum_{i=1}^{m} E_i(P_{gi}), \tag{9}$$

$$E_i(P_{gi}) = d_i + e_i P_{gi} + f_i P_{gi}^2 \text{ Kg/hr} \tag{10}$$

where, d_i, e_i and f_i are the pollution coefficients of the ith generating unit subjected to demand constraint and generating capacity constraints, as stated in eqn.2 and eqn.3; E is the total emission release; $E_i(P_{gi})$ is the emission release of ith generating in Kg/hr.

2.3. Economic Environmental Dispatch

The non linear constrained multi-objective optimization problem can be converted to a single objective optimization problem (EED) as follows:

$$T = \alpha(\text{cost of generation}) + (1-\alpha)^{**}(\text{cost of pollution control}) \tag{11}$$

where, T is total cost in Rs/hr; *PPF* is price penalty factor in Rs/kg and $0 < \alpha < 1$ is a compromise factor which is primarily a function of the preference of the decision maker. The total cost can be calculated as

$$T = \sum_i^m \begin{bmatrix} \alpha\{a_i P_{gi}^2 + b_i P_{gi} + c_i\} \\ +(1-\alpha) PPF\{d_i P_{gi}^2 + e_i P_{gi} + f_i\} \end{bmatrix} \text{Rs/hr.}$$

$$T = \sum_i^m P_{gi}^2 [\alpha a_i + (1-\alpha) PPF.d_i] + P_{gi}[\alpha b_i + (1-\alpha) PPF.e_i] + [\alpha c_i + (1-\alpha) PPF.f_i] \text{Rs/hr.}$$

$$T = \sum_i^m (a_i^* P_{gi}^2 + b_i^* P_{gi} + c_i^*) \text{Rs/hr.} \tag{12}$$

where,

$$a_i^* = \alpha a_i + (1-\alpha) PPF.d_i$$
$$b_i^* = \alpha b_i + (1-\alpha) PPF.e_i$$
$$c_i^* = \alpha c_i + (1-\alpha) PPF.f_i$$

The power system emission limit is expressed as

$$\sum_{i=1}^{m} E(P_{gi}) \le E_{lt} \tag{13}$$

where, E_{lt} is the total emission limit over the operating zones.

Therefore, the transformed single objective optimization problem can be described as:

$$\text{Minimize } T(P_{gi}) \tag{14}$$

$$\text{Subject to}: g(P_{gi}) = 0 \tag{15}$$

$$h(P_{gi}) \le 0 \tag{16}$$

where, g is the equality constraint representing the power balance equation, while h is the inequality constraint representing the generation capacity constraint.

3. Proposed Solution Methodology

3.1. Gravitational Search Algorithm

The gravitational search algorithm (GSA), is one of the newest metaheuristic search algorithm developed by Rashedi et al. in 2009 [12]. GSA is followed by the physical law of gravity and the law of motion. The gravitational force between two particles is directly proportional to the product of their masses and inversely proportional to the square of the distance between them. In the proposed algorithm, agents are considered as objects and their performance is measured by their masses. M_i is the mass of the of the object i, $G(t)$ is the gravitational constant at time t, The gravitational constant, G, is initialized at the beginning and will be decreased with the time to control the search accuracy. In other words, G is function of the initial value (G_0) and time (t):

$$G(t) = G(G_0, t) \quad (17)$$

$$G(t) = G_0 e^{-\alpha t/T}. \quad (18)$$

The masses of the agents are calculated using fitness evaluation. A heavier mass means a more efficient agent. This means that better agents have higher attractions and moves more slowly. Supposing the equality of the gravitational and inertia mass, the values of masses are calculated using the map of fitness.

$fit_i(t)$ represents the fitness value of the agent i at time t, and the $best(t)$ and $worst(t)$ in the population, respectively indicate the strongest and the weakness agent according to their fitness value.

For a *minimization problem*:

$$best(t) = \min_{j \in \{l, \cdots m\}} fit_j(t) \quad (19)$$

$$worst(t) = \max_{j \in \{l, \cdots m\}} fit_j(t) \quad (20)$$

For a *Maximization problem:*

$$best(t) = \max_{j \in \{l, \cdots m\}} fit_j(t) \quad (21)$$

$$worst(t) = \min_{j \in \{l, \cdots m\}} fit_j(t) \quad (22)$$

The GSA consists of following steps such as i) Initialization, ii) Fitness evaluation (objective function), iii) Agent force calculation, iv) Evaluation of acceleration of an agent, v) Update the agents' position and velocity and vi) Stopping criterion.

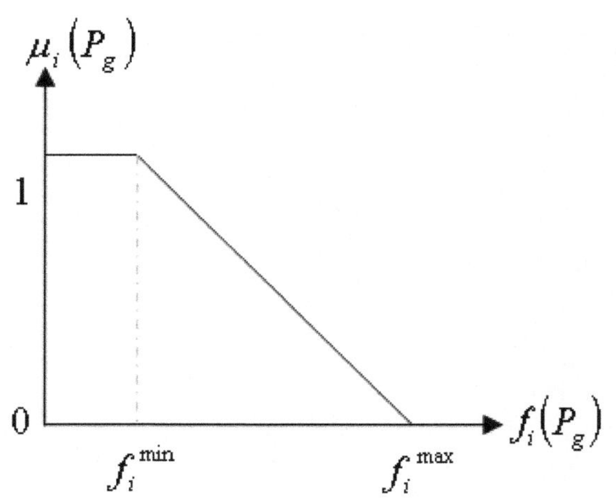

Figure 1. Membership function of fuzzy fuel cost

3.2. Best Compromising Solution

The fuzzy set theory has been implemented to derive efficiently a solution from the set of non-dominated solutions. The fuzzy decision making function is represented by the membership function to replace each variable as a precise value. Figure 1. depicts the membership function μ_i for the fuzzy variable signifying total fuel cost $f_i(P_{gi})$. The decision maker is fully satisfied with the cost if $\mu_i = 1$, and not satisfied at all if $\mu_i = 0$. Therefore, the value of membership function indicates the adaptability of the economy index. Due to the imprecise nature of decision maker's judgment, the ith objective function of a solution in the non-dominated set $f_i(P_{gi})$ is represented by a membership function $\mu_i(P_{gi})$ defined as:

$$\mu_i(P_{gi}) = \frac{f_i^{max} - f_i(P_{gi})}{f_i^{max} - f_i^{min}} \quad (23)$$

where, f_i^{max} and f_i^{min} are the maximum and minimum values of ith objective, respectively.

$$FDM_i(P_{gi}) = \begin{cases} 0 & \mu_i(P_{gi}) \le 0 \\ \mu_i(P_{gi}) & 0 < \mu_i(P_{gi}) < 1 \\ 1 & \mu_i(P_{gi}) \ge 1 \end{cases} \quad (24)$$

For each non-dominated solution k, the normalized membership function FDM^k is defined as:

$$FDM^k = \frac{\left[\sum_{i=1}^{2} FDM_i^k(P_{gi})\right]}{\left[\sum_{j=1}^{M}\sum_{i=1}^{2} FDM_i^j\right]} \quad (25)$$

The best compromising solution of EED problem is the one having the maximum value of FDM^k as fuzzy decision making function. Where, M is the total number of non dominated solutions. Then, all the solutions are arranged in descending order according to their membership function values which will guide the decision maker with a priority list of non-dominated solutions in view of the current operating conditions.

3.3. Computational Procedure of Proposed Fuzzy Based GSA Technique for EED

The computational procedure of the proposed fuzzy based GSA technique for the EED problem can be described as follows.

Step-1: Initialize Input parameters of the system.

Step-2: Initialize randomly the agents of the population. These agents must be feasible candidate solutions that satisfy the given constraints.

Step-3: Let $P_{gj} = [P_{g1} \quad P_{g2} \ldots P_{gm}]$ be the trial vector denoting the particles of population to be evolved. The elements of P_i are the real power outputs of the committed m generating units subjected to their respective capacity constraints in (3). To meet exactly the load demand in (2), a dependent unit is randomly selected from among the committed m units. Let P_d be the power output of the dependent unit (slack generator), then P_d is calculated as $P_d = P_D - \sum_{\substack{i=1 \\ i \neq d}}^{m} P_{gi}$, $i = 1,2,\ldots m$.

Step-4: If the equation (2) is satisfied, go to step-6, otherwise go to step-2.

Step-5: Calculate the fitness value of each agent.

Step-6: Update gravitational constant $G(t)$, $best(t)$ and $worst(t)$ and update the mass of the object $M_i(t)$.

Step-7: Calculate Force in different directions.

Step-8: Calculate acceleration and velocity of each agent.

Step-9: Update the position of each agent.

Step-10: Repeat Step-5 to Step-9 until the stop criterion is satisfied.

Step-11: The agent that generates the latest $best(t)$ is the optimal value.

Step-12: Fuzzy set theory is applied to get the best compromise solution and stop.

The flow chart of the overall computational procedure for EED using Fuzzy based GSA algorithm is shown in Figure 2.

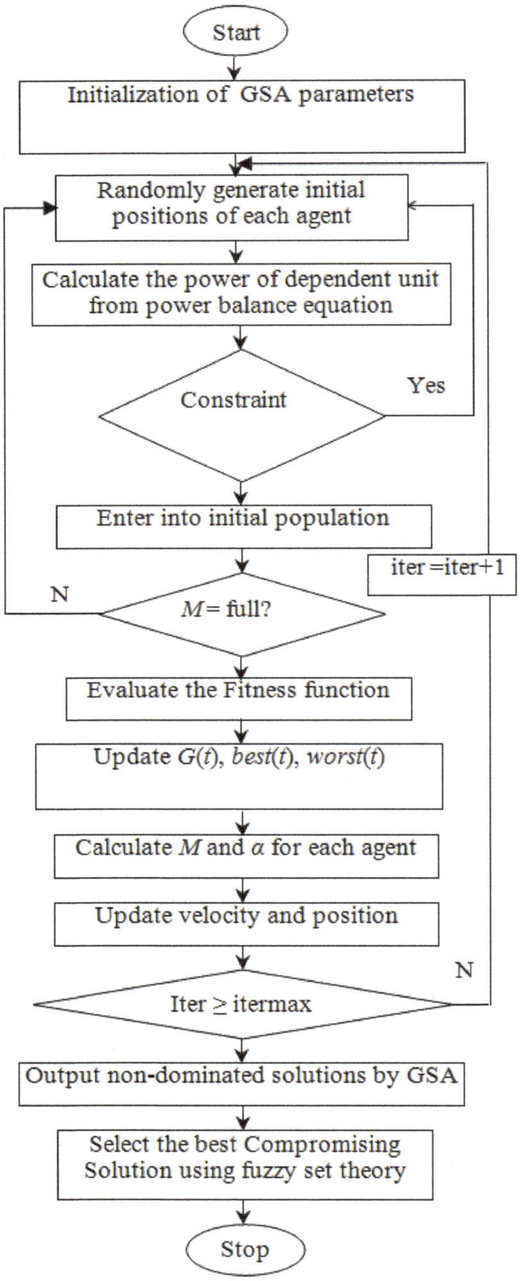

Figure 2. Flow chart for the proposed fuzzy based GSA method

4. Simulation Results

The proposed fuzzy set based GSA methodology is applied to the 6-generator system. The input data of the test system is adopted from [8] and system load demands are taken as 500 MW, 700MW and 900MW for comparison point of view.

The following key parameters are selected such as: M = 100, G_0 = 120, α = 8, $iter_{\max}$ (T) =50, etc.

Table 1. Economic load dispatch results (Load Demand=700 MW)

Methods	Fuel cost (FC), Rs/h	Emission Kg/h	Losses P_{loss} MW	Execution time, s
Classical Method	37288.700	495.3480	26.570	0.25
Real coded GA	37137.9600	489.5590	23.124	14.61
Hybrid GA	37137.9600	489.5590	23.124	1.21
PSO	36921.5274	494.9329	19.164	1.16
Proposed GSA	36912.326	498.683	19.405	0.54

Table 2. Emission dispatch results (Load Demand=700 MW)

Methods	Fuel cost (FC), Rs/h	Emission Kg/h	Losses P_{loss}, MW	Execution time, s
Classical Method	38364.500	437.9660	20.2400	0.26
Real coded GA	38186.400	435.0750	17.3660	14.61
Hybrid GA	38186.400	435.0750	17.3660	1.21
PSO	38099.352	434.138	16.5517	1.32
Proposed GSA	38081.946	433.178	16.552	0.54

Table 3. Economic Environmental dispatch (EED) results

P_D, MW	500 MW	900 MW
PPF Rs/kg	43.898	47.822
Performance	Fuel cost, Rs/h Emission, kg/h P_{loss}, MW Total cost, Rs/h	Fuel cost, Rs/h Emission, kg/h P_{loss}, MW Total cost, Rs/h
Classical Method Rs/hr	27692.10 262.472 8.830 39159.50	48567.90 701.428 35.230 82436.58
Real coded GA Rs/hr	27692.10 263.370 10.172 39258.10	48567.70 694.169 29.725 81764.50
Hybrid GA Rs/hr	27695.00 263.370 10.135 39257.50	48567.50 694.172 29.718 81764.40
PSO Rs/hr	27619.237 264.714 9.002 39239.737	48556.034 692.229 27.959 81660.014
Fuzzy based GSA Rs/hr	27586.663 262.948 8.996 39129.564	48410.851 691.360 27.890 81473.090

For the load demand of 700MW, the EED problem is solved separately by proposed fuzzy based GSA and the results are compared with the conventional method, real coded GA, hybrid GA and PSO method in Table 1 and Table 2, respectively. The results proved that the proposed method is capable of providing non-dominated quality solutions within less execution time. It is evident from

Table 3 that the solution obtained by the proposed method has minimum fuel cost, minimum emission and minimum loss.

Table 4 shows the optimal power output of generators for the EED problem. After getting many optimal solutions with varying degree of compromise with a load demand of 700 MW, the fuzzy set theory has been implemented to derive efficiently the best compromising solution in presence of conflicting objectives for the decision makers. The best compromising solution (10th one) is obtained with maximum FDM value as 0.1191 out of eleven non-dominated solutions as shown in Table 5.

Table 4. Optimal power dispatch results obtained by proposed fuzzy based GSA method for EED problem

P_D, MW	P_1	P_2	P_3	P_4	P_5	P_6
500	30.1938	25.4906	90.0264	90.9107	138.3767	133.9984
900	91.7054	102.0342	150.3813	148.4540	219.0278	216.2881

Table 5. Non Dominated Solutions of Two Objectives with Fuzzy Decision Making Function

Sr. no.	α	$1-\alpha$	F (Rs/h)	E (Kg/h)	$\mu(f_1)$	$\mu(f_2)$	FDM	P_{loss} MW
1	1.0	0.0	36912.326	498.683	1.000	0.0000	0.0814	19.30413
2	0.9	0.1	36912.487	498.557	0.9999	0.0020	0.08154	19.04899
3	0.8	0.2	36912.571	498.079	0.9998	0.0092	0.08212	19.36825
4	0.7	0.3	36912.955	496.724	0.9995	0.0299	0.08379	19.24657
5	0.6	0.4	36913.992	495.339	0.9986	0.0510	0.08542	19.28452
6	0.5	0.5	36915.448	493.755	0.9973	0.0752	0.0873	19.25140
7	0.4	0.6	36922.356	488.362	0.9914	0.1576	0.09352	19.06470
8	0.3	0.7	36930.870	483.857	0.9841	0.2263	0.09852	18.89331
9	0.2	0.8	36952.879	476.717	0.9653	0.3353	0.10585	18.64566
10	**0.1**	**0.9**	**37060.594**	**459.983**	**0.8732**	**0.5908**	**0.1191**	**18.00375**
11	0.0	1.0	38081.946	433.178	0.0000	1.0000	0.0814	16.50262

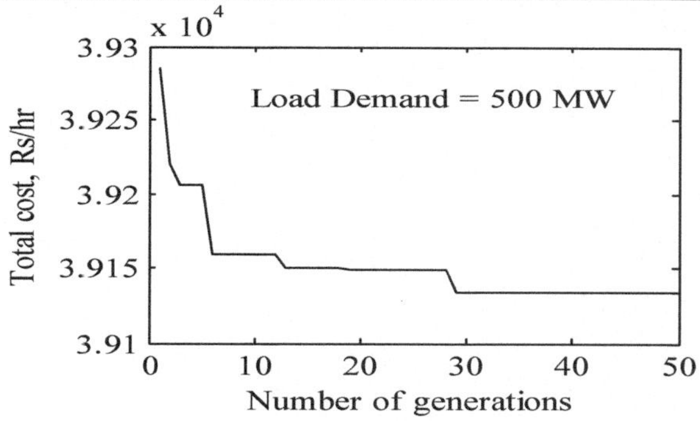

Figure 3. Convergence characteristics of proposed method with P_D = 500 MW (EED problem)

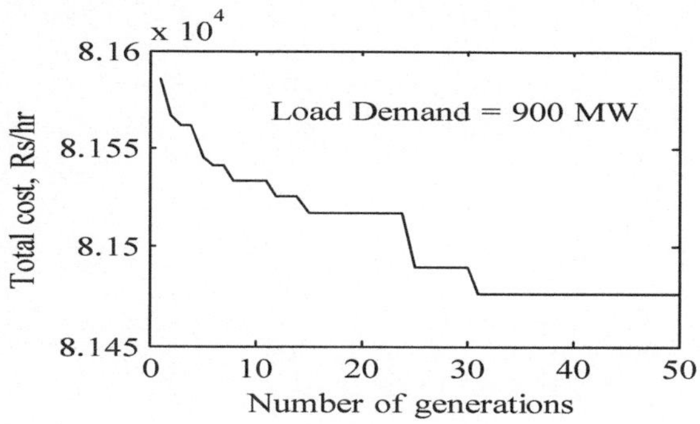

Figure 4. Convergence characteristics of proposed method with P_D = 900 MW (EED problem)

Table 6. Generation Schedules corresponding to Non Dominated Solutions

Sr. no.	P_1 (MW)	P_2 (MW)	P_3 (MW)	P_4 (MW)	P_5 (MW)	P_6 (MW)
1	28.465	10.0	119.9194	119.3457	230.3894	211.0886
2	28.4016	10.0012	119.7185	120.5502	230.4752	210.2322
3	28.9907	10.0	120.4949	118.5121	227.767	213.6226
4	29.6001	10.0	120.8020	120.1669	227.4199	211.3475
5	29.8483	10.0199	122.648	120.5831	224.8195	211.3670
6	31.3968	10.8484	119.5816	121.9910	225.3194	210.1236
7	31.1025	15.5821	121.5623	120.3979	220.4680	209.9601
8	32.6022	16.9158	124.3681	122.4443	215.86	206.7076
9	36.0081	20.6317	125.0190	123.8452	210.0	203.1505
10	**44.1872**	**34.5814**	**124.0012**	**123.0279**	**202.1902**	**190.0372**
11	79.2633	82.2247	114.0173	113.3536	162.9846	164.7091

Table 6 shows the best generation schedules for all non-dominated solutions. The best generation schedule corresponding to best compromising solution is highlighted in Table 6. The convergence characteristics of a six-unit system obtained by the proposed method for the load demand of 500 MW and 900 MW are shown in Figure 3 and Figure 4, respectively. The required number of iteration are found to be 28 and 32, respectively.

Further, using the load flow program, the security constraints are checked and all the transmission lines are found to carry the real power in MW less than or equal to their corresponding maximum real power limits.

5. Conclusion

The proposed fuzzy based GSA technique has been implemented in the security constrained economic environmental load dispatch problems which, appears to be a simple and efficient optimization method. The obtained compromising solution permits the system operator to decide on different preferences for the objectives according to system operating conditions, thus resulting more flexible operation of generating units with improved man machine interface. The comparison of results for the test cases clearly identifies the superiority of the proposed method and is capable of obtaining reliable and efficient solutions in a short CPU time. Further, other practical objective functions such as reliability, etc. can also easily be incorporated in the proposed algorithm.

Acknowledgements

The support from the Electrical Engineering Department of Veer Surendra Sai University of Technology, Burla, India, extended to the second author for the work reported in this paper is gratefully acknowledged.

References

[1] Nanda J, Kothari D P, Linga Murthy K S, "*Economic emission load dispatch through Goal Programming Techniques*", IEEE Trans. On Energy Conversion, 1988; Vol. 13, No 1, pp. 26-32.

[2] Palanichamy C, Srikrishna K, "*Economic thermal power dispatch with emission constraint*", Journal of the Institution of Engineers (India), pt. EL, 1991, Vol. 72, pp. 11-18.

[3] Nanda J, Lakshman H, Kothari M L, "*Economic emission load dispatch with line flow constraints using a Classical Technique*", IEE proceedings–Generation, Transmission and Distribution, January, 1994, Vol.141, No.1, pp. 1-10.

[4] Venkatesh P, Kannan P S, Sudhakaran M, "*Application of computational intelligence to economic load dispatch*", Journal of the Institution of Engineers (India), pt. EL, 2000, Vol. 81, pp. 39-43.

[5] Kamaraj N, Sivakamasundari C, Kannan P S, "*Economic dispatch with Line flow and emission constraints using genetic algorithm*", Journal of the Institution of Engineers (India), pt. EL, 2001, Vol. 82, pp. 195-199.

[6] Baskar S, Subbaraj P, Rao M V C, "*Hybrid genetic algorithm solution to emission and economic dispatch problems*", Journal of the Institution of Engineers (India), pt. EL, March, 2002, Vol. 82, pp. 343-249.

[7] Sudhakaran M, Slochanal S M R, Sreeran R, Sekhar N C, "*Application of refined genetic algorithm to combined economic and emission dispatch*", Journal of the Institution of Engineers (India), pt. EL, September, 2004, Vol. 85, pp. 115-119.

[8] Sudhakaran M, Sivakumar G, Vengatachalapathy P, Latchumi K, "*Particle swarm optimization for economic and emission dispatch problems*", Journal of the Institution of Engineers (India), pt. EL, 2007, Vol. 88, pp. 39-45.

[9] Zimmermann H J, "*Fuzzy Set Theory and Its Applications*", Kluwer-Nijhoff Publishing, 1985.

[10] Hota P K, Chakrabarti R, Chattopadhyay P K, "*A fuzzy-set based optimization technique for economic emission load dispatch*", Journal of the Institution of Engineers (India), pt. EL, November, 1999, Vol. 80, pp. 99-103.

[11] Hota P K, Chakrabarti R, Chattopadhyay P K, "*Economic emission load dispatch through an interactive fuzzy satisfying method*", Electric Power Systems Research, 2000, Vol. 54, No.3, pp. 151-157.

[12] Rashedi E, Nezamabadi-pour H, Saryazdi S, "GSA: A Gravitational Search Algorithm", Information Science, 2009, Vol. 179, pp. 2232-2248.

[13] Swain R K, Sahu N C, Hota P K, "*Gravitational Search Algorithm for Optimal Economic Dispatch*", Procedia Technology, Science Direct, Elsevier, ICCCS-2012, pp. 1-9.

[14] Dugman S, Guvenc U, Yorukeren N, "*Gravitational Search Algorithm for Economic Dispatch with Valve Point Effects*", International Review of Electrical Engineering, 2010, Vol. 5, No. 6, pp. 2890-2895.

Synchronization of Diffusively Coupled Oscillators: Theory and Experiment

B. Nana[1,*], **P. Woafo**[2]

[1]Department of Physics, Higher Teacher Training College, University of Bamenda, PO Box 39 Bamenda, Cameroon
[2]Laboratory of Modelling and Simulation in Engineering, Biomimetics and Prototypes, Faculty of Science, University of Yaounde I, PO Box 812 Yaounde, Cameroon
*Corresponding author: na1bo@yahoo.fr

Abstract In this paper complete synchronization of diffusively coupled oscillators is considered. We present the results of both, theoretical and experimental investigations of synchronization between two, three and four almost identical oscillators. The method of linear difference signal has been applied. The corresponding differential equations have been integrated analytically and the synchronization threshold has been found. Hardware experiments have been performed and the measured synchronization error of less than 1% has been determined. Good agreement is found between theoretical and experimental results.

Keywords: *Chaos, Synchronization, Oscillator*

1. Introduction

In the last few years, several researchers have focused their attention on the problems related to the synchronization of chaotic systems [1-5]. Today the potential of chaos theory is recognized in the world-wide with research groups actively working on this topic [6-10]. One of the great achievements of the chaos theory is the application in secure communications. The chaos communication fundament is the synchronization of two chaotic systems under suitable conditions if one of the systems is driven by the other. Since Pecora and Carrol [11] have demonstrated that chaotic systems can be synchronized, the research in synchronization of couple chaotic circuits is carried out intensively and some interesting applications such as communications with chaos have come out of that research.

There are several methods for synchronizing chaotic oscillators described in literature [12-17]. The simplest one employs the feedback in the form of linear difference between the output of the transmitter $u_1(t)$ and the output of the receiver $u_2(t)$. The difference signal $K(u_2 - u_1)$ when applied with a certain weight $K > K_{\min}$ to an appropriate input of the receiver synchronizes the latter to the transmitter [16]. Although there are many papers describing global synchronization of a network of coupled oscillators, less attention has been devoted to experimental results for bidirectional coupled systems.

In this paper, attention will be drawn to complete synchronization of two, three and four coupled chaotic oscillators. The remainder of this paper is organized as follows. In section 2, we present the used oscillator and study its dynamic behavior as a function of control parameter. In section 3, synchronization of two, three and four diffusively coupled oscillators are analyzed theoretically. Section 4 deals with the experimental setup and where experimental results are compared to the numerical ones. Finally conclusions are drawn in section 5.

2. Circuit Model and Description

2.1. Circuit Model

The circuit diagram of the oscillator is shown in Figure 1.

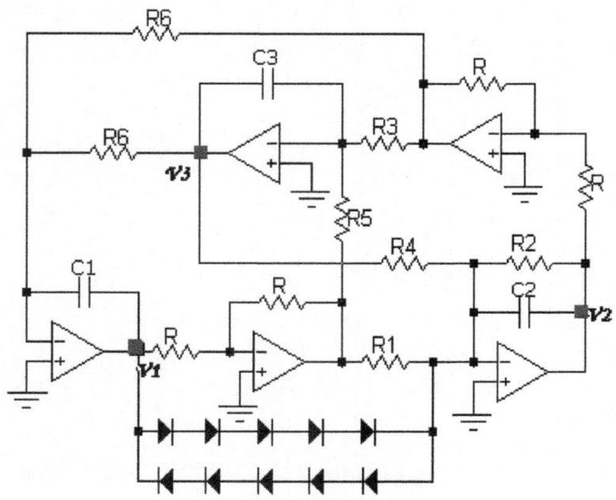

Figure 1. Electronic scheme of the oscillator

The circuit is a third order nonlinear oscillator containing five operational amplifiers. We assume that all the operational amplifiers operate in their linear domain. In our model, the diode acts like non linear component and we model its voltage-current characteristic with an exponential function, namely

$$i = I_0\left[\exp\left(\frac{v}{V_0}\right)-1\right] \tag{1}$$

where i is the current through the diode, v is the voltage across the diode, I_0 is the inverse saturation current and $V_0 \approx 26\ \text{mV}$ at the room temperature. The dynamics of the oscillator is given by the following set of differential equations:

$$\frac{dv_1}{dt'} = \frac{1}{R_6C_1}v_2 - \frac{1}{R_6C_1}v_3, \tag{2}$$

$$\frac{dv_2}{dt'} = \frac{1}{R_1C_2}v_1 - \frac{1}{R_2C_2}v_2 - \frac{1}{R_4C_2}v_3 - \frac{2I_0}{C_2}\sinh\left(\frac{v_1}{5V_0}\right), \tag{3}$$

$$\frac{dv_3}{dt'} = \frac{1}{R_5C_3}v_1 + \frac{1}{R_3C_3}v_2, \tag{4}$$

Introducing the following dimensionless variables and parameters

$$x = \frac{v_1}{5V_0},\ y = \frac{v_2}{5V_0},\ z = \frac{v_3}{5V_0},\ t = \omega_0 t',\ a = \frac{1}{R_1C_2\omega_0},$$
$$b = \frac{1}{R_2C_2\omega_0},\ c = \frac{1}{R_4C_2\omega_0},\ d = \frac{1}{R_3C_3\omega_0} \tag{5}$$
$$\text{where } \omega_0 = \frac{1}{R_6C_1} \text{ and } 5V_0C_2 = 2I_0R_6C_1,$$

we come to the set of differential equations convenient for numerical integrations

$$\dot{x} = y - z, \tag{6}$$

$$\dot{y} = ax - by - cz - \sinh(x), \tag{7}$$

$$\dot{z} = x + dy. \tag{8}$$

2.2. Dynamic behavior

Assuming that in the oscillatory state, variables x, y and z can be replaced by their corresponding virtual orbits x_0, y_0 and z_0 respectively, we find x_0, y_0 and z_0 in the following form:

$$\begin{cases} x_0 = P_1\cos\omega t - Q_1\sin\omega t \\ y_0 = P_2\cos\omega t - Q_2\sin\omega t \\ z_0 = P_3\cos\omega t - Q_3\sin\omega t \end{cases} \Rightarrow \begin{cases} x_0 = Ae^{j\omega t} + \overline{A}e^{-j\omega t} \\ y_0 = Be^{j\omega t} + \overline{B}e^{-j\omega t} \\ z_0 = Ce^{j\omega t} + \overline{C}e^{-j\omega t} \end{cases}$$

with (9)

$$A = \frac{1}{2}(P_1 + jQ_1),\ B = \frac{1}{2}(P_2 + jQ_2),\ C = \frac{1}{2}(P_3 + jQ_3).$$

Using the above expressions and neglecting the higher harmonics, we can show that

$$(x_0)^{2n+1} = \frac{(2n+1)!}{n!(n+1)!}|A|^{2n}x_0 \text{ then}$$
$$\sinh(x_0) \approx \sum_{n=0}^{N}\left[\frac{(x_0)^{2n+1}}{(2n+1)!}\right] \approx x_0\sum_{n=0}^{N}\left[\frac{|A|^{2n}}{n!(n+1)!}\right]. \tag{10}$$

By substituting system (8) and equation (9) in the set of equations (5), (6) and (7) and equating the coefficients of $e^{j\omega t}$ and $e^{-j\omega t}$ separately to zero, we obtain the radian frequency of oscillatory orbit by. While the amplitude of oscillation is obtained by solving the following polynomial equation:

$$\sum_{n=0}^{N}\left[\frac{|A|^{2n}}{n!(n+1)!}\right] - r = 0 \text{ with } r = a + \frac{c(1-bd)}{b+d}. \tag{11}$$

In view to derive the analytical expression of amplitude A, we fixe $N = 3$ and we obtain the following result:

$$|A| = \left[-4 - 2\left(-1 - 9r + 3\sqrt{1+2r+9r^2}\right)^{1/3} + 2\left(1 + 9r + 3\sqrt{1+2r+9r^2}\right)^{1/3}\right]^{1/2} \tag{12}$$

This system presents stationary, periodic and chaotic attractors depending on the value of the parameters (a, b, c, d). The bifurcation diagram as well as the Lyapunov exponent drawn in Figure 2 show that for certain sets of parameters, the system exhibits chaotic oscillations. We used c as control parameter and other used parameters are the following: $a = 3.5$, $b = 0.5$, $d = 1.2$.

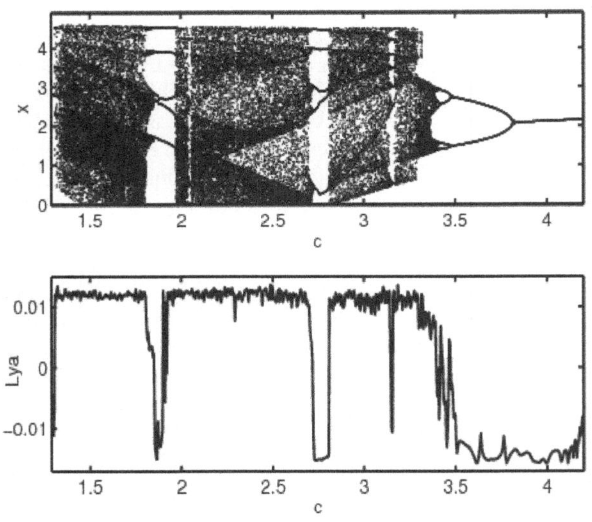

Figure 2. a) One parameter bifurcation diagram in the (c, x) plane and b) maximal Lyapunov spectrum Lya

The bifurcation diagram consists of quasiperiodicity, chaos, windows, period adding sequences and the familiar period doubling bifurcation sequence, intermittency and so on. As shown, there is a good agreement between the bifurcation diagram and its corresponding maximal Lyaponov exponent.

3. Synchronization of Coupled Oscillators

Recently, Woafo and Kraenkel [18] considered the problem of stability and duration of the synchronization process between classical Van der Pol oscillators and showed that the critical slowing-don behavior of the synchronization time and the boundaries of the synchronization domain can be estimated by analytical investigations. The next subsections extend the calculations of Ref. [18] to two, three and four diffusely coupled oscillators. Only in our analytical treatment, we assume that oscillators have identical coefficients.

3.1. Synchronization in a Case of two Oscillators

Here, we aim to determine the threshold value for synchronization (the minimal value K such that practical synchronization occurs) of two oscillators. The two oscillators are diffusely coupled with a buffer and a variable resistor R_c, which give the coupling constant $K = (R_c C_1 \omega_0)^{-1}$. The dynamics of two diffusely coupled oscillators can be described by the following set of equations:

$$\begin{cases} \dot{x}_j = y_j - z_j + K(x_{i\neq j} - x_j), \\ \dot{y}_j = ax_j - by_j - z_j - sh(x_j), \\ \dot{z}_j = x_j + dy_j, \end{cases} \tag{13}$$

where index i and j represent the oscillator number ($i, j \in \{1,2\}$). Assuming the three following vectors defined as $u_1(x_1, y_1, z_1)$, $u_2(x_2, y_2, z_2)$ and $\varepsilon(\varepsilon_1, \varepsilon_2, \varepsilon_2) = u_1 - u_2$, the stability of synchronization manifold is decided by the asymptotic behavior of $\varepsilon_1 = x_1 - x_2$, $\varepsilon_2 = y_1 - y_2$ and $\varepsilon_3 = z_1 - z_2$. At a linear approximation, ε_1, ε_2 and ε_3 obey to

$$\begin{cases} \dot{\varepsilon}_1 = \varepsilon_2 - \varepsilon_3 - 2K\varepsilon_1, \\ \dot{\varepsilon}_2 = [a - ch(x_1)]\varepsilon_1 - b\varepsilon_2 - c\varepsilon_3, \\ \dot{\varepsilon}_3 = \varepsilon_1 + d\varepsilon_2. \end{cases} \tag{14}$$

By replacing the chaotic variable x_1 by its virtual orbit x_0, the dynamics of the synchronization errors is then described by the linear system

$$\begin{pmatrix} \dot{\varepsilon}_1 \\ \dot{\varepsilon}_2 \\ \dot{\varepsilon}_3 \end{pmatrix} = \begin{pmatrix} -2K & 1 & -1 \\ \alpha & -b & -c \\ 1 & d & 0 \end{pmatrix} \begin{pmatrix} \varepsilon_1 \\ \varepsilon_2 \\ \varepsilon_3 \end{pmatrix},$$

with (15)

$$\alpha = a - ch(x_0) = a - \sum_{n=0}^{N} \left[\frac{|A|^{2n}}{(n!)^2} \right].$$

Using the Lyapunov criteria, the synchronization is stable if the real part of all eigenvalues is negative. Assuming that λ is the eigenvalue of system (15), then it obeys the following algebraic third order equation:

$$\lambda^3 + (b+2K)\lambda^2 + (2Kb + cd - \alpha + 1)\lambda + (c + b + 2Kcd + \alpha d) = 0. \tag{16}$$

The determination of signs of the real parts of the root λ may be carried out by making use of the Routh-Hurwitz criterion. In applying this criterion, we find that the real parts of the roots are negative if

$$\begin{cases} b + 2K > 0, \\ 2Kb + cd - \alpha + 1 > 0, \\ c + b + 2Kcd + \alpha d > 0, \\ 4K^2 b + (2b^2 - 2\alpha + 2)K + (bcd - b\alpha - \alpha d - c) > 0. \end{cases} \tag{17}$$

Then the synchronization is said to be stable if $K > K_{2\min}$, where the new parameter $K_{2\min}$ is defined as follow:

$$K_{2\min} = -\frac{c + b + \alpha d}{2cd}. \tag{18}$$

As shown in Figure 3, the result obtained from equation (18) is verified by a direct numerical simulation of system (13).

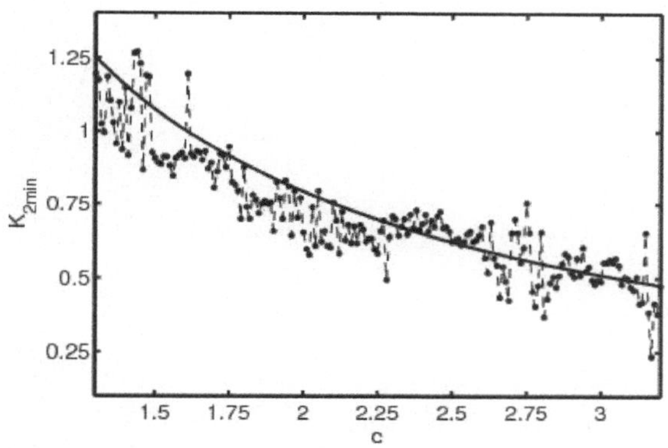

Figure 3. Synchronization boundaries in the case of two coupled oscillators

Numerically, we use the fourth order Runge Kutta algorithm and we find the first value of K for which quantity $|x_1 - x_2| < 0.05$ for $t \geq 400$. In Figure 3, the numerical result (points and dashed-line) fluctuates around the analytic curve (solid line).

3.2. Synchronization in a Case of three Oscillators

In this case, the dynamics of three diffusely coupled oscillators can be described by the following set of equations:

$$\begin{cases} \dot{x}_j = y_j - z_j + K(x_{j-1} - 2x_j + x_{j+1}), \\ \dot{y}_j = ax_j - by_j - z_j - sh(x_j), \\ \dot{z}_j = x_j + dy_j. \end{cases} \tag{19}$$

Index j (with $1 \leq j \leq 3$) represents the oscillator number and the two periodic boundary conditions $x_0 = x_3$ and $x_4 = x_1$ are used. Assuming the five following vectors defined as $u_1(x_1, y_1, z_1)$, $u_2(x_2, y_2, z_2)$, $u_3(x_3, y_3, z_3)$, $\eta(\eta_1, \eta_2, \eta_2) = u_1 - 2u_2 + u_3$, and $\varepsilon(\varepsilon_1, \varepsilon_2, \varepsilon_2) = u_1 - u_3$, the stability of synchronization manifold is decided by the asymptotic behavior of η and ε. At a linear approximation, the components of η and ε obey to

$$\begin{cases} \dot{\eta}_1 = \eta_2 - \eta_3 - 3K\eta_1, \\ \dot{\eta}_2 = \left[a - ch(x_1)\right]\eta_1 - b\eta_2 - c\eta_3, \\ \dot{\eta}_3 = \eta_1 + d\eta_2. \end{cases} \tag{20}$$

$$\begin{cases} \dot{\varepsilon}_1 = \varepsilon_2 - \varepsilon_3 - 3K\varepsilon_1, \\ \dot{\varepsilon}_2 = \left[a - ch(x_1)\right]\varepsilon_1 - b\varepsilon_2 - c\varepsilon_3, \\ \dot{\varepsilon}_3 = \varepsilon_1 + d\varepsilon_2. \end{cases} \tag{21}$$

The form of systems (20) and (21) brings the following comment: the three oscillators fall together in the synchronization. Proceeding in the same manner as the above subsection, we obtain the synchronization boundary as $K > K_{3\min}$ where

$$K_{3\min} = -\frac{c + b + \alpha d}{3cd}. \tag{22}$$

Figure 4 shows comparison between analytical result (full line) and numerical result (points and dashed-line). Numerically, we find the first value of K for which quantity $0.5(|x_1 - x_2| + |x_1 - x_3|) < 0.05$ for $t \geq 400$.

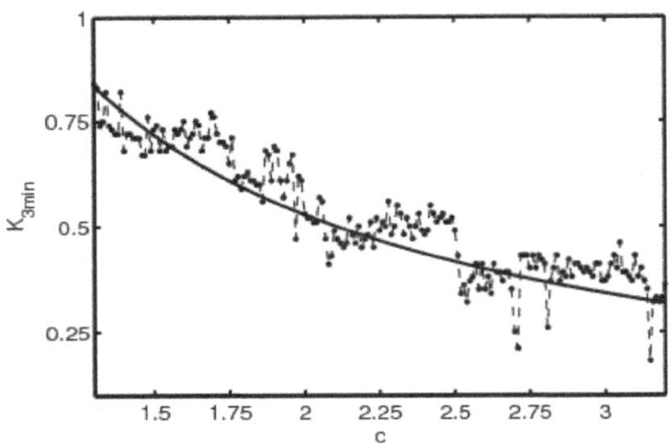

Figure 4. Synchronization boundaries in the case of three coupled oscillators

3.3. Synchronization in a Case of four Oscillators

Four systems are diffusely coupled in a ring structure with a coupling constant K. The dynamics of four diffusely coupled oscillators can be described by the following set of equations:

$$\begin{cases} \dot{x}_j = y_j - z_j + K\left(x_{j-1} - 2x_j + x_{j+1}\right), \\ \dot{y}_j = ax_j - by_j - z_j - sh\left(x_j\right), \\ \dot{z}_j = x_j + dy_j. \end{cases} \tag{23}$$

Index j (with $1 \leq j \leq 4$) represents the oscillator number and the two periodic boundary conditions $x_0 = x_4$ and $x_5 = x_1$ are used. Assuming the three following vectors defined as

$$\eta = (u_1 + u_3) - (u_2 + u_4), \ \varepsilon = u_1 - u_3 \ \text{and} \ \xi = u_2 - u_4. \tag{24}$$

The stability of synchronization manifold is decided by the asymptotic behavior of η, ε and ξ. At a linear approximation, the components of η, ε and ξ obey to

$$\dot{\eta} = M_1\eta, \quad \dot{\varepsilon} = M_2\varepsilon, \quad \dot{\xi} = M_2\xi \quad \text{where}$$

$$M_1 = \begin{pmatrix} -4K & 1 & -1 \\ \alpha & -b & -c \\ 1 & d & 0 \end{pmatrix} \text{and } M_2 = \begin{pmatrix} -2K & 1 & -1 \\ \alpha & -b & -c \\ 1 & d & 0 \end{pmatrix}. \tag{25}$$

Choosing $K_{4min} = -\dfrac{c + b + \alpha d}{4cd}$, if $K_{4min} < K < 2K_{4min} = K_{2min}$ the ring falls in the cluster synchronization ($u_1 \simeq u_3$, $u_2 \simeq u_4$ while $u_1 \neq u_2$). If $K > 2K_{4min}$ the complete synchronization ($u_1 \simeq u_2 \simeq u_3 \simeq u_4$) occurs in the ring. To verify our assumption, we plot as shown in Figure 5 our analytical result ($2K_{4min}$) as function of c and our numerical result obtained while recording the first value of K for which $\frac{1}{3}(|x_1 - x_2| + |x_1 - x_3| + |x_1 - x_4|) < 0.05$ (dashed lines).

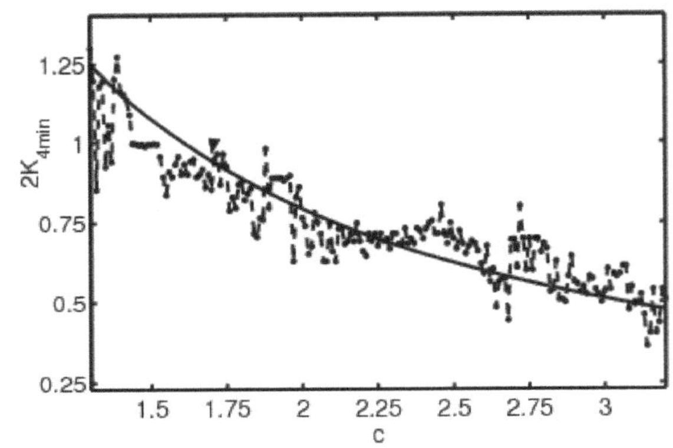

Figure 5. Synchronization boundaries in the case of four coupled oscillators

Although the cluster domain obtained analytically is verified numerically, the oscillators lost their chaotic state.

4. Experimental and Numerical Results

4.1. Experimental Setup

An experimental setup consisting of a network of four oscillators is shown in Figure 6. The networks of two and three oscillators can be derived from Figure 6 by choosing the suitable connections. In the set of equations (13), (19) and (23) the variables x_j, y_j and z_j are the voltages across the capacitors, C_{1j}, C_{2j} and C_{3j} respectively. The coupling strength between oscillators is controlled by four variable resistors R_{cj}. The circuits are built using (TL082) Operational Amplifiers, (BBY40) Diodes,

Capacitances and Resistances. The nominal values of the components can be found in Table 1. Due to the tolerances of the components, oscillators are slightly different. Therefore synchronization in the sense that $|u_i(t)-u_j(t)|=0$ is not possible and practical synchronization is defined as $|u_i(t)-u_j(t)|\le\delta$ with $\delta \ll 1$.

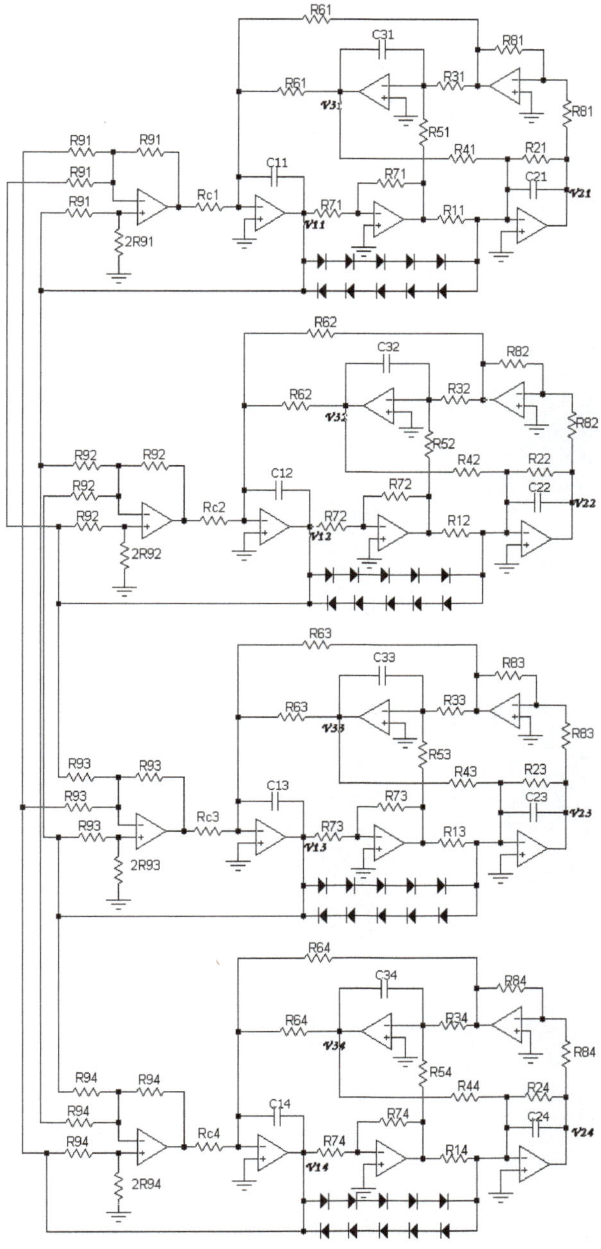

Figure 6. Electronic schematic of overall system

4.2 Experimental and numerical results

Before the coupling, Figure 7(a) shows a phase portrait of the first oscillator, which corresponds to a chaotic phase portrait (Figure 7(b)) obtained numerically with the following normalized parameters: $a_1 = 3.522$, $b_1 = 0.513$, $c_1 = 2.511$ and $d_1 = 1.264$.

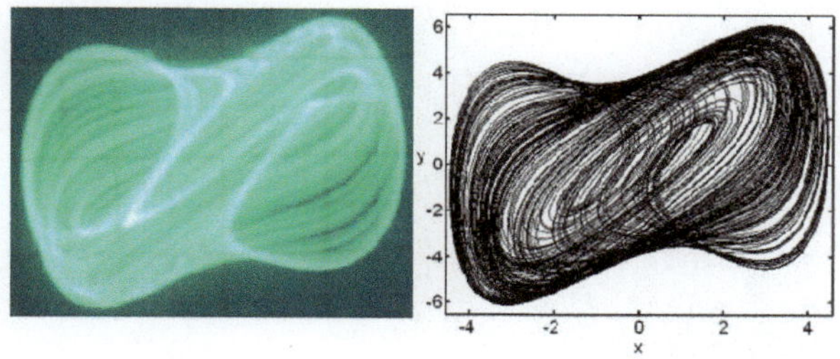

a) b)

Figure 7. Phase portrait of one oscillator before the coupling. a) Experimentally obtained, b) Numerically plot

Maintaining the same parameters used in Figure 7, we show in Figure 8 the phase portraits (x_1, x_2) to illustrate the absence of synchronization in the system before the coupling.

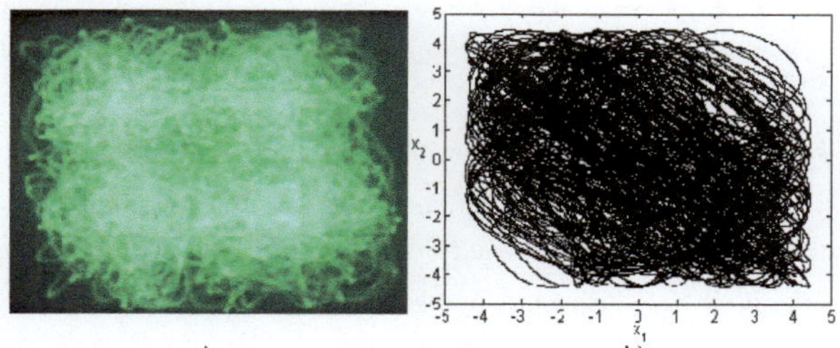

a) b)

Figure 8. Phase portrait in the plane (x_1, x_2) before the coupling. a) Experimental result, b) Numerical result

After setting the coupling, we decrease the values of the resistances R_{cj} to find experimentally the synchronization domain.

Table 1. Values of capacitors and resistances used in the overall electronic circuit

Resistances (Ω) and Capacitances (nF)	First Oscillator ($j=1$)	Second Oscillator ($j=2$)	Third Oscillator ($j=3$)	Fourth Oscillator ($j=4$)
C_{1j}	12.0	12.5	12.6	12.3
C_{2j}	192.8	201.2	202.6	198.0
C_{3j}	10.1	10.2	10.2	10.2
R_{1j}	177.4	170.2	168.8	173.1
R_{2j}	1218	1158	1154	1170
R_{3j}	9437	9329	9366	9337
R_{4j}	248.9	238.7	238.7	242.3
R_{5j}	11920	11810	11810	11820
R_{6j}	10040	10060	10050	10060
R_{7j}	35100	35140	35170	35110
R_{8j}, R_{9j}	$18\cdot10^4$	$18\cdot10^4$	$18\cdot10^4$	$18\cdot10^4$

4.2.1. Two Oscillators

In this case, we find that for $R_{cj} \le 400\Omega$, $(j=1,2)$, the system falls in the synchronization. Figure 9(a) and 9(b) illustrate the complete synchronization state of the ring for $R_{c1} = 315.0\Omega$, $R_{c2} = 318.7\Omega$ and $K_1 = 0.813$, $K_2 = 0.815$. Another used parameters are the following: $a_1 = 3.522, b_1 = 0.513, c_1 = 2.511, d_1 = 1.264$, $a_2 = 3.518,\ b_2 = 0.517, c_2 = 2.508$ and $d_2 = 1.266$.

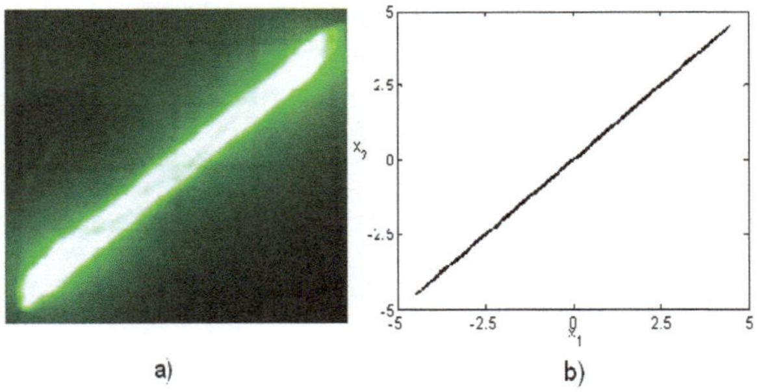

Figure 9. Phase portraits in the plane (x_1 , x_2) illustrating synchronization in the ring of two coupled oscillators. a) Experimental result, b) Numerically plot

4.2.2. Three oscillators

In the case of three coupled oscillators, when the coupling resistances satisfy $R_{cj} \leq 298\Omega$, $(j = 1,2,3)$, the ring is in the complete synchronization. To prove it, we plot in Figure 10 some phase portraits: Figure 10(a) (resp. Figure 10(c)) represents our experimental result in the plane (x_1, x_2) (resp. (x_1, x_3)) while Figure 10(b) (resp. Figure 10(d)) is its numerical analogous. The coupling resistances are $R_{c1} = 237.2\Omega$, $R_{c2} = 240.5\Omega$, $R_{c3} = 237.8\Omega$ and $K_1 = 0.53$, $K_2 = 0.52$, $K_3 = 0.55$. Other parameters are keep constant: $a_1 = 3.522, b_1 = 0.513,$ $c_1 = 2.511,$ $d_1 = 1.264$, $a_2 = 3.518,$ $b_2 = 0.517,$ $c_2 = 2.508$, $d_2 = 1.266$, $a_3 = 3.525,$ $b_3 = 0.515,$ $c_3 = 2.513$ and $d_3 = 1.261$.

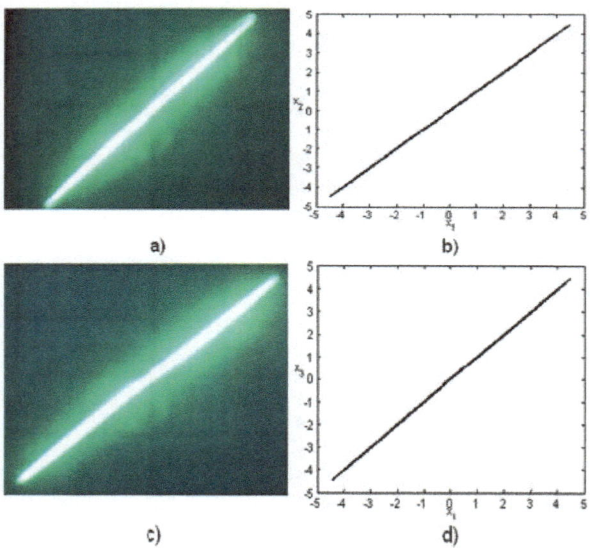

Figure 10. Phase portraits illustrating the complete synchronization in the ring of three coupled oscillators. a) Experimental result in the plane (x_1 , x_2), b) Numerically plot in the plane (x_1 , x_2), c) Experimental result in the plane (x_1 , x_3), d) Numerically plot in the plane (x_1 , x_3)

4.2.3. Four Oscillators

Although the analytical study foresaw a cluster synchronization, the numerical and the experimental studies showed only a complete synchronization state. This is obtain experimentally when $R_{cj} \leq 345\Omega$, $(j = 1,2,3,4)$. Proceeding as the above subsection, we plot in Figure 11 different phase portraits: Figure 11(a), Figure 11(c) and Figure 11(e) represent our experimental results drawn in the planes (x_1, x_2), (x_1, x_3) and (x_1, x_4) respectively, while Figure 11 (b), Figure 11 (d) and Figure 11 (f) are their numerical corresponding. These following coupling parameters are used: $R_{c1} = 327.2\Omega$, $R_{c1} = 325.5\Omega$, $R_{c1} = 328.4\Omega$, $R_{c1} = 324.8\Omega$ and $K_1 = 8.03$, $K_2 = 8.025$, $K_3 = 8.027$, $K_4 = 8.031$, $a_1 = 3.522,$ $b_1 = 0.513,$ $c_1 = 2.511,$ $d_1 = 1.264$, $a_2 = 3.518,$ $b_2 = 0.517,$ $c_2 = 2.508$, $d_2 = 1.266$, $a_3 = 3.525, b_3 = 0.515, c_3 = 2.513$, $d_3 = 1.261$, $a_4 = 3.515,$ $b_4 = 0.520,$ $c_4 = 2.511$ and $d_4 = 1.265$.

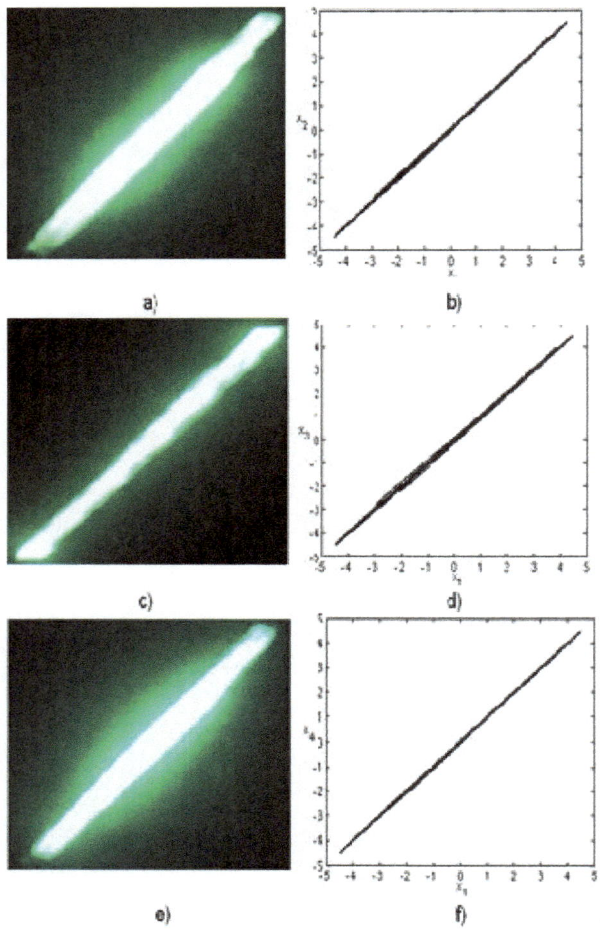

Figure 11. Phase portraits illustrating the complete synchronization in the ring of four coupled oscillators. a) Experimental result in the plane (x_1 , x_2), b) Numerically obtained in the plane (x_1 , x_2), c) Experimental result in the plane (x_1 , x_3), d) Numerically obtained in the plane (x_1 , x_3), e) Experimental result in the plane (x_1 , x_4), f) Numerically obtained in the plane (x_1 , x_4)

5. Conclusion

In this paper, theoretical and experimental complete synchronization of diffusively two, three and four coupled oscillators are presented. With the experimental setup it is impossible to achieve a zero synchronization error due to the tolerances of the electrical components. We obtain that three and four diffusely coupled oscillators synchronized or desynchronized together, provided initial values are chosen in the vicinity of the synchronization manifold. Despite the fact that the single harmonic response give in equation (9) may be questionable, the analytical treatment gives a good indication on the boundary of K for synchronization to be achieved. The presented experimental results are qualitative comparable with numerical simulations.

Acknowledgement

This work is supported by the Academy of Sciences for the Developing World (TWAS) under Research grant N.03-322 RG/PHYS/AF/AC.

References

[1] Ling, L. and Le, M, "Parameter identification and synchronization of spatiotemporal chaos in uncertain complex network," Nonlinear Dynamics 66 (4). 489-495. Dec. 2011.

[2] Carroll, T.L. and Pecora, L.M, "Synchronizing chaotic circuits," IEEE Trans Circ Syst 38 (2). 453-456. Apr. 1991.

[3] Yu P.E. and Kuznetsov A.P, "Synchronization of coupled van der pole and Kislov-Dmitriev self-oscillators," Technical Physics 56, (4). 435-442. Apr. 2011.

[4] Peng J.H., Ding, E.J., Ding, M. and Yang, W, "Synchronizing hyperchaos with a scalar transmitted signal," Phys Rev Lett 76 (6). 904-907. 1996.

[5] Nijmeijer H, "Control of chaos and synchronization," Systems and Control Letters 31 (5). 259-262. Oct. 1997.

[6] Kennedy M.P., Rovatti R. and Setti, G, "Chaotic Electronics in Telecommunications," CRC Press, USA. 2000.

[7] Gamal M.M., Shaban A.A. and AL-Kashif M.A, "Dynamical properties and chaos synchronization of a new chaotic complex nonlinear system," Nonlinear Dynamics. 51, 171-181 Oct. 2008.

[8] Tanmoy B., Debabrata B. and Sarkar B.C, "Complete and generalized synchronization of chaos and hyperchaos in a coupled first-order time-delayed system," Nonlinear Dynamics. 71 (1). 279-290. Jan. 2013.

[9] Buric N. and Vasovic N, "Global stability of synchronization between delay-differential systems with generalized diffusive," Chaos Solitons and Fractals. 31 (2). 336-342. 2007.

[10] Fotsin H. and Bowong S, "Adaptive control and synchronization of chaotic systems consisting of Van der Pol oscillators coupled to linear oscillators," Chaos Solitons and Fractals. 27 3). 822-835. Apr. 2006.

[11] Carroll T.L. and Pecora L.M, "Nonlinear Dynamics in Circuits," World Scientific Publishing, Singapore. 1995.

[12] Juan C., Jun-an L. and Xiaoqun W, "Bidirectionally coupled synchronization of the generalized Lorenz systems," Journal of Systems Science and Complexity 24 (3). 433-448. Jun. 2011.

[13] Pecora L.M. and Carroll T.L, "Driving systems with chaotic signals," Physl Rev A. 44 (4). 2374-2383. Aug. 1991.

[14] Cuomo K.M. and Oppenheim A.V, "Circuit implementation of synchronized chaoswith application to communications," Phys Rev Lett. 71 (1). 65-68. Jul. 1993.

[15] Nana B. and Woafo P, "Synchronization in a ring of four mutually coupled Van der pol oscillator: Theory and experiment," Phys Rev E 74, 1-8. 2006.

[16] Yu Y.H., Kwak K., Lim T.K, "Synchronization via small continuous feedback," Phys Lett A. 191 (3). 233-237. 1994.

[17] Li-feng Z., Xin-lei A. and Jian-gang Z, "A new chaos synchronization scheme and its application to secure communications," Nonlinear Dynamics. 73 (2). 705-722. Jan. 2013.

[18] Woafo P. and Kraenkel R.K, "Synchronization stability and duration time," Phys Rev E. 66. 25-31. 2002.

A Digital Filter for Electrical Drive with Elastic Shaft

Jumber Dochviri*, Oleg Khachapuridze, Nana Beradze

Department of Electrical Engineering, Faculty of Power Engineering and Telecommunications, Georgian Technical University, M. Kostava str., Tbilisi, Georgia

*Corresponding author: Jumber_Dochviri@yahoo.com

Abstract In the paper analytically is constructed digital filter for DC-electrical drive with elastic junctions. It could be realized with two discrete integrators and five operational amplifiers. The filter would be insert sequentially on the output of the speed's regulator. As results transient processes similar to transients of dc-electric drive with rigid mechanical shaft are obtained.

Keywords: *DC -thyristor electrical drive, elastic shaft, digital filter*

1. Introduction

On the power technological machines, including metallurgical rolling mills and paper machines with the purpse of the reliability mainly are used thyristor dc electric drives. Because mentioned technological machines have long mechanical shafts in their dynamics frequently arise strong elastic vibrations. In the continuous control systems in practice are widely applied soft differential feedbacks with R-C circuits to damping oscillations in the dynamics, as well as filters with R-L-C elements [1,2]. Such filters are not suitable for digitally controled elastic dc drives and we get naturally raised problem.

2. Analytical Design of a Digital Filter

To solve the problem we suggest digital filter based on integrators and operational amplifiers similar to the Kalman-Bucy filter [3,4]. To construct analitically such a filter first we write equations of movement of elastic electrical drive due to D'alambert's principle:

$$\begin{cases} M - M_e = J_1 \dfrac{d\omega_1}{dt}; \\ M_e - M_M = J_2 \dfrac{d\omega_2}{dt}; \\ M_e = c_{12} \int (\omega_1 - \omega_2) dt + b(\omega_1 - \omega_2), \end{cases} \quad (1)$$

where M, M_e and M_M are moment of rotation of the dc motor, elastic moment of mechanical transmission shaft and resistance moment of the mechanism reduced on the shaft of the motor, respectively; J_1 and J_2 are moments of inertia of the drive and mechanism; ω_1 and ω_2 are angular speeds of the motor and mechanism; c_{12} is coefficient of elasticity of long mechanical shaft; b is damping coefficient of torsional vibrations of inner viscous friction of long shaft.

If we write the variables in the relative increments and use simple transformations in the (1) we get:

$$\begin{cases} \dfrac{d\Delta\nu_1}{dt} = \dfrac{1}{T_1}(\Delta\mu - \Delta\mu_e); \\ \dfrac{d\Delta\nu_2}{dt} = \dfrac{1}{T_2}(\Delta\mu_e - \Delta\mu_M); \\ \dfrac{d\Delta\mu_e}{dt} = \dfrac{1}{T_c}(\Delta\nu_1 - \Delta\nu_2) - \dfrac{T_d}{T_c} \cdot \left(\dfrac{1}{T_1} + \dfrac{1}{T_2}\right)\Delta\mu_e, \end{cases} \quad (2)$$

where $T_1 = \dfrac{J_1\omega_{ST}}{M_{ST}}$ and $T_2 = \dfrac{J_2\omega_{ST}}{M_{ST}}$ are mechanical time constants of the motor and mechanism, respectively; $T_c = \dfrac{M_{ST}}{c_{12} \cdot \omega_{ST}}$ and $T_d = \dfrac{b}{c_{12}}$ are time constants of long (elastic) shaft; M_{ST} and ω_{ST} are base (static) values of the moment of resistance and angular speed of the drive of mechanism.

In the Cauchy form the system of equations (2) can be write as following:

$$\begin{cases} \dfrac{dx}{dt} = A \cdot x + B \cdot u; \\ y = Cx, \end{cases} \quad (3)$$

where y is output signal of the object (angular speed of the motor);

$$A = \begin{bmatrix} 0 & \dfrac{1}{T_2} & 0 \\ -\dfrac{1}{T_c} & -\dfrac{T_\Sigma T_d}{T_1 T_2 T_C} & \dfrac{1}{T_C} \\ 0 & -\dfrac{1}{T_1} & 0 \end{bmatrix};$$

$$B = \begin{bmatrix} 0 \\ \frac{T_d}{T_1 T_C} \\ \frac{1}{T_1} \end{bmatrix}; x = \begin{bmatrix} x_1 \\ x_2 \\ x_3 \end{bmatrix}; \quad (4)$$

$C = [0\ 0\ 1]$; x_1, x_2 and x_3 - are state variables of the object, namely x_1-angular speed of the motor; x_2-elastic moment of the long shaft; x_3 - angular speed of the mechanism; $T_\Sigma = T_1 + T_2$-total mechanical time constant of the drive.

To obtain transfer function of regulator first of we should define transfer function of the object via input signal u and output signal $y = x_3$. With this purpose let us write the system of equations given by (3) in the operational form ($s = d/dt$) as following:

$$\begin{cases} sX(s) = A \cdot X(s) + B \cdot U(s); \\ Y(s) = C \cdot X(s), \end{cases} \quad (5)$$

From the first equation of (5) we can write:

$$X(s) = (s \cdot E - A)^{-1} B \cdot U(s), \quad (6)$$

Then from the (5) we get

$$Y(s) = C(s \cdot E - A)^{-1} \cdot B \cdot U(s). \quad (7)$$

Using inner current feedback and matrices from (4), an approximate transfer function of the object of control system could be given as following:

$$\theta_{OB}(s) = \frac{Y(s)}{U(s)} = \frac{X_3(s)}{U(s)} \approx \frac{1}{k_t \cdot T_\Sigma \cdot s} \cdot \frac{T_2^{/2} \cdot s^2 + T_d \cdot s + 1}{T_1^{/2} \cdot s^2 + T_d \cdot s + 1}, \quad (8)$$

where $T_1^{/} = (T_1 \cdot T_2 \cdot T_C / T_\Sigma)^{1/2}$; $T_2^{/} = (T_2 \cdot T_C)^{1/2}$.

By substituting numerical values of the parameters $T_1 = 1.5$ sec.; $T_2 = 10$ sec; $T_d = 0.002$ sec; $T_c = 0.0004$ sec; $T_1 = 1.5$ sec and $k_t = 0.1$ for the electrical drive of press of paper machine we get:

$$\theta_{OB}(s) = \frac{0.004 \cdot s^2 + 0.002 \cdot s + 1}{s \cdot (0.0006 \cdot s^2 + 0.0025 \cdot s + 1.15)}. \quad (9)$$

Expressing transfer function of (9) in the discrete z-transform form we use well-known [5] approximation $z = e^{T_0 s} \Rightarrow z \approx 1 + T_0 s \Rightarrow s = (z-1)/T_0$, where sample time $T_0 = 0.01$ sec. and then we have:

$$\theta_{OB}(z) = \frac{0.06367 \cdot z^2 - 0.1254 \cdot z + 0.06334}{z^3 - 2.774 \cdot z^2 + 2.734 \cdot z - 0.9592}. \quad (10)$$

On Figure 1,a are given Bode diagrams simulated corresponding to the transfer function (10).

In the expression (8) second multiplier denotes transfer function of elastic link of the object of dc drive. Compensation of its action could be realized via inversely given digital filter:

$$\theta_F(s) = \frac{T_1^{/2} \cdot s^2 + T_d \cdot s + 1}{T_2^{/2} \cdot s^2 + T_d \cdot s + 1} = \frac{0.00053 \cdot s^2 + 0.002 \cdot s + 1}{0.004 \cdot s^2 + 0.002 \cdot s + 1}. \quad (11)$$

z-transform of the (11) similarly to (10), has following view:

$$\theta_f(z) = \frac{0.1325 \cdot z^2 - 0.2459 \cdot z + 0.1383}{z^2 - 1.97 \cdot z + 0.995}. \quad (12)$$

On Figure 1,b are given Bode diagrams corresponding to transfer function (12). To purge resonant peak on the L_f -logarithmic amplitude frequency response, it is necessary second summand in denominator increase 20-times. Therfore, adjusted transfer function of digital filter could be write as following:

$$\theta_F(s) = \frac{0.00053 \cdot s^2 + 0.002 \cdot s + 1}{0.004 \cdot s^2 + 0.04 \cdot s + 1}. \quad (13)$$

Transfer function (13) in z-transform form looks as following:

$$\theta_F(z) = \frac{U_2(z)}{U_1(z)} = \frac{0.1325 \cdot z^2 - 0.2466 \cdot z + 0.1378}{z^2 - 1.881 \cdot z + 0.9048}, \quad (14)$$

where $U_1(z)$ and $U_2(z)$ are input and output signals of the filter.

On Figure 1,c Bode diagrams corresponding to (14) are presented, where resonant peak is suppressed.

To plot Bode diagrams (Figure 1,d) of the external open loop of speed we should use following transfer function:

$$\vartheta_{OP.} = \theta_{OB}(z) \cdot \vartheta_F(z) \cdot \vartheta_{SR}(z) = \frac{\begin{pmatrix} 0.06 \cdot z^5 - 0.28 \cdot z^4 + 0.56 \cdot z^3 \\ -0.91 \cdot z^2 + 0.28 \cdot z - 0.06 \end{pmatrix}}{\begin{pmatrix} z^6 - 5.68 \cdot z^5 + 13.56 \cdot z^4 - 17.48 \cdot z^3 \\ +12.84 \cdot z^2 - 5.12 \cdot z + 0.86 \end{pmatrix}}, \quad (15)$$

where $\vartheta_{SR}(z) = \beta \cdot \frac{z-a}{z-1} = 5 \cdot \frac{z-0.975}{z-1}$ is PI-type digital transfer function of speed regulator which is obtained from $\vartheta_{SR}(s) = \beta \cdot \frac{\tau s + 1}{\tau s} = 5 \cdot \frac{0.4 \cdot s + 1}{0.4 \cdot s}$ via Matlab. On Bode diagrams L_{OP} and ϕ_{OP} are not resonant peaks that usually are features of the elesaticities.

To construct electrical scheme of the filter from (14) we should rewrite it as following:

$$U_2 = \frac{1}{z}\begin{bmatrix} (0.1378 \cdot U_1 - 0.9048 \cdot U_2)\frac{1}{z} \\ +(1.881 \cdot U_2 - 0.2466 \cdot U_1) \end{bmatrix} + 0.1325 \cdot U_1. \quad (16)$$

Digital block diagram with two digital integrators and five operational amplifiers are presented on the computer based block digram (Figure 2) as separate part, which is involved in the output of digital speed regulator (SR) of the electrical drive.

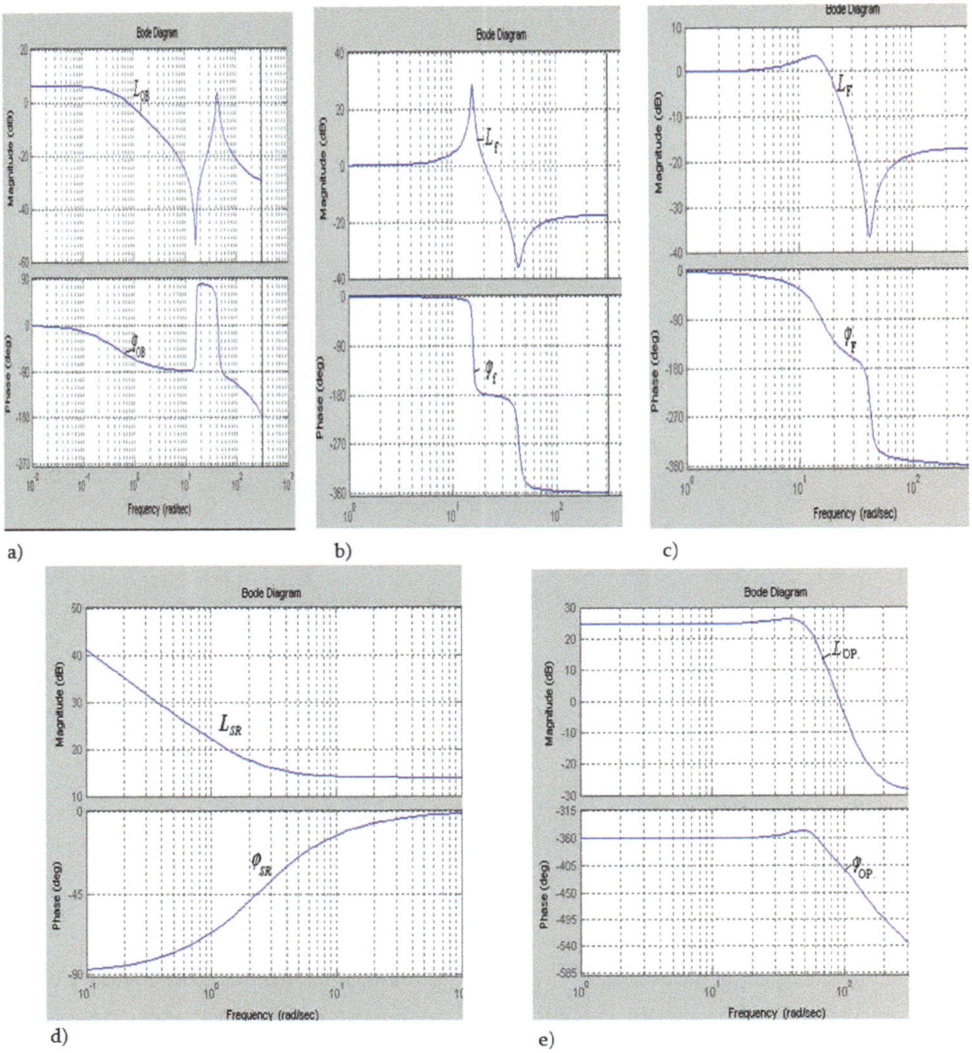

Figure 1. Bode diagrams: a) of the object of control system; b) of unadjusted filter according to (12); c) of adjusted filter according to (14); d) of the open loop of speed contour (together with regulator)

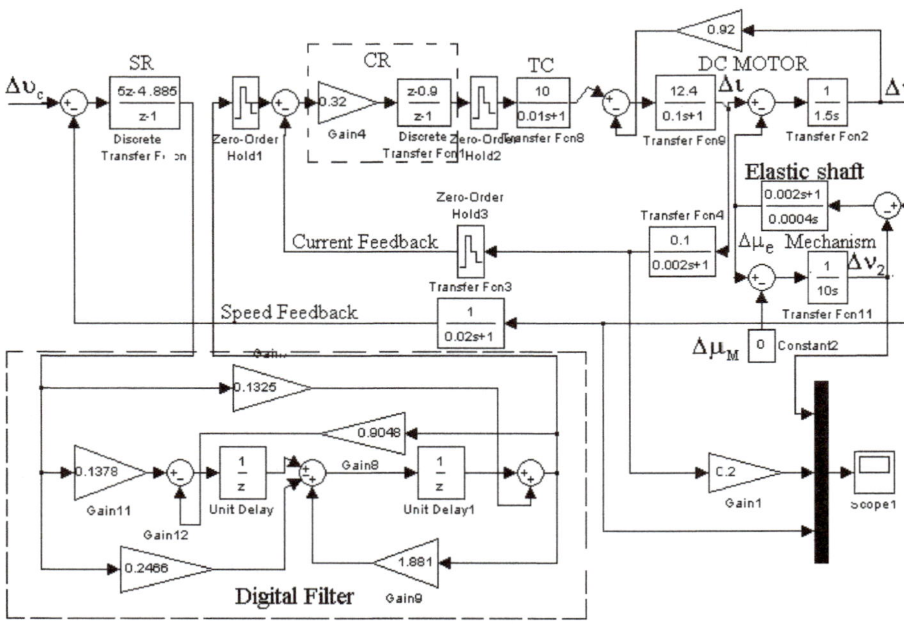

Figure 2. Numerical block diagram of elastic thyristor dc electric drive under digital control

On Figure 2 following notations are used: $\Delta\upsilon_c$ denotes relative increment of control (input) signal; SR and CR are digital regulators of speed and current; TC is thyristor converter; $\Delta\iota$ denotes relative increment of motor anchor; $\Delta\mu_M$ is moment of resistance of the drive from mechanism.

Investigations of dynamical modes using programm MatLab shows efficiency of suggested digital filter. Qualititive characteristics of transients of constructed digital filter ensures work of elastic drive system almost similar to the drive system with short (hard) shaft: $\sigma = 45\%$, $t_T \approx 1\sec$ and $\Delta\nu_{dyn} = 0.015$. It must noticed that constructed filter is better than feedbacks on the base R-C circuits. Such type of discrete filter completely cleans the power mechanical part of the drive systems from adverse impulsive strikes, for this reason we explore beginings of the speed transient curves (Figure 4, a,b,c - [6]).

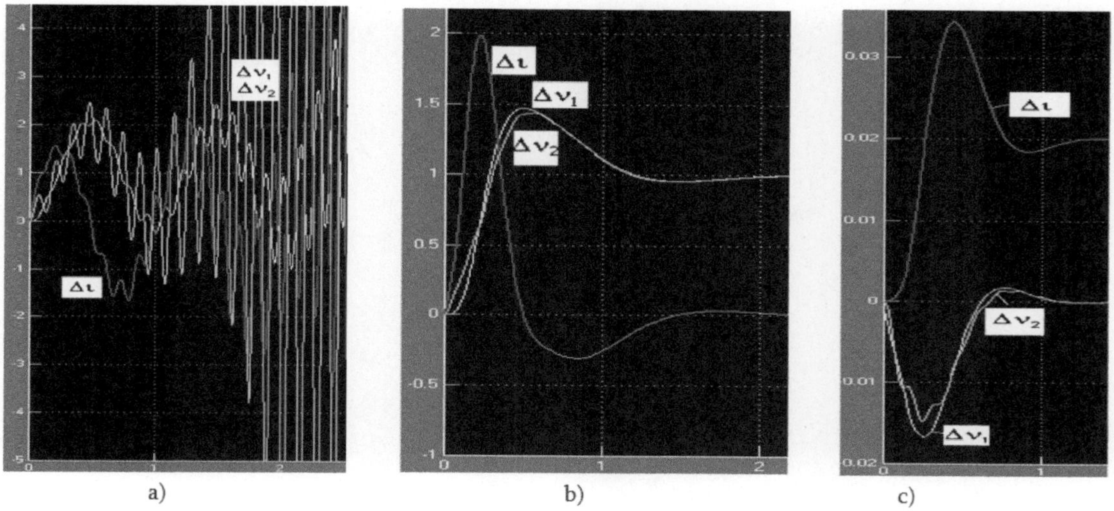

Figure 3. Transient curves of thyristor electrical drive with elastic shaft under digital control: a) without digital filter; With digital filter: b) under stepwise change of $\Delta\upsilon_c$ control signal; c) under stepwise change of loading $\Delta\mu_M$

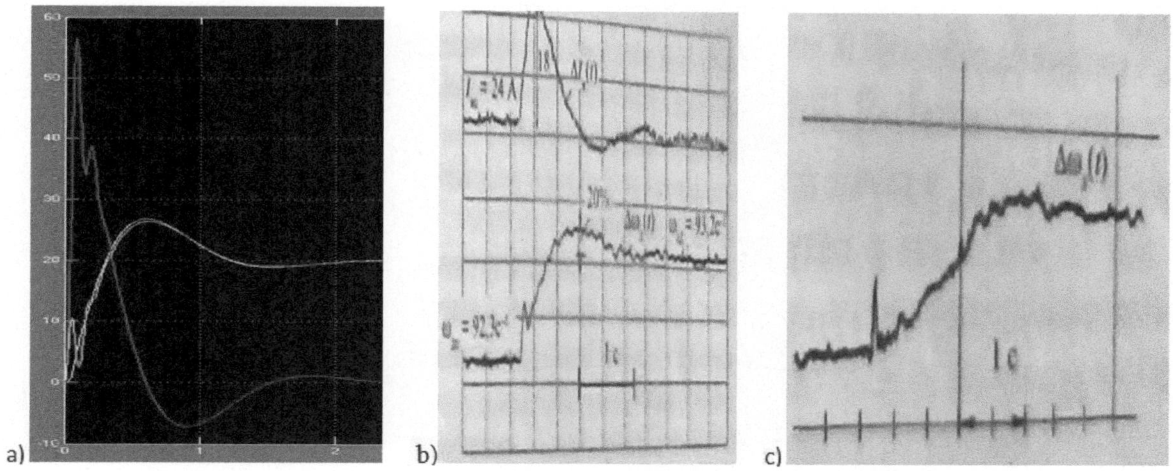

Figure 4. Transient processes of dc elastic electrical drive with feedback using R-C circuits: a) obtained via Matlab simulations; b) and c) are experimental results on the acting paper machine

3. Conclusions

In the work we suggest new type digital filter based on discrete integrators to damping elastic vibrations of signals in dynamical modes of dc drives. For tuning of optimal parameters frequency characteristic method is used. Transient curves similar to non-elastic (with short, i.e. hard shaft) electrical drive are obtained $\sigma = 45\%$, $t_T \approx 1.0\sec$, $\Delta\nu_{dyn} = 0.015$.

Acknowledgements

The first author has been supported by the grant FR/7/3-160/14 from Shota Rustaveli National Science Foundation of Georgia.

References

[1] V.D. BARYSHNIKOV, Y.A. BORTSOV et al. -Dynamics of thyristor electrical drives of the sections of paper machines // *J. Russian Electrical Technology.* 1973, vol.1, pp. 43-48.

[2] E. RAATZ -Regelung von antrieben mit Elastischer verbindung zur arbeitsmaschinen // ENZ, 1981, A92, N4, s. 211-216

[3] A. ANTONIOU -Digital Filters: Analysis, Design, and Applications // McGraw-Hill, New York, NY, 1993.

[4] J. DOCHVIRI -Kalman-Frobenius dynamic optimization of a thyristor drive with an elastic component //Allerton press, USA, *J. Russian Electrical Engineering, 2004,vol.75(5),pp.7-17.*

[5] K. OGATA -Discrete-time control systems. Upper Saddle River, N.J.: Prentice hall, 1994.

[6] J. DOCHVIRI -The optimization of transients in multi-motor thyristor electric drives with elastic links for the press mechanisms of continuous technological machines // *J. Russian Electrical Technology*. 2006, v. 2, pp. 34-42.

26

Design and Performance Evaluation of Microstrip Antenna for Ultra-Wideband Applications Using Microstrip Feed

Gourav Misra[1,*], Arun Agarwal[1], Kabita Agarwal[2]

[1]Department of ECE, Faculty of Engineering & Technology, Siksha 'O' Anusandhan University, Khandagiri Sqare, Bhubaneswar, Odisha, India
[2]CV Raman College of Engineering, Department of ETC, Bhubaneswar, Odisha, India
*Corresponding author: gourav.misra.ima@gmail.com

Abstract In the modern days the development in communication systems engineering requires the development of low profile antennas that are capable of maintaining high performance over a wide spectrum of frequencies. In this review, we have designed an elementary Microstrip patch Antenna for ultra-wideband application. This technological trend has focused much effort on the basic concepts, characteristics and design of a Microstrip patch antenna. This paper explains the pattern of designs of the Microstrip patch antenna along with detailed study. A thorough analysis and observations of the problem begins with the application of the equivalence principle that introduces the unknown electric and magnetic surface current densities on the dielectric surface. Here in this technical review, our main goal is to design an elementary Microstrip antenna, which will work in the ultra-wide band range and also it should work in the desired operating frequency. Ansoft High Frequency Structured Simulations (HFSS) software is used for the elementary design of a rectangular patch Microstrip antenna and also some of the basic calculations has been done by using MATLAB. This review paper also comprises of basic parameters, types of antenna, working, fundamental concepts and characteristics as well as feeding techniques and simulation, results of Microstrip patch antenna.

Keywords: *low profile, microstrip patch antenna, electric and magnetic surface current densities, radiation problems, HFSS, MATLAB, feeding techniques, simulations, results*

1. Introduction

An Antenna is an impedance matching device or in other words we can say that an Antenna is a transitional structure between free space and guiding device. The transmission line may take the form of a coaxial line or a hollow pipe i.e. called a wave guide, and it is used to transport electromagnetic energy from the transmitting source to the antenna or from the antenna to the receiver. Antenna converts electrical signal to electromagnetic waves or radio waves. So we can also say that antenna is a microwave transducer because it deals with microwave frequencies (300MHz-300GHz). In transmission, a radio transmitter supplies an electric current oscillating at radio frequency i.e. a high frequency alternating current to the antenna's terminals. The antenna emits the energy from the current as electromagnetic waves or radio waves or microwaves. In reception, an antenna intercepts some of the power of an electromagnetic wave in order to produce a voltage at its terminals.

Antennas are very much essential components of all equipment that are using radio. They are widely used in different types of communication systems such as TV broadcasting, radar, cellphones, and satellite communications, as well as other devices.

Antenna consists of an arrangement of number of conductors and they are connected electrically to the transmitter or receiver. An oscillating current of electrons imposed through the antenna by a transmitter will create an oscillating magnetic field around the antenna elements, while the charge of the electrons will also create an oscillating electric field along the elements and these time-varying fields discharge away from the antenna into space as a moving transverse electromagnetic field wave. Again at the time of reception, the oscillating electric fields and magnetic fields of an incoming radio wave or microwave apply force on the electrons in the antenna elements, prompting them to move back and forward, creating oscillating currents in the antenna [1].

Antennas could be designed to transmit or receive radio waves or microwaves or electromagnetic waves in all horizontal directions equally or partially in a distinct direction. In the next step, it may happen that the antenna may include some supplementary elements or surfaces with no any kind of electrical connection to the transmitter or receiver and this will serve to direct the electromagnetic waves into a beam or other desired radiation pattern. In [1], the authors have described that an antenna is defined by Webster's Dictionary as "a usually metallic device (as a rod or wire) for radiating or receiving radio waves." In this review paper, we have mentioned some of the basic parameters, types of antenna as well as basic characteristics and simulations, results of an elementary Microstrip patch antenna which has been designed by using HFSS software.

2. Basic Parameters of Antenna

2.1. Introduction

To explain the performance and characteristics of an antenna, various fundamental parameters are required. Some of the parameters are affiliated and not all of them need be specified for complete explanation of the antenna performance. Parameter definitions will be given in this chapter. Many of those in citation marks are from the IEEE Standard Definitions of Terms for Antennas (IEEE Std 145-1983). This is a revision of the IEEE standard 145-1973. Beamwidth, Directivity, Gain, Radiation Pattern, Return loss etc. are some if the basic or fundamental parameters of an Antenna.

2.2. Beamwidth

The beamwidth of a pattern is defined as the angular separation between two identical points on opposite side of the maximum pattern of the lobe. In an antenna pattern, there are a number of beamwidths. One of the most commonly used beamwidths is the Half-Power Beamwidth (HPBW). Another important beamwidth is the angular separation between the first nulls of the pattern, and it is known as the First-Null Beamwidth (FNBW). In a radio antenna pattern, the half power beamwidth is an angel between the half-power points of the main lobe. In below Figure 1 the beamwidth of an antenna's mentioned.

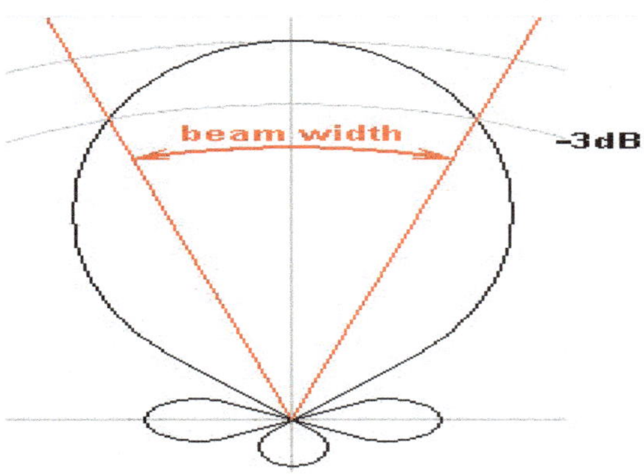

Figure 1. Beamwidth of an Antenna [26]

2.3. Directivity

Directivity of an antenna is defined as the ratio of the radiation intensity in a given direction from the antenna to the radiation intensity averaged over all directions. The radiation intensity is equal to the total power radiated by the antenna divided by 4π. Directivity is the figure of merit of an antenna. It measures the power density an antenna radiates in the direction of its powerful or robust emission versus the power density radiated by a faultless or supreme isotropic radiator radiating the same total power.

2.4. Gain

A comparative measure of an antenna's ability to direct radio frequency energy in a particular direction. Basically the measurement is done in dB

2.5. Radiation Pattern

Radiation pattern is a graphical illustration of the relative field strength transmitted from the antenna or received by the antenna. The radiation patterns of Antennas are taken at one frequency, one polarization, and one cut.

2.6. Return Loss

In [1], it is mentioned that the return loss is the loss of signal power resulting from the reflection caused at a disruption in a transmission line. This disruption can be miss-match with the terminating load or with a device inserted in the line.

3. Types of Antenna

3.1. Wire Antenna

Wire antennas are very much well known to the layman because they are seen effectively everywhere— on automobiles, buildings, aircraft, various spacecraft, and so on. There are various shapes of wire antennas such as a straight wire (dipole), loop, and helix. Loop antennas need not only be circular. The circular loop is the most common because of its simplicity in design. A wire antenna (loop antenna) is drawn below (Figure 2).

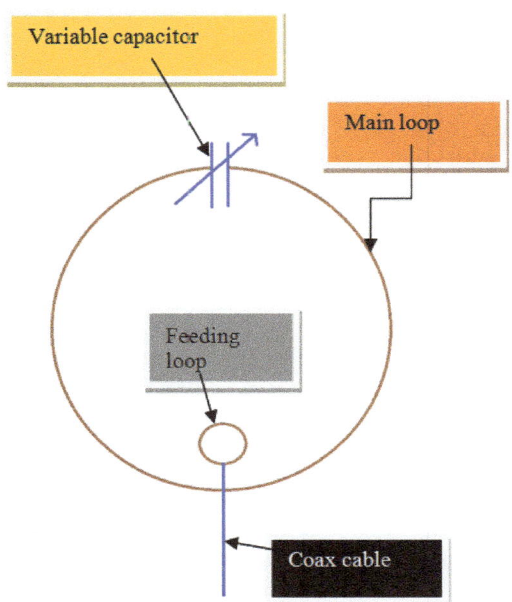

Figure 2. Wire Antenna (Loop Antenna)

3.2. Aperture Antenna

Aperture antenna (mentioned below in Figure 3) may be more accustomed to the common man today than in the past because of the increasing demand for more enlightened forms of antennas and the utilization of very high frequencies. Antennas of this type are very useful for Aircraft and spacecraft applications, because they can be mounted on the skin of the aircraft or spacecraft. In addition, they are covered with a dielectric material to protect them from various hazardous conditions.

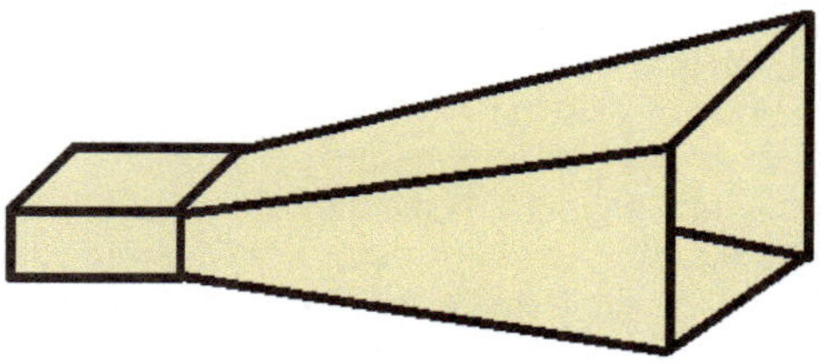

Figure 3. Pyramidal Horn Antenna

3.3. Microstrip Antenna

In [1], it is mentioned that these antennas consist of a metallic patch on a grounded substrate. The micro strip antennas are low profile, compatible to planar and non-planar surfaces, simple and very less expensive to manufacture using modern printed-circuit technology, mechanically strong and long-lasting when scaled on hard or inflexible surfaces, well matched with MMIC designs. The Microstrip antennas (mentioned below in Figure 4) can be scaled on the surface of very high-performance spacecraft, satellites, missiles, cars, and even handheld mobile phones or cellphones.

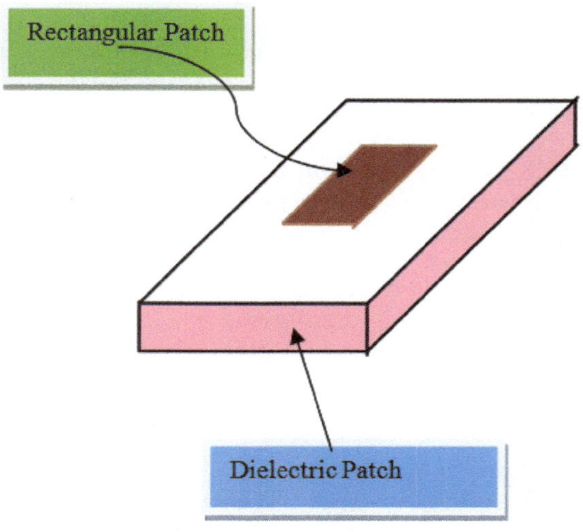

Figure 4. Rectangular Patch Microstrip Antenna

3.4. Array Antenna

Many applications require radiation characteristics that may not be achievable by a single element. Array antenna (mentioned below in Figure 5) is the combination of multiple antennas. The arrangement of the array may be such that the radiation from the elements sums up to give a radiation maximum in a particular direction, minimum in other directions.

3.5. Reflector Antenna

Reflector antennas or parabolic reflector antennas (mentioned below in Figure 6) has largest applications. Reflector antenna is the combination of a feed antenna and a metallic plate which is parabolic in structure, in which the metallic plate acts as the primary antenna and the feed antenna or horn antenna acts as the secondary antenna. These antennas are mainly used in TV broadcasting, radio broadcasting, satellite communications etc.

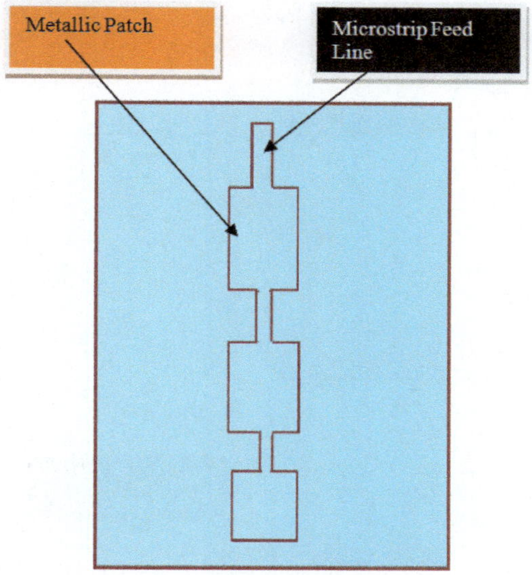

Figure 5. Top View of Microstrip Antenna (Array of rectangular patch)

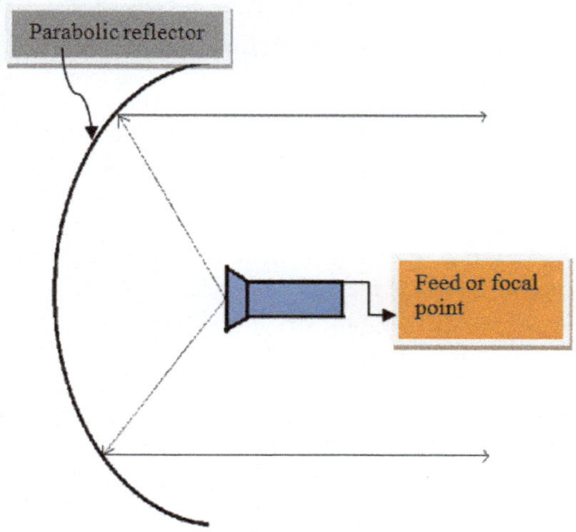

Figure 6. Parabolic Reflector Antenna

4. Microstrip Patch Antenna

The microstrip antenna concept was first proposed in the 1950s. Due to the development of printed-circuit board (PCB) technology, many practical applications of microstrip antennas scaled on missiles and aircraft were indicated in the early 1970s. In [2] it is mentioned that, Microstrip antennas can be used to meet some specific requirements in commercial applications, for example in the field of wireless communication and mobile radio etc. Since then, the study of microstrip antennas has resonated, giving birth to a new antenna industry.

Figure 7 shows the basic geometry of a microstrip antenna: a metallic patch is printed on a grounded dielectric substrate. The metallic patch could be of any shape or size, but in practical applications, rectangular and circular patches are most common. Because of its simple geometry, the microstrip antenna offers many striking advantages, such as low profile, and easy manufacturing etc.

Some major disadvantages of microstrip patch antennas are their low efficiency, low power, high Q, poor polarization purity, poor scan performance, false feed radiation and very narrow frequency bandwidth, which is typically only a fragment of a percent.

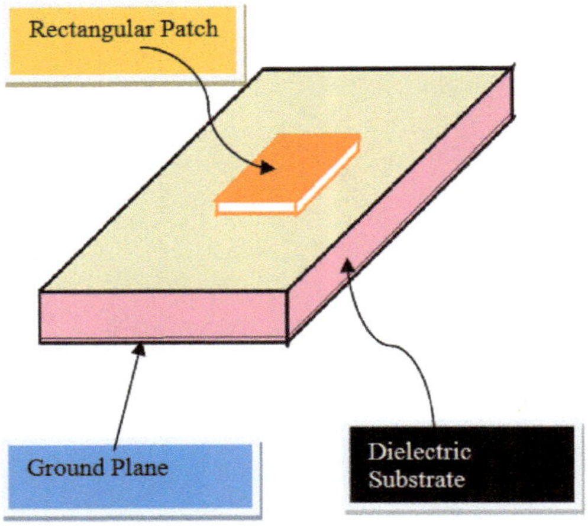

Figure 7. Microstrip patch Antenna

5. Working, Fundamental Concepts, and Characteristics of Micropstrip Antenna

The working of Microstrip patch antenna includes a simple concept. In Micristrip antennas the microstrip feed lines are used for feeding purpose. So when we give a supply (current) through the feed, the electrons or current is accelerated in the feed line and reach in the metallic patch. After that they are radiated in the edge of the metallic plate because there is a property of electron is that, electron will start radiating after the breaking of path i.e. after a discontinuity in the conducting surface, it will radiate. So here also after breaking of path (at the edge) the electrons are radiating (as shown in the Figure 8 which is mentioned below).

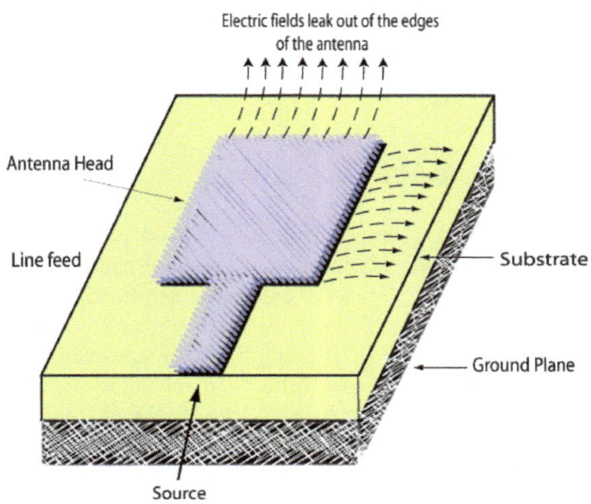

Figure 8. Radiation of Microstrip patch Antenna [27]

Figure 8 consists of very thin ($t << \lambda 0$) metallic patch placed a small fraction of wavelength above the ground plane. Here $\lambda 0$ is the free space wavelength. The microstrip antenna is designed so its radiation pattern maximum is perpendicular to the patch and in the rectangular patch antenna, the length L of the rectangular patch should be with in $\lambda 0/3 < L < \lambda 0/2$.

The patch and the ground plane are separated by a dielectric sheet or substrate. The dielectric substrate is the electrical insulators and they are serving as a base for other materials. There are a lot of substrates that can be used for the design or manufacturing of micro-strip patch antennas. Their dielectric constants are in the range of 2.2 to 12. In [4], it is written that if the dielectric constant will be low i.e. it ranges in the lower range then a thick substrate will be required. If the dielectric constant will be high i.e. it ranges in the higher range then a thin substrate will be needed and if the substrate is thick then it will produce large bandwidth and also it will produce good efficiency. Thin substrates with higher dielectric constants are sensible for microwave circuitry because they need tightly bound fields to minimize unacceptable or unwanted radiation and coupling, and it will lead to smaller element sizes; and also because of their various losses, they are very less efficient and also comparatively they have smaller bandwidths [4]. Patches could be of many shapes that means it could be of any shape and size but practically rectangular and circular patches are used because of easy fabrication on the substrate.

6. Feeding Techniques

In [9], the authors have mentioned that microstrip patch antenna has many feeding methodologies. As these antennas are having dielectric substrate on one side and the radiating element on the other side. So these feeding techniques are being put as two different classifications contacting and non-contacting technique. Non-contacting is a technique or method where an electromagnetic magnetic coupling is done to transfer the power between the Microstrip line and the radiating patch. There are many new methods of feed techniques are available but some of the common techniques are

A) Microstrip line
B) Coaxial probe
C) Aperture coupling
D) Proximity coupling
E) Co planner wave guide feed etc.

7. Simulation and Results (12 GHz Operating Frequency)

In this topic the HFSS design model of Microstrip Antennas has been mentioned below in Figure 9 and also the side view as well as the top view of Microstrip patch antenna is mentioned below in Figure 10 and Figure 11 respectively.

Simulation and results are one of the most important things in our research. Here in this research in order to design a Microstrip antenna we have used two software and those are such as-: Ansoft High Frequency Structured Simulations (HFSS), and MATLAB. Here we HFSS has been used for the designing of the Microstrip antenna and for MATLAB has been used for the calculations like length, breath etc. of a patch. Here we have designed an Antenna which will operate at 12 GHz. That means here the cut-off frequency is 12 GHz.

So below the computer aided design has been mentioned.

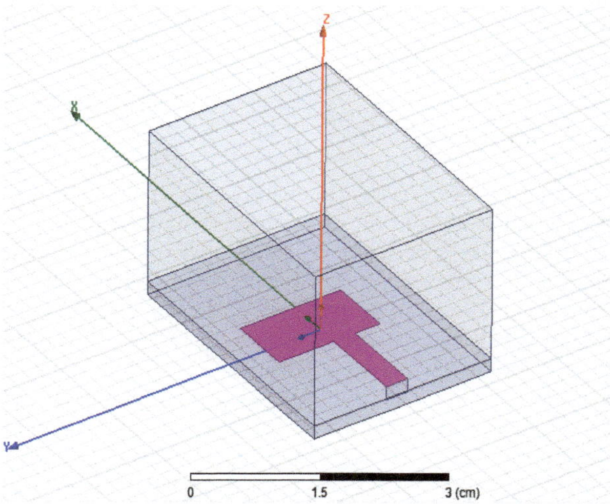

Figure 9. Microstrip patch Antenna (HFSS Design)

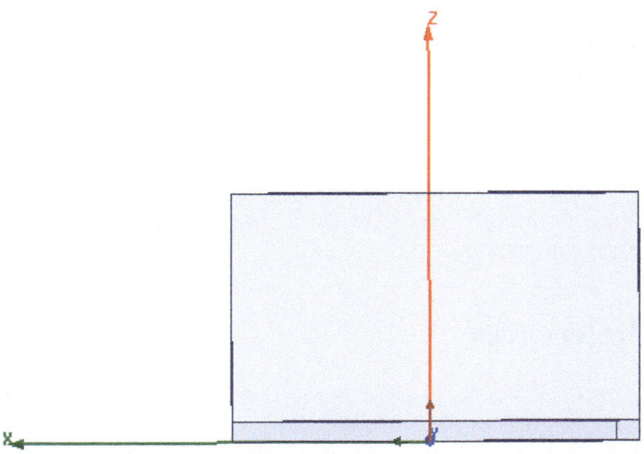

Figure 10. Microstrip patch Antenna (HFSS Side View)

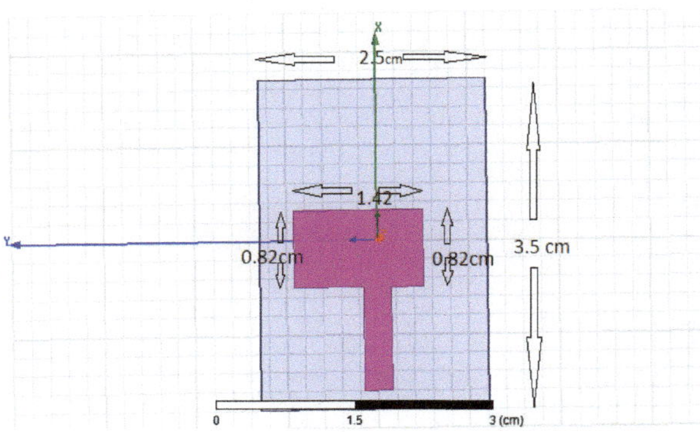

Figure 11. Microstrip patch Antenna (HFSS top View)

Now below in Figure 12, we have mentioned the s_{11} plot or frequency response and in Figure 13, Figure 14 we have mentioned the radiation pattern as well as the 2D rectangular plot of the Microstrip antenna respectively.

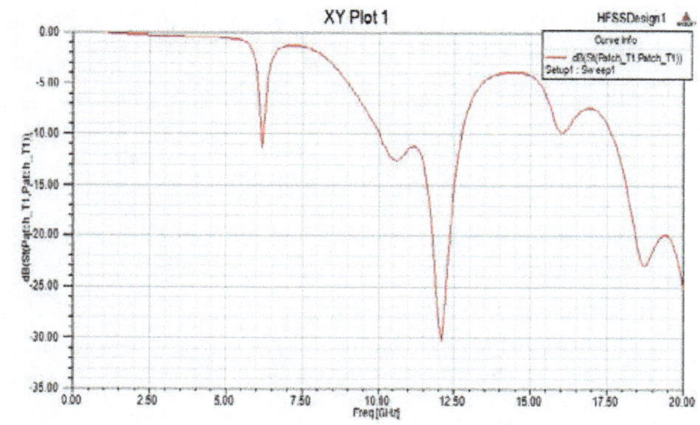

Figure 12. s_{11} plot of Microstrip Antenna (HFSS Result)

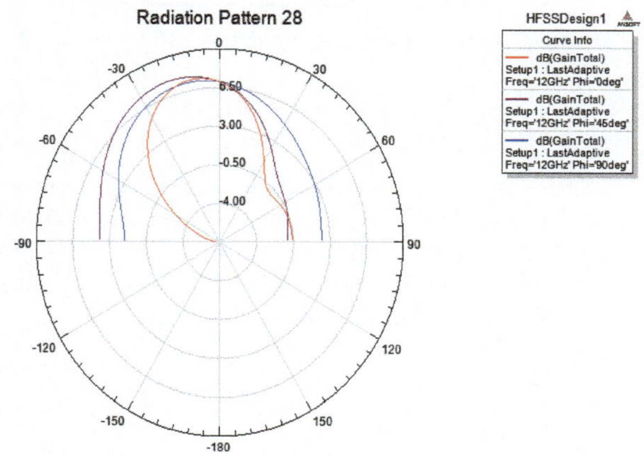

Figure 13. Radiation Pattern of Microstrip Antenna (HFSS Result)

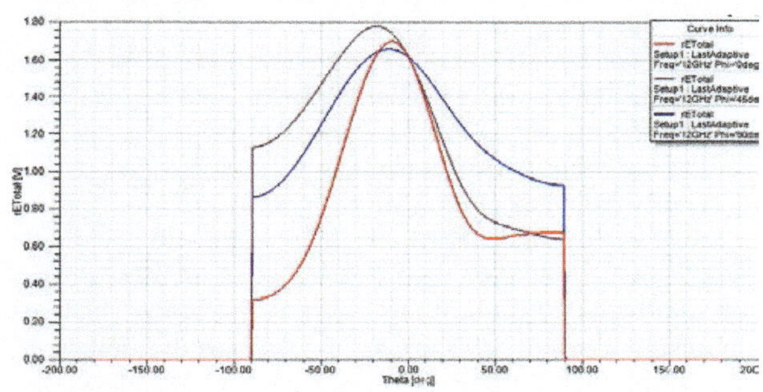

Figure 14. 2D rectangular plot (HFSS Result)

Here in this design first we decided the operating frequency and here the operating frequency is 12 GHz. After that we used some design equations to calculate the length, breath of the rectangular patch.

Here we have taken the height or thickness of the substrate as 0.16 cm. here we calculated the length, breath by using MATLAB that means here we wrote a MATLAB code to find out these parameters. Below some design equations have been mentioned.

Formula for effective dielectric constant [1]:

$$\epsilon_{reff} = \frac{\epsilon_r + 1}{2} + \frac{\epsilon_r - 1}{2}\left[1 + 12\frac{h}{w}\right]^{-1/2} \tag{1}$$

Where ϵ_{reff} = effective dielectric constant

ϵ_r = dielectric constant= 2.2

h = thickness of the substrate

w = width of the patch

Due to fringing fields the change in dimensions in length is given by [1]

$$\frac{\Delta L}{h} = 0.412 \frac{(\epsilon_{reff} + 0.3)(\frac{w}{h} + 0.264)}{(\epsilon_{reff} - 0.258)(\frac{w}{h} + 0.8)} \quad (2)$$

Where ϵ_{reff} = effective dielectric constant

h = thickness of the substrate

w = width of the patch.

Formula for the width of the patch is [3]

$$w = \frac{1}{2f_r\sqrt{\mu_0 \epsilon_0}}\sqrt{\frac{2}{\epsilon_r + 1}} = \frac{v_0}{2f_r}\sqrt{\frac{2}{\epsilon_r + 1}} \quad (3)$$

Formula for the length of the patch is [3]

$$L = \frac{1}{2f_r\sqrt{\epsilon_{reff}}\sqrt{\mu_0 \epsilon_0}} - 2\Delta L \quad (4)$$

So here in our review, the above design equations are used for the designing of an elementary Microstrip Antenna. Now below we have mentioned the calculated values of the above parameters and these calculated values are calculated by using MATLAB.

ϵ_{reff} =2.1960

ϵ_r =2.2

f_r =12 GHz

h=0.16 cm

L = 0.6738 cm, W = 1.42 cm, L (Substrate) = 3.5 cm,

b (Substrate) = 2.5 cm

Now we have mentioned the coordinate of the patch, substrate, and port of the Microstrip Antenna below.

The coordinates of the substrate is: (-1.8,-1.2, 0)

The coordinates of the patch is: (-0.5,-0.5, 0.16)

The coordinates of the port is: (-1.6,-0.15, 0)

Here in this fundamental design theoretically we got the length of the patch as 0.6738 cm. but when we designed our antenna by putting this value we didn't get the frequency response exactly at 12 GHz. So in order to get our desire output that means the frequency response at 12 GHz, we tuned the length because while designing the Microstrip patch antenna we have neglected many losses like fringing losses, current losses, capacitive losses, and metal loss. So here we tuned the length up to 0.82 cm. and after putting the tuned value we got the response at 12 GHz which is our desired output. Generally we consider the operating frequency below 10 dB line in the s_{11} plot because 10 dB is approximately equal to 90% that means, 90% of total power should be delivered from the generator to the antenna through the transmission line. It would be better if more power will be transmitted to the antenna because our objective is to transmit more information in the form of energy from the transmitter to the receiver. The s_{11} plot (Figure 12) is otherwise called as return loss of an antenna i.e. because of reflection, the loss of signal power occurred due to the discontinuity in the transmitted line. In this case we got our desired output with minimum tuning that means the tuning was from 0.6738 cm to 0.82 cm. Here also the material of the substrate that we have used is Rogers RT/ duroid 5880 (tm).

We consider the frequency response below 10 dB and also here the bandwidth is nearly about 2.7 GHz that means it is in ultra-wide band range i.e. the antenna that we have designed is an ultra-wide band antenna and also here also we got a radiation pattern which shows a positive result. In the radiation pattern (Figure 13), the radiations of the elementary antenna was in the broader side and also here any kind of side lobes does not appear in the graph. So the radiation pattern is good in this case.

8. Conclusion

In this paper the design and simulation of an elementary Rectangular patch Microstrip antenna with microstrip feed technique was presented. The objective or goal was to design an antenna for ultra-wide band range and the antenna should work in the target operating frequency (12 GHz), which was calculated by using MATLAB. In order to achieve target operating frequency (12 GHz) with the calculated value, the length of the rectangular patch was increased and after that the design was simulated again by Ansoft HFSS and simulation result gave a large bandwidth which is in the ultra-wide band range. The rectangular patch antenna was designed by changing the feed position, size of the substrate, and shape of the patch to get our objective or target output. This elementary design was based on changing the dimension of length because here various losses are not considered.

Microstrip antenna has founded many applications in the real world due to its low profile feature. Some of the applications are GPS, Bluetooth, Wi-Fi, missiles, and also in radio communication system etc. Many other applications are under research like multiband antenna, array antenna etc.

References

[1] Constatine A. Balanis, "Antenna Theory – Analysis and Design", John Wiley & Sons. Inc. 1997.

[2] G.A. Deschamps, "Microstrip Microwave Antennas," Presented at the third USAF Symposium on Antennas, 1953.

[3] Kin-Lu Wong, Compact and Broadband Microstrip Antennas, Jon Wiley & Sons, Inc., 2002.

[4] D. M. Pozar, "Microstrip Antennas," Proc. IEEE, Vol. 80, No. 1, pp. 79_81, January 1992.

[5] T. Taga and K, Tsuna Kawa , "Performance Analysis of a Built-Planar Inverted-F Antennafor 800 MHz Band Portable Radio Units", IEEE Journal on selected areas in communication, Vol.5, Nop.5, pp. 921-29, June 1987.

[6] Caver K, and Mini J, "Microstrip Antenna Technology", IEEE transactions on antenna and propagation, vol .29, no 1, January 1981.

[7] J.Q Howll, "Microstrip Antennas", IEEE trans. antenna propagation, vol .ap-23, no.1, pp.90-9, Jan 1975.

[8] Lee, H.M. Electron. Eng. Dept., Kyonggi Univ., Suwon, South Korea, "Pattern reconfigurable microstrip patch array antenna using switchable feed-network", Microwave Conference Proceedings (APMC), Issue Date : 7-10 Dec. 2010.

[9] T. Durga Prasad, et.al.,"comparisons of Circular and Rectangular Microstrip Patch Antennas" International Journal of Communication Engineering Applications-IJCEA, Vol. 02, Issue 04; pp. 187-97, July 2011

[10] J. T. Aberle and F. Zavosh, "Analysis of Probe-Fed Circular Microstrip Patches Backed by Circular Cavities," Electromagnetics, Vol. 14, pp. 239-58, 1994.

[11] H. F. Pues and A. R. Van de Capelle, "An Impedance Matching Technique for Increasing the Bandwidth of Microstrip Antennas,"

IEEE Trans. Antennas Propagat., Vol. AP-37, No. 11, pp. 1345-1354, November 1989.

[12] R. E. Munson, "Conformal Microstrip Antennas and Microstrip Phased Arrays," IEEE Trans. Antennas Propagat, Vol. AP-22, No. 1, pp. 74-78, January 1974.

[13] J. W. Howell, "Microstrip Antennas," IEEE Trans. Antennas Propagat., Vol. AP-23, No. 1, pp. 90-93, January 1975.

[14] L. C. Shen, S. A. Long, M. R. Allerding, and M. D. Walton, "Resonant Frequency of a Circular Disc, Printed-Circuit Antenna," IEEE Trans. Antennas Propagat, Vol. AP-25, No. 4, pp. 595-96, July 1977.

[15] A. G. Derneryd, "Analysis of the Microstrip Disk Antenna Element," IEEE Trans. Antennas Propagat., Vol. AP-27, No. 5, pp. 660_4, September 1979.

[16] S. A. Long and M. D. Walton, "A Dual-Frequency Stacked Circular-Disc Antenna," IEEE Trans. Antennas Propagat., Vol. AP-27, No. 2, pp. 270-273, March 1979.

[17] I. J. Bahl and P. Bhartia, Microstrip Antennas, Artech House, Dedham, MA, 1980.

[18] E. H. Newman and P. Tylyathan, "Analysis of Microstrip Antennas Using Moment Methods,"IEEE Trans. Antennas Propagat, Vol. AP-29, No. 1, pp. 47-53, January 1981.

[19] D. C. Chang, "Analytical Theory of an Unloaded Rectangular Microstrip Patch," IEEE Trans. Antennas Propagat., Vol. AP-29, No. 1, pp. 54-62, January 1981.

[20] T. Itoh and W. Menzel, "A Full-Wave Analysis Method for Open Microstrip Structures," IEEE Trans. Antennas Propagat, Vol. AP-29, No. 1, pp. 63-68, January 1981.

[21] P. B. Katehi and N. G. Alexopoulos, "On the Modeling of Electromagnetically Coupled Microstrip Antennas-The Printed Strip Dipole," IEEE Trans. Antennas Propagat, Vol. AP- 32, No. 11, pp. 1179-86, November 1984.

[22] C. C. Liu, A. Hessel, and J. Shmoys, "Performance of Probe-Fed Rectangular Microstrip Patch Element Phased Arrays," IEEE Trans. Antennas Propagat, Vol. AP-36, No. 11, pp. 1501-09, November 1988.

[23] E. O. Hammerstad, "Equations for Microstrip Circuit Design," Proc. Fifth European Microwave Conf., pp. 268-72, September 1975.

[24] R. E. Collinan d F. J. Zucker, Antenna Theory, Part I, Chapter 5, McGraw-Hill Book Co.,New York, 1969.

[25] N.T. Markad, Dr. D.G. Wakade," Probe Feed Microstrip Patch Antenna Computer Aided Design Methodology" International Journal of Scientific and Research Publications, Volume 2, Issue 5, , pp. 1-6, May 2012.

[26] http://www.wirelesstut.com/images/knowledge/beamwidth.jpg.

[27] http://www.amitsingh.co.in/blog/wp-content/uploads/2012/10/new.jpg.

Combined Effect of Block interleaving and FEC on BER Performance of OFDM based WiMAX (IEEE 802.16d) System

Arun Agarwal[1,2,*], **Saurabh N. Mehta**[3]

[1]Department of ECE, Institute of Technical Education & Research, Siksha 'O' Anusandhan University, Khandagiri Square, Bhubaneswar-751030, Odisha, India
[2]Ph.D Scholar, Department of Information Technology, AMET University, Tamil Nadu, Chennai, India
[3]Department of Electronics and Telecommunication Engineering, Vidyalankar Institute of Technology, Mumbai, Maharashtra, India
*Corresponding author: arunagrawal@soauniversity.ac.in

Abstract In this World of Digital era the demand of mobile Internet and wireless multimedia applications are growing faster than ever. To satisfy the user requirements and to overcome the limitations of existing wireless technologies, have led the researchers to come up with more advanced and efficient technology. Orthogonal Frequency Division Multiplexing (OFDM) based WiMAX (Worldwide Interoperability for Microwave Access) is the outcome in this direction which promises to solve the last mile access technology to provide high speed internet access in the residential as well as small and medium sized enterprise sectors. In this paper we have analyzed the effect of Block Interleaving on the Bit Error Rate (BER) performance of the WiMAX Physical layer baseband system conforming to the parameters established by IEEE 802.16 standards for different digital modulation schemes. From the analysis it was observed that addition of interleaving with forward error correction (FEC) improves the system performance by reducing the burst errors during transmission.

Keywords: *BER, burst error, digital modulation, interleaving, FEC, OFDM, WiMAX*

1. Introduction

WiMAX known as Worldwide Interoperability for Microwave Access is a wireless digital communications system designed to provide 30 to 40 megabit-per-second data rates, with the 2011 update providing up to 1 Gbit/s for fixed stations. It is also known as IEEE 802.16 that is intended for Wireless Metropolitan Area Networks (WMAN). WiMAX can provide broadband wireless access (BWA) typically 50 km for fixed stations, and 5 - 15 km for mobile stations. Whereas in contrast, the Wi-Fi/802.11 wireless local area network standard is limited in most cases to only 30 – 100 m. WiMAX is an IP based, wireless broadband access technology that provides performance similar to 802.11/Wi-Fi networks with the coverage and QOS (quality of service) of cellular networks. WiMAX can be used to provide a wireless alternative to cable and DSL for broadband access, and to provide high-speed data and telecommunications services.

WiMAX is based upon IEEE 802.16e-2005, approved in December 2005. It is a supplement to the IEEE 802.16-2004, and so the actual standard is 802.16-2004 as amended by 802.16e-2005. The original version of the standard on which WiMAX is based (IEEE 802.16) specified a physical layer operating in the 10 to 66 GHz range. 802.16a updated in 2004 to 802.16-2004, added specifications for the 2 to 11 GHz range. 802.16-2004 was updated by 802.16e-2005 in 2005 and uses scalable orthogonal frequency-division multiple access (SOFDMA) which is a method of encoding digital data on multiple carrier frequencies.

OFDM has developed into a popular scheme for wideband digital communication, whether wireless or over copper wires, used in applications such as digital television and audio broadcasting. SOFDMA as opposed to the fixed orthogonal frequency-division multiplexing (OFDM) version with 256 sub-carriers (of which 200 are used) in 802.16d. More advanced versions, including 802.16e, also bring multiple antenna support through MIMO (multiple input multiple output). This brings potential benefits in terms of coverage, self installation, power consumption, frequency re-use and bandwidth efficiency. WiMAX is the most energy-efficient pre-4G technique among LTE and HSPA+.

The medium access control (MAC) and the physical (PHY) layers [8] have been included as the air interface defined by WiMAX standard. The MAC layer ensures efficient bandwidth usage by use of TDM/TDMA scheduled uplink/downlink frames. Since IEEE 802.16 supports ATM, IPv4, IPv6, Ethernet and Virtual Local

Area Network (VLAN) services therefore it can support a variety of high bandwidth broadband wireless applications such as Cellular Backhaul, Residential Broadband etc. The interoperability of WiMAX products are certified by WiMAX Forum, which was formed in June 2001 to promote conformity and interoperability of the standard. There are two version of the IEEE 802.16 standard [3] to provide different types of access which have been adopted by WiMAX Forum, one is IEEE802.16-2004 standard for fixed and nomadic access which uses OFDM to provide connection in Line of Sight (LOS) and Non Line of Sight (NLOS) environment and the other is IEEE 802.16e standard for Portable/Mobile Access which incorporates Scalable OFDM Access and provides solutions to handoffs and roaming. Table 1 below from the WiMAX Forum summarizes the 802.16 standards. In this work we have implemented the physical layer of fixed WiMAX and evaluated the Bit-Error-Rate (BER) performance using different guard time intervals.

Following this introduction the remaining part of the paper is organized as follows. Section II presents detailed description of physical layer of the WiMAX. In Section III the details of the modeling and simulation of the system using MATLAB is presented. Then Simulation results have been discussed in Section IV. Finally Section V provides the conclusion.

Table 1. IEEE Wimax standard for BWA [8]

WiMAX Standard & Completion year	IEEE 802.16 Dec 2001	IEEE 802.16a Jan 2003 & Revised in 2004	IEEE 802.16e-2005
Modulation Type	BPSK, QPSK 16-QAM 64-QAM	QPSK 16-QAM 64-QAM	QPSK 16-QAM 64-QAM
Spectrum Allocated	10-66 GHz	2-11 GHz	2-6 GHz
Duplex type	TDD/FDD	TDD/FDD	TDD/FDD
Bandwidth range	1.25 to 28.0 MHz	1.25 to 28.0 MHz	1.25 to 28.0 MHz
Bit Rate	Up to 134 Mbps	Up to 75 Mbps	Up to 15Mbps
Propagation/channel conditions	LOS	NLOS	NLOS
Mobility	Fixed	Fixed	Fixed/Nomadic (Pedestrian Mobility)
Guard period	1/4, 18, 1/16, 1/32	1/4, 18, 1/16, 1/32	1/4, 18, 1/16, 1/32
FFT size	-	256	128, 256, 512, 1024, 2048
MIMO	Yes [2X2]	Yes[2X2]	Yes[2X2]
aerA egarevoC	1 to 3 miles	3 to 5 miles, Max. 30 miles depending upon antenna gain and transmitted power	1 to 3 miles

Table 2. WiMAX OFDM PhyParameters

Parameter	Fixed WiMAX OFDM-PHY	Mobile WiMAX Scalable OFDMA-PHY
FFT size (N_{FFT})	256	128 512 1,024 2,048
Number of data subcarriers	192	72 360 720 1,440
Number of pilot subcarriers	8	12 60 120 240
Number of null/guard subcarrier	56	44 92 184 368
Guard Time (Tg)	1/4	1/32, 1/16, 1/8,1/4
Channel bandwidth (MHz)	3.5	1.25 5 10 20
Subcarrier frequency spacing, Δf (kHz)	15.625	10.94
Useful symbol time Tb(μs)	64	91.4
OFDM symbol duration Ts(μs)	72	102.9

2. IEEE 802.16 WiMAX PHY Layer

This section discusses about the OFDM based IEEE 802.16 PHY layer which can be used in conjunction with the MAC layer to provide a reliable end-to-end link.

Orthogonal Frequency-Division Multiplexing (OFDM) is the key development in the WiMAX PHY layer that makes it resistant to multipath fading effects and inters symbol interference (ISI). Addition of cyclic prefix is the key operation for ISI removal. OFDM is derived from the fact that the high serial bit stream data is transmitted over large (parallel) number sub-carriers (obtained by dividing the available bandwidth), each of a different frequency and these carriers are orthogonal to each other. OFDM converts frequency selective fading channel into N flat fading channels, where N is the number of sub-carriers. Othogonality is maintained by keeping the carrier spacing multiple of 1/Ts by using Fourier transform methods, where Ts is the symbol duration. Since channel coding is applied prior to OFDM symbol generation which accounts for the term 'coded' in COFDM. Figure 1 below presents the detailed block diagram of an OFDM Transceiver.

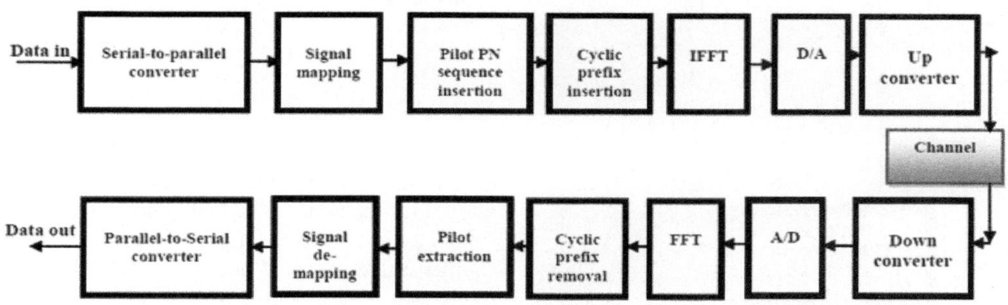

Figure 1. Block diagram of an OFDM Transceiver

2.1. Fixed WiMAX OFDM-PHY

This is based on 256-point Fast Fourier Transform (FFT) OFDM multiplexing scheme in which 192 subcarriers used for carrying data, 8 used as pilot subcarriers for channel estimation and synchronization purposes, and the rest 56 used as guard band subcarriers. This PHY is for 2-11 GHz frequency band and uses TDD mode of access. The channel bandwidth can be an integer multiple of 1.25 MHz, 1.5 MHz, 1.75 MHz, 2MHz and 2.75 MHz with a maximum of 20 MHz. Forward Error Correction (FEC) is done on two phases through the outer Reed-Solomon (RS) code and inner Convolutional code (CC). The RS coder corrects burst error at the byte level. It is particularly useful for OFDM links in the presence of multipath propagation. The CC corrects independent bit errors [8].

2.2. Mobile WiMAX OFDM-PHY

This is based on 2048-point FFT OFDMA scheme but the FFT size can vary from 128 to 2,048. Here, when the available bandwidth increases, the FFT size is also increased to keep the subcarrier spacing always 10.94 kHz [8]. A scalable design also keeps the costs low. This PHY is below 11 GHz frequency band. The details of Parameters for both the IEEE 802.16 PHY layer is shown in Table 2.

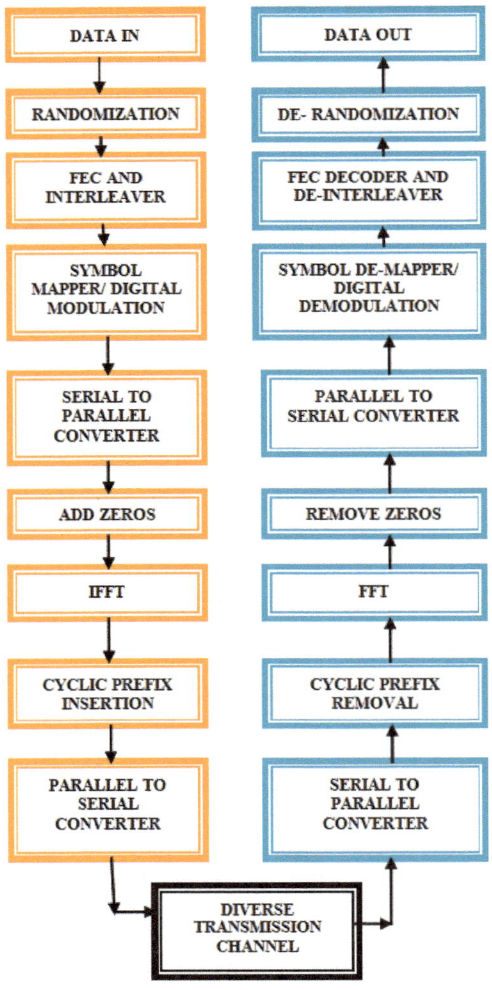

Figure 2. Simulation model of WiMAX-PHY layer

3. Simulation Model

This section discusses the simulation model implemented by us in MATLAB 8.0 environment to evaluate BER performance analysis of IEEE 802.16 OFDM based WiMAX system having 20 MHz bandwidth conforming to the parameters as specified in Table I and II. The functional blocks of baseband WiMAX transceiver is depicted in Figure 2. In the transmitter part the data from the source is randomized first and then subjected to forward error correction (FEC) Encoder employing punctured convolutional codes with code rate 1/2. The coded data is then Block interleaved and mapped into QPSK/QAM symbols. Previously, multi-carrier systems were implemented through the use of separate local oscillator. This was both inefficient and costly. With the advent of cheap powerful DSP processors, the sub-carriers can now be implemented by the fast Fourier transform algorithms (FFT) which keep sub-carriers orthogonal with each other. The symbol is modulated onto sub carriers by applying the Inverse Fast Fourier Transform (IFFT). The output is converted to serial data and a cyclic prefix of $\frac{1}{32}$ is added to make the system robust to inter-symbol interference (ISI). In channel, additive white Gaussian noise characteristics are taken along with the fading Rayleigh channel. The receiver performs the reverse operations of the transmitter. The important blocks of the simulation model is discussed in detail as follows:

3.1. Randomizer

To avoid long sequence of continuous ones and zeros the information bits is randomized before the transmission. This ensures appropriate energy dispersal in the transmitted signal minimize the possibility of transmissions of non-modulated sub carriers. This is implemented with a Pseudo Random Binary Sequence (PRBS) generator which uses a 15-stage shift register with a generator polynomial of $1 + x^{14} + x^{15}$ with XOR gates in feedback configuration as shown in Figure 3.

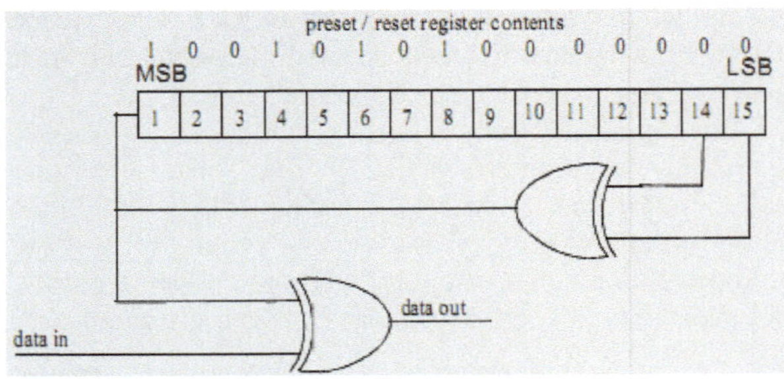

Figure 3. Randomizer [3]

3.2. FEC Encoder

This block consists of a concatenation of an outer Reed Solomon (RS) code and an inner convolutional code (CC) as a FEC scheme. The last block of the encoder is an interleaver to avoid long burst errors. The RS encoder implemented is a systematic RS (N=255, K=239, T=8) code using Galois field GF (28), where N is the number of bytes after encoding, K is the number of data bytes before encoding and T is the number of bytes which can be corrected.

After the RS encoding process, the data bits are further encoded by a binary Convolutional encoder, which has a mother code rate of $\frac{1}{2}$ and a constraint length of 7. The generator polynomials used to derive its two output code bits, denoted X and Y, are $G1 = 171_{OCT}$ & $G2 = 133_{OCT}$ respectively. This convolutional encoder is shown in Figure 4.

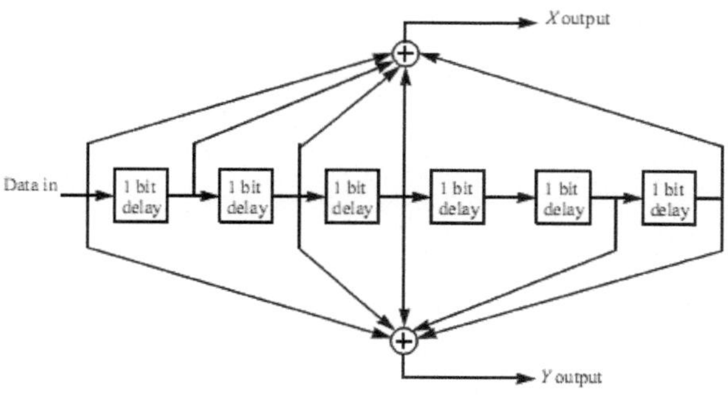

Figure 4. Convolutional Encoder of mother code rate $\frac{1}{2}$ [3]

3.3. Block Interleaving

Data interleaving is performed prior to symbol mapping to scatter burst errors so that it converts in few scarcely spaced single symbol errors, which are more easily correctable, thereby increasing the efficiency of FEC. The RS-CC encoded data are interleaved by a block interleaver whose block size corresponds to the number of coded bits per the allocated sub channels per OFDM symbol, N_{cbps}.

In IEEE 802.16 standard [3], the interleaver is defined by two step permutation. The first ensures that adjacent coded bits are mapped onto non-adjacent subcarriers. The second permutation ensures that adjacent coded bits are mapped alternately onto less or more significant bits of the constellation, thus avoiding long runs of unreliable bits.

The index value of the bits after first and second permutation will be calculated using following Equation (1) and (2) respectively.

$$x_k = \left(N_{cbps}/12\right).k_{\text{mod}12} + floor\left(k/2\right) \qquad (1)$$

where $k = 0,1,2,\ldots\ldots..N_{cbps}-1$, known as the index of the coded bit before the first permutation.

$$y_k = s.floor\left(x_k/s\right) + \left(m_k + N_{cbps} - floor\left(12.m_k/N_{cbps}\right)\right)_{\text{mod}(s)} \qquad (2)$$

where $s = ceil\left(N_{cpc}/2\right)$, and N_{cpc} stands for the number of coded bits per subcarrier, i.e., 1,2,4 or 6 for BPSK, QPSK 16-QAM, or 64-QAM, respectively.

3.4. Digital Modulation Mapper

After channel coding and interleaving data are then entered serially to the digital modulation mapper that support BPSK, grey-mapped QPSK, 16-QAM, and 64-QAM as specified in the standard [3]. The bits are mapped to a subcarrier amplitude and phase, which is represented by a complex in-phase and quadrature-phase (IQ) vector. The IQ plot for a modulation scheme shows the transmitted vector for all data word combinations. Gray coding has been used for this allocation so that adjacent points in the constellation only differ by a single bit. This coding helps to minimize the overall bit error rate as it reduces the chance of multiple bit errors occurring from a single symbol error [10]. The constellation plots for BPSK, 4-QAM, and 16-QAM modulations are depicted in Figure 5.

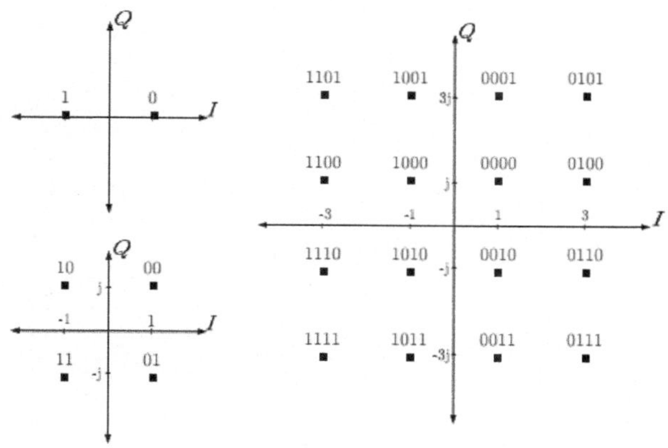

Figure 5. BPSK, 4-QAM, and 16-QAM constellation diagrams [10]

3.5. Channel

Finally to evaluate the performance of the developed WiMAX communication system, the transmitted signal is exposed to six different types of channel conditions [8,10] which are three paths fading channels. The wireless channel designed use

Multipath delay spread, Fading characteristics, Doppler spread, the effect of Path loss (including shadowing), Co channel and adjacent channel interference.

4. Result and Discussion

In this section we have presented the simulation results along with the bit error rate (BER) curve analysis for Rayleigh and Rician three path fading channel. The simulation parameters for WiMAX system are as per Table 1 and Table 2. This section gives the effect of bit interleaving on the performance of different digital modulation.

Figure 6 and Figure 7 below, presents the BER performance improvement due to bit interleaver for BPSK modulation on channel 1 and channel 2.

It can be seen from these BER curves that inclusion of bit interleaver helps FEC to work properly. It can evaluated from the Figure 6 and Figure 7 that bit there is a interleaver-coding gain of 5 dB at BER level of 10^{-4} on channel 1 and 10.3 dB SNR improvement on channel 2.

After analyzing effect of interleaver for BPSK modulation, we will now investigate the performance improvement for QPSK, 16QAM and 64QAM on the same channel conditions, as depicted from Figure 8 to Figure 13 below. The SNR improvement observed from these all figures are tabulated in Table 3.

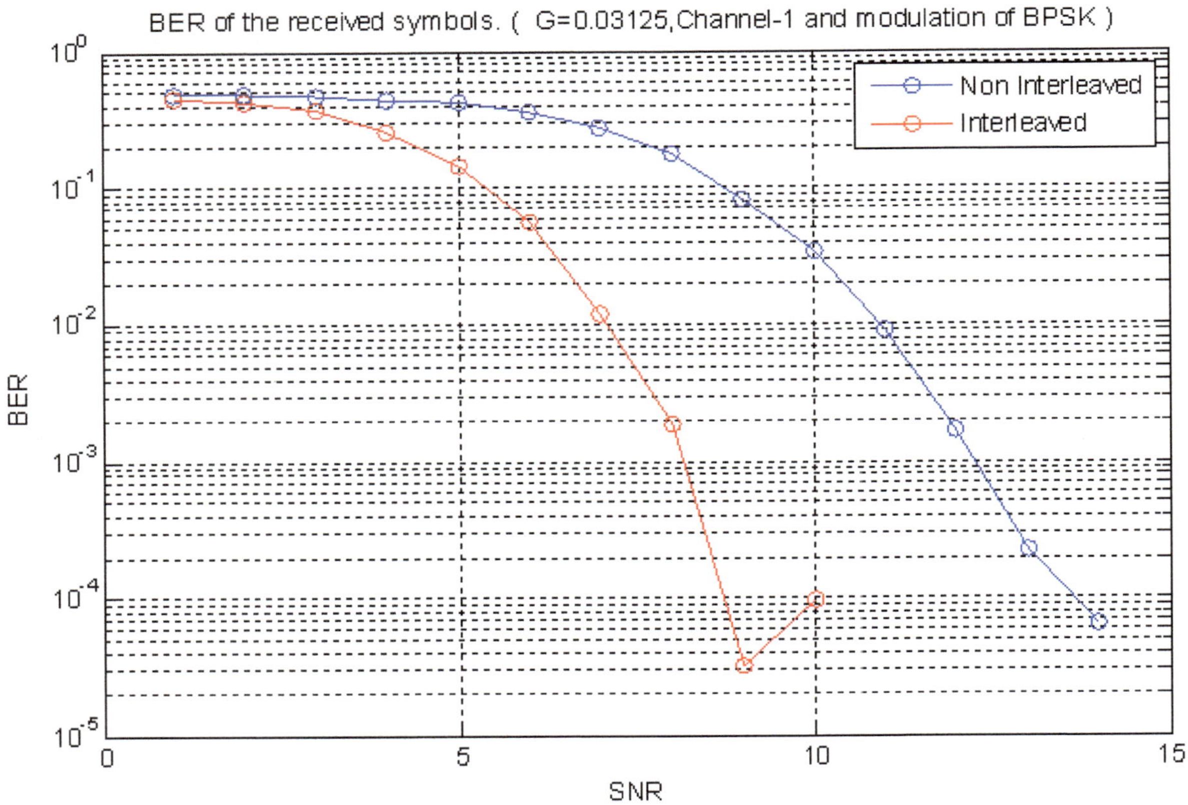

Figure 6. Effect of Block interleaver in BPSK on Channel-1

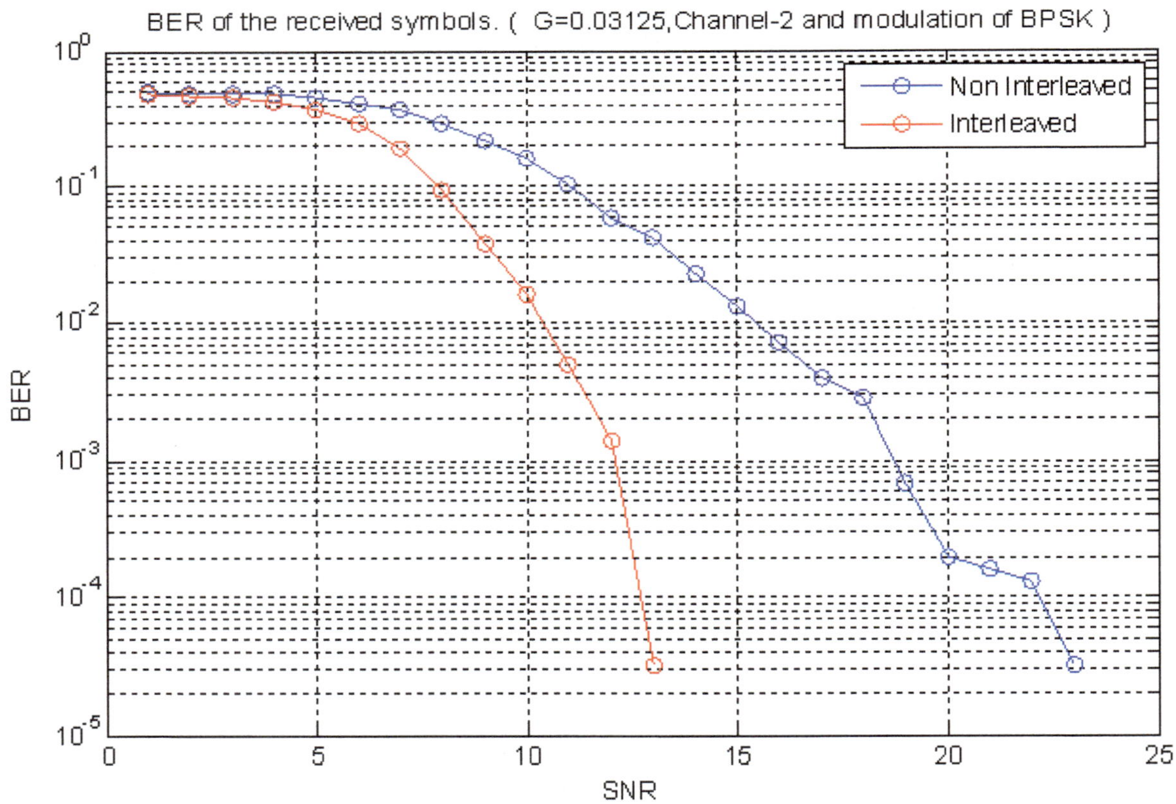

Figure 7. Effect of Block interleaver in BPSK on Channel-2

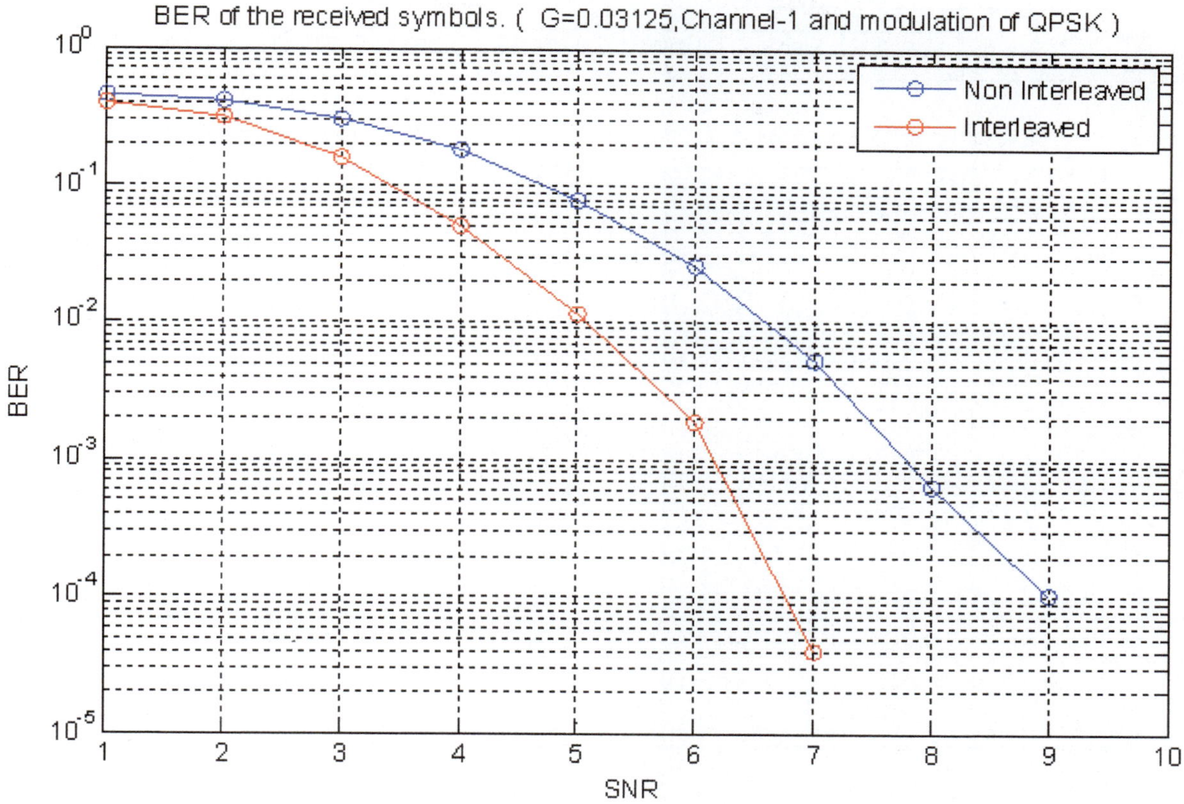

Figure 8. Effect of Block interleaver in QPSK on Channel-1

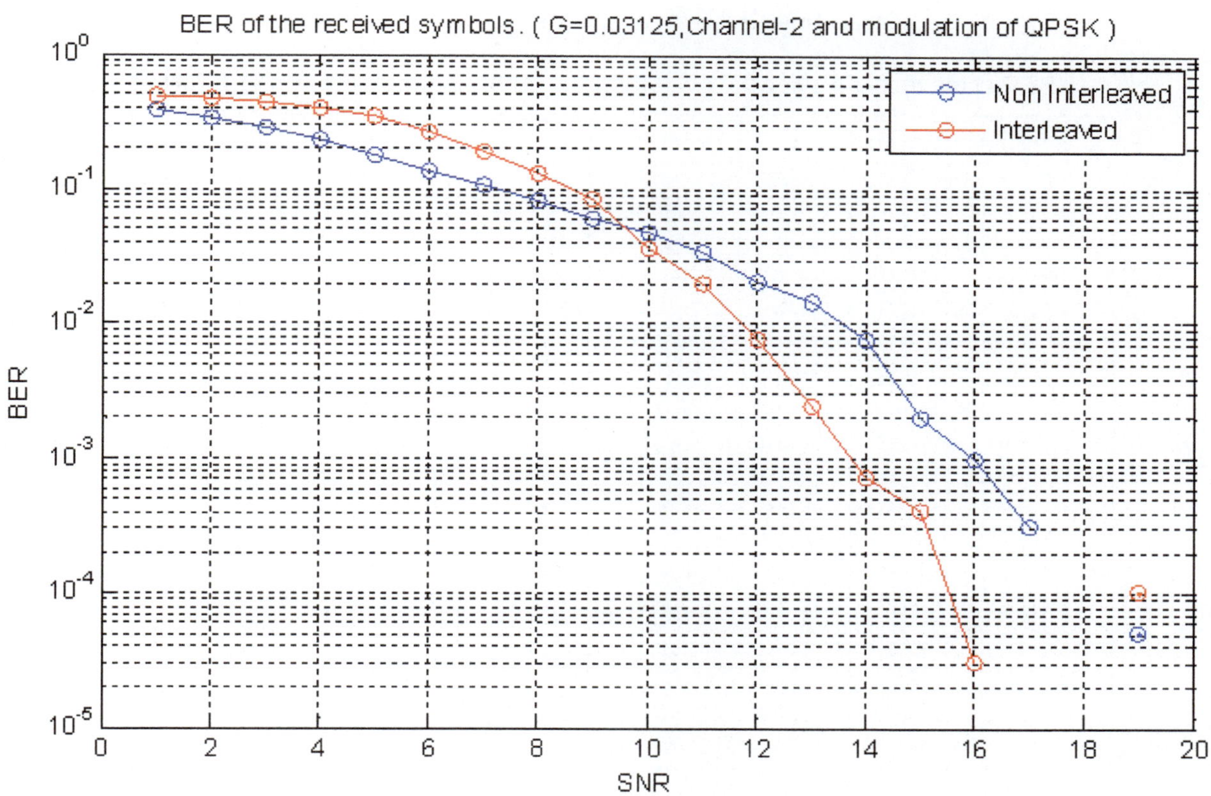

Figure 9. Effect of Block interleaver in QPSK on Channel-2

Table 3. Performance improvement due to block interleaving

Modulation	BPSK		QPSK		16QAM		64QAM	
	Channel 1	Channel 2	Channel 1	Channel 2	Channel 1	Channel 2	Channel 1	Channel 2
Coding gain (dB) at BER 10^{-4}	5	10.3	3.2	2.5	1.9	1.8	2.4	3

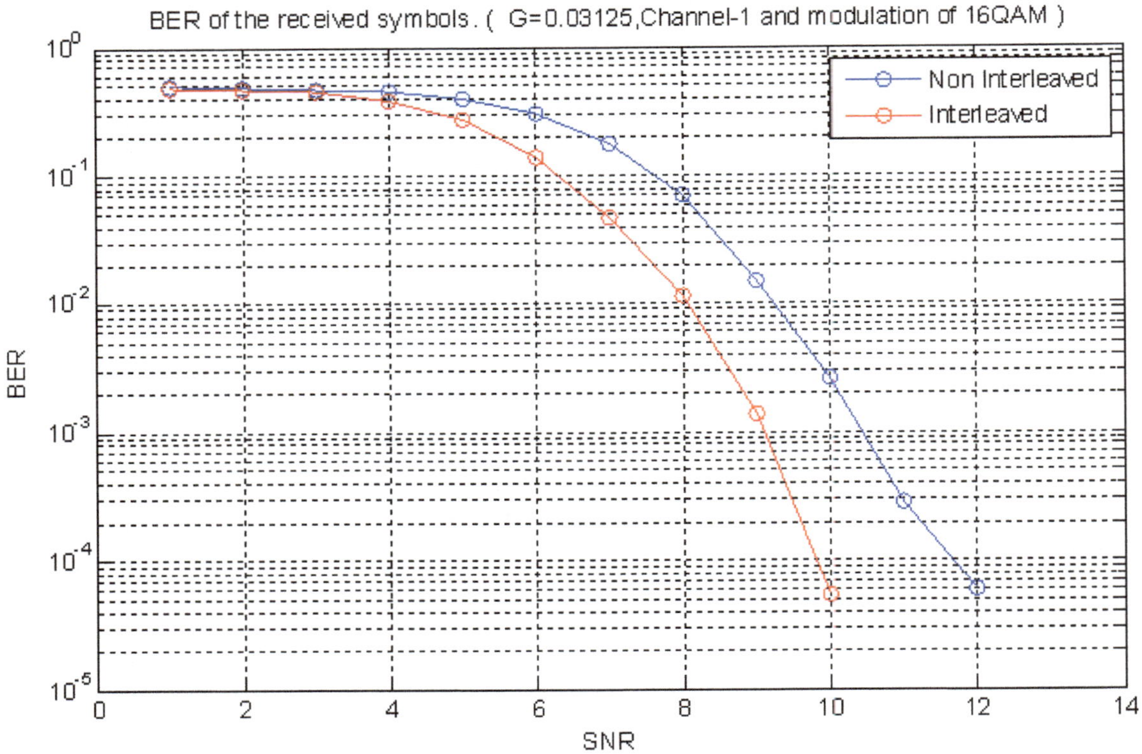

Figure 10. Effect of Block interleaver in 16-QAM on Channel-1

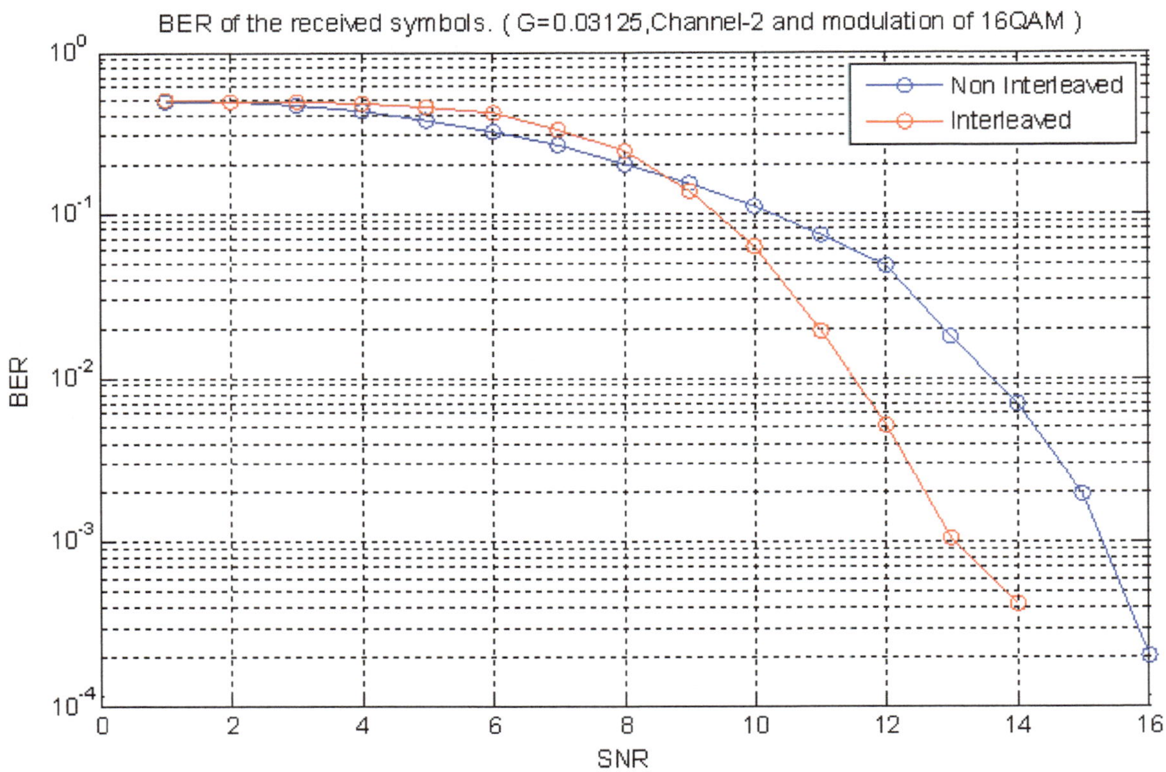

Figure 11. Effect of Block interleaver in 16-QAM on Channel-2

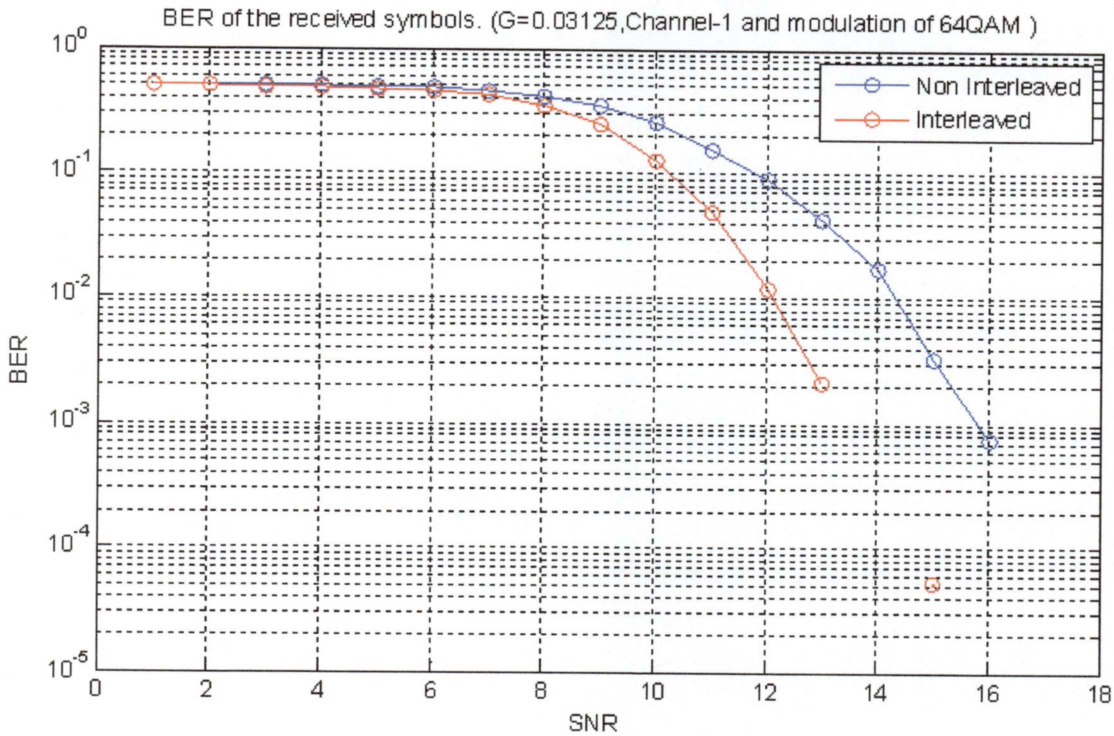

Figure 12. Effect of Block interleaver in 64-QAM on Channel-1

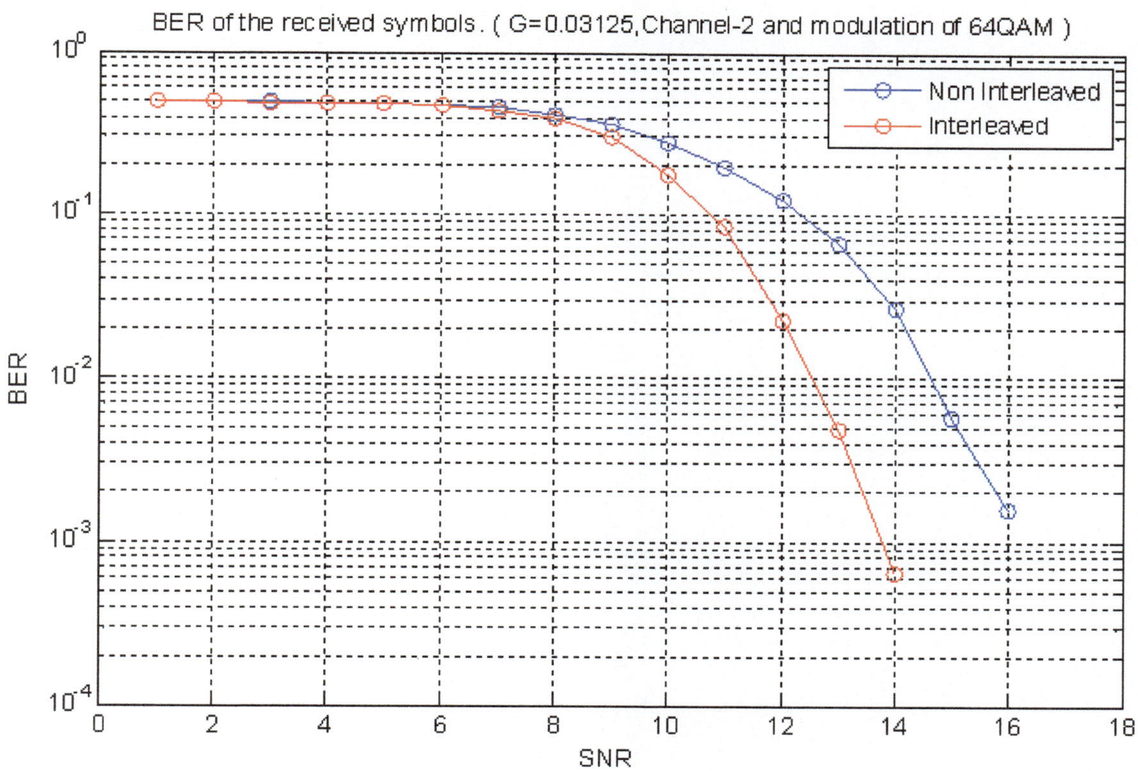

Figure 13. Effect of Block interleaver in 64-QAM on Channel-2

5. Conclusion

The key contribution of this paper was the implementation PHY layer of IEEE 802.16 OFDM based WiMAX system to evaluate the effect of block interleaver on the BER performance. It was shown clearly that interleaver improves the performance of the WiMAX system. Use of interleaving techniques increases latency which should be taken care when designing the communication system. The work can be successfully extended with addition of other interleaving techniques such as convolutional interleaving, frequency interleaving for further improvement in performance.

References

[1] Mobile WiMAX – Part I: A Technical Overview and Performance Evaluation, http://www.wimaxforum.org/technology/downloads/Mobile_WiMAX_Part1_Overview_and_Performance.pdf.

[2] WiMAX Forum: "Mobile WiMAX. Part I: A Technical Overview and Performance Evaluation," August 2006. http://www.intel.com/netcomms/technologies/wimax

[3] IEEE 802.16-2004, "IEEE Standard for Local and Metropolitan Area Networks -Part 16: Air Interface for Fixed Broadband Wireless Access Systems", 1 October, 2004.

[4] IEEE Std 802.16-2001," IEEE Std. 802.16-2001 IEEE Standard for Local and Metropolitan area networks Part 16: Air Interface for Fixed Broadband Wireless Access Systems", December 2001.

[5] Arun Agarwal, S.K. Patra, "Performance prediction of Eureka-147 DAB system Using Interleaving and different Coding rates", in IEEE International Conference on MEMS, Nano & Smart Systems, Proceedings, ICMENS, 2011, vol 403-408, pp. 4119-4125,.

[6] Ghosh, A.; Wolter, D.R.; Andrews, J.G.; Chen, R., "Broadband wireless access with WiMax/802.16: current performance benchmarks and future potential", IEEE Communications Magazine, Feb. 2005, Vol. 43, Iss. 2, Pages: 129-136.

[7] Arun Agarwal, Kabita Agarwal, "Design and Simulation of COFDM for high speed wireless communication and Performance analysis", in IJCA-International Journal of Computer Applications, Oct-2011, Vol-2., pp 22-28, ISBN: 978-93-80865-49-3.

[8] Hasan, Mohammad Azizul," Performance Evaluation of WiMAX/IEEE 802.16 OFDM Physical Layer", Master's Thesis at Helsinki University of Technology, Espoo, June 2007.

[9] Tariq, Umar, Umer Naeem Jilani and Tauseef Ahmad Siddiqui, "*Analysis on Fixed and Mobile WiMAX*", Master's Thesis at Blekinge Institute of Technology, 2007.

[10] Roca, Amalia, "Implementation *of a WiMAX simulator in Simulink*", Diploma Thesis, Vienna, February 2007.

[11] John. G. Proakis, "Digital Communications", 3rd edition, McGraw-Hill, 1995.

[12] Walter Tuttlebee, "Software Defined Radio", 2002 John Wiley & Sons, Ltd.

[13] H. Harada & Ramjee Prasad, Simulation and Software Radio for mobile communications.: Artech House, 2003.

[14] Ohrtman, Frank,"WiMAX Hand Book Building 802-16 Wireless Network". McGraw Hill Publishers, 2005.

[15] The Matlab help: "Communications Blockset," http://www.mathworks com

[16] http://www.en.wikipedia.org/wiki/WiMAX

[17] http://www.wimax.com/general/what-is-wimax

[18] Arun Agarwal, Chinmayee Behera, Kabita Agarwal, "Effect of Guard Time on BER Performance of OFDM based WiMAX (IEEE 802.16d) System", in International Journal of Electronics and Communication Engineering. 2013, Volume 6, Number 1, pp. 41-55.

28

Different Methods of Speed Control of Three-Phase Asynchronous Motor

G. C. Diyoke*, Okeke C., Uchechi Aniagwu

Department of Electrical and Electronic Engineering, Michael Okpara University of Agriculture, Umudike Abia State, Nigeria
*Corresponding author: geraldiyoke@mouau.edu.ng

Abstract Asynchronous motors are the most widely used electrical motors due to their reliability, low cost, ruggedness and robustness. However, asynchronous motors do not inherently have the capability of variable speed operation. Due to this reason, earlier dc motors were applied in most of the electrical drives. But the recent developments in speed control methods of the asynchronous motor have led to their large scale use in almost all electrical drives. In this paper, various methods of speed control of induction (asynchronous) motors such as variable rotor resistance, variable stator voltage, constant Volts/hertz control, frequency variation, pole changing, etc., were carried out and MATLAB software was used to develop the codes for the analysis of these methods.

Keywords: *asynchronous motor, speed control, mode of operation*

1. Introduction

An induction or asynchronous motor is a type of AC motor where power is supplied to the rotor by means of electromagnetic induction, rather than a commutator or slip rings as in other types of motor. These motors are widely used in industrial drives, particularly poly-phase induction motors, because they are rugged and have no brushes. But they require much more complex methods of control, more expensive and higher rated power converters than DC and permanent magnet machines [1]. Single-phase versions are used in small appliances. Their speed is determined by the frequency of the supply current, so they are most widely used in constant-speed applications, although variable speed versions, using variable frequency drives are becoming more common. The most common type is the squirrel cage motor, and this term is sometimes used for induction motors generally [2,3,4,5]. But it is desirable to replace the single phase induction motor drives by three phase induction motor drives in residential appliances, farming and low power industrial applications. Induction motors have performed the main part of many speed control systems and found usage in several industrial applications [6]. The benefit of improvement in the motor drive industry has touched varied applications, from heavy and large industrial equipment such as rolling mills in steel making plants, paper mills, etc. to "Mechatronics" equipment used in machine tools and semiconductor fabrication machines [7].

Three phase induction machines are synchronous speed machines, operating below synchronous speed when motoring and above synchronous speed when generating [8]. They are comparatively less expensive to equivalent size synchronous or dc machines and range in size from a few watts to 10,000hp. As motors, they are rugged and require very little maintenance. However, their speeds are not as easily as with dc motors. They draw large starting currents, typically, about six to eight times their full load values, and operate with a poor lagging power factor when lightly loaded.

This paper is structured as: In section I, the concept of asynchronous motor and its applications. Section II presents different speed control techniques. Section III details the principle of operation of induction motor. Here in section IV presents in details the equivalent circuit analysis. In section V, Matlab/Simulink simulation results are displayed. Finally, in section VI conclusion is presented.

2. Different Speed Control Methods

The speed control of asynchronous motor (AM) or induction motor (IM) can be varied by varying the slip 'S' or number of poles 'p' or frequency 'f' of the supply. The ability of varying any one of the above three quantities will provide methods of speed control of an induction motor. Constant V/F method is commonly used for constant and variable speed control of induction motor. The different methods of speed control of IM can be broadly classified into scalar and vector control methods. In this paper, scalar control methods are used. The scalar methods of speed control can be classified as

A. *Rotor Resistance Control*

This method is applicable only to the wound rotor induction motors. Speed variation can be obtained by inserting external resistance in the rotor circuit. External resistances are placed in series with the rotor windings during starting to limit the starting current. Without the external resistances, the starting currents are many times the rated currents. Depending on the size of the machine, it can draw 300% to over 900% of full-load current. Once the motor is started, the external resistance can be cut out to obtain high torque throughout the accelerating range.

B. *Supply Frequency Control*

By changing the supply frequency, the motor synchronous speed can be altered and thus the torque-speed of a three- phase induction motor can be controlled. The synchronous speed of the motor at rated conditions is known as the base speed. By using variable frequency control, it is possible to adjust the speed of the motor either above or below the base speed. Increase in frequency increases the torque-speed relation and a decrease in frequency decreases the torque-speed relation of the motor.

C. *Supply Voltage Control*

As can be seen from Equation 12, the torque developed by an induction motor varies as square of the voltage applied to its stator terminals. Thus by varying the applied voltage, the electromagnetic torque developed by the motor can be varied. This method is generally used for small squirrel-cage motors where cost is an important criterion and efficiency is not. However, this method has rather limited range of speed control.

D. Stator *Voltage and Frequency Control i.e. Volts per Hertz Control*

The stator voltage and frequency control is to control the induction motor speed and torque by varying the ratio of voltage to frequency. When low voltage and low frequency is applied to the motor, the maximum torque available decreases at reduced speeds. If the ratio of voltage to frequency is kept constant, this technique allows the induction motor to deliver its rated torque at speeds up to its rated speed. This is because; the air gap flux is reduced due to the drop in the stator impedance while motor operates at a low frequency. We vary the stator voltage in such a way that the flux remains constant by simultaneously varying the supply frequency such that the ratio V/f remains constant [9].

E. *Pole Changing*

The primary factor in determining the speed of an induction motor is the number of poles, given by the formula;

$$\omega_s = \frac{120f}{P} \text{ rpm or } n_s = \frac{2f}{p} \text{ rps}$$

ω_s = Synchronous speed, in revolutions per minute,
f = AC power frequency
p = Number of poles per phase winding.

From the formula above, Pole changing in induction machine can be done using a Dahlander motor also known as a pole changing motor. Pole changing can be used to achieve different speeds in induction machine by switching the configuration of the electrical stator windings in the ratio of 2:1, indirectly adding or removing poles and thus varying the rotor speed. The number of stator poles can be changed by (a) multiple stator winding, (b) method of consequent poles, (c) pole amplitude modulation.

3. Operating Principle

A three-phase power supply provides a rotating magnetic field in an induction motor. In both induction and synchronous motors, the AC power supplied to the motor's stator creates a magnetic field that rotates in time with the AC oscillations. Whereas a synchronous motor's rotor turns at the same rate as the stator field, an induction motor's rotor rotates at a slower speed than the stator field. The induction motor stator's magnetic field is therefore changing or rotating relative to the rotor. This induces an opposing current in the induction motor's rotor, in effect the motor's secondary winding, when the latter is short-circuited or closed through external impedance. The rotating magnetic flux induces currents in the windings of the rotor in a manner similar to currents induced in a transformer's secondary winding(s). The currents in the rotor windings in turn create magnetic fields in the rotor that react against the stator field. Due to Lenz's Law, the direction of the magnetic field created will be such as to oppose the change in current through the rotor windings. The cause of induced current in the rotor windings is the rotating stator magnetic field, so to oppose the change in rotor-winding currents; the rotor will start to rotate in the direction of the rotating stator magnetic field. The rotor accelerates until the magnitude of induced rotor current and torque balances the applied load. Since rotation at synchronous speed would result in no induced rotor current, an induction motor always operates slower than synchronous speed. Thus, the operation of induction motor results to losses and power output as show in Figure 1 below.

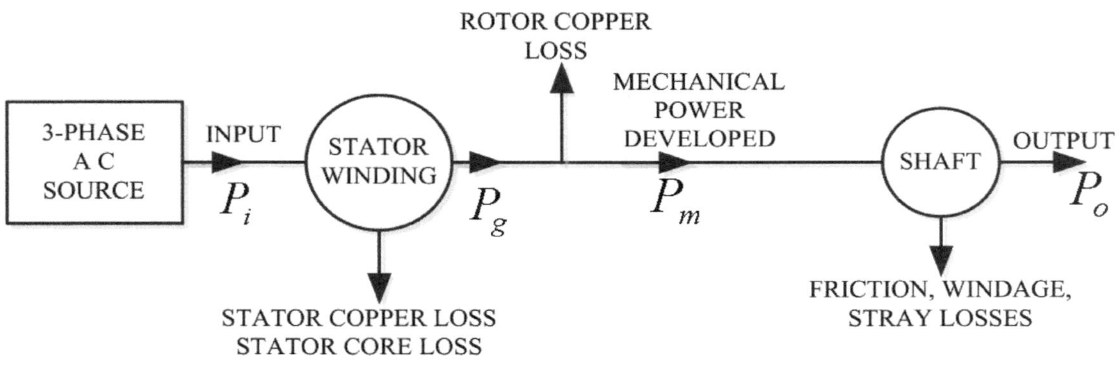

Figure 1. Power flow diagram of 3-phase induction motor

The torque-slip/speed curve is drawn in Figure 2 below. This curve has the following characteristic features:

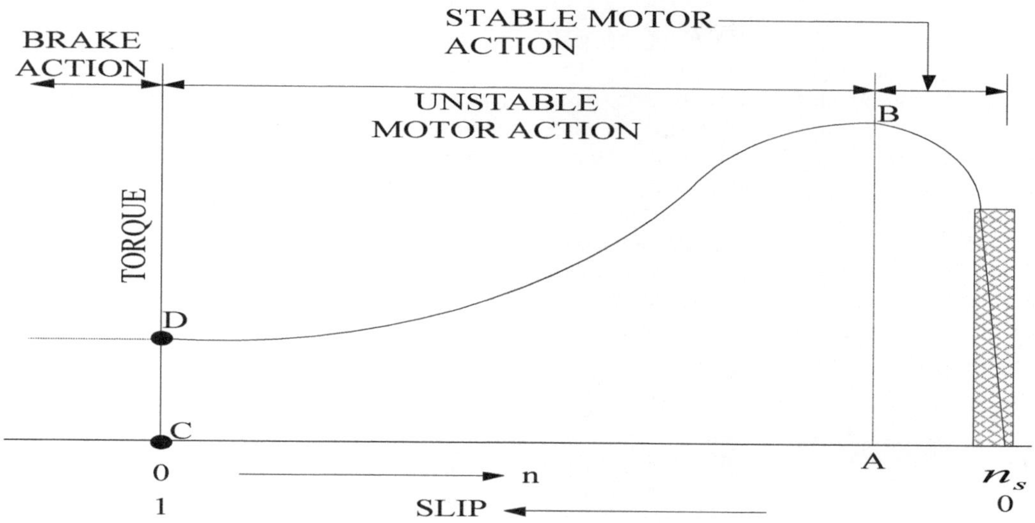

Figure 2. Torque slip/speed curve of 3 phase induction motor

(1) When n=ns, and s=0, torque is zero. This is because at synchronous speed there is no relative motion because rotor conductors and stator rotating field, so that e. m. f and current are zero.

(2) The hatched area shows the normal operating range for motor action. In this region torque increases linearly with slip.

(3) CD represents the torque at s=1, i.e. at standstill conditions. If load torque is less than CD, the motor will accelerate till the torque developed by motor is equal to load torque. This happens at a speed which is close to but less than synchronous speed.

(4) AB is the maximum or pull-out torque. If the load torque is more than this torque, the motor will come to standstill. The region BD is the unstable region.

(5) The maximum torque occurs at a low value of slip. This is typical of all induction motors and is a desired feature, because the change in speed with load is small, making it almost a constant speed motor.

(6) The value of slip greater than 1 are also shown in Figure 2. To achieve a value of slip more than 1, the rotor must be coupled to a prime mover driven in the direction opposite to that of stator rotating field, the stator still being connected to the three-phase supply. In this region the motor is receiving both electrical and mechanical input, all the power being dissipated as loss. This mode of operation is known as brake action.

4. Circuit Analysis

In developing the equivalent circuit of an induction motor, the circuit models for stator and rotor will be combined together so as to give the equivalent circuit model.

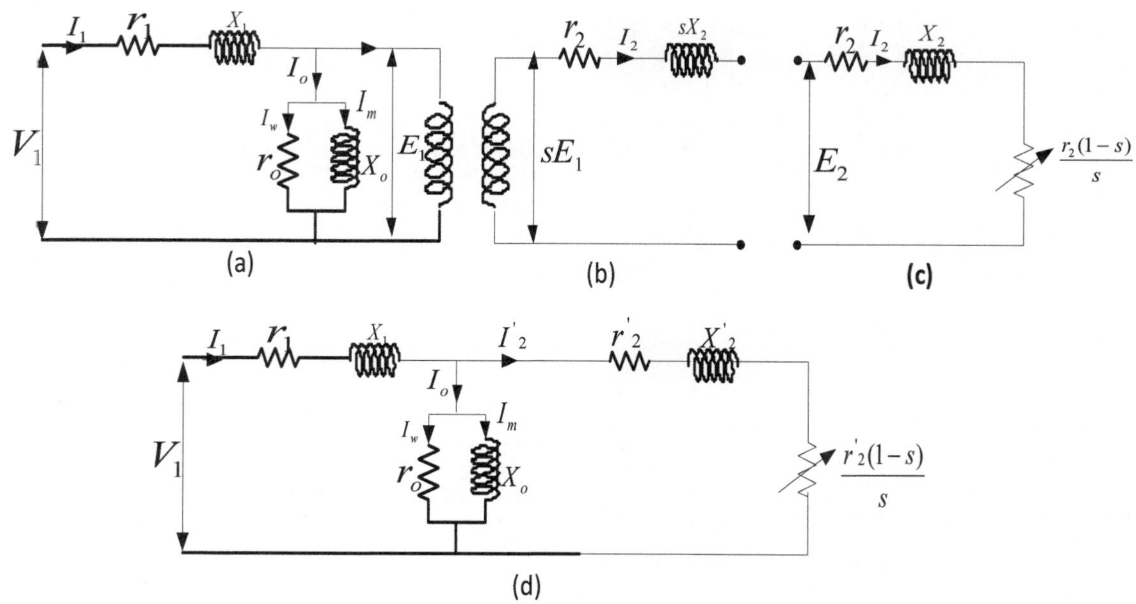

Figure 3. Development of equivalent circuit of 3-phase induction motor (a) stator circuit model (b and c) rotor models (d) complete equivalent circuit model

Figure 3(a) shows the stator circuit model, when a voltage V_1 is applied to stator terminals, a flux $\emptyset_m$ is step up. This flux induces e. m. f E_1 in the stator winding. The flow of current I_1 causes voltage drops $I_1 r_1$ and $I_1 x_1$ in the stator winding. E_1, V_1, $I_1 r_1$ and $I_1 x_1$ are related to give motor equation. Figs. 3(b and c) are rotor circuit models. The mutual flux $\emptyset_m$ induces an e. m. f sE_2 in the rotor and a current I_2 flows. r_2 and x_2 are rotor resistance and leakage reactance at standstill (i. e. s =1). When the motor is operating at slip s, the leakage inductance is sx_2 as shown in Figure 3b. Figure 3c shows alternative rotor circuit model with mechanical load and I_2 in this model represents the line frequency current.

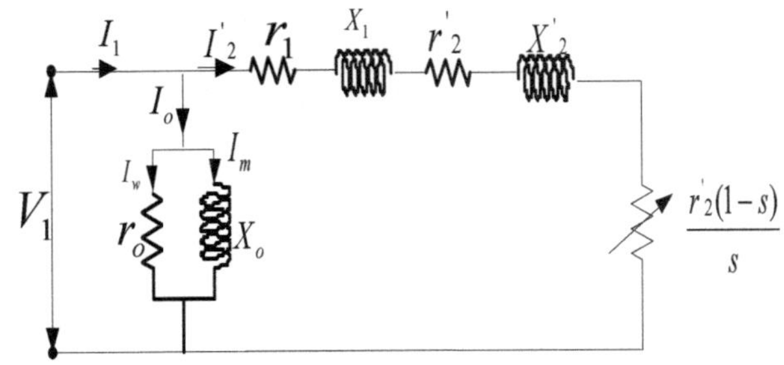

Figure 4. Approximate equivalent circuit of 3phase induction motor

The 3-phase power input pi to an induction motor is given by

$$P_i = 3V_1 I_1 cos\theta_1 \quad (1)$$

Where V_1, I_1 and $\cos\theta$ are the per phase values of stator voltage and current and stator power factor respectively. A part of this power is consumed in stator core losses stator copper losses. The remainder is P_g, the air gap power.

Stator copper losses is given by

$$P_{cu} = 3I_1^2 r_1 \quad (2)$$

$$P_g = P_i - P_{cu} \quad (3)$$

P_g is the power transferred, across the air gap, to the rotor. After rotor copper losses are subtracted from P_g, the mechanical power developed, P_m is obtained. A part from P_m of power output P_o is consumed as friction and windage and stray losses and the remainder is the power output P_o as can be seen in Figure 1.

Now,

$$P_g = 3E_2 I_2 cos\theta_2 = \frac{3I_2^2 r_2}{s} \quad (4)$$

Also,

$$P_g = 3E_2 I_2' cos\theta_2 = \frac{3I_2'^2 r_2'}{s} \quad (5)$$

Rotor Copper-loss

$$P_{cu} = 3I_1^2 r_1 = 3I_2'^2 r_2' \quad (6)$$

From equations (4) and (6), Rotor Copper-loss = sP_g

$$P_m = P_g - Rotor\,Copperloss = P_g(1-s) \quad (7)$$

$$P_o = P_m - \text{Friction, windage and stray losses} \quad (8)$$

The frequency of rotor current is very low. Therefore, power loss is very small and hence negligible.

The torque developed T is defined as the torque generated due to mechanical power developed by electro-mechanical power conversion process in the motor. The actual load torque available at the shaft is the torque developed less the windage and friction and stray losses.

Thus,

$$T = \frac{P_m}{\omega_m} = \frac{P_m}{2\pi n} \quad (9)$$

Or

$$T = \frac{3I_2'^2 r_2'}{2\pi n}\left[\frac{1-s}{s}\right] \quad (10)$$

Hint, n = $n_s(1-s)$

The torque developed is proportional to the air gap power P_g. The air gap power P_g is usually known as torque measured in synchronous watts. This indicates that if this power is divided by synchronous angular velocity $2\pi n_s$, Torque developed is obtained.

It can be ascertained from the approximate equivalent circuit in Figure 4 that

$$I_2' = \frac{V_1}{\left[(r_1 + r_2'/s)^2 + (x_1 + x_2')^2\right]^{0.5}} \quad (11)$$

Putting the value of I_2' from Eq. (11) into Eq. (10) yields

$$T = \left[\frac{3}{2\pi n_s}\right]\left[\frac{r_2'}{s}\right]\left[\frac{V_1^2}{(r_1 + r_2'/s)^2 + (x_1 + x_2')^2}\right]. \quad (12)$$

5. Simulation Results

(a) MACHINE DATA

Table 1 below gives the machine data used for the simulation of the machine model for the analysis of speed control schemes.

Table 1. Machine details used in MATLAB codes execution for the various methods of speed control

supply voltage (star connected)	400 Volts
Number of poles	4
Stator resistance	0.6Ω
Rotor resistance referred to stator	0.3Ω
Frequency	50 Hertz
Stator leakage reactance	1.1Ω
Rotor leakage reactance referred to stator	0.5Ω
V/f ratio	10
Magnetizing inductance	5.834H

(b) STEADY STATE BEHAVIOUR OF INDUCTION MACHINE

The steady state behaviour of the induction motors are graphically represented in Figure 5, Figure 6, Figure 7, Figure 8 and Figure 9 depending on the following variables: rotor resistance, line frequency, number of poles, stator voltage and stator voltage/line frequency.

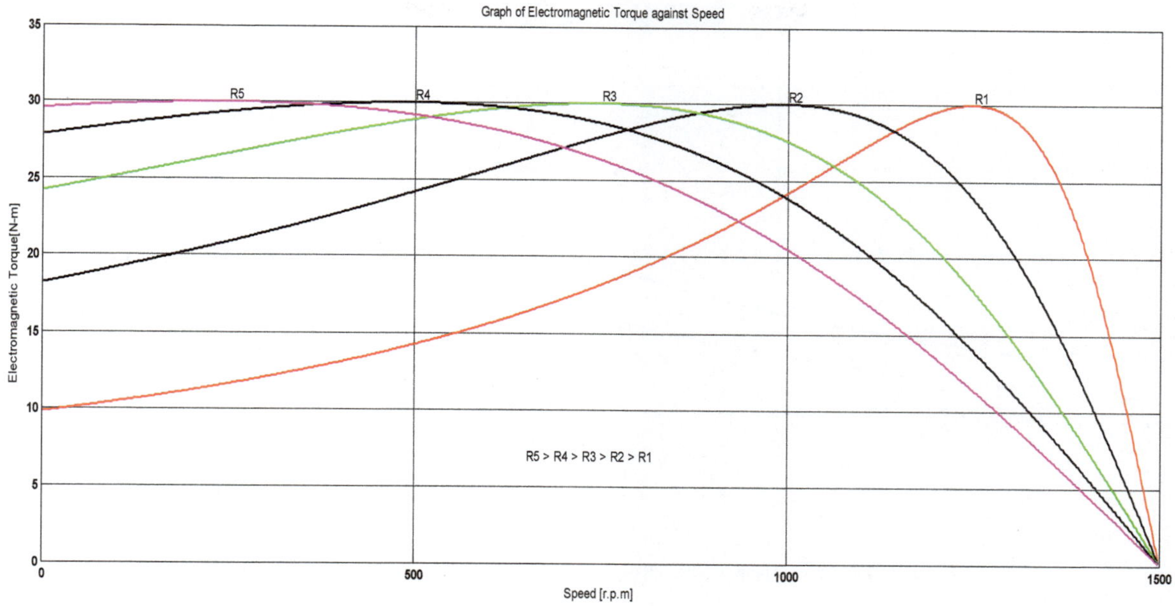

Figure 5. Torque-speed curves of 3 phase induction motor for different values of rotor resistance

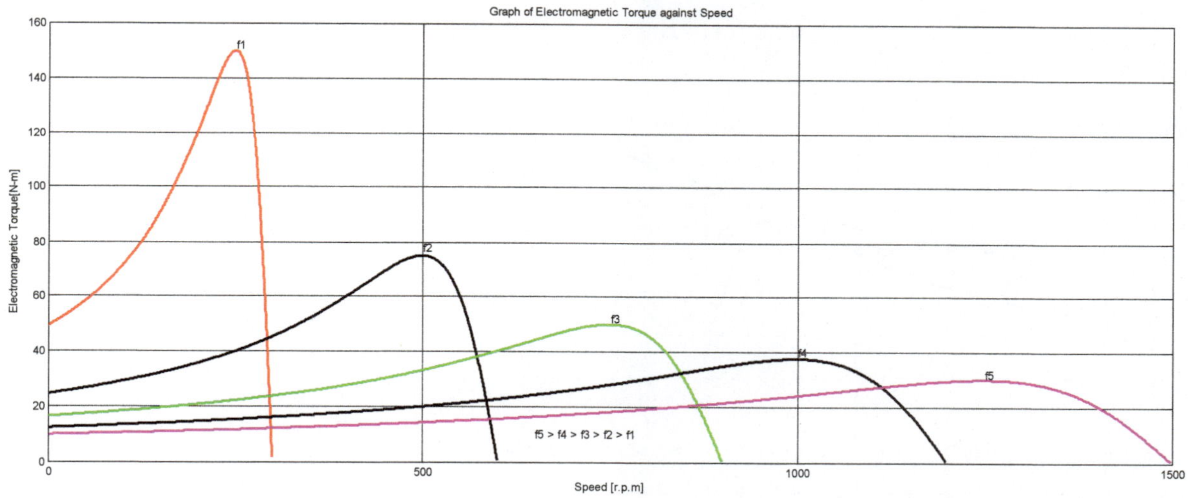

Figure 6. Torque-speed curves of 3-phase induction motor by varying line frequency

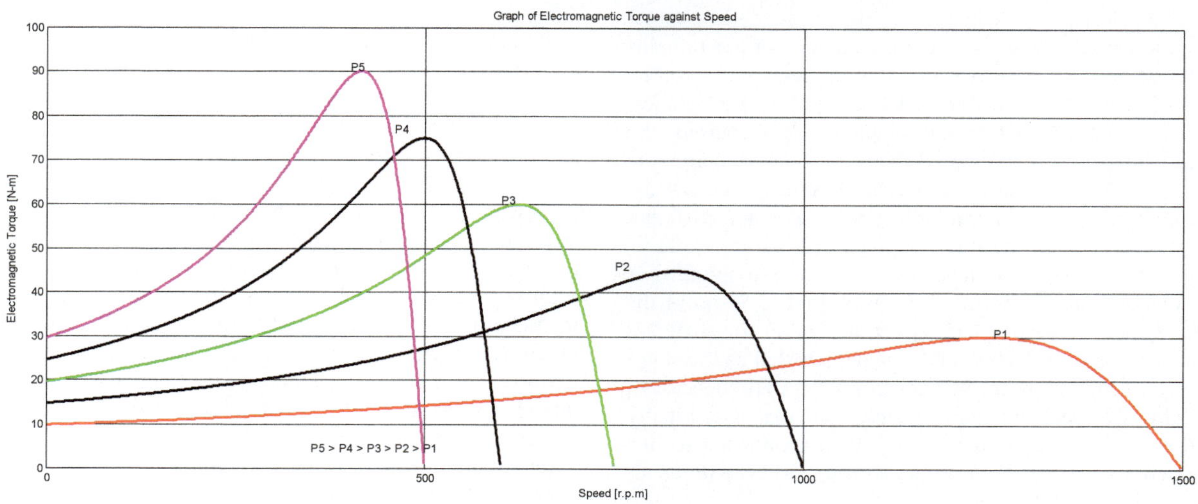

Figure 7. Torque-speed curves of 3 phase induction motor by varying the number of poles.

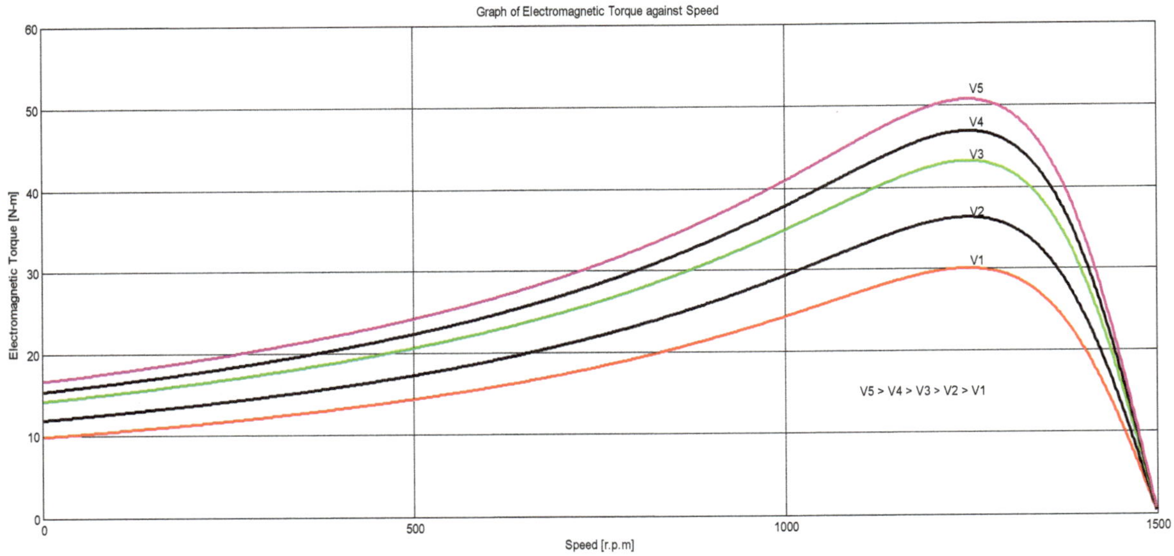

Figure 8. Torque-speed curves of 3-phase induction motor by varying the supply voltage

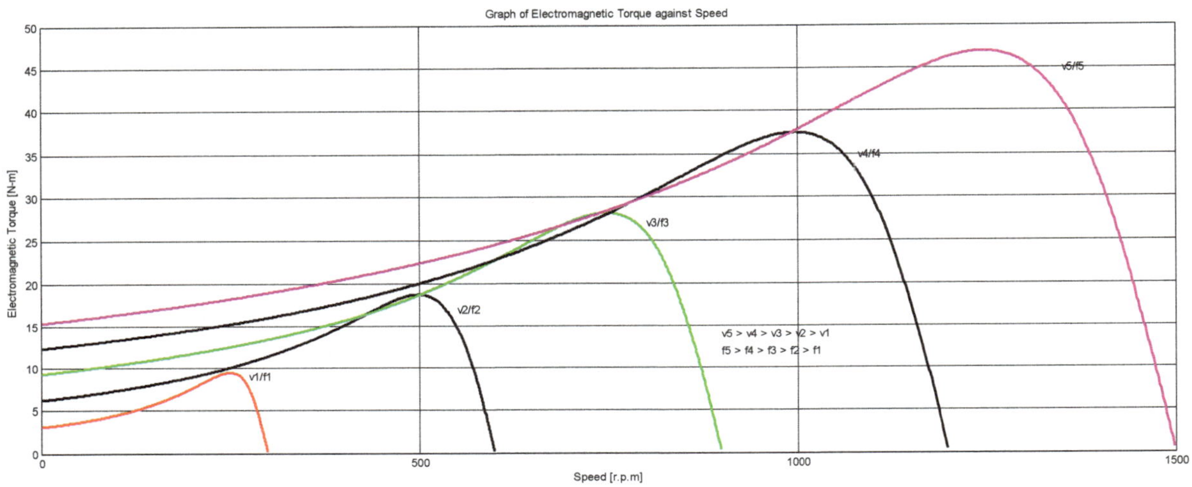

Figure 9. Torque-speed curves of 3-phase induction motor by varying stator voltage/line frequency

Figure 5 shows the torque-speed curve of an induction motor for different values of rotor resistances $(R_1, R_2, R_3, R_4, R_5)$. From the graph it is observed that as the rotor resistance is increased, the speed decreases (or slip increases) while the electromagnetic torque remains constant. Figure 6 shows the torque-speed curve of an induction motor for different values of line frequencies. The plot depicts that as the frequency is increased, the speed increases (or slip deceases), consequently the electromagnetic torque decreases. Figure 7 shows the torque-speed curve of an induction motor for different values of poles. The plot depicts that as the pole is increased, the speed reduces (or slip increases), and thus the electromagnetic torque increases. Figure 8 shows the torque-speed curve of an induction motor for different values of stator voltages. The plot depicts that as the stator voltage is increased, the speed increases (or slip deceases), and thus the electromagnetic torque increases also. Finally, if the line voltage is changed in direct proportion to line frequency, the resulting torque-speed curves will be as shown in Figure 9.

6. Conclusion

Induction motors are not good machines for applications requiring considerable speed control. The normal operating range of a typical induction motor is confined to less than 5% slip, so the need for different speed controls arises. Torque-Speed characteristics for different methods of speed control of a three-phase Induction Motor were obtained and analyzed by developing MATLAB codes. In variable rotor resistance, the maximum electromagnetic torque is independent of rotor resistance, so, when the rotor resistance is increased, the maximum electromagnetic torque is unaffected but the speed at which it occurs can be directly controlled. In variable stator frequency, Increase in frequency increases the torque-speed relation and a decrease in frequency decreases the torque-speed relation of the motor. In variable stator voltage, varying the terminal voltage will vary the operating speed but with also a variation of operating torque. In terms of the range of speed variations, it is not significant hence this method is only suitable for small motors only. In constant volts/hertz control, the supply voltage as well as the supply frequency can be varied such that the ratio remains constant and the flux remains constant too. So, different operating zones for various speeds and torques can be got and also we can get different synchronous speed with almost same maximum torque. Thus the motor is completely utilized and also we have a good range of speed control. In pole changing method of speed control, to achieve higher speed and torque, lesser number of poles is required and to achieve lower speed and torque, the number of poles ought to be increased. Finally, it is recommended that any of the speed control method be implemented in a prototype.

References

[1] Rashidi, F., Rashidi, M.,"Design of robust sliding mode speed control with fuzzy approach for induction motor", IEEE international conference on industrial technology (ICIT03), pp. 27-30.

[2] Stephen J. Chapman; Electric Machinery Fundamental; Second Edition; McGraw-Hill companies; 2012.

[3] O. I. Okoro; Introduction to Matlab/Simulink for Engineers and Scientists; Second Edition; John Jacob's Classic Publishers Ltd; 2008.

[4] P. S. Bimbhra; Power Electronics; 3rd Edition; Khanna Publishers; 2005.

[5] A. E. Fitzgerald et al; Electric machinery; Sixth Edition; McGraw publishers; 2003.

[6] Shilpa V. Kailaswar, R. A. Keswani, "Speed Control of Three Phase Induction Motor by V/f Method for Batching Motion System" International Journal of Engineering Research and Application (IJERA) Vol. 3, Issue 2, March-April 2013, pp. 1732-1736.

[7] Gourambika, M. S. Aspalli, "Speed Control of Three Phase Induction Motor by VVVF method using G7/A-1000 Drive" International Journal of Recent Technology and Engineering (IJRTE) Vol.-3, Issue-4, Sept 2014, pp. 43-49.

[8] Chee- Mun Ong; Dynamic Simulation of Electric Machinery using Matlab and Simulink; Prentice hall PTR; 1998.

[9] M.S.Aspalli, R. Asha, P.V. Hunagund, "Three Phase Induction Motor Drive Using IGBTs and Constant v/f Method" International Journal of Advanced in Electrical, Electronics and instrumentation Engineering vol. 1, Issue 5, November 2012, pp. 463-469.

Appendix

MATLAB Code for Speed Control of three-phase Induction motor by varying the stator voltage. See Figure 7.

```
%===========================================================================
% M-file for Speed Control of three-phase Induction motor
% by varying the stator voltage.
%===========================================================================
s= 0:0.001:1; % slip
R1= 0.6; % stator resistance
X1= 1.1; % stator leakage reactance
R2p= 0.3; % rotor resistance referred to stator
X2p= 0.5; % rotor leakage reactance referred to stator
f= 50;P=4; % Supply frequency
Ws= 120*f/P; % synchronous speed
%Wm= -100:20:900; % rotor speed
Wm= (1-s).*Ws; % rotor speed
for Vs=400
  A= (R1+(R2p./s).^2);
  B= (X1+X2p).^2;
  V= Vs./sqrt(3); % supply voltage
  I1=(V.^2./(A+B)); % stator current
  A1= 3./Wm;
  A2= (1-s)./s;
  A3= R2p.*I1;
  Tem1= A1.*A2.*A3; % Electromagnetic torque
end
for Vs=440
  A= (R1+(R2p./s).^2);
  B= (X1+X2p).^2;
  V= Vs./sqrt(3); % supply voltage
  I1=(V.^2./(A+B)); % stator current
  A1= 3./Wm;
  A2= (1-s)./s;
  A3= R2p.*I1;
  Tem2= A1.*A2.*A3; % Electromagnetic torque
end
for Vs=480
  A= (R1+(R2p./s).^2);
  B= (X1+X2p).^2;
  V= Vs./sqrt(3); % supply voltage
  I1=(V.^2./(A+B)); % stator current
  A1= 3./Wm;
  A2= (1-s)./s;
  A3= R2p.*I1;
  Tem3= A1.*A2.*A3; % Electromagnetic torque
end
for Vs=500
  A= (R1+(R2p./s).^2);
  B= (X1+X2p).^2;
  V= Vs./sqrt(3); % supply voltage
  I1=(V.^2./(A+B)); % stator current
  A1= 3./Wm;
  A2= (1-s)./s;
  A3= R2p.*I1;
  Tem4= A1.*A2.*A3; % Electromagnetic torque
end
for Vs=520
  A= (R1+(R2p./s).^2);
  B= (X1+X2p).^2;
  V= Vs./sqrt(3); % supply voltage
  I1=(V.^2./(A+B)); % stator current
  A1= 3./Wm;
  A2= (1-s)./s;
  A3= R2p.*I1;
  Tem5= A1.*A2.*A3; % Electromagnetic torque
end
figure(1)
plot(Wm,Tem1,'r',Wm,Tem2,'b',Wm,Tem3,'g',Wm,Te
m4,'k',Wm,Tem5,'m','linewidth',2.5)
grid on
xlabel('Speed [r.p.m]')
ylabel('Electromagnetic Torque [N-m]')
title('Graph of Electromagnetic Torque against Speed')
text(1250,31,'V1')
text(1250,37,'V2')
text(1250,44,'V3')
text(1250,48,'V4')
text(1250,52,'V5')
text(1100,15,'V5 > V4 > V3 > V2 > V1')
```

Electric Circuit Analysis Using a Simple Assumption-Based Technique

Mousaab M. Nahas*

Electrical and Computer Engineering Department, Faculty of Engineering, University of Jeddah, Jeddah, Saudi Arabia
*Corresponding author: mnahas1@uj.edu.sa

Abstract This paper demonstrates a simple technique to analyze linear electric circuits. The technique is based on assuming some voltage or current values according to the given resistance/impedance values such that a relative circuit performance is obtained. The actual circuit results are achieved directly by normalizing the assumed values using a specific formula. In practice, this method either reduces the number of simultaneous equations or simplifies the mathematical formulas; hence the overall analysis procedure is reasonably accelerated. It is highly recommended that this method is taught in electric circuit courses especially in the beginning chapters so that students can have more variety of circuit analysis techniques. It is also possible to apply such a method to verify an existing solution or vice versa.

Keywords: *electric circuits, circuit analysis techniques, circuit theory, DC and AC circuits, methods of analysis*

1. Introduction

It is already recognized that the electric circuit theory is one of the main fundamental theories on which all branches of electrical engineering are based. Therefore, electric circuit analysis techniques are very important to be taught to electrical engineering students at the beginning of electrical engineering program [1]. They are also useful for students in other specializations such as physics and applied mathematics [2,3,4]. It is well-known that there are different circuit analysis techniques which are all based on three main laws: Ohm's law, KCL and KVL. The techniques can involve combining resistors/impedances in series or parallel, voltage division or current division. For more complex circuits, nodal or mesh analysis are advised to be used where the solution procedure is significantly facilitated [1,5-12]. In fact, the above techniques along with the Y-Δ transformations are fairly sufficient to solve all linear electric circuits where the appropriate procedure is decided according to the given circuit design.

In this research, we propose additional simple technique that is never presented elsewhere such that a student can have more options either to analyze a given circuit or to verify an existing solution. This technique is again based on the three famous laws mentioned above but it works backwards since we begin by assuming circuit values for some elements located far from the source. After complete analysis, these values are normalized to obtain the actual circuit results. This is unlike original methods in which the analysis usually starts from the given source and moves forwards to reach the actual circuit values. The technique proposed here can apply easily to most linear circuit designs including resistive DC circuits, charging *RC*, *RL* and *RLC* circuits, and AC circuits. It is not claimed that this technique can replace the traditional methods, especially the nodal and mesh techniques, but in many cases it can be the best option where the number or complexity of equations is reasonably reduced.

For simplicity and as done in all references, we start with simple resistive DC circuits upon which our proposed method is fully explained. Various cases (examples) are given to ensure the validity of the technique over different designs. These examples are mostly taken from one reference [1] just for consistency and rapid verification. Other circuits including AC ones are then presented in the later sections.

2. Simple Resistive Circuits

As known, simple resistive circuits are composed of resistors that are connected in series, parallel or both. The new technique is mainly based on Ohm's law which states that the voltage across a resistor is directly proportional to the current passing through the resistor [5-12]. Mathematically,

$$v = Ri \tag{1}$$

It is therefore possible to say that the voltage across a resistor is directly proportional to the given resistance assuming constant current. On the other hand, the current through a resistor is inversely proportional to the given resistance assuming constant voltage. This would imply that, for any resistor, the voltage value can be initially assumed where it is either a multiple or fraction of the given resistance. This is of course assuming that i is the proportionality constant in (1). For simplicity, the voltage and resistance can be assumed to be equal where i is assumed to be 1. Therefore, equation (1) temporarily becomes

$$V = R \quad (2)$$

where capital letter is used to distinguish from the actual voltage value.

In contrast, if the voltage is fixed, it is possible to predict a current value according to the resistance where the proportionality constant is the voltage in this case.

The above two situations are applicable in case that there are several resistors in the circuit. However, both situations cannot be applied simultaneously since they are based on assumptions hence relative results, and dual assumption would corrupt the relativity. Therefore, the choice of the right approach must be dependent on the circuit design which mostly involves series and/or parallel connections. The following sections discuss the different possible options and their application.

2.1. Series Resistors

Suppose we have resistors connected in series as shown in Figure 1. The original approach is to combine them into one equivalent resistor thus the current i is obtained directly by Ohm's law hence v_1 and v_2. Alternatively, the voltage division principle can be used where the source voltage v is divided among the resistors in direct proportion to their resistances [5,6,8].

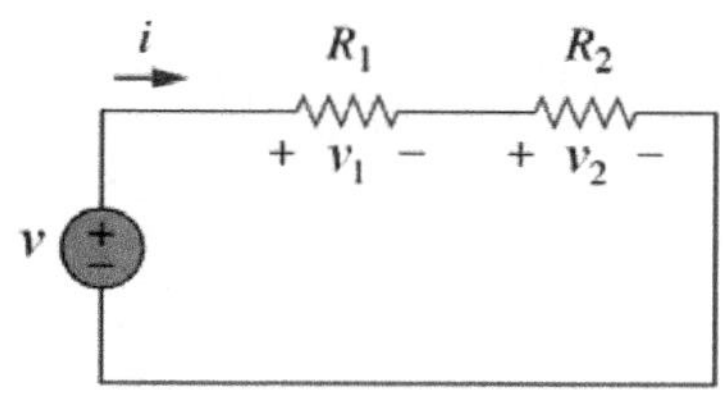

Figure 1. Series resistors

In our proposed method, since the current is already constant through both resistors, it is possible to apply equation (2) as

$$V_1 = R_1, V_2 = R_2$$

where V_1 and V_2 are relative voltages assuming $i = 1$.

According to KVL, the total assumed voltage V is

$$V = V_1 + V_2 = R_1 + R_2 \quad (3)$$

Certainly, this value is relative and is most probably a multiple of the given source voltage v. Thus it is possible to define a *multiplicity factor* (m) that is the ratio of the relative voltage obtained by (3) to the actual given voltage, i.e.

$$m = \frac{V}{v} \quad (4)$$

Therefore, the actual voltages across the resistors are

$$v_1 = \frac{V_1}{m}, v_2 = \frac{V_2}{m} \quad (5)$$

In practice, the above solution procedure can be done very fast where simple calculations are involved.

The unit prefix of the assumed values must be compatible with the given source unit. Typically, the assumed voltages are in V while the currents are either in A or mA. Basically, if the resistances are given in kΩ, the currents are assumed in mA.

2.2. Parallel Resistors

Suppose we have resistors connected in parallel as shown in Figure 2. Again, the original approach is to combine them into one equivalent resistor thus the current i is obtained directly by (1). The sub-currents i_1 and i_2 are obtained either by (1) or by current division where the total current i is divided among the resistors in direct proportion to the other resistance in parallel [6,8,10].

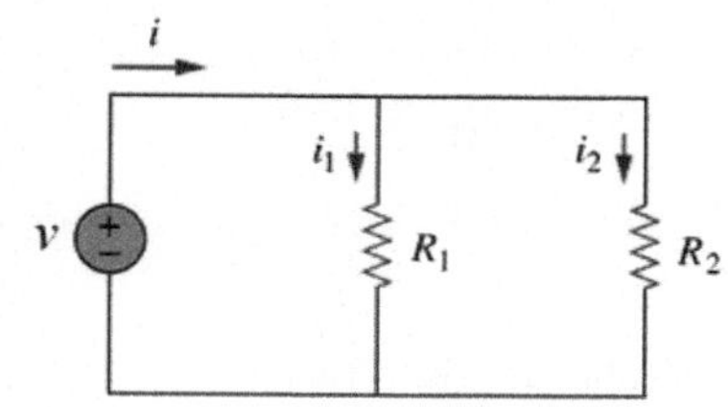

Figure 2. Parallel resistors

In our approach, since the voltage is already constant across both resistors and the current through each resistor is directly proportional to the other resistance, it is possible to assume the current values as

$$I_1 = R_2, I_2 = R_1 \quad (6)$$

According to KCL, the total assumed current I is

$$I = I_1 + I_2 = R_2 + R_1 \quad (7)$$

and

$$V = V_1 = V_2 = R_1 I_2 = R_2 I_1$$

From (6) and (7), the voltage is then

$$V = R_1 R_2 \quad (8)$$

Obviously, the current and voltage values obtained above are all relative and are multiples (or fractions) of the actual currents and voltages, respectively. The actual values are then achieved by dividing all assumed values by the multiplicity factor m defined in equation (4). Therefore,

$$v = \frac{V}{m}, i = \frac{I}{m}, i_1 = \frac{I_1}{m}, i_2 = \frac{I_2}{m}. \quad (9)$$

If the circuit has a current source rather than a voltage source, the multiplicity factor m is then defined as the ratio of the relative current obtained by (7) to the actual current i given in the circuit. That is

$$m = \frac{I}{i}$$

In general, the multiplicity factor can be re-defined as the ratio of the assumed value to the actual value and is determined based on the source type given in the circuit, i.e.

$$m = \frac{\text{Assumed}}{\text{Actual}} \quad (10)$$

3. The Technique Application

3.1. Series Resistors Example

Consider that we are required to find v_1, v_2 and i in the circuit shown in Figure 3.

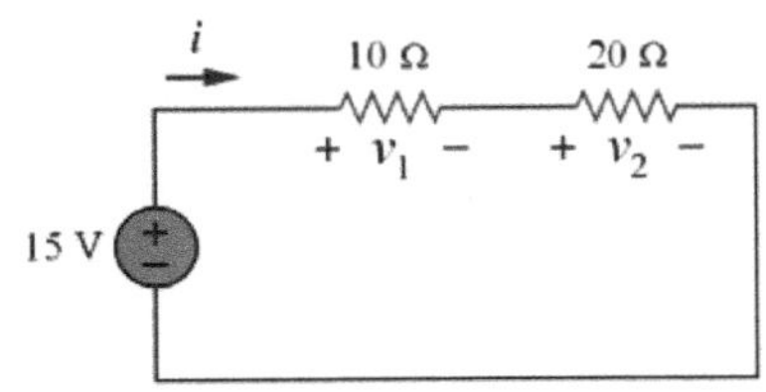

Figure 3. Series resistors example

From (2), we assume that

$$V_1 = 10\text{V}, V_2 = 20\text{V}$$

Thus, the total relative voltage is

$$V = 30 \text{ V}$$

From (4), the multiplicity factor is

$$m = \frac{30}{15} = 2$$

Therefore,

$$v_1 = \frac{10}{2} = 5\text{V}, v_2 = \frac{20}{2} = 10\text{V}.$$

Hence

$$i = 0.5\text{A}.$$

3.2. Parallel Resistors Example

Consider we are required to determine all currents and voltages in the circuit shown in Figure 4 [1].

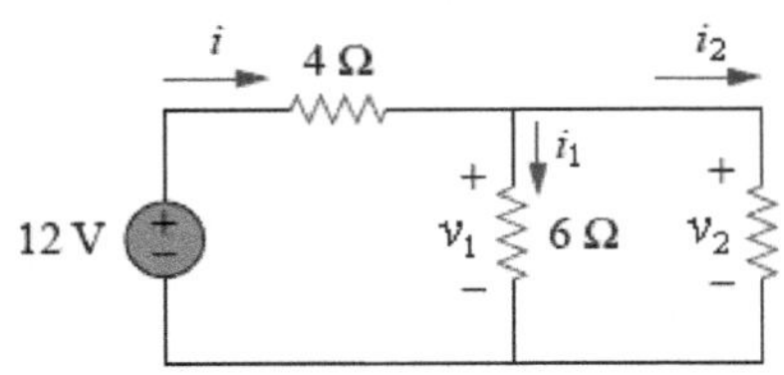

Figure 4. Parallel resistors example

From (6), we can assume that

$$I_1 = 3\text{A}, I_2 = 6\text{A}.$$

Thus, the total relative current is

$$I = 9\text{A}$$

and the voltages are

$$V_1 = V_2 = 3 \times 6 = 18\text{V}.$$

Now, the total relative voltage of the circuit is the sum of V_1 and the 4-Ω resistor's relative voltage which is basically $4I$ that is 36 V. Therefore,

$$V = 36 + 18 = 54\text{V}.$$

Thus, the multiplicity factor is

$$m = \frac{54}{12} = 4.5$$

As a result,

$$v_1 = v_2 = \frac{18}{4.5} = 4\text{V}, i = \frac{9}{4.5} = 2\text{A},$$

$$i_1 = \frac{3}{4.5} = 0.67\text{A}, i_2 = \frac{6}{4.5} = 1.33\text{A}.$$

Obviously, the above solutions are excessively detailed and can be shortened into fewer steps if the solver is familiar with the technique.

3.3. Series Resistors with Dependent Source Example

Consider the circuit shown in Figure 5 [1], in which we need to find i and v_o.

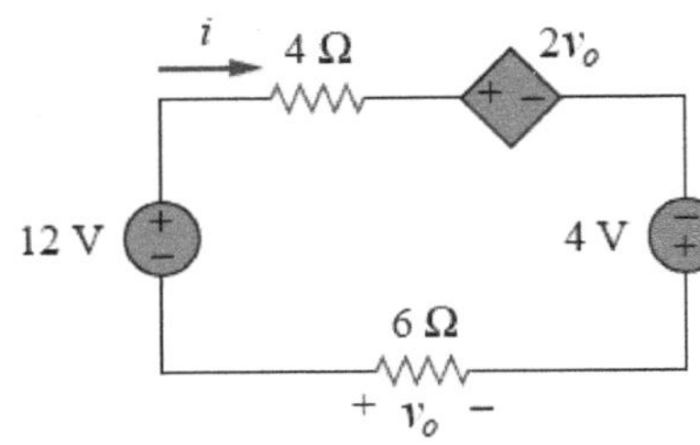

Figure 5. Series resistors with dependent source

The total actual supplied voltage is

$$v = 12{+}4 = 16 \text{ V}.$$

The assumed voltages are $V_{4\Omega}$ = 4 V and V_o = −6 V (according to the given polarity)

Since, the dependent source has a function of v_o which is assumed to be V_o, it is added to the relative voltages, thus equation (3) becomes

$$V = |V_{4\Omega}| + |V_o| + 2V_o = 4 + 6 + 2(-6) = -2\text{V} \quad (11)$$

Note that the resistor voltages are always positive in (11) as passive sign convention is always satisfied. However, the dependent source voltage is not.

The multiplicity factor is

$$m = \frac{-2}{16} = -\frac{1}{8}$$

Hence

$$i = \frac{1}{-1/8} = -8\text{A}, v_2 = \frac{-6}{-1/8} = 48\text{V}.$$

3.4. Parallel Resistors with Dependent Source Example

Consider that we want to determine i_o and v_o in the parallel circuit shown in Figure 6 [1].

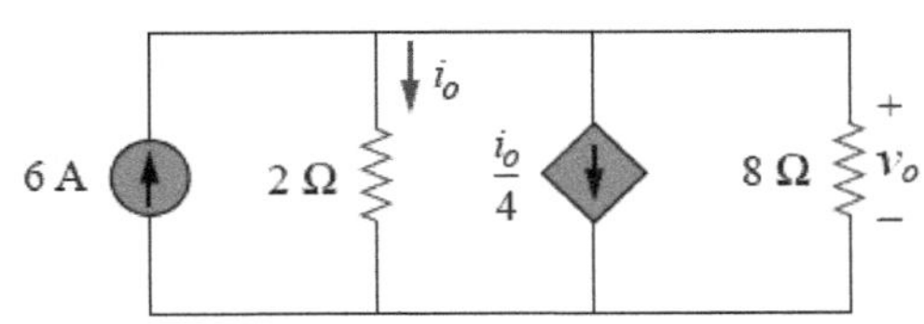

Figure 6. Parallel resistors with dependent source

According to (6), the relative (assumed) values are

$$I_o = 8\text{A},\ I_{8\Omega} = 2\text{A} \Rightarrow V_o = 2\times 8 = 16\text{V}.$$

Since, the dependent source is a function of i_o which is assumed to be I_o, it is added to the relative currents, thus equation (7) becomes

$$I = I_o + I_{8\Omega} + \frac{I_o}{4} = 8 + 2 + \frac{8}{4} = 12\text{A}. \tag{12}$$

Note that if the dependent source was in the opposite direction, it would be subtracted in (12).

However, the multiplicity factor is

$$m = \frac{12}{6} = 2$$

and hence

$$i_o = \frac{8}{2} = 4\text{A},\ v_o = \frac{16}{2} = 8\text{V}.$$

In fact, the solution of the above circuits in Figure 5 and Figure 6 is much shorter than that obtained by traditional approaches.

3.5. Charging *RLC* Circuit Example

Consider the circuit in Figure 7 [1]. We would like to find i, v_C and i_L under dc conditions.

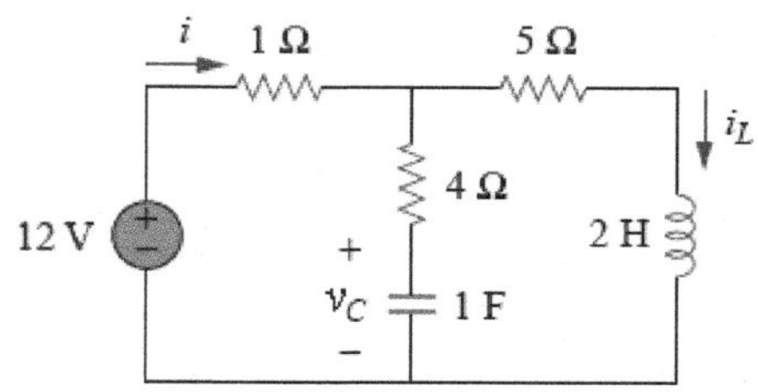

Figure 7. Charging *RLC* circuit

Under dc conditions, the circuit becomes as shown in Figure 8, which is a simple resistive circuit with series connection.

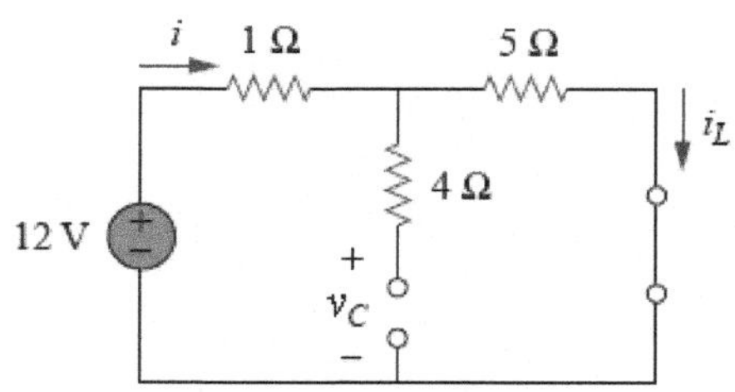

Figure 8. *RLC* circuit under dc conditions

We assume

$$V_{1\Omega} = 1\text{V}, V_{5\Omega} = 5\text{V} \Rightarrow I = 1\text{A}.$$

Using equation (3),

$$V = 6\text{V}.$$

Thus,

$$m = \frac{6}{12} = \frac{1}{2}$$

As a result,

$$i = i_L = \frac{1}{1/2} = 2\text{A}, v_C = \frac{5}{1/2} = 10\text{V}.$$

4. AC Circuits

As known, AC circuits are usually composed of impedances that are connected in series, parallel or both. These impedances are related to resistors, capacitors and inductors. Since Ohm's and Kirchhoff's laws work in phasor domain with all kinds of passive elements in the same way used with DC circuit resistors, it is possible to apply our technique on AC circuits as well. To do so, we shall now base on Ohm's law in phasor form which states that the phasor voltage across a passive element is directly proportional to the phasor current passing through the element [5-12]. Mathematically,

$$\mathbf{V} = \mathbf{ZI} \tag{13}$$

where **Z** is the element impedance. Again, it is possible to state that the phasor voltage is directly proportional to the given impedance assuming constant current. Also, the phasor current through an element is inversely proportional to its impedance assuming constant voltage. This would imply that the phasor voltage value can be initially assumed as long as it is in direct proportion to **Z**, or the phasor current is assumed such that it is inversely proportional to **Z**. Doing this would allow finding a relative circuit values that are eventually normalized to reach the actual values. However, the application of the above two assumptions is based on the AC circuit design as discussed in the next sections.

4.1. Series Impedances

Suppose we have AC circuit with passive elements connected in series as shown in Figure 9. The traditional approach is to combine the impedances into one equivalent impedance thus the phasor current **I** is obtained directly by (13) hence $\mathbf{V}_1$ and $\mathbf{V}_2$. Alternatively, the voltage division principle can be used where the source voltage **V** is divided among the elements in direct proportion to their impedances [1,7,8,9].

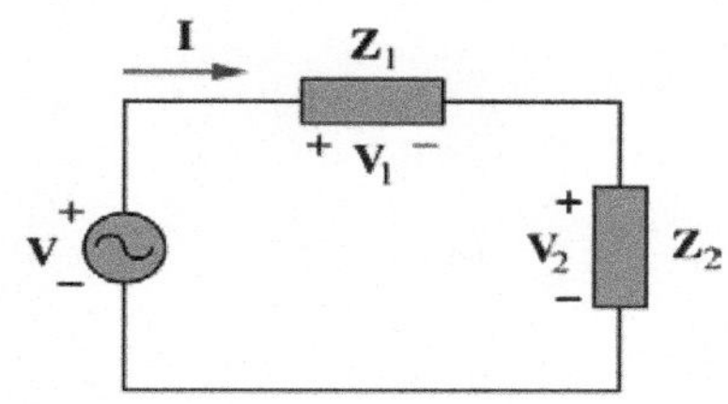

Figure 9. Series impedances

In our technique, we solve the circuit by assuming the element voltage value in direct proportion to its given impedance as mentioned above. For simplicity, we can assume that the voltage across any element is

$$\mathbf{V}' = \mathbf{Z} \tag{14}$$

where the current is assumed to be 1, and prime is used to distinguish from the real voltage value.

Applying (14) on both elements, we get the relative voltages as

$$\mathbf{V}_1' = \mathbf{Z}_1, \mathbf{V}_2' = \mathbf{Z}_2$$

According to KVL, the total assumed voltage $\mathbf{V}'$ is

$$\mathbf{V}' = \mathbf{V}_1' + \mathbf{V}_2' = \mathbf{Z}_1 + \mathbf{Z}_2 \tag{15}$$

Clearly, this value is relative and is simply a multiple or fraction of the given source voltage **V**. Thus it is possible to define the multiplicity factor in phasor domain as the ratio of the relative phasor value (whether voltage or current) to the source phasor value given in the circuit. In this case,

$$\mathbf{M} = \frac{\mathbf{V}'}{\mathbf{V}} \tag{16}$$

Note that the multiplicity factor is now complex, and that is why it is represented by capital-bold letter.

Therefore, the actual phasor voltages across the elements are basically

$$\mathbf{V}_1 = \frac{\mathbf{V}_1'}{\mathbf{M}}, \mathbf{V}_2 = \frac{\mathbf{V}_2'}{\mathbf{M}}. \tag{17}$$

We believe that the above procedure is simpler than the conventional procedure for such circuit.

4.2. Parallel Impedances

Suppose we have impedances connected in parallel as shown in Figure 10. Again, the conventional approach is to combine them into one impedance thus the current **I** is obtained directly by Ohm's law hence $\mathbf{I}_1$ and $\mathbf{I}_2$. Alternatively, the current division principle can be used where the total phasor current **I** is divided among the elements in direct proportion to the other impedance in parallel [6,9,12].

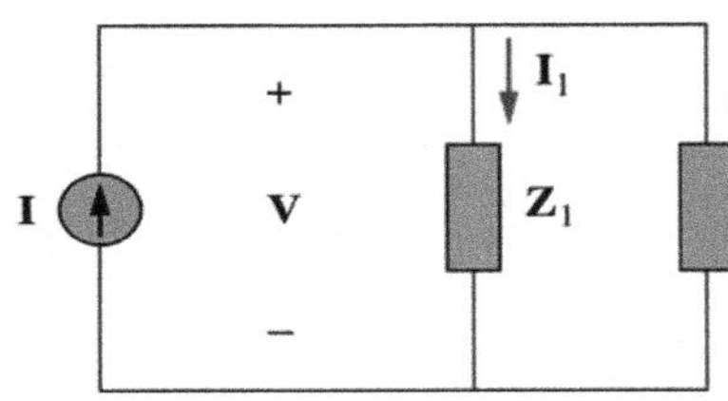

Figure 10. Parallel impedances

In our approach, since the voltage is constant across both elements and the current through each one is directly proportional to the other impedance, it is possible to assume the phasor current values as

$$\mathbf{I}_1' = \mathbf{Z}_2, \mathbf{I}_2' = \mathbf{Z}_1 \tag{18}$$

According to KCL, the total assumed current $\mathbf{I}'$ is

$$\mathbf{I}' = \mathbf{I}_1' + \mathbf{I}_2' = \mathbf{Z}_2 + \mathbf{Z}_1 \tag{19}$$

and the voltage across both elements is

$$\mathbf{V}' = \mathbf{V}_1' = \mathbf{V}_2' = \mathbf{Z}_1\mathbf{I}_2' = \mathbf{Z}_2\mathbf{I}_1'$$

Using (18), the parallel voltage is then

$$\mathbf{V}' = \mathbf{Z}_1\mathbf{Z}_2 \tag{20}$$

The phasor current and voltage values obtained by (19) and (20) are obviously relative and are again multiples or fractions of the actual current **I** and voltage **V**, respectively. The actual values are then achieved by dividing all assumed values by the multiplicity factor **M** which is in this case

$$\mathbf{M} = \frac{\mathbf{I}'}{\mathbf{I}} \tag{21}$$

Therefore,

$$\mathbf{I} = \frac{\mathbf{I}'}{\mathbf{M}}, \mathbf{V} = \frac{\mathbf{V}'}{\mathbf{M}}, \mathbf{I}_1 = \frac{\mathbf{I}_1'}{\mathbf{M}}, \mathbf{I}_2 = \frac{\mathbf{I}_2'}{\mathbf{M}}. \tag{22}$$

5. The Technique Application in AC Circuits

5.1. Series Impedances Example

Assume we need to find $v(t)$ and $i(t)$ in the circuit shown in Figure 11 [1].

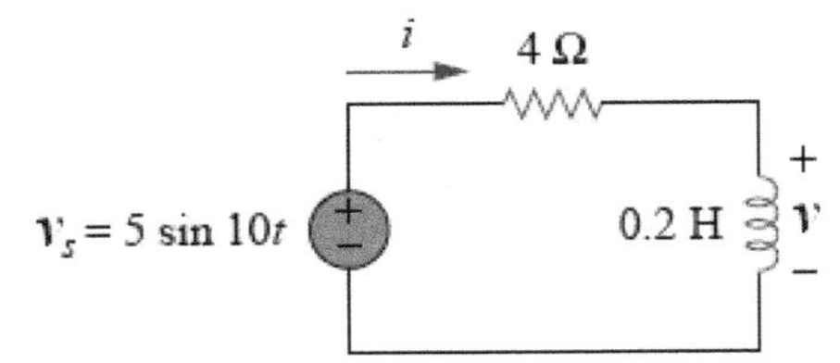

Figure 11. Series impedances example

$$\mathbf{V}_S = 5\angle -90°\text{V}, \omega = 10$$

From (15), the total assumed (relative) AC voltage and current are

$$\mathbf{V}_S' = 4 + j2\text{V}, \ \mathbf{I}' = 1\text{A}$$

Thus,

$$\mathbf{M} = \frac{\mathbf{V}_S'}{\mathbf{V}_S} = \frac{4+j2}{5\angle\text{-}90°} = 0.894\angle 116.57°$$

Therefore, the actual phasor voltage and current are

$$\mathbf{V} = \frac{j2}{0.894\angle 116.57°} = 2.236\angle\text{-}26.57°\text{V}$$

$$\mathbf{I} = \frac{1}{0.894\angle 116.57°} = 1.118\angle\text{-}116.57°\text{A}$$

Hence,

$$v(t) = 2.236\sin(10t + 63.43°)\text{V}$$

$$i(t) = 1.118\sin(10t - 26.57°)\text{V}$$

5.2. Parallel Impedances Example

Consider we are required to determine $v_o(t)$ in the circuit shown in Figure 12 [1].

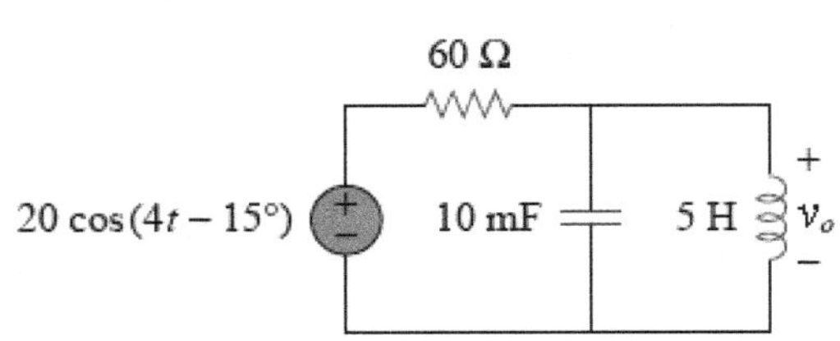

Figure 12. Parallel impedances example

Since the inductor and capacitor are in parallel, we can use (18) to assume that

$$\mathbf{I}_L' = \mathbf{Z}_C, \ \mathbf{I}_C' = \mathbf{Z}_L$$

That is

$$\mathbf{I}'_L = -j25\text{A}, \mathbf{I}'_C = j20\text{A}.$$

Therefore, the total assumed current according to (19) is

$$\mathbf{I}' = \mathbf{I}'_L + \mathbf{I}'_C = -j5\text{A}$$

and the relative voltage across the inductor and capacitor according to (20) is

$$\mathbf{V}'_o = \mathbf{I}'_L \times \mathbf{I}'_C = 500 \text{ V}.$$

From the above, the relative voltage across the 60-Ω resistor can be obtained using 60**I′** which is –*j*300 V. Therefore, the total relative voltage for the whole circuit is

$$\mathbf{V}' = 500 - j300.$$

The multiplicity factor is

$$\mathbf{M} = \frac{500 - j300}{20\angle\text{-}15°} = 29.15\angle\text{-}15.96°$$

As a result,

$$\mathbf{V}_o = \frac{500}{29.15\angle\text{-}15.96°} = 17.15\angle 15.96°\text{V}$$

In time domain,

$$v_o(t) = 17.15\cos(4t + 15.96°)\text{V}.$$

6. Conclusion

In this paper, we introduce a simple technique to analyze linear DC and AC electric circuits. The technique is based on assuming element values according to the given resistances/impedances such that a relative circuit results are attained. The actual performance is achieved by normalizing the assumed values using a specific formula. It is practically shown that this technique significantly facilitates the solution procedure in many cases where it either simplifies the calculations or reduces the number of equations. However, it is not claimed that such a method is proposed to replace the conventional methods, but it is highly recommended to be considered for circuit analysis where some complex steps can effectively be saved. In addition, the technique is advised to be used with the Y-Δ transformations and the circuit theorems including superposition, source transformation, Thevenin's and Norton's theorems. Furthermore, we suggest that this method is taught in electric circuit courses especially in the introductory chapters to give the students more variety of circuit solving techniques.

References

[1] C. Alexander and M. Sadiku, *Fundamentals of Electric Circuits*, McGraw-Hill, January 2012.

[2] A. Fischer-Cripps, *The Electronics Companion: Devices and Circuits for Physicists and Engineers*, CRC Press, September 2014.

[3] J. Walker, *Physics*, Addison-Wesley, January 2009.

[4] C. Marinov, P. Neittaanmäki, *Mathematical Models in Electrical Circuits: Theory and Applications*, Springer, September 2012.

[5] T. Floyd, *Electric Circuits Fundamentals*, Prentice Hall, July 2009.

[6] T. Floyd, *Principles of Electric Circuits: Conventional Current Version*, Prentice Hall, March 2009.

[7] R. Boylestad, *Introductory Circuit Analysis*, Prentice Hall, January 2010.

[8] C. Steinmetz, *Theory and Calculation of Electric Circuits*, Watchmaker Publishing, April 2010.

[9] R. Thomas, A. Rosa and G. Toussaint, *The Analysis and Design of Linear Circuits*, Wiley, December 2011.

[10] J. Svoboda and R. Dorf, *Introduction to Electric Circuits*, Wiley, March 2013.

[11] J. Nilsson and S. Riedel, *Electric Circuits*, Prentice Hall, January 2014.

[12] H. Jackson, D. Temple, and B. Kelly, *Introduction to Electric Circuits*, Oxford University Press, July 2015.

30

A Review on Hybrid Controller Using Soft Computing Algorithms

Lokesh Kumar Agrawal[1,*], Bhavesh Kumar Chauhan[2], G. K. Benerjee[1]

[1]Department of Electrical Engineering, IFTM University, Moradabad(U.P.), India
[2]Department of Electrical Engineering, ABESIT, Ghaziabad (U.P.), India
*Corresponding author: lokesh_7575@rediffmail.com

Abstract Aadaptability and self-organization of a system is two key factors, when it comes to how well the system is surviving for the changes to the environment and how these work within the plant. Different tuning methods and soft computing techniques improve these two factors in controllers. Considering the increasing complexity of dynamic systems along with their need for feedback controls, using more complicated controls has become necessary and these techniques can be a suitable response to this necessity. This paper briefly describes a review on different techniques used for PID tuning as well as soft computing algorithm for hybrid controllers. This paper provides a comprehensive reference source for people working with hybrid controllers.

Keywords: *PI Controller, PID Controller, FLC, Neuro-fuzzy system, Tuning, ANFIS*

1. Introduction

Conventional proportional-integral-derivative (PID) controllers are extensively used for industrial automation and process control. PID controllers are simple, easy to understand and implement in hardware and software and do not require a process model for operation [1]. Many nonlinear processes can be satisfactorily controlled using PID controller parameters provided that controller parameters are tuned well. According to a survey in 1989, 90% of process industries use them. This widespread use of PID in industry can be attributed to their simplicity and ease of re-tuning on-line [2].

The PID controller is so named because its output sum of three terms, proportional, integral and derivative term.

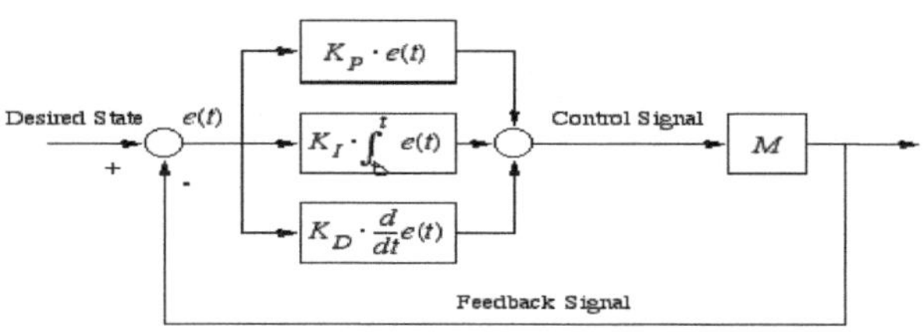

Figure 1. Block diagram of PID controller

Each of these terms is dependent on the error value e(t) between the input and the output,

$$output = K_P e(t) + K_I \int_0^t e(t)dt + K_D \frac{de(t)}{dt} \qquad (1)$$

Where K_P, K_I and K_D are the P, I and D parameters respectively. By tuning these constants in PID algorithm, the controller provides control action designed for specific requirements [3].

The Fuzzy Logic is closer as to human reasoning and in language than conventional logical systems [3]. Fuzzy hardware systems have been developed, including fuzzy rules, fuzzy interface systems and optical fuzzy inference systems [3]. Fuzzy logic is successful applications for control engineering field and can be used for various parameters control of the real time systems. The main problem to overcome in applications is the uncertainty in fuzzy modeling of the linguistic structure for process [3].

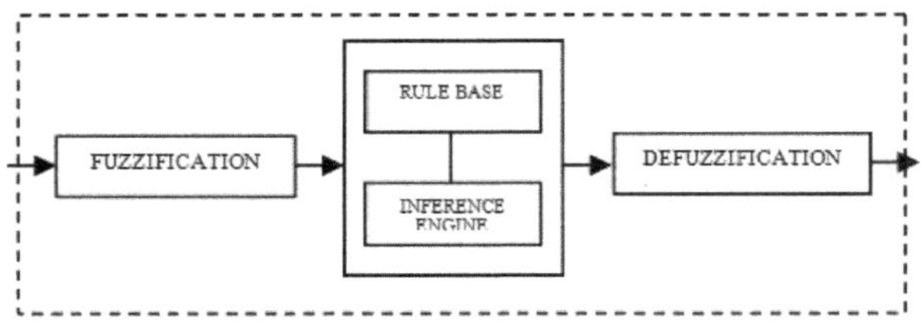

Figure 2. Block diagram of Fuzzy logic controller

While Artificial Neural Networks are mathematical models inspired from biological nervous system [4]. They can get large number of inputs and learn solely from training samples [4]. As mathematical models for biological nervous system, artificial neural networks are useful in establishing relationships between input and output of the system [4]. Neural network is collection of artificial neuron. Neural network can learn from data. However, the knowledge learned by neural networks has been difficult to understand while Fuzzy models are easy to understand due to use linguistic terms and the structure of IF-THEN rules.

However, fuzzy logic by itself cannot learn. Since neural networks can learn.

Neuro-fuzzy system is the combination of artificial neural network and fuzzy logic [4]. Fuzzy system provide a framework to represent imprecise information and to reason with this kind of information, while neural networks enhance fuzzy systems with the capability of learning capability of input-output samples, learning is used to adapt parameters of the fuzzy system as a membership function or rule [4]. By these two methodologies, improve the performance of the system and get advantage of the each methodology and at the same time overcome the limitations of individual techniques [4].

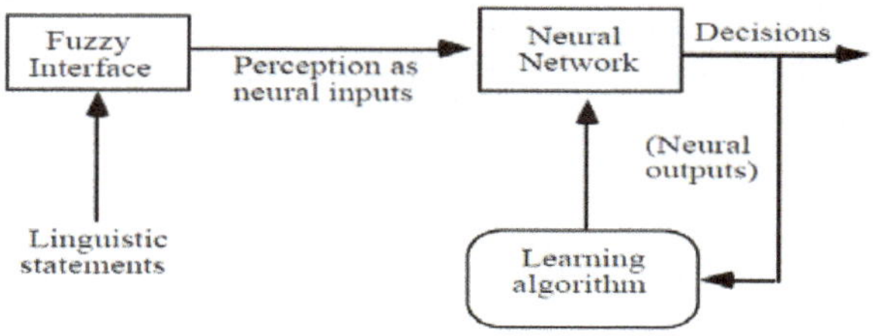

Figure 3. First model of fuzzy neural system

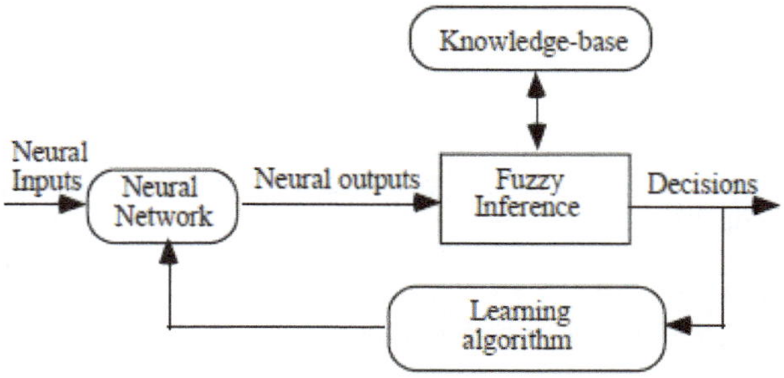

Figure 4. Second model of fuzzy neural system

2. Literature Review

The literature survey carried out related to impact in the study of Fuzzy system, neural network, Neuro-Fuzzy system and PID controller.

Amit Kr. Singh et al.[4] described the Self Tuning Fuzzy PID controller developed to improve the performance of the plants having 2nd order, 3rd order and 5th order system. Firstly, maintained the modeling for the higher order plant and designed the adaptive fuzzy-PID controller and simulated using MATLAB/Simulink. The response of the plants (i.e.2nd, 3rd and 5th order) using adaptive fuzzy-PID controller has improved significantly as compared to conventional PID controller.

Figure 5 shows the comparative result of PID Controller and Adaptive Fuzzy PID Controller for 5th order plant. A simulation result shows that the response of Adaptive Fuzzy PID Controller is superior in comparison to conventional PID Controller.

Y.M.Zho et al. [5] investigate performance-based parameter tuning method of model-driven Two-Degree-of- Freedom PID (MD TDOF PID) control system to enhance the control performances of a process. Tuning methods for the MD TDOF PID are based on internal model control (IMC) method instead of optimizing the performance indices. In this paper, an Integral of Time Absolute Error (ITAE) zero-position-error optimal tuning and noise effect minimizing method is proposed for tuning two parameters in MD TDOF PID control system to achieve the desired regulating and disturbance rejection performance.

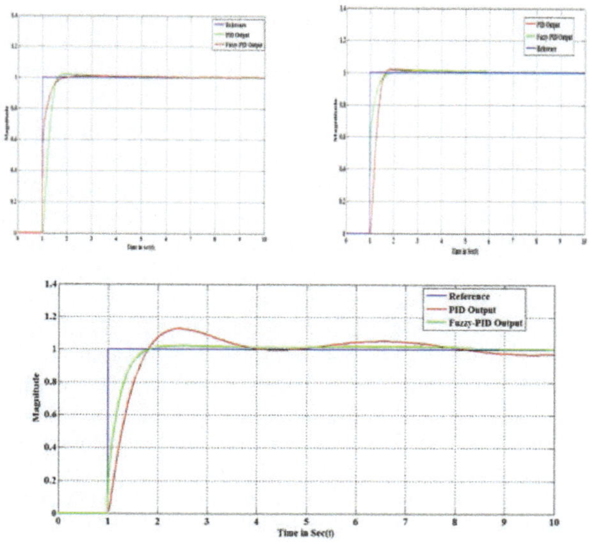

Figure 5. Comparative result of 2nd , 3rd and 5th order plant

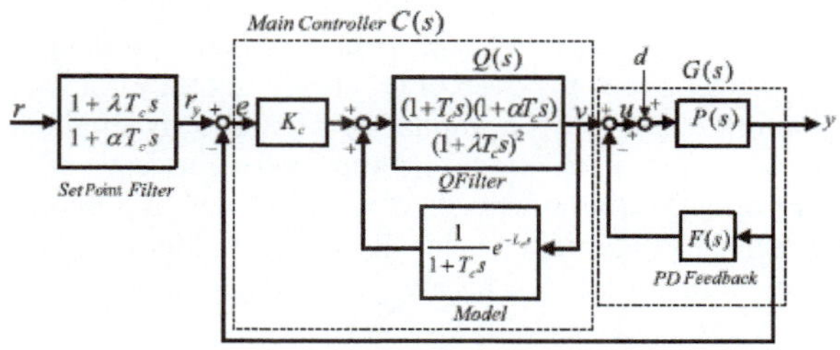

Figure 6. MD TDOF PID control system

In this paper, the function of each block of MDTDOFPID is analyzed. Two performance decisive factors – variance control and variance tracking – are used to design the tuning method. The variance tracking is the ability for output to follow the input signal and the variance control consists in keeping the output of the system at a set point while recovering as quickly as possible from disturbances.

MDTDOFPID control system functionally having three parts: PD feedback mechanism, main controller and set point. Main controller further decomposed into gain block, Q filter block and first order model with time delay. PD feedback is used to stabilizing the plant and make a transfer function of the process including PD feedback itself to first order system with time delay. Main controller is designed by adjusting the parameters of Q filter to meet design specifications and improves the speed response ability of the system as well as also helps in rejecting noise from entering system in low frequency. Set point filter is used to cancel one zero and one pole in the system and reduce the order of the system from second order to first one.

After tuning the both parameters of Q filter, there be two cases in this paper: for unstable first order plus time delay process and for pure integrating process plus time delay process. By simulation, the results have been compared with ITMDTDOFPID, PTMDOFPID and TDOFCSMSP. The results show, by tuning of parameters of PD feedback compensator, most of many processes can be converted to a first order with dead time process which makes it easy to design MDTDOFPID. The main controller improves the set point tracking and disturbance rejection properties of control system. Set point filter helps system reduce oscillations and make system more stable.

Mohamed S. Zaky [6] presented a self tuning PI controller for the speed control of electrical motor drives. The performance of the proposed self tuning PI controller using field oriented control of hybrid stepper motor is simulated and compared with conventional PI controller under tracking performance, parameters variation and load torque disturbances. Firstly maintained the dynamic modeling of hybrid stepper motor with d-q model using Park transformation and apply field oriented control principles in which direct current i_d set zero for simplify the control and then design the self tuning PI controller in which gains are varied over a pre-determined range for varying operating condition. Proportional term (K_p) is responsible for improving overshoots, rise time response while integral term (K_i) reduces steady state error. For large speed error, high value proportional gain is required for speed up the transient response and small integral gain is used to eliminate the undesirable oscillations and overshoot while for small seed error, high value of integral gain is required to overcome steady state error and minimum value of proportional gain is used to eliminate overshoots and oscillations. So that two gains are varied online as a function of speed error e(t). Hence in proposed self tuning PI controller using Root Locus Method gives better performance for highly under damped system in comparison to other control.

Dao Thanh Liem et al. [7] Presented a new method for estimating the load torque of a DC motor shaft by using a novel modelling method based on an adaptive control technique, named as online tuning grey fuzzy PID (OTGFPID). A test rig using a DC motor is setup to investigate the torque behavior as well as to evaluate the developed estimator. Firstly, mathematical model is developed for the motor. Secondly, the experimental speed-torque data and the optimized motor model is used to optimize the torque estimator. Then the optimized estimator is used to estimate accurately the load torque. This paper proposes to separate the actual speed of the DC motor ω into two distinct parts for estimating load torque: speed due to the armature current (ω_n) and speed variation (ω_e) due to the current generated from the unknown load torque which needs to be estimated. Therefore, it is necessary to estimate the ideal speed ω_n and subsequently, to determine the speed difference ω_e caused by the load torque. A control algorithm is then proposed to drive the second DC motor model to track this speed different. If the controller is able to make the model track the desired speed, ω_e, then the load torque can be precisely estimated.

Firstly physical model of DC motor is developed and optimized through an identification process using least square algorithm and iterative procedure. OTGFPID model having three main blocks: the online tuning fuzzy PID controller, the optimized motor model and the Fuzzy Grey Predictor (FGP). The OTFPID controller is to produce the control input for the optimized model which is also the estimated load torque. Then model creates a speed corresponding to the voltage and the estimated torque. For actual working condition, a noise source is added to the model output. By subtracting to the ideal speed, the estimated value of the speed difference is achieved. At the same time, the input and output MFs of the fuzzy structure inside the OTFPID are optimized online with respect to the control error minimization by using the learning mechanism while the PID gains are updated through the robust checking conditions, consequently, improving the control quality. To improve the control quality, the FGP predictor with online tuning ability of its prediction step size is designed to estimate the motor speed difference in a near future. This estimated speed is finally fed-back to the controller to create the control input and subsequently, performs the close-loop control system.

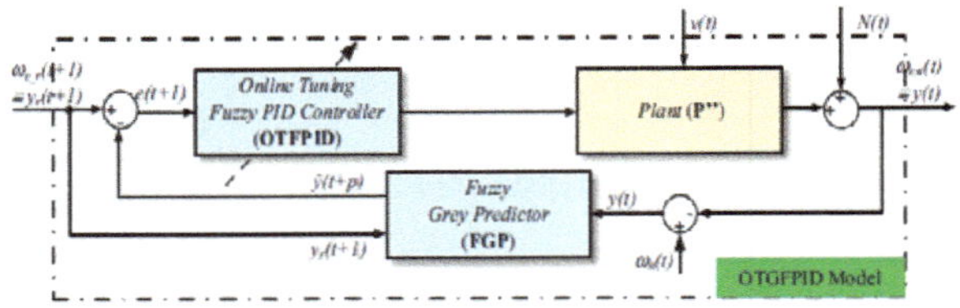

Figure 7. Structure of the OTGFPID model

Hence, the proposed method using FPID-based estimator investigates the effectiveness of the purpose estimation technique. The OTGFPID was successfully design as the combination of online tuning fuzzy PID controller and fuzzy gray predictor. Here, OTFPID controller based on PID algorithm is to estimate the motor torque from motor speed. Learning mechanism with robust checking conditions was implemented into the OTFPID in order to optimize online its parameters with respect to control error minimization. Meanwhile, the FGP predictor techniques takes part in estimating the system output in near future to optimize the controller parameters in advance and consequently, improve estimation performance.

Juing-Shian Chiou et al. [8] presents a novel approach to determining the parameters of fuzzy PID controller using practical swarm optimization (PSO) reinforcement evolutionary algorithm. PSO with Q- learning developed an adaptive fuzzy PID controller scheme for automotive active suspension system.

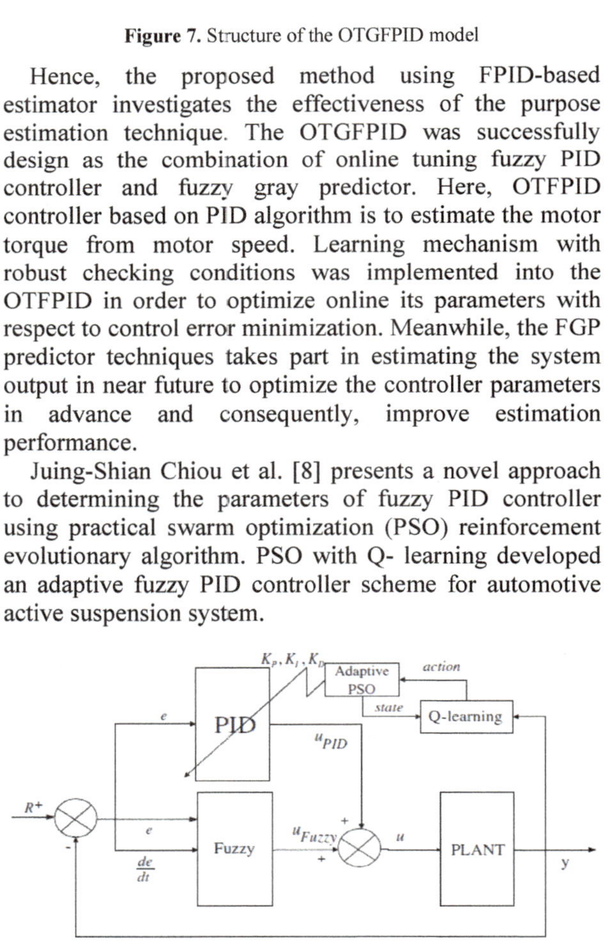

Figure 8. Hybrid structure of fuzzy-PID controller

For improving the efficiency and accuracy, PSO reinforcement evolutionary algorithm is employed to adjust the swarm of Q-learning to obtain global optimum solution. By this scheme, reduces the suspension deflection, spring mass acceleration and beating distance between the tire and the ground relative and also improves ride comfort and vehicle stability.

Rajani K. Mudi et al. [9] investigate an improved auto tuning scheme for PI controller which proposes an augmented Zieglar Nichols PI controller (AZNPIC). This proposed controller is tested with several high order linear and non-linear dead time processes under both set point change and load disturbance. It shows better performance, if compared with ZNPIC and RZNPIC (Refined Zieglar Nichols PI controller).

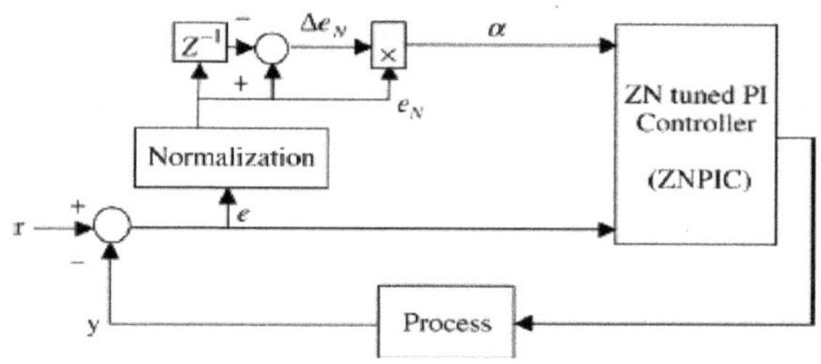

Figure 9. Block diagram of proposed AZNPIC

AZNPIC introduce an online gain updating factor which continuously modifies its proportional and integral gains based only on recent process trend.

Transfer function of the second-order linear process is

$$G_p(s) = \frac{e^{-Ls}}{(1+s)^2} \tag{2}$$

Response characteristics for the process in (2) with L = 0.2 s and L= 0.3 s under ZNPIC, RZNPIC, and AZNPIC are as under

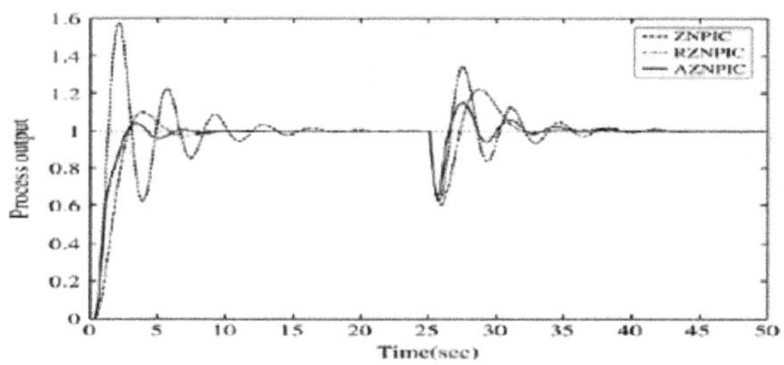

Figure 10. Response of the second order linear process with L=0.2 s

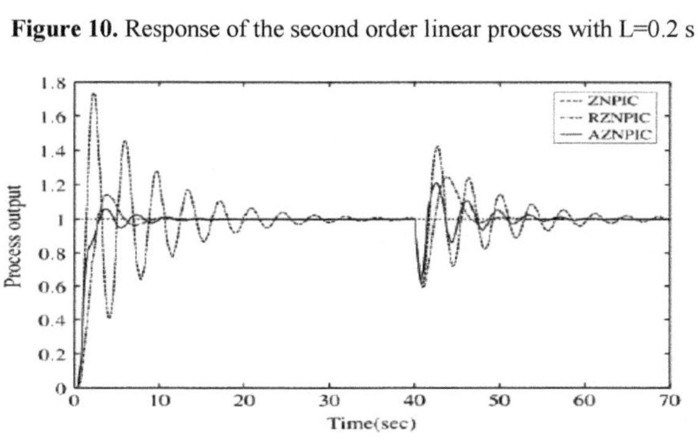

Figure 11. Response of the second order linear process with L=0.3 s

Figure 10 indicates that the performance of AZNPIC is better than that of RZNPIC. To study the robustness of the proposed scheme a 50% higher value of dead time i.e., L=0.3 sec is considered with same controller setting as that of L=0.2 sec and the corresponding responses are shown in Fig 11also shows that AZNPIC is superior over RZNPIC and ZNPIC.

The performance of the proposed auto-tuner is tested for different nonlinear processes which is described by

$$\frac{d^2y}{dt^2} + \frac{dy}{dt} + 0.2y^2 = u(t-L) \tag{3}$$

Response characteristics for the nonlinear process in (3) with L = 0.2 s and L= 0.3 s under ZNPIC, RZNPIC, and AZNPIC are as under

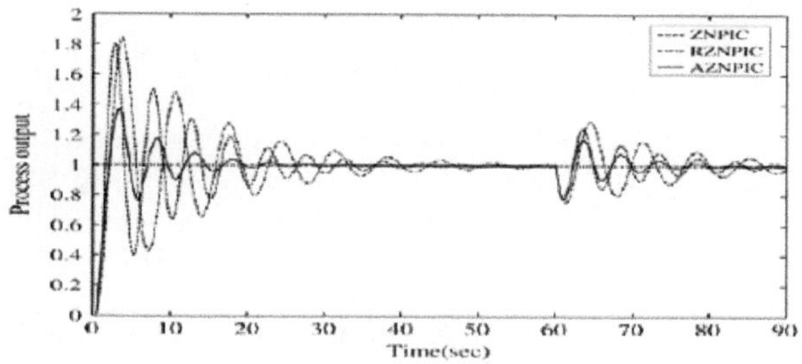

Figure 12. Response of the second order nonlinear process with L=0.2 s

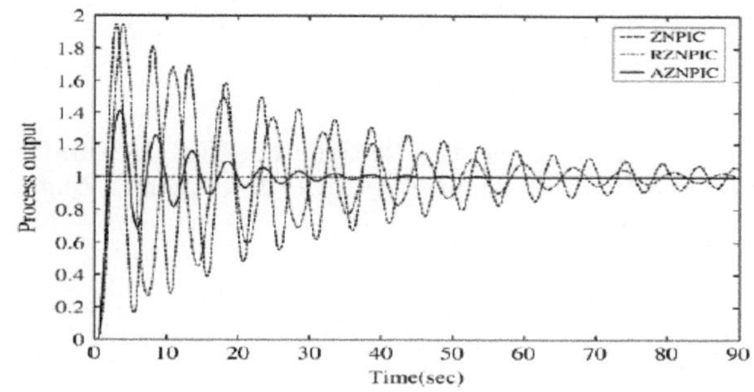

Figure 13. Response of the second order nonlinear process with L=0.3 s

From Figure 12 and Figure13, it is seen that in case of AZNPIC, % overshoot is considerably reduced, and steady state is reached faster, which reveal that unlike ZNPIC and RZNPIC.

Transfer function of the second-order marginally stable process is

$$G_p(s) = \frac{e^{-Ls}}{s(s+1)} \tag{4}$$

Response characteristics for the integrating process in (4) with L = 0.2 s and L= 0.3 s under ZNPIC, RZNPIC, and AZNPIC are as under

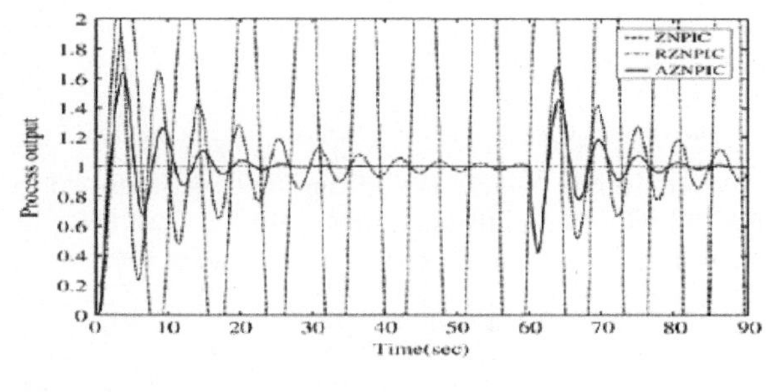

Figure 14. Response of the marginally stable process with L=0.2 s

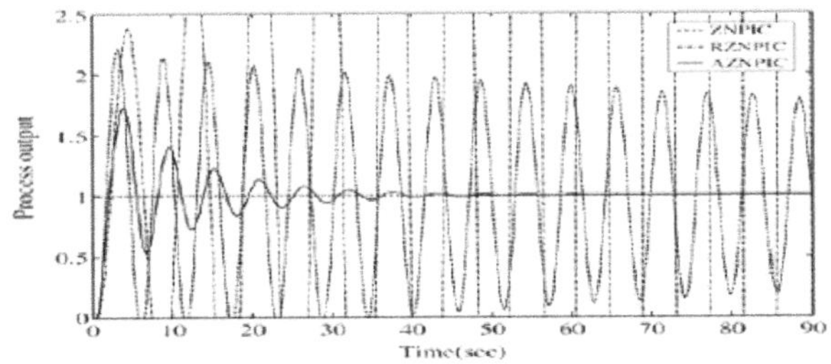

Figure 15. Response of the marginally stable process with L=0.3 s

ZNPIC provides a very poor performance due to excessively large overshoot and oscillation. The %OS of ZNPIDC is also very large (about 80%). Whereas, AZNPIC is found to reduce the %OS and t_s by about 40% and 60%, respectively compared to ZNPIC. When the process dead-time is increased from 0.2 to 0.3 s without changing the controller parameters, the system goes to the verge of instability under ZNPIC. However, AZNPIC still provides a highly stable performance. In both cases (L = 0.2 s and 0.3 s) the integrating process in (4) becomes completely unstable under RZNPIC.

In this study third-order linear as well as nonlinear process models are also tested to justify the effectiveness of the proposed scheme. Such a linear system is expressed by the following transfer function

$$G_p(s) = \frac{e^{-Ls}}{s^3 + s^2 + s + 0.2} \tag{5}$$

Response characteristics for the process in (2) with L = 0.4 s and L= 0.5 s under ZNPIC, RZNPIC, and AZNPIC are as under

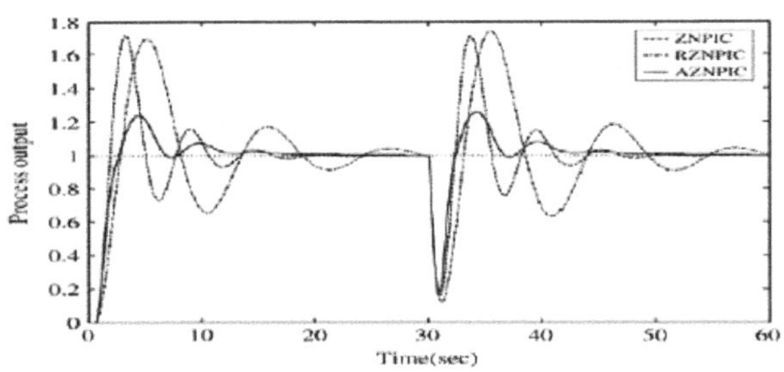

Figure 16. Response of the third order linear process with L=0.4 s

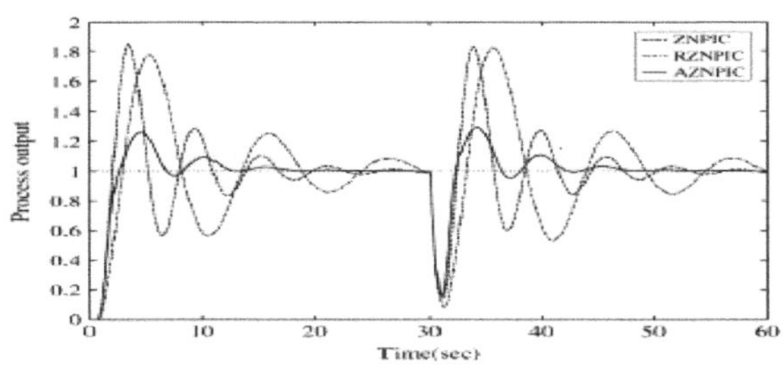

Figure 17. Response of the third order linear process with L=0.5 s

Figure 16 and Figure 17 indicate that the performance of AZNPIC is better than that of RZNPIC and ZNPIC.

Third order non-linear process is considered as:

$$\frac{d^3y}{dt^3} + \frac{d^2y}{dt^2} + \frac{dy}{dt} + 0.5y = u(t - L) \tag{6}$$

Response characteristics for the nonlinear process in (3) with L = 0.3 s and L= 0.4 s under ZNPIC, RZNPIC, and AZNPIC are as under:

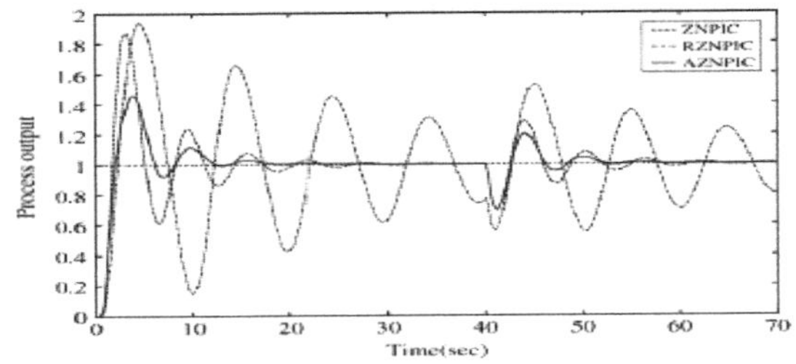

Figure 18. Response of the third order nonlinear process with L=0.3 s

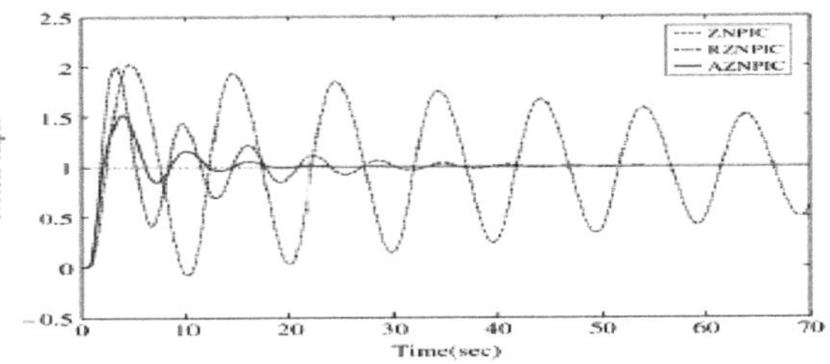

Figure 19. Response of the third order nonlinear process with L=0.4 s

The proposed controller (AZNPIC) exhibits better performance compared to ZNPIC for nonlinear process in (6). Also, RZNPIC shows a very poor performance.

The performance of AZNPIC is also evaluated for practical process, pH neutralization. For this process both ZNPIC and RZNPIC are not suitable.

From above results for various processes it is evident that the proposed AZNPIC shows consistently improved performance over ZNPIC and RZNPIC.

Chanchal Dey and Rajani K Mudi [10] investigate an improved auto tuning scheme for PID controller which proposes an augmented Zieglar Nichols PID controller (AZNPIDC) through an online gain updating factor which continuously modifies its proportional, integral and derivative gains based only on recent process trend and compared with ZNPIDC and RZNPIDC, gives better performance.

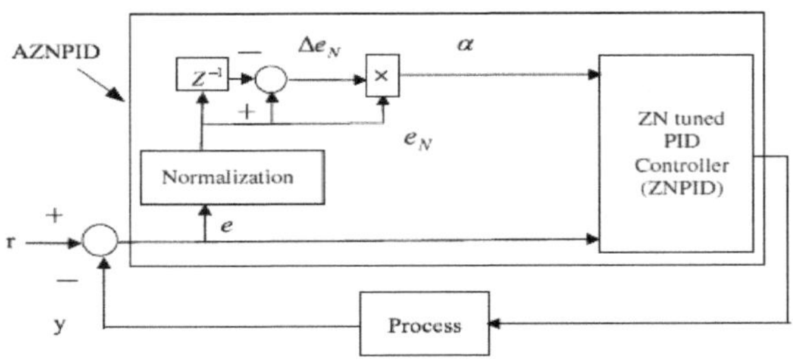

Figure 20. Block diagram of proposed AZNPID

For various processes (linear process and nonlinear process) it is evident that the proposed AZNPIDC shows consistently improved performance over ZNPIDC and RZNPIDC.

Muguo Li and Da Liu [11] proposed recurrent wavelet neural network PID controller [RWNNPID] for PMSM speed servo drive system to adjust the parameters online to be adaptive to the parameters variety and load disturbance.

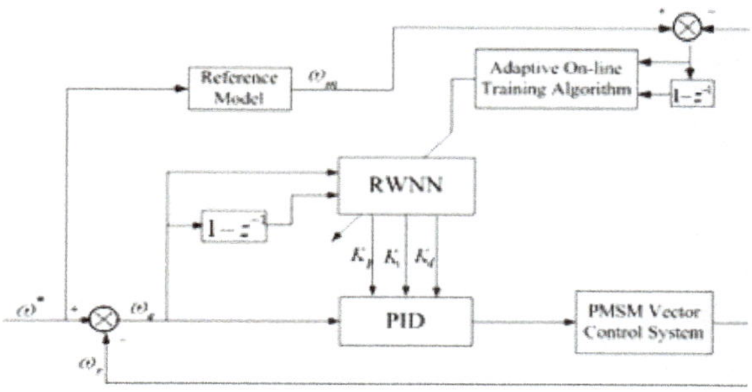

Figure 21. Recurrent wavelet neural network PID control system

The algorithm uses WNN to turn PID parameters adaptively. The recurrent nodes in addition to the network increase the network response speed and ensure the network has stable outputs. Network algorithm is working online for obtaining optimal parameters to realize an adaptive system with strong learning ability and good robustness.

From Figure 22 and Figure 23, results show that proposed controller gives improve performance if compared with conventional PID controller for PMSM speed regulation.

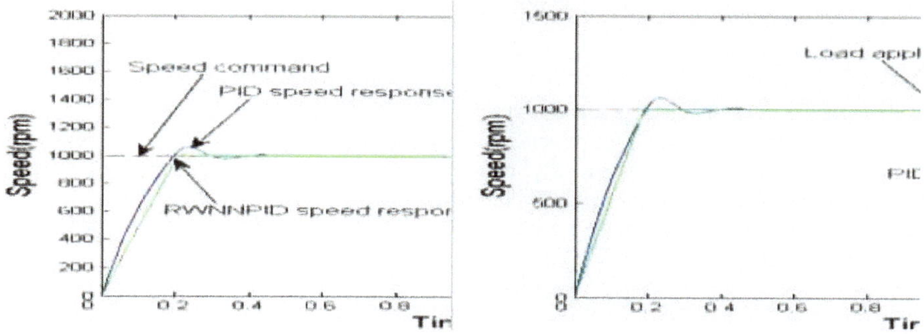

Figure 22 (a). Simulation investigation of conventional PID and RWNNPID speed response on condition that step speed command is changed; (b): Simulation investigation of conventional PID and RWNNPID speed response on condition that load is applied

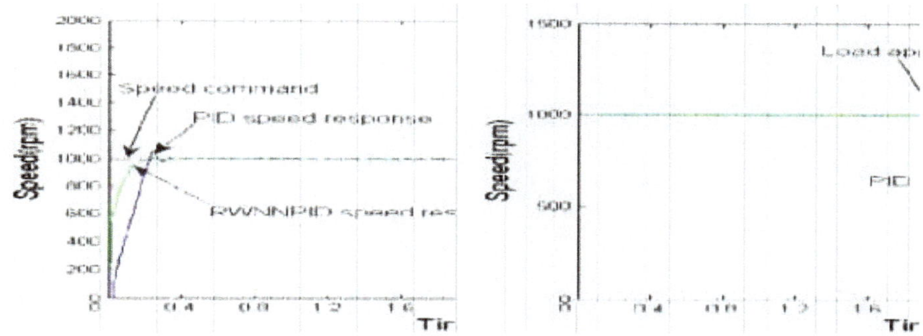

Figure 23. (a): Experimental results of conventional PID and RWNNPID speed response on condition that step speed command is changed; (b): Experimental results of conventional PID and RWNNPID speed response on condition that load is applied

Muawia A. Magzoub et al. [12] investigate the efficiency improvement of induction motor variable speed drive using a hybrid fuzzy-fuzzy controller to overcome drawback of field oriented method. There are two fuzzy controller used as fuzzy frequency controller during accelerate-decelerate stage and fuzzy stator current magnitude controller for steady state stage.

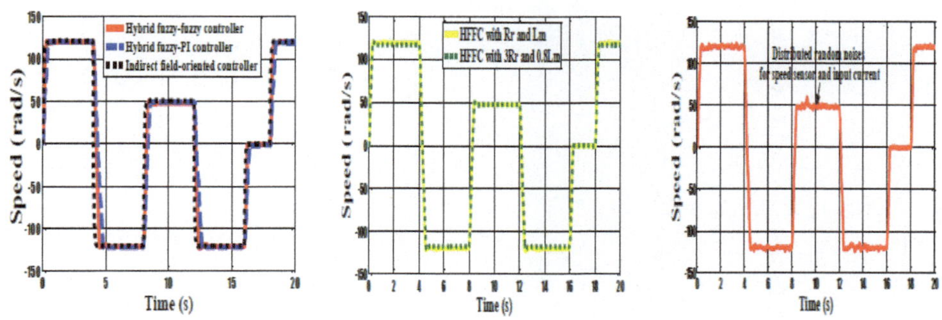

Figure 24. (a): Speed response of HFFC; (b): With parameters variation; (c): With noise

After simulation, compared with FOC results faster speed and the large changes can be accommodated in load torque. The minimization of current losses which contribute to energy efficiency system and efficiency is improved by HFFC. Henceforth, from the exploitation of the two features of field-oriented control, the fuzzy fuzzy controller has been found to be more efficient than scalar controller.

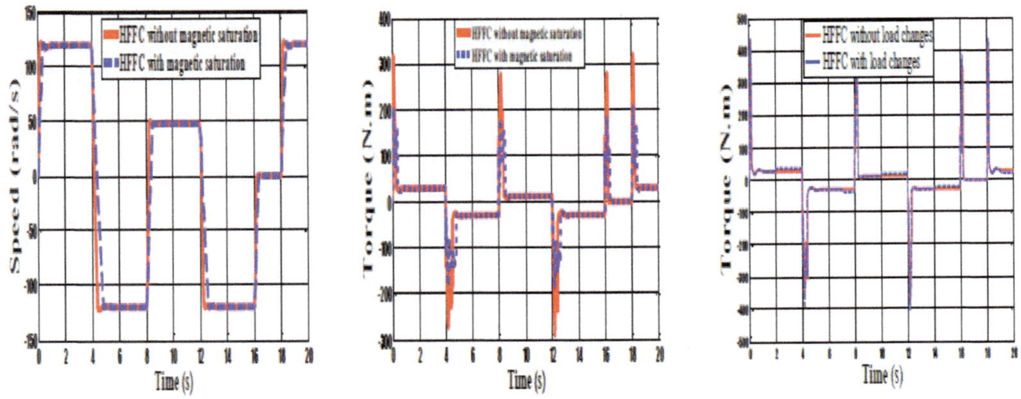

Figure 25. (a): With the effect of magnetic saturation; (b): Torque with the effect of magnetic saturation; (c): With load changes

Sakuntala Mahapatra et al. [13] developed a prototype model using PSO-ANFIS (Particle Swarm Optimization-Adaptive Neuro-Fuzzy Inference System) hybrid technique to control the speed of an induction motor. The speed of an Induction Motor drive is controlled by the hybrid of PSO- ANFIS algorithms.

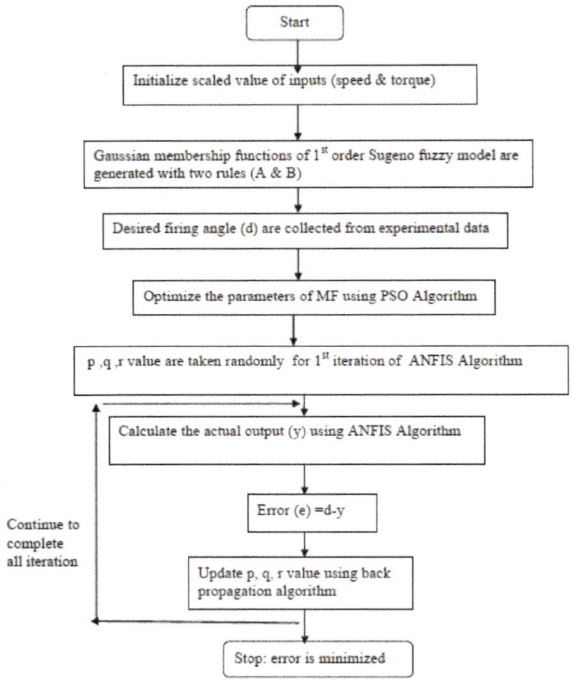

Figure 26. Flowchart of proposed model

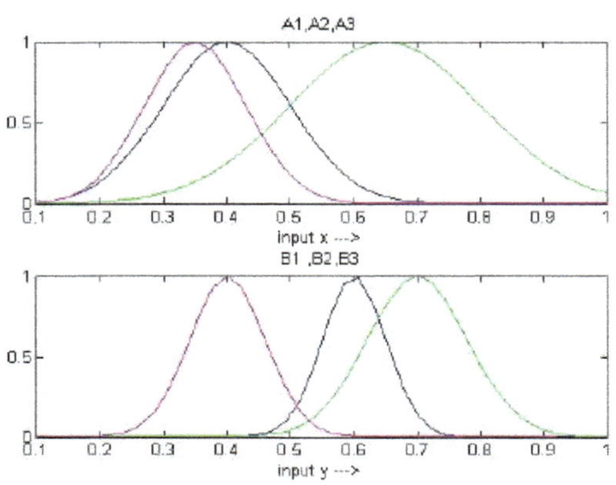

Figure 27. Plot of membership function using 2 inputs 3 rules

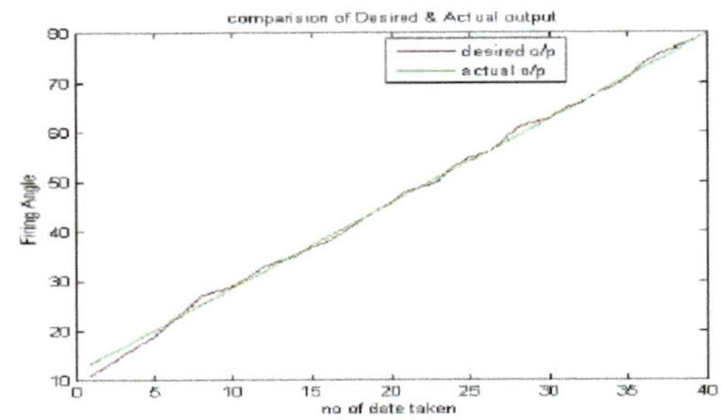

Figure 28. Comparison of actual and desired output plot

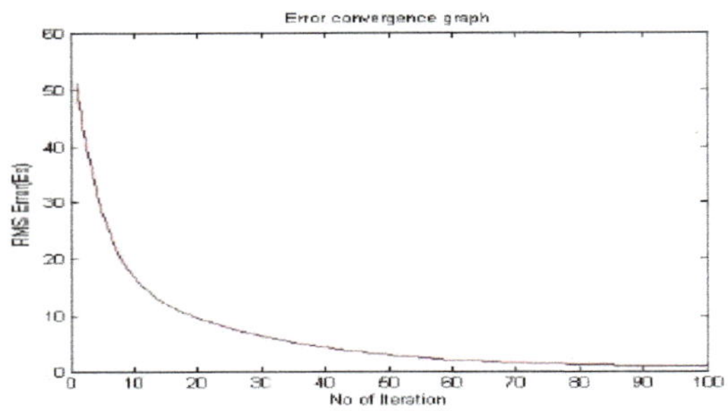

Figure 29. Error convergence plot for ANFIS

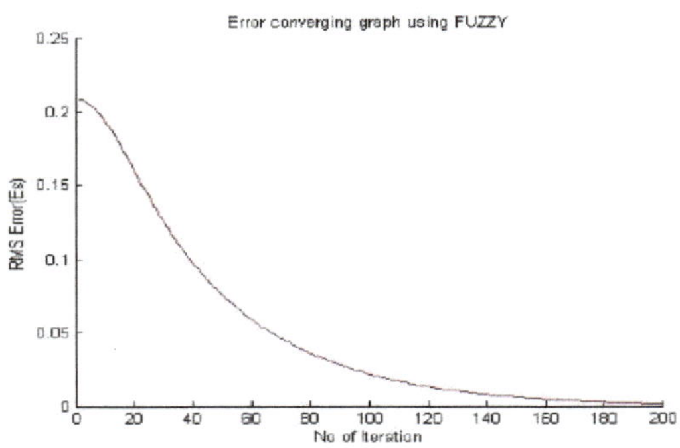

Figure 30. Error convergence plot of fuzzy back propagation

According to MATLAB simulation, the Adaptive Neuro Fuzzy (ANFIS) controller efficiently is better than the traditional FLC. The ANFIS-PSO is the best controller which presented satisfactory performances. Therefore, chose with the use of the Neural Networks and Particle Swarm Optimization for the optimization of this controller in order to control Induction motor speed. Finally, the proposed controller (PSO-ANFIS) gives a very good result. The ANFIS-PSO is the best controller which presented satisfactory performances.

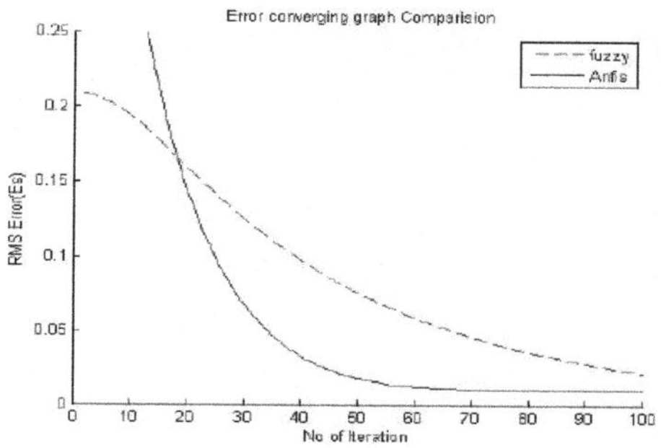

Figure 31. Error convergence comparison between ANFIS and fuzzy back propagation

Sharda Patwa [14] investigated the fuzzy logic controller for controlling induction motor parameters such as starting current, flux, torque, and speed. The results of FLC are also compared the controlling of induction motor same parameters with PI controller. The results showed that FLC has given the better performance than PI controller for controlling the induction motor parameters with improved rise time with less starting current.

P.M.Menghal and A.Jaya Laxmi [15] proposed an adaptive Neuro-fuzzy controller based dynamic performance of induction motor to overcome the problems developed in conventional fuzzy controllers. The main problem with conventional FLC is that the parameters associated with the membership functions and rules depend on intuition of the experts.

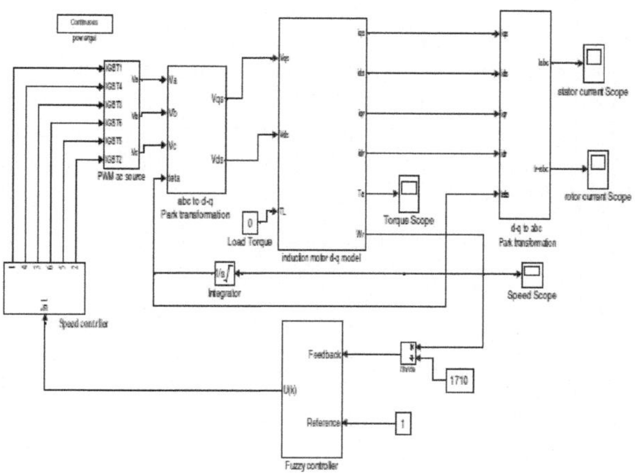

Figure 32. Adaptive neuro-fuzzy controller simulation model

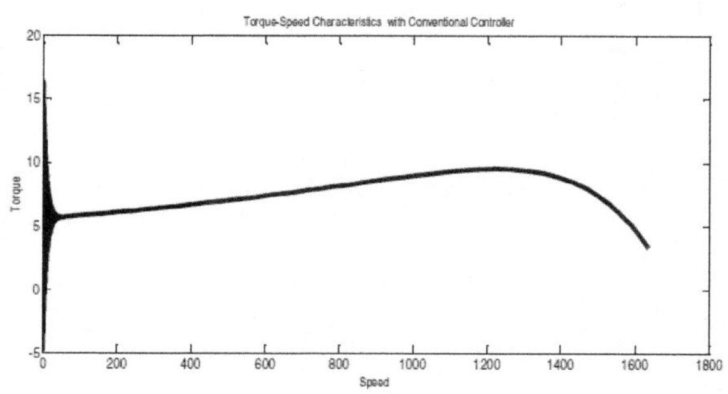

Figure 33. (a): Torque-speed characteristics with Conventional Controller

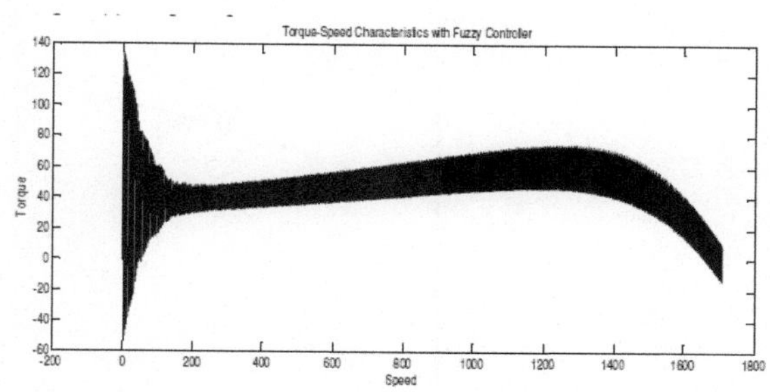

Figure 33(b). Torque-speed characteristics with Fuzzy controller

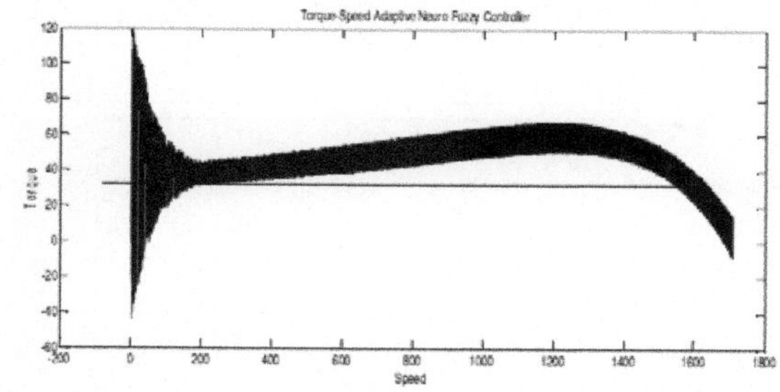

Figure 33. (c): Torque-speed characteristics with Adaptive Neuro Fuzzy controller

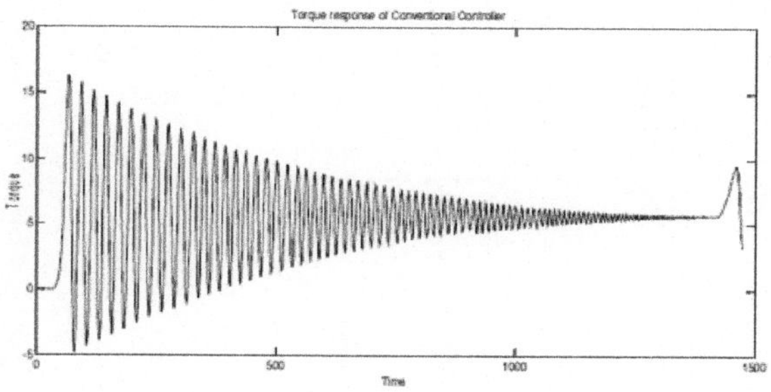

Figure 34. (a): Torque response of Conventional Controller

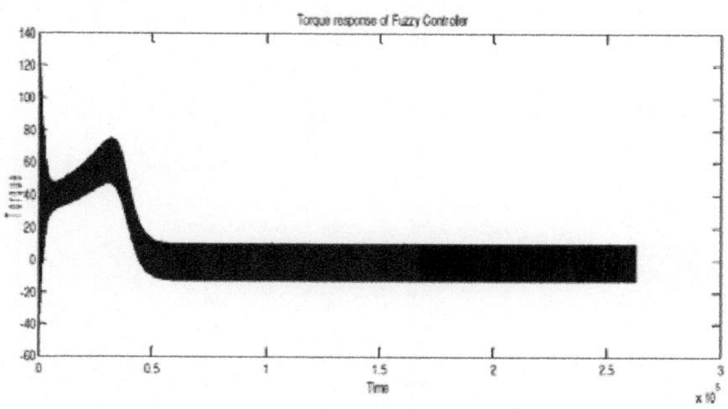

Figure 34. (b): Torque response of Fuzzy Controller

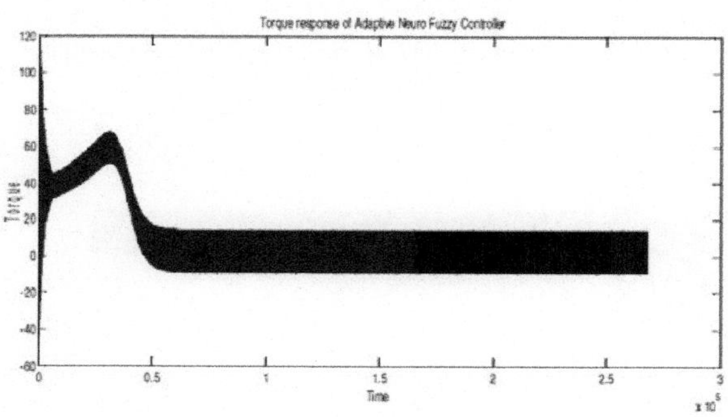

Figure 34(c). Torque response of Adaptive Neuro Fuzzy Controller

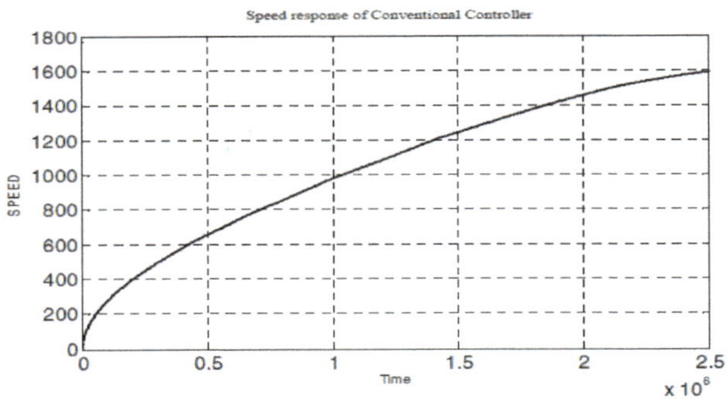

Figure 35. (a): Speed response of Conventional Controller

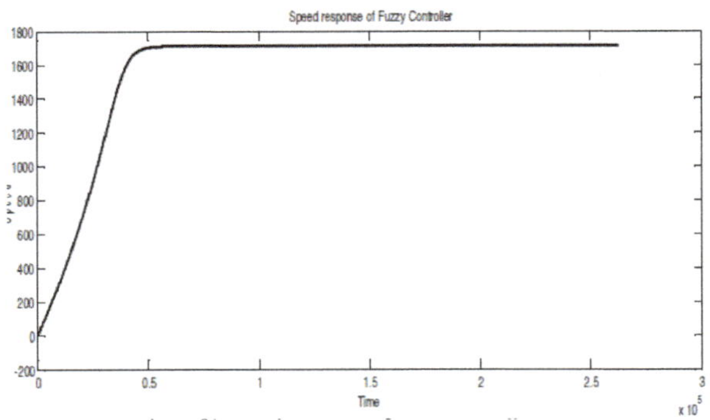

Figure 35. (b): Speed response of Fuzzy Controller

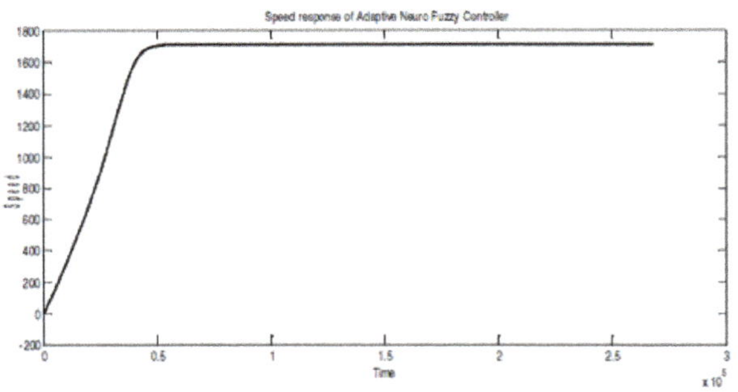

Figure 35. (c): Speed response of Adaptive Neuro Fuzzy Controller

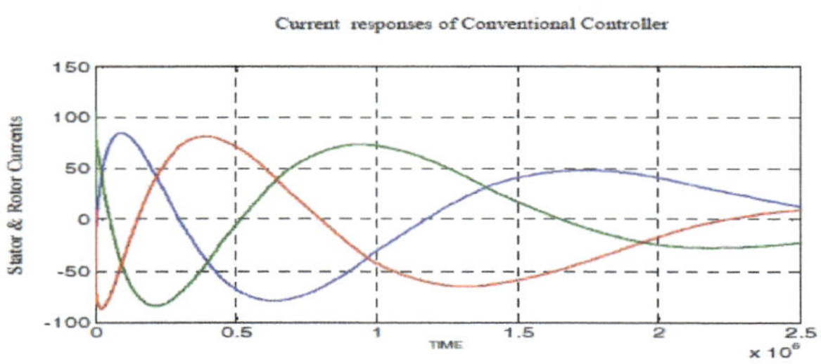

Figure 36. (a): Stator & Rotor Currents of Conventional Controller

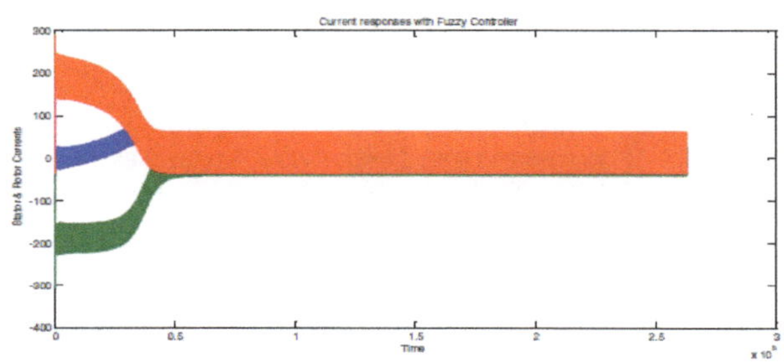

Figure 36. (b): Stator & Rotor Currents of Fuzzy Controller

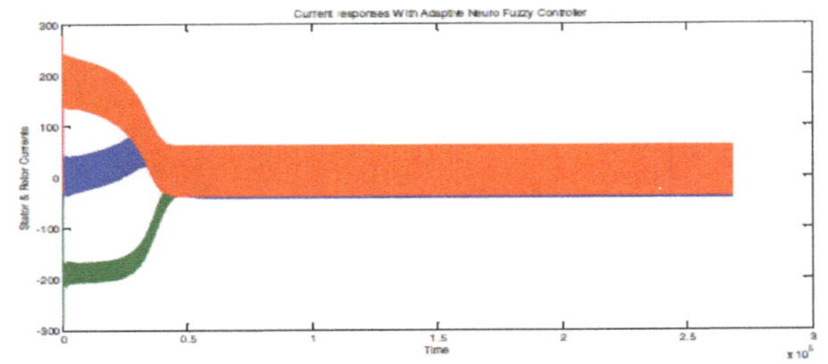

Figure 36. (c): Stator & Rotor Currents of Adaptive Neuro Fuzzy controller

The proposed adaptive neuro-fuzzy controller simulation model is designed and simulated in MATLAB/Simulation and compare with conventional and fuzzy controller. Results show that proposed adaptive neuro-fuzzy controller give better performance like speed-torque characteristics, torque response, speed response, rotor and stator currents compared to conventional and fuzzy controller.

P.M.Menghal et al. [16] present dynamic simulation of induction motor drive using neuro controller and the performance of fuzzy logic and artificial neural network based controller's are compared with that of the conventional proportional integral controller. The most popular supervised learning algorithm (i.e. back propagation) is used. On comparing neuro controller with FLC model, by adding learning algorithm to the control system will decrease rise time more than expectation and it proves neuro controller has better dynamic performance as compared to FLC and conventional controller.

Ahmed J. Fattah and Ikhlas Abdel-Qader [17] investigated an improved hybrid PID fuzzy controller for performance analysis of speed control induction motor and also compare their performance with conventional PI controller. They proposed a hybrid controller with vector control to control the speed of induction motor. PI-fuzzy, PD-fuzzy and PID-fuzzy controllers are designed and simulated with MATLAB/simulink.

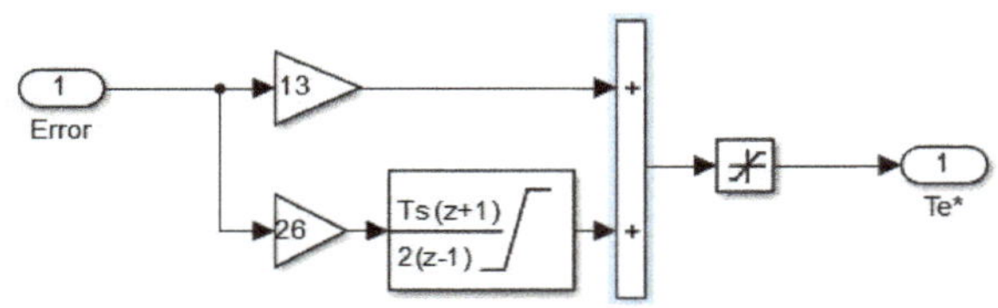

Figure 37. The PI Controller MATLAB model

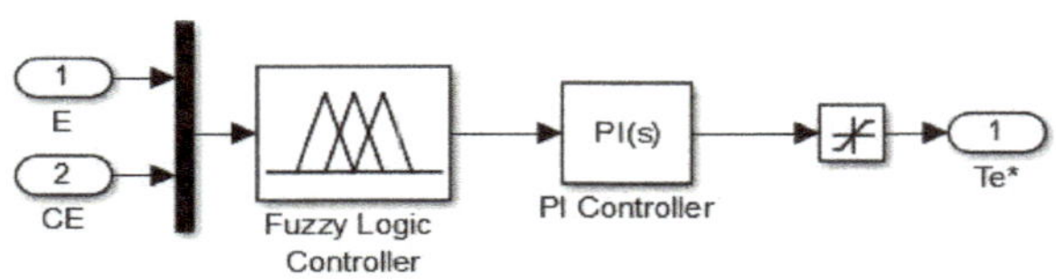

Figure 38. The PI-Fuzzy Controller MATLAB model

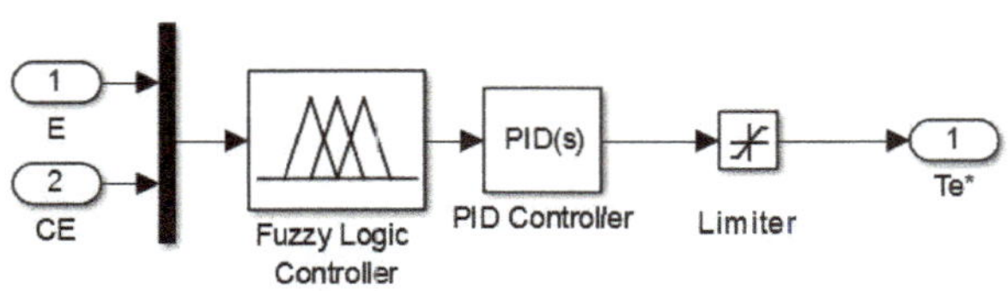

Figure 39. The PID-Fuzzy Controller MATLAB model

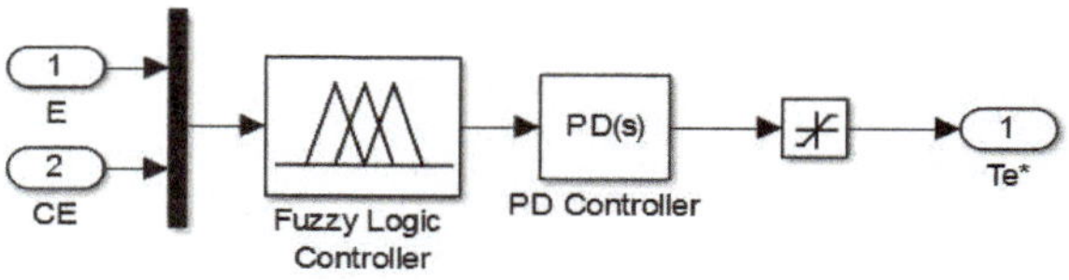

Figure 40. The PD-Fuzzy Controller MATLAB model

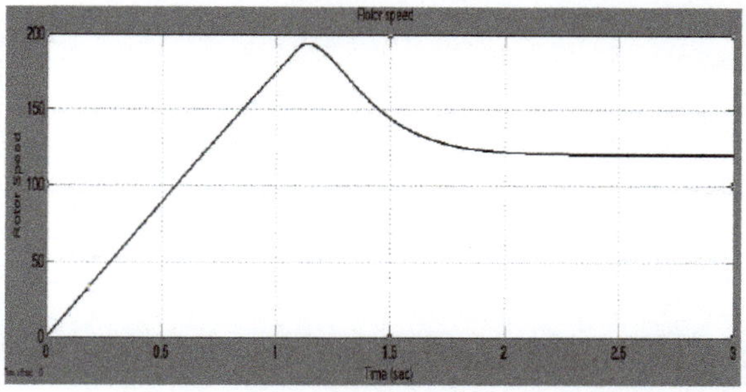

Figure 41. Speed response curve of SCIM at 0 N.m load and 120 reference speed using PI Controller

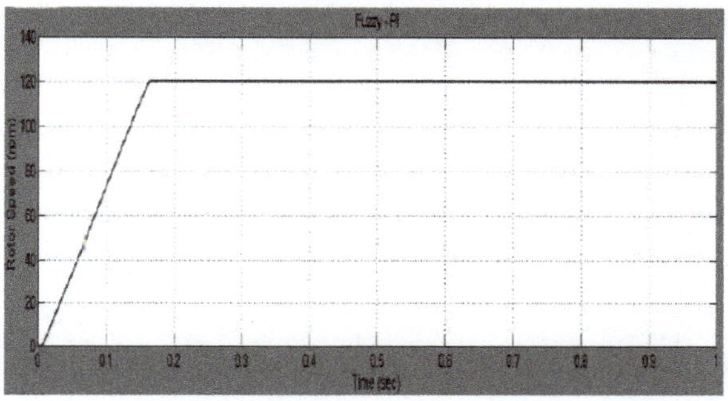

Figure 42. Speed response curve of SCIM at 200 N.m load and 120 reference speed using PI-Fuzzy Controller

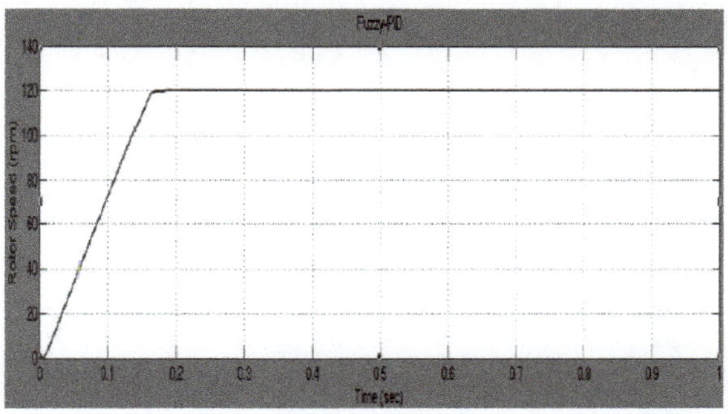

Figure 43. Speed response curve of SCIM at 200 N.m load and 120 reference speed using PID-Fuzzy Controller

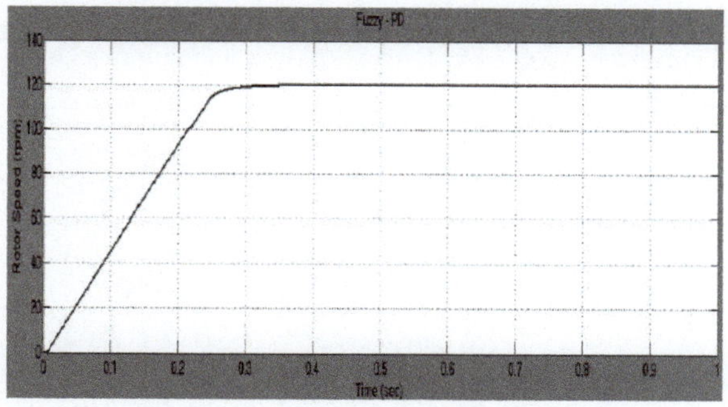

Figure 44. Speed response curve of SCIM at 200 N.m load and 120 reference speed using PD-Fuzzy Controller

Table 1. Performance analysis of different speed controllers for SCIM at different load

Load (N.m)	Controller	t_r (sec)	t_s (sec)	M_p	E_{ss}
0	PI	0.5439	1.9326	61.2028	0.015
	PD-Fuzzy	0.1649	0.2246	0	0.004
	PI-Fuzzy	0.1071	0.1386	0.1423	0.1600
	PID-Fuzzy	0.1071	0.1386	0.0763	0.005
100	PI	0.8264	2.6355	69.8537	0.005
	PD-Fuzzy	0.1815	0.2444	0	0.0570
	PI-Fuzzy	0.1146	0.1484	0.0994	0.1100
	PID-Fuzzy	0.1146	0.1484	0.0384	0.0420
150	PI	1.1171	3.3759	74.3692	0.0040
	PD-Fuzzy	0.1915	0.2563	0	0.0835
	PI-Fuzzy	0.1189	0.1539	0.0792	0.0880
	PID-Fuzzy	0.1189	0.1539	0.0200	0.0250
200	PI	1.7215	4.9398	77.3644	0.007
	PD-Fuzzy	0.2027	0.2698	0	0.1098
	PI-Fuzzy	0.1234	0.1598	0.0591	0.0650
	PID-Fuzzy	0.1234	0.1598	0.0022	0.0005

With different load conditions, comparison with conventional PI controller, the hybrid controller is increased dynamic performance such as rise time, peak overshoot, settling time and steady state error of induction motor and also to provide good stabilization.

Shamseldin and Adel A. EL-Samahy [18] presented three different controllers for speed control of BLDC motor. In which, first is conventional controller. The second controller is to use to GA technique to adjust the three cost functions. Finally, third controller hybrid fuzzy PID controller is development. All controllers are tested for speed regulation and speed tracking.

Firstly, the parameters (K_p, K_i, K_d) of PID controller is achieved by Ziegler-Nichols method and three cost function genetic algorithm techniques. Comparing among the controllers, the performance of the three cost function controller is best.

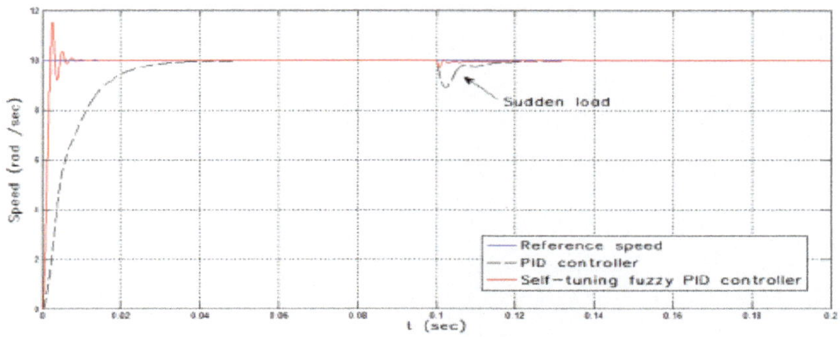

Figure 45. PID Control and Self –tuning Fuzzy-PID Controller response at sudden load

Finally, they proposed self tuning fuzzy PID controller and simulated with MATLAB/Simulink and compare with PID controller for speed regulation at sudden load, sinusoidal load, parameter variation and speed tracking of BLDC motor.

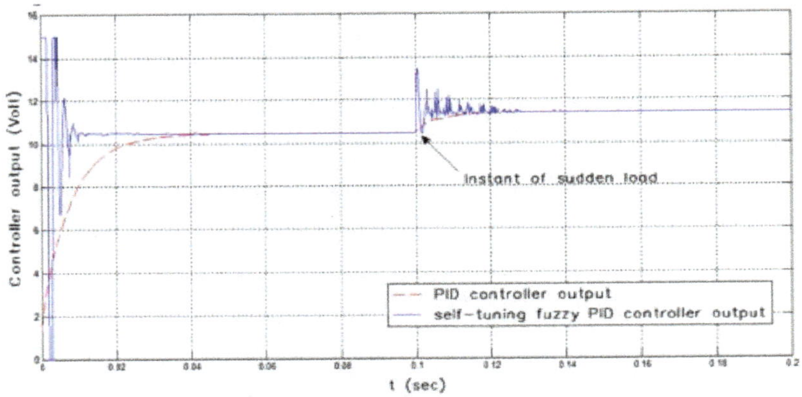

Figure 46. PID Control and Self –tuning Fuzzy-PID Controller output at sudden load

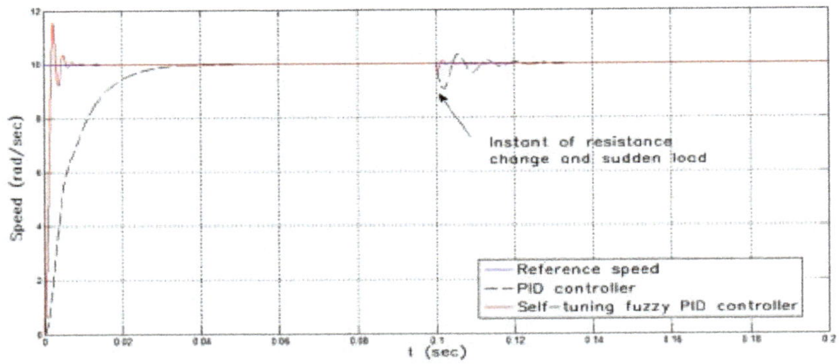

Figure 47. PID Controller and Self –tuning Fuzzy-PID Controller response at resistance change and sudden load

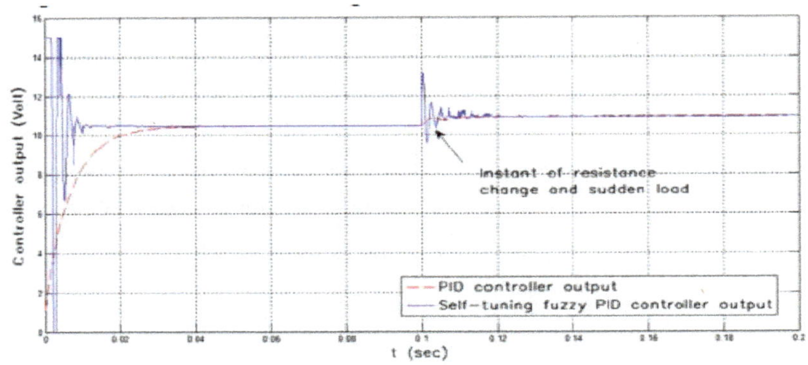

Figure 48. PID Controller and Self –tuning Fuzzy-PID Controller output at resistance change and sudden load

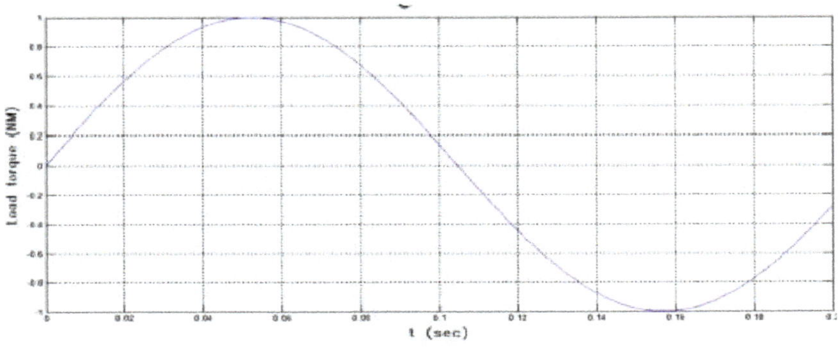

Figure 49. Sinusoidal load torque that exposed to the BLDC motor during simulation

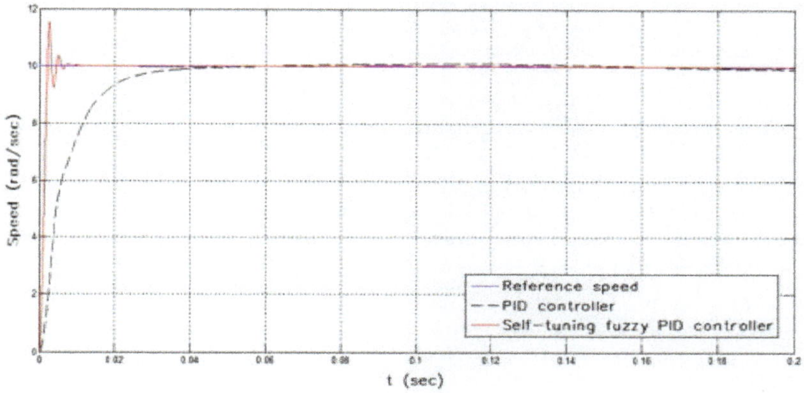

Figure 50. PID Controller and Self –tuning Fuzzy-PID Controller response at Sinusoidal load

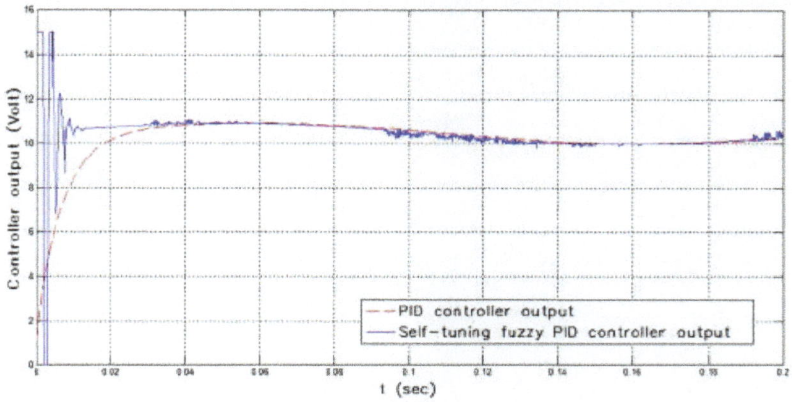

Figure 51. PID Controller and Self –tuning Fuzzy-PID Controller output at Sinusoidal load

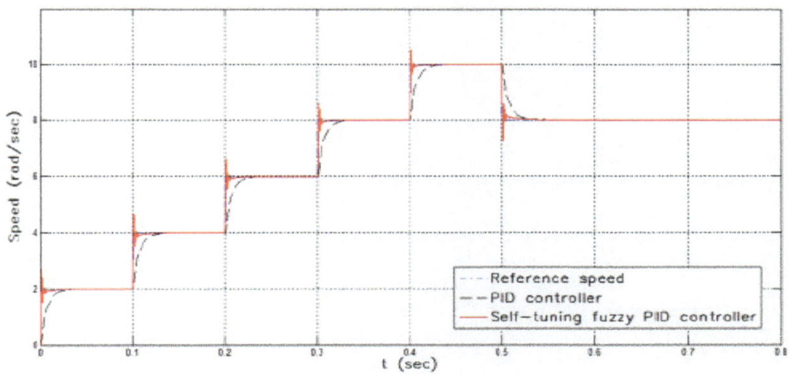

Figure 52. PID Controller and Self –tuning Fuzzy-PID Controller response at speed tracking

Results show that proposed self tuning fuzzy PID controller give better performance compared to other two techniques.

3. Conclusion

This review presents a overall summary on the various implementation techniques of fuzzy systems, neural network, PID-fuzzy system, neuro-fuzzy system, PID controller. This overview represents the information about Neuro-Fuzzy system and PID Controller focuses on its usability and challenges. This study suggests that the future development of Neuro-Fuzzy PID controllers using different soft computing techniques will improve its flexibility and re-usability of the controller.

Acknowledgements

Authors are thankful to all the staff of Electrical Engineering Department of KSVCEM, ABESIT and IFTM University for their support directly or indirectly.

References

[1] Emad Ali, "Automatic Fuzzy Tuning of Proportional-Integral Controller Based on Time-Domain Specification", *Journal of King Saud University, 17, Eng. Sci. (2), 171-196, 2005.*

[2] Hari Om Bansal, Rajamayyoor Sharma, P. R. Shreeraman, "PID Controller Techniques: A Review", *Journal of Control Engineering and Technology, Volume 2, Issue 4, October 2012, Pages 168-176.*

[3] Amit Kr Singh, Manjari Mehrotra, and AK Pandey, "Improvement of the Performance of Different Order System with Adaptive Fuzzy PID Controller", *VSRD-IJEECE, Volume 2, Issue 2, 27-38, 2012.*

[4] Prajakta S.Taywade, K.N.Kasat, "A Review on Implementation of Neuro-fuzzy PID Controller using FPGA", *International Journal of Engineering Research and General Science, Volume 3, Issue 1, January-February 2015, Pages 1145-1152.*

[5] Y.M. Zhao, W.F. Xie, X.W. Tu, "Performance-based parameter tuning method of model-driven PID control systems", *ISA Transactions, Volume 51, Issue 3, May 2012, Pages 393-399.*

[6] Mohamed S. Zaky, "A self-tuning PI controller for the speed control of electrical motor drives", *Electric Power Systems Research, Volume 119, February 2015, Pages 293-303.*

[7] Dao Thanh Liem, Dinh Quang Truong, Kyoung Kwan Ahn, "A torque estimator using online tuning grey fuzzy PID for applications to torque-sensorless control of DC motors", *Mechatronics, Volume 26, March 2015, Pages 45-63.*

[8] Juing-Shian Chiou, Shun-Hung Tsai, Ming-Tang Liu, "A PSO-based adaptive fuzzy PID-controllers", *Simulation Modelling Practice and Theory, Volume 26, August 2012, Pages 49-59.*

[9] Rajani K. Mudi, Chanchal Dey, Tsu-Tian Lee, "An improved auto-tuning scheme for PI controllers", *ISA Transactions, Volume 47, Issue 1, January 2008, Pages 45-52.*

[10] Chanchal Dey, Rajani K. Mudi, "An improved auto-tuning scheme for PID controllers,"*ISA Transactions, Volume 48, Issue 4, October 2009, Pages 396-409.*

[11] Muguo Li, Da Liu, "A Novel Adaptive Self-turned PID controller based on Recurrent-Wavelet-Neural-Network for PMSM Speed Servo Drive System", *Procedia Engineering, Volume 15, 2011, Pages 282-287.*

[12] Muawia A. Magzoub, Nordin B. Saad, Rosdiazli B. Ibrahim, "Efficiency Improvement of Induction Motor Variable Speed Drive Using a Hybrid Fuzzy-fuzzy Controller", *Energy Procedia, Volume 75, August 2015, Pages 1529-1535.*

[13] Sakuntala Mahapatra, Raju Daniel, Deep Narayan Dey, Santanu Kumar Nayak, "Induction Motor Control Using PSO-ANFIS", *Procedia Computer Science, Volume 48, 2015, Pages 753-768.*

[14] Sharda Patwa, "Control of starting current in three phase induction motor using fuzzy logic controller," *International Journal of Advanced Technology in Engineering and Science, Volume 1,Issue 12, pages 27-32, December 2013.*

[15] P.M.Menghal, A.Jaya Laxmi, "Adaptive Neuro Fuzzy based dynamic simulation of induction motor drives", *IEEE International Conference Fuzzy Systems (FUZZ-IEEE 2013) 7-10 July 2013.*

[16] P.M.Menghal, A.Jaya Laxmi, N.Mukhesh, "Dynamic simulation of induction motor drive using neuro controller", *International Journal on Recent Trends in Engineering and Technology, Volume 10, Issue 2, January 2014.*

[17] Ahmed J. Fattah, Ikhlas Abdel-Qader, "Performance and Comparison Analysis of Speed Control of Induction Motors using Improved Hybrid PID-Fuzzy Controller", *Electro/Information Technology (EIT), 2015 IEEE International Conference on Year 2015, Pages 575-580.*

[18] Mohammed Abdelbar Shamseldin and Adel A. EL-Samahy, "Speed Control of BLDC Motor by using PID Control and Self-tuning Fuzzy PID Controller", *15th International Workshop on Research and Education in Mechatronics (REM), September 9-11, 2014, IEEE Conference Publication.*

[19] Bemal K. Bose, "Modern Power Electronics and Ac Drives", *1st ed., Bernard Goodwin, Ed. USA: Prentice Hall, Inc., 2002.*

[20] Raul Rojas, "Neural Network a Systematic Introduction", Springer-Verlag, Berlin 1996.

Modeling and Simulation of a New Duty-Cycle Modulation Scheme for Signal Transmission Systems

Leandre Nneme Nneme*, **Jean Mbihi**

Department of Electrical and Electronics Engineering, ENSET (University of Douala), Douala, Cameroon
*Corresponding author: leandren@gmail.com

Abstract This paper studies a new signal transmission scheme founded on duty-cycle modulation. The proposed signal transmission system consists of a low cost duty-cycle modulation circuit, a transmission line model and a linear demodulator. The study of the signal transmission system is conducted using analytical developments and virtual simulation models. Then, the results obtained under a variety of modulating waveforms including standard signals and arbitrary waveforms, show the high quality of the proposed duty-cycle modulation scheme for signal transmission systems.

Keywords: *modeling, Duty-Cycle-Modulation, lossless transmission line, Low-Pass Filter, simulation*

1. Introduction

The duty-cycle modulation (DCM) is a class of signal conversion techniques in which an input signal x is transformed into a periodic switching wave $x_m(t)$ with period $T_m(x)$, and duty-cycle $R_m(x) = T_{on}(x)/T_m(x)$, where $T_{on}(x)$ is the pulse width of $x_m(t)$. Generally speaking, $R_m(x)$ is a monotonous function of x since it should be possible to extract the modulating x from $R_m(x)$ using a simple demodulation circuit if any.

As reported in [1], although the general DCM principle is not new, over years many types of implementation techniques with specific characteristics have been built for application in a variety of application areas. In the pulse width modulation (PWM) technique where $T_m(x) = T_m$ is kept constant, a basic PWM circuit requires a triangle wave oscillator and a comparator [2], and a whole minimal implementation requires three integrated operational amplifiers and four passive components. In the sigma-delta modulation (SDM) policy where both $T_{on}(x)$ and $T(x)$ depend on x, a standard SDM circuit outlined in [3] consists of a summing amplifier, an integrator, a controlled latch, a feedback mechanism, a separate clock source, and more. Furthermore, a few novel DCM configurations are encountered in the space of modern patents ([4,5]). In the patent [4], the novel DCM circuit presented structurally behaves as a closed loop system, consisting of a controlled Schmitt trigger and a summing integrator, in which case a whole realization requires at least two integrated circuits and seven passive elements. Whereas in the patent [5], the new modulation circuit presented consists of a number of building blocks organized according to a feedback configuration, including a summing amplifier, an accumulator, a binary comparator, a delay circuit, a ROM device and more.

Surprisingly, the hardware architecture of existing DCM schemes are notoriously complicated. Unfortunately, according to our best knowledge, a clear proof of the significant profit arising from the structural complexity of the modulation circuit at the expense of its simplicity, appears to be missing in the DCM literature.

Thus, the active need of DCM schemes providing both structural simplicity and high quality, has motivated the development of a class of low cost DCM circuits presented in [6], consisting of the use of a single integrated operational circuit and a few passive components. Following [6], further research works have successfully developed low cost DCM schemes for a variety of application areas in industrial electronics ([7,8,9]). In [7], a novel analog-to-digital conversion (ADC) technique is designed and well tested, and its multichannel model is also implemented in [8]. In [9], a novel digital-to-analog conversion (DAC) technique using low cost duty-cycle modulation, is designed and successfully implemented. Furthermore, it is shown in [10] that a duty-cycle modulation (DCM) driver for a buck power converter offers best characteristics compared to that of a traditional pulse width modulation (PWM). However, the literature review related to the application of low cost DCM schemes presented above, is mainly limited to instrumentation and industrial electronics engineering.

The new application area of the low cost DCM scheme studied in this paper, is a signal transmission system. The architecture and analytical models of the transmission system are presented in depth in Section 2. Then, Matlab/Simulink model and the simulated results obtained when simulating a prototyping transmission system, are presented in Section 3, whereas the conclusion of the paper is outlined in Section 4.

2. Architecture and Analytical Models of DCM-Based Signal Transmission System

The architecture of the proposed DCM-based signal transmission system is presented in Fig.1, where its main subsystems are numbered from 1 to 4. It consists of a modulating signal source (N°1), a simple DCM circuit (N°2), a transmission line (N°3) and a demodulation filter (N°4).

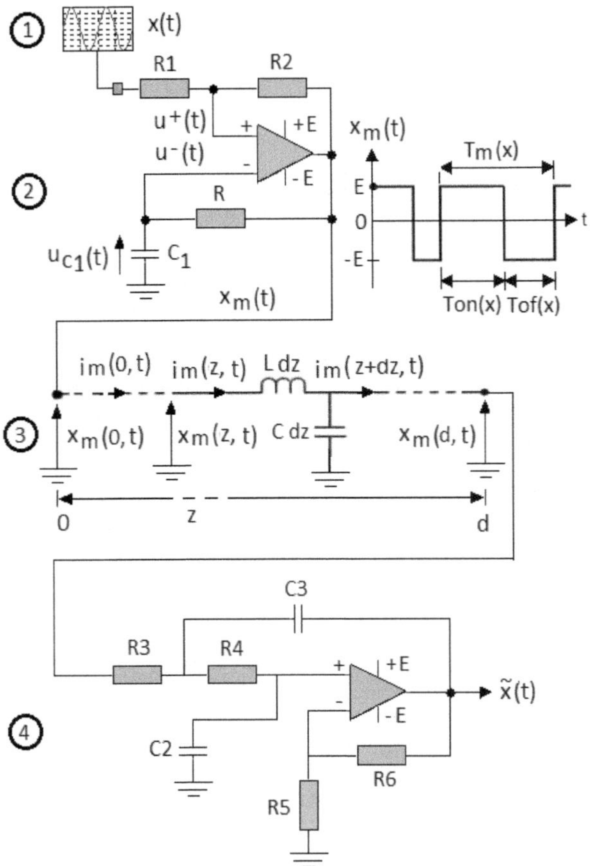

Figure 1. Schematic diagram of the signal transmission system

The modulating signal source (N°1) might be any testing instrument with a variety of selectable output signals including standard waveforms and white noises.

The simple DCM circuit (N°2) behaves as a controlled oscillator with embedded clock. For a modulating $x(t)$, the Fourier series of the resulting periodic modulated wave could be computed at a given time t as follows [10]:

$$x_m(t) = \underbrace{\left(2\,R_m(x(t)) - 1\right)E}_{Low\ \ frequency} + \underbrace{\left(\frac{4\,E}{\pi}\right)\sum_{n=1}^{\infty}\left(\frac{\dfrac{\sin\left(n\pi\,R_m(x(t))\right)}{n}}{\cos\left(2\,\pi\,n\dfrac{t}{T_m(x(t))}\right)}\right)}_{High\ \ frequencie\ \ s} \qquad (1)$$

where $R_m(x)$ could be described in its linear range by the following relation ship :

$$R_m(x(t)) = \frac{T_{on}(x(t))}{T(x(t))} = p_m\ x(t) + \frac{1}{2} \qquad (2)$$

with,

$$p_m = \frac{\dfrac{\alpha_1\alpha_2}{E(1-\alpha_1^2)}}{\log\left(\dfrac{1+\alpha_1}{1-\alpha_1}\right)},\quad \alpha_1 = 1-\alpha_2 = \frac{R_1}{R_1+R_2} \qquad (3)$$

Thus, Equation (1) given (2) shows that the modulating signal $x(t)$ could be recovered from (1) using a suitable low-pass filter with static gain K_f defined as follows:

$$K_f = \frac{1}{2\ p_m E} \qquad (4)$$

Unfortunately, in the DCM-based signal transmission system considered in this paper, there is a long transmission line between the remote DCM source and the demodulation process, might have significant impacts on the quality of the overall DCM scheme.

The parameters of the transmission line (N°3) between the DCM source and the demodulation filter are defined as follows: d (length), L (inductance per unit length) and C (capacitance per unit length). Assuming that the line is lossless, then the signal $x_m(z,\ t)$ propagating along the line at distance z from the origin $z = 0$ could be as the following Telegraph's Equations [11] :

$$\begin{cases} \dfrac{\partial x_m(z,t)}{\partial z} = -L\dfrac{\partial i_m(z,t)}{\partial t} \\ \dfrac{\partial i_m(z,t)}{\partial z} = -C\dfrac{\partial x_m(z,t)}{\partial t} \end{cases} \qquad (5)$$

or equivalently,

$$\frac{\partial^2 x_m(z,t)}{\partial z^2} - \frac{1}{\left(\dfrac{1}{\sqrt{LC}}\right)^2}\frac{\partial^2 x_m(z,t)}{\partial t^2} = 0 \qquad (6)$$

Then, the general solution of Equation (6) at the end of the line where $z = d$ is given as follows [12]:

$$\begin{aligned} x_m(d,t) = &\underbrace{x_m^{+}(t-\sqrt{LC}\ d)}_{Forward-travelling\ \ wave}\ \overbrace{u(t-\sqrt{LC}\ d)}^{Step\ \ with\,(+)\ \ delay} \\ + &\underbrace{x_m^{-}(t+\sqrt{LC}\ d)}_{Backword-travelling\ \ wave}\ \overbrace{u(t+\sqrt{LC}\ d)}^{Step\ \ with\,(-)\ \ delay} \end{aligned} \qquad (7)$$

As a conclusion at time t, the modulated 1-bit signal $x_m(d,t)$ available for filtering at the end of the lossless transmission line with length d, behaves as a delayed image of $x_m(t) = x_m(0,t)$.

The demodulation unit (N°3) is a standard second order active filter, with transfer function given as follows :

$$\begin{aligned} F(s) &= \frac{\tilde{X}(s)}{X_m(s)} \\ &= \frac{K_f}{R_3R_4C_2C_3\ s^2 + \left(R_3C_2 + R_4C_2 + (1-K_f)\,R_3C_3\right)s + 1} \end{aligned} \qquad (8)$$

where K_f = 1+ R6/R5 (static gain). As an implication, from Equation (4, the following constraint should be satisfied :

$$\frac{1}{2\ p_m E} = 1 + \frac{R_6}{R_5} \qquad (9)$$

For the sake of simplicity, it is easy to show that Equation (8) could be written as follows:

$$F_c(s) = \frac{Y_f(s)}{X_M(s)} = \frac{K_f\ \omega_n^2}{s^2 + 2\ \xi\ \omega_n\ s + \omega_n^2} \qquad (10)$$

In which case, ω_n is the natural frequency (in rad/s), and ξ being the damping coefficient. In addition, for a given cut off frequency f_c (in Hz), the related natural frequency could be computed using the following formula [13]:

$$\omega_n = \frac{2\ \pi\ f_c}{\sqrt{1 - 2\xi^2 + \sqrt{4\xi^4 - 4\xi^2 + 2}}} \qquad (11)$$

Equations (1)-(11) stands for a complete set of analytical characteristics of the proposed DCM-based signal transmission system.

3. Matlab/Simulink Model and Simulation Results of the DCM Transmission System

3.1. MatLab/Simulink Model

In engineering, the use of Matlab/Simulink as a suitable framework for rapid simulation of dynamic systems ([14,15,16]). Following this common practice, the complete Matlab/Simulink model of the DCM-based transmission system is presented in Figure 2, where the subsystems are numbered according to the same order used in the general DCM architecture shown earlier in Figure 1.

The remote modulating subsystem (N°1) consists of a low frequency virtual instrument with selectable standard waveforms (sine, square, saw teeth) and a chirp signal generator. The readers interested to signal generation algorithms for standard signals and arbitrary waveforms including wavelets, can find more details in ([17,18,19]). The selection of signals is conducted using a couple of bus create/select objects. Then, the DCM subsystem is built using the Kirchhoff law applied to the behavior of the modulation circuit observed on Figure 1. These equations are given as follows:

$$u^+(t) = \alpha_1 x_m(t) + \alpha_2 x(t) \qquad (12)$$

$$u^-(t) = u_{C1}(t) \qquad (13)$$

$$e(t) = u^+(t) - u^-(t) \qquad (14)$$

$$x_m(t) = RC_1 \frac{du_{C1}(t)}{dt} + u_{C1}(t) = \begin{cases} E \ if\ e(t) \geq 0 \\ -E\ if\ e(t) > 0 \end{cases} \qquad (15)$$

From (4), it is obvious to write an equivalent integral,

$$u_{C1}(t) = \frac{1}{RC_1} \int_0^t \left(x_m(\tau) - u_{C1}(\tau) \right) d\tau \qquad (16)$$

As an implication, a set of nine basic operators available in Simulink toolbox are sufficient for building a complete model of the proposed DCM process.

The lossless transfer line (N°3) is available in Simulink toolbox, and the critical parameter to be specified by the user, is the propagation time delay along the transmission line. It corresponds in Equation (14) to the factor $t_d = \sqrt{LC}\ d$. Finally, the model of the demodulation low-pass filter has been implemented with ease using the transfer function objet available in Simulink tool box.

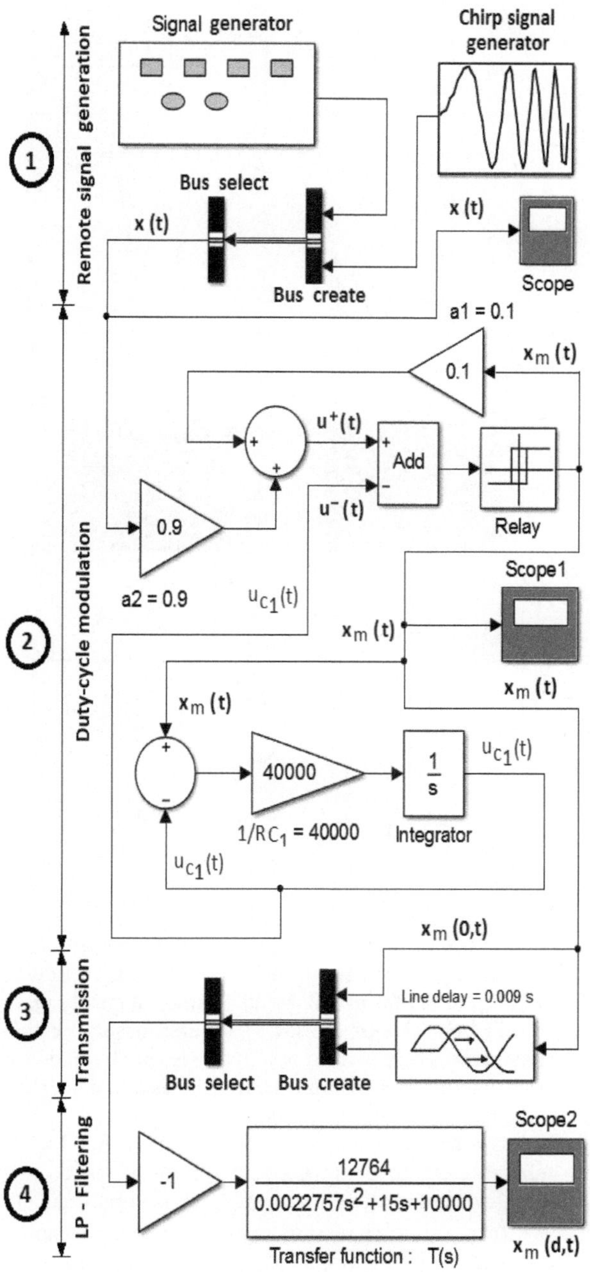

Figure 2. Simulink models of the DCM transmission system

3.2. Simulation Results

The main data used for numerical simulation of the prototyping transmission, is summarized in Table 1.

Table 1. Summary of data used for the simulation

N°	Main data used for numerical simulation		
	Subsystem	**Data type**	**Parameters**
1	Modulating signal generation	Sine	2 V, 20 Hz
		Square	
		Saw teeth	
		Chrirp	10 Hz to 40 Hz
		Bus create/select	2 input sources
2	DCM circuit	α_1	0.1
		α_2	0.9
		1/RC1	40000
		E	9 V
3	Transmission line	Line delay T_d	0.005 s
		Bus create/select	2 input lines
		Padé order	2
4	Demodulation circuit	K_f	1.2764
		ω_n	2096.3 rad/s
		ξ	1.5726

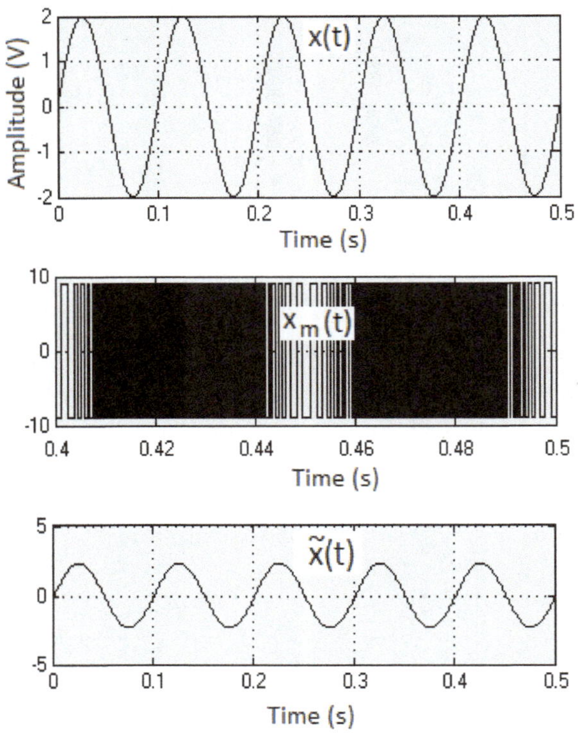

Figure 3. Transmission of a sine wave (Ideal transmission line)

For the sake of clarity, the simulation results presented in this Section are organized into set of m coupled figures (Fig. m, Fig. m+1) for $m = 2, 3, \ldots, 9$. In each case, Fig. m reports the behavior of the DCM scheme assuming that the transmission line is ideal, whereas Fig. m+1 stands for the case where the transmission line is assumed lossless under the same modulating input.

As shown from (Figure 5, Figure 6) to (Figure 9, Figure 10) related to the transmission of a variety of waveforms including an arbitrary chirp wave, it is important to observe that, the transmission result $\tilde{x}(t)$ obtained under the assumption of ideal transmission line, is almost perfect since the overall output $\tilde{x}(t)$ appears to be an exact copy of the modulating input $x(t)$. Furthermore, in all cases where the dynamic model of the lossless transmission line is taken into account, the resulting wave $\tilde{x}(t)$ obtained is quite closed to $x(t)$. However, a narrow gap observed between the two, is due to the fact that Simulink model of a lossless transmission line is computed from a given time delay T_d using a $[n/m]$ Padé approximant [20]. As an implication, the model of the lossless transmission line behaves as an additional rational transfer function (characterized by m zeros and n poles), placed in series upstream the demodulation filter. These findings emerging from the simulation results lead to conclusion of this paper.

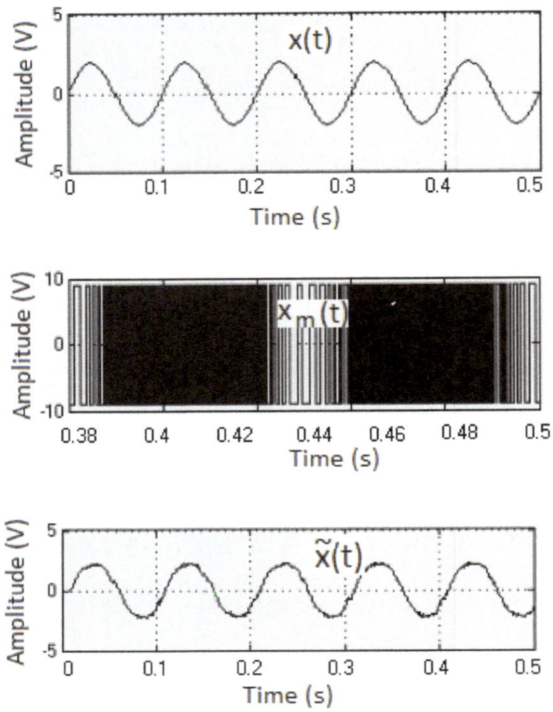

Figure 4. Transmission of a sine wave (Lossless transmission line)

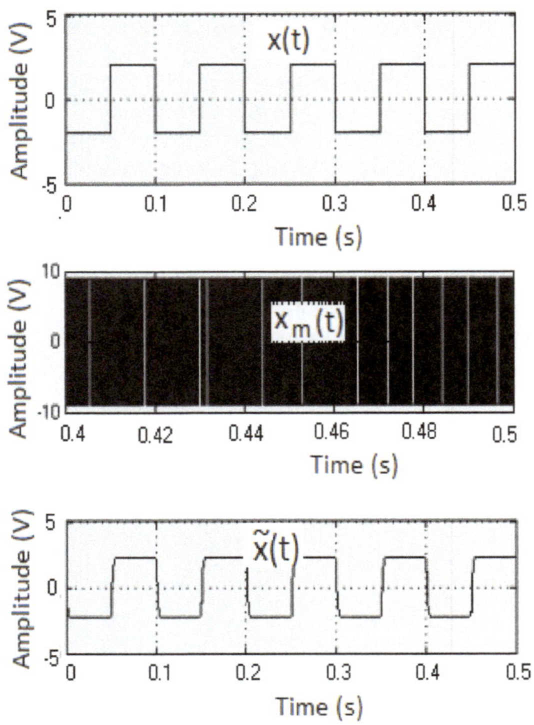

Figure 5. Transmission of a square wave (Ideal transmission line)

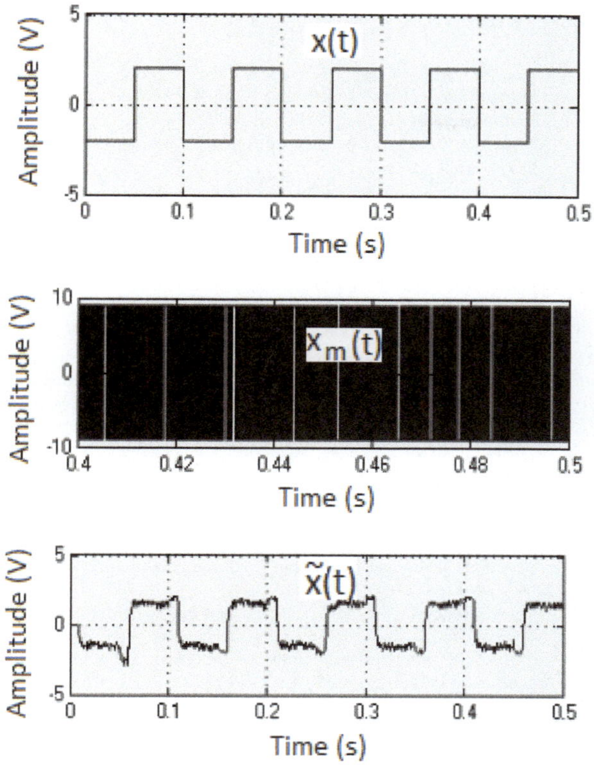

Figure 6. Transmission of a square wave (Lossless transmission line)

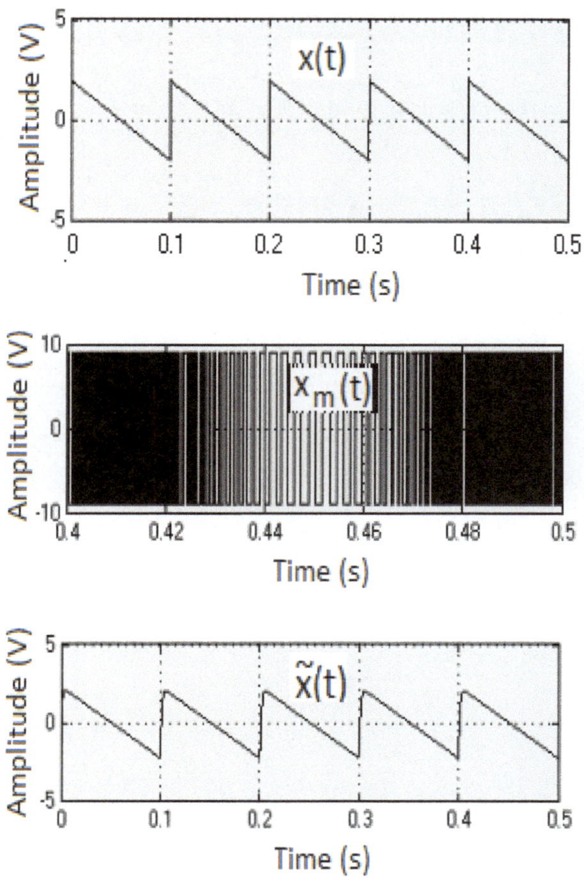

Figure 7. Transmission of a saw teeth wave (Ideal transmission line)

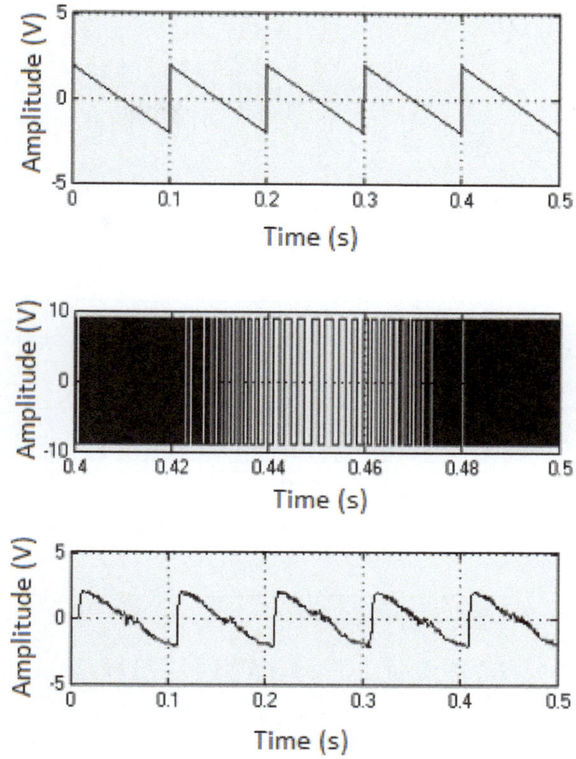

Figure 8. Transmission of a saw teeth wave (Lossless transmission line)

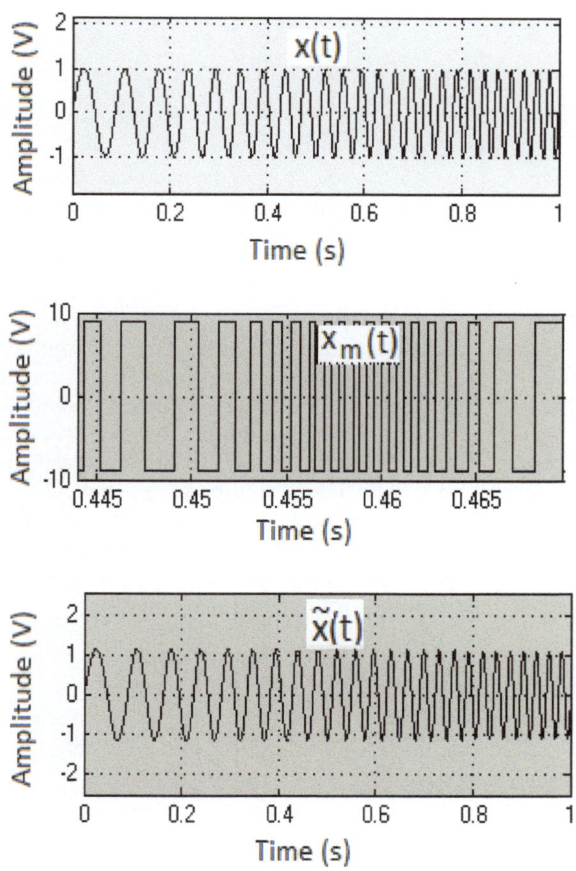

Figure 9. Transmission of a chirp signal (Lossless transmission line)

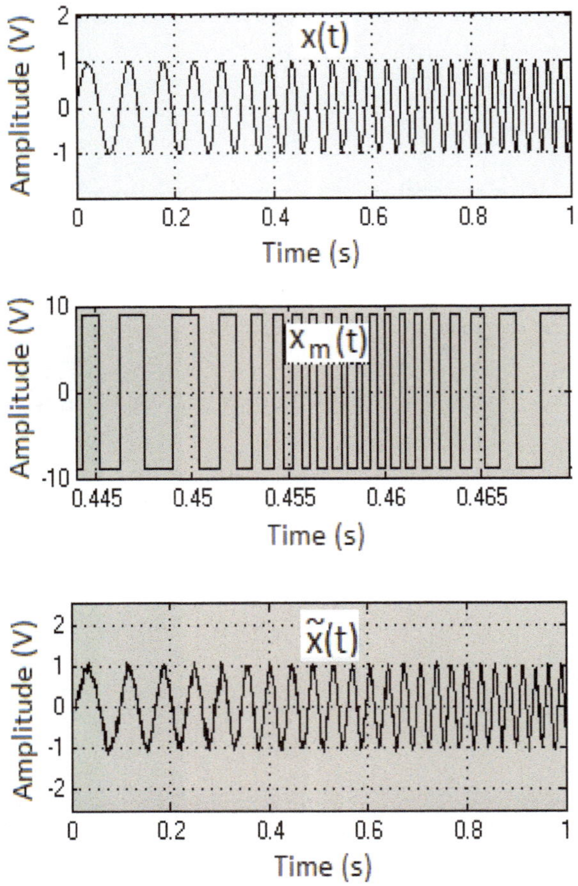

Figure 10. Transmission of a chirp signal (Lossless transmission line)

4. Conclusion

The analytical developments and the related simulation results presented in this paper, shows the high quality of the proposed DCM scheme for signal transmission systems. Although the simulation has been conducted for low frequency modulating waves, a higher modulating frequency bandwidth might be covered for an optimal choice of the design parameters associated with both the duty-cycle modulator and the low-pass filter. In addition, it would be interesting to study the behavior of the proposed DCM scheme under a more realistic transmission line with losses. Furthermore, it would be also a challenge to build a real prototyping DCM system, in order to show the realistic nature of the proposed signal transmission scheme. These unsolved problems appear to be good opportunities for future research works.

References

[1] Roza, E. "*Analog-to-digital conversion via duty-cycle modulation*", IEEE transactions on circuits and systems II: Analog and digital signal processing, 44 (11) (1997), pp. 907-914.

[2] Millman, I., Grabel, A. Micro Electro,ics, 2nd Edition, 10001 pages, 1987, © Mc Graw Hill.

[3] Sangil Park, Motorola digital signal processors – Principles of sigma-delta modulation for analog-digital (February, 2010), http://digitalsignallabs.com/SigmaDelta.pdf, pp 1.1-9.6.

[4] Harvey, P., C. Transmitting a signal using duty-cycle modulation, United State Patent N° 6 111 680, August 29, 2000.

[5] Takamuki, H. Sigma-Delta modulation circuit. United State Patent N° 6 057 794, May 2, 2000.

[6] Mbihi, J., Ndjali Beng, F., Mbouenda, M., *Modeling and simulation of a class of duty-cycle modulators for industrial instrumentation*, Iranian Journal of Electrical and Computer engineering, 4 (2)2 (2005) pp. 121-128

[7] Mbihi, J., Ndjali Beng, F., Kom, M., and Nneme Nneme, L. *A novel analog-to-digital conversion technique using nonlinear duty-cycle modulation*, International Journal of electronics and computer science engineering, 1(3), 2012, pp. 818-825.

[8] Mbihi, J., Nneme Nneme, L. A multichannel analog-to-digital conversion technique using parallel duty-cycle modulation, 1(3), 2012, pp. 826-833.

[9] Moffo Lonla, B., Mbihi, J., Nneme Nneme, L. and Kom, M. *A novel digital-to-analog conversion technique using Duty-cycle modulation*, International Journal of circuits, systems and signal processing, 1(7) 2013, pp. 47-49.

[10] Mbihi, J., Nneme Nneme, L. A novel control scheme for buck power converters using duty-cycle modulation, International Journal of power electronics, 5(3), 2013, pp. 185-199.

[11] Nitsch, J., Tkachenko, S. *Telegrapher equations for arbitrary frequencies and modes*: Radiation of an infinite, lossless transmission line, Radio Science, 30(2), 2004.

[12] *Engineering Electromagnetics*, 4th edition by William Hayt, 1981. Chapter 12, section 1, "The Transmission Line Equations".

[13] OGATA, K. "Modern control engineering – Second Edition", pp 526-528, prentice hall, 1990.

[14] Tilak Thakur, and Kumar Gupta S., Modeling of a Stand-Alone Induction Generator on Load Using MATLAB Simulink, *Journal of Emerging Trends in Engineering and Applied Sciences*, Vol. 3, No. 4, 2012, pp. 729-733.

[15] A. Aldair A. A., Hardware Implementation of the Neural Network Predictive Controller for Coupled Tank System, *American Journal of Electrical and Electronic Engineering*. 2014, 2(2), pp. 40-47.

[16] Karthikeyan, V., Vijayalakshmi, V.J. and Jeyakumar P., Selective Harmonic Elimination (SHE) for 3-Phase Voltage Source Inverter (VSI), *American Journal of Electrical and Electronic Engineering*. 2014, 2(1), pp. 17-20.

[17] Hung-Chen, C. Square/Triangular wave generator using Single DVCC and three grounded passive components, *American Journal of Electrical and Electronic Engineering, 1(2) 2013, pp. 32-36.*

[18] Mbihi, J. and Motto, A. *Instrumentation virtuelle assistée par ordinateur - Principes et techniques – Cours et exercice*, 240 p., Ocober 2012, © Ellipse, Paris.

[19] J. Mbihi, J., Tene, N. and Takoufet, K., Etude et prototypage d'un générateur de signaux arbitraires à vocation didactique, *Journal sur l'Enseignement des Sciences et des Technologies de l'Information et des Systèmes, Vol 9, N°2, 2010, © EDP Sciences*

[20] Baker, G. A., and Graves-Morris, P., *Padé Approximants*, second Edition, Encyclopedia of Mathematics and its Applications, March 11, 2010.

Permissions

All chapters in this book were first published in AJEEE, by Science and Education Publishing; hereby published with permission under the Creative Commons Attribution License or equivalent. Every chapter published in this book has been scrutinized by our experts. Their significance has been extensively debated. The topics covered herein carry significant findings which will fuel the growth of the discipline. They may even be implemented as practical applications or may be referred to as a beginning point for another development.

The contributors of this book come from diverse backgrounds, making this book a truly international effort. This book will bring forth new frontiers with its revolutionizing research information and detailed analysis of the nascent developments around the world.

We would like to thank all the contributing authors for lending their expertise to make the book truly unique. They have played a crucial role in the development of this book. Without their invaluable contributions this book wouldn't have been possible. They have made vital efforts to compile up to date information on the varied aspects of this subject to make this book a valuable addition to the collection of many professionals and students.

This book was conceptualized with the vision of imparting up-to-date information and advanced data inthis field. To ensure the same, a matchless editorial board was set up. Every individual on the board wentthrough rigorous rounds of assessment to prove their worth. After which they invested a large part of theirtime researching and compiling the most relevant data for our readers.

The editorial board has been involved in producing this book since its inception. They have spent rigoroushours researching and exploring the diverse topics which have resulted in the successful publishing of this book. They have passed on their knowledge of decades through this book. To expedite this challenging task, the publisher supported the team at every step. A small team of assistant editors was also appointed to further simplify the editing procedure and attain best results for the readers.

Apart from the editorial board, the designing team has also invested a significant amount of their time in understanding the subject and creating the most relevant covers. They scrutinized every image to scout for the most suitable representation of the subject and create an appropriate cover for the book.

The publishing team has been an ardent support to the editorial, designing and production team. Theirendless efforts to recruit the best for this project, has resulted in the accomplishment of this book. They are a veteran in the field of academics and their pool of knowledge is as vast as their experience in printing. Their expertise and guidance has proved useful at every step. Their uncompromising quality standards have made this book an exceptional effort. Their encouragement from time to time has been an inspiration for everyone.

The publisher and the editorial board hope that this book will prove to be a valuable piece of knowledge for researchers, students, practitioners and scholars across the globe.

List of Contributors

Anireh V.I.E.
Department of Electrical/Computer Engineering, Rivers State University of Science and Technology, Rivers State, Nigeria

John Tarilanyo Afa
Department of Electrical/Electronic Engineering, Niger Delta University, Bayelsa State, Nigeria

Gada Mahmood Faisal and Kaydar Majeed Quboa
Elec. & Com. Dept, College of Engineering, Mosul University, Mosul /Iraq

Dia Mohamad Ali
Com. Dept, College of Electronic Engineering, Mosul University, Mosul /Iraq

Bilal Arif, Mohd. Usama Ismail and Ale Imran
Department of Electronics Engineering, Aligarh Muslim University, Aligarh, India

Ammar A. Aldair and Electrical Engineering
Department, University of Basrah, Basrah, Iraq

D. K. Sambariya
Department of Electrical Engineering, Rajasthan Technical University, Kota, India

R. Prasad
Department of Electrical Engineering, Indian Institute of Technology Roorkee, Roorkee, India

Laith Ali Abdul-Rahaim
Electrical Engineering Department, Babylon University, Babil, Iraq

D. K. Srivastava and V. K. Singh
Electronics Engineering Department, Institute of Engineering and Technology, Lucknow, 226021 India

R. Senani
Division of Electronics and Communication Engineering, NetajiSubhas Institute of Technology, Sector3, 4 Dwarka, New Delhi, 110078, India

F. F. Mende
B.I. Verkin Institute for Low Temperature Physics and Engineering, NAS Ukraine, 47 Lenin Ave., Kharkov, Ukraine

Mohammad Mahdi Share Pasand
Department of Electrical and Electronics Engineering, Standard Research Institute- SRI, Alborz, Iran

Manoj Kumar Jain and V. K. Singh
Department of Electronics and Communication Engineering, Institute of Engineering and Technology, Lucknow, India

Ali H. Ahmad
Department of Electrical, Eng. College, University of mosul, Mosul, Iraq

Nashwan Saleh Sultan
Technical Eng.College of Mosul, Mosul, Iraq

Emmanuel U. Oleka
Electrical and Computer Engineering Department, North Carolina A&T State University, Greensboro NC, USA

Samuel N. Ndubisi
Electrical Engineering Department, University of Port Harcourt, PortHarcourt, Nigeria

Gerald K. Ijemaru
Electrical and Electronic Engineering Department, Federal University of Oye-Ekiti, Ekiti State Nigeria

N. Schinas and A. Moullas
Department of Electrical Engineering, Technological Educational Institute of Western Greece, Patras, Greece

V.G. Kudrya
Odesa National Polytechnic University, Ukraine

Gourav Misra, Vivek Kumar and Arun Agarwal
Department of Electronics & Communication Engineering, Institute of Technical Education & Research, Faculty of Engineering & Technology, Siksha 'O' Anusandhan University, Khandagiri Square, Bhubaneswar-751030, Odisha, India

Kabita Agarwal
Department of Electronics and Telecommunication Engineering, CV Raman College of Engineering, Bhubaneswar-751054, Odisha, India

Jyothilal Nayak Bharothu and G Venkatesh
Associate professor of Electrical & Electronics Engineering, Sri Vasavi Institute of Engineering & Technology, Nandamuru, India

Y. Wang
IREENA Ecole polytechnique - de l'université de Nantes, Nantes, France

B. Feuvrie
IREENA Ecole polytechnique - de l'université de Nantes, Nantes, France
IUT GEII Nantes Site Fleuriaye, Av. du Prof J. Rouxel, Carquefou, France

M. Diop
Ecole Supérieure Polytechnique, UCAD, BP 5085, Dakar-Fann, Sénégal

Andre YOUMSSI
Electric Energy Laboratory (LEE), Montreal, Canada
Ecole Polytechnique de Montreal, Succursale Centre Ville, Montreal, Quebec, Canada Electromagnetic Laboratory & Process Engineering Laboratory, The University of Ngaoundere, Ngaoundere, Cameroon

Emmanuel Patrick FEUDJIO
Electromagnetic Laboratory & Process Engineering Laboratory, The University of Ngaoundere, Ngaoundere, Cameroon

B. Nana and S. B. Yamgoué
Department of Physics, Higher Teacher Training College, University of Bamenda, PO Box 39 Bamenda, Cameroon

R. Tchitnga
Laboratory of Electronics and Signal Processing, Faculty of Science, University of Dschang, PO Box 67 Dschang Cameroon

P. Woafo
Laboratory of Modelling and Simulation in Engineering, Biomimetics and Prototypes, Faculty of Science, University of Yaounde I, PO Box 812 Yaounde, Cameroon

Arun Agarwal
Department of Electronics & Communication Engineering, Institute of Technical Education and Research, Siksha 'O' Anusandhan University, Khandagiri Square, Bhubaneswar, India

Kabita Agarwal
Department Electronics & Telecommunication Engineering, CV Raman College of Engineering, Bhubaneswar, India

Gang Ma, Guchao Xu, Linru Jiang and Tao Qian
School of Electric & Automation Engineering, Nanjing Normal University, Nanjing, China

D. R. Bhaskar
Department of Electronics and Communication Engineering, Jamia Millia Islamia, New Delhi, India

R. Senani
Devision of Electronics and Communication Engineering, Netaji Subhas Institute of Technology, New Delhi, India

P. K. Hota
Department of Electrical Engineering Depart, Veer Surendra Sai University of Technology (VSSUT), Burla, India

N. C. Sahu
Department of Electrical Engineering, ITER, SOA University, Bhubaneswar, India

B. Nana
Department of Physics, Higher Teacher Training College, University of Bamenda, PO Box 39 Bamenda, Cameroon

P. Woafo
Laboratory of Modelling and Simulation in Engineering, Biomimetics and Prototypes, Faculty of Science, University of Yaounde I, PO Box 812 Yaounde, Cameroon

Jumber Dochviri, Oleg Khachapuridze and Nana Beradze
Department of Electrical Engineering, Faculty of Power Engineering and Telecommunications, Georgian Technical University, M. Kostava str., Tbilisi, Georgia

Gourav Misra and Arun Agarwal
Department of ECE, Faculty of Engineering & Technology, Siksha 'O' Anusandhan University, Khandagiri Sqare, Bhubaneswar, Odisha, India

Kabita Agarwal
CV Raman College of Engineering, Department of ETC, Bhubaneswar, Odisha, India

Arun Agarwal
Department of ECE, Institute of Technical Education & Research, Siksha 'O' Anusandhan University, Khandagiri Square, Bhubaneswar-751030, Odisha, India

Ph.D Scholar, Department of Information Technology, AMET University, Tamil Nadu, Chennai, India

Saurabh N. Mehta
Department of Electronics and Telecommunication Engineering, Vidyalankar Institute of Technology, Mumbai, Maharashtra, India

G. C. Diyoke, Okeke C. and Uchechi Aniagwu
Department of Electrical and Electronic Engineering, Michael Okpara University of Agriculture, Umudike Abia State, Nigeria

Mousaab M. Nahas
Electrical and Computer Engineering Department, Faculty of Engineering, University of Jeddah, Jeddah, Saudi Arabia

Lokesh Kumar Agrawal and G. K. Benerjee
Department of Electrical Engineering, IFTM University, Moradabad(U.P.), India

Bhavesh Kumar Chauhan
Department of Electrical Engineering, ABESIT, Ghaziabad (U.P.), India

Leandre Nneme Nneme and Jean Mbihi
Department of Electrical and Electronics Engineering, ENSET (University of Douala), Douala, Cameroon

Index